W9-AEJ-594

Introduction to Dynamic Systems

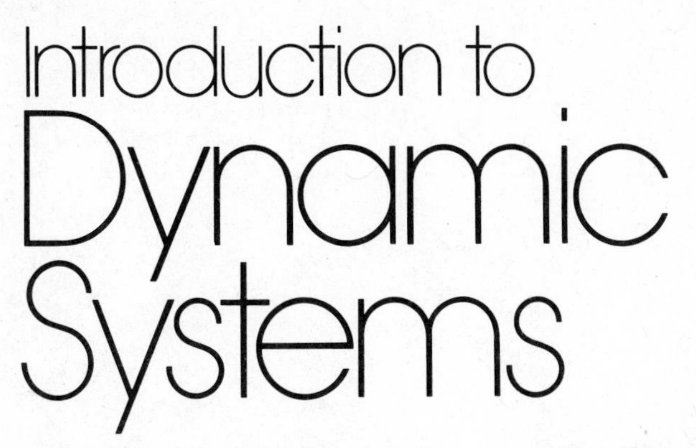

Theory, Models, and Applications

David G. Luenberger

Stanford University

John Wiley & Sons

New York Chichester Brisbane Toronto

Library of Congress Cataloging in Publication Data

Luenberger, David G., 1937–
Introduction to dynamic systems.

Includes bibliographical references and index.
1. System analysis. 2. Differential equations.
3. Control theory. I. Title. II. Title: Dynamic Systems.

QA402.L84 003 78-12366
ISBN 0-471-02594-1

Printed in the United States of America

10

To my parents

Preface

This book is an outgrowth of a course developed at Stanford University over the past five years. It is suitable as a self-contained textbook for second-level undergraduates or for first-level graduate students in almost every field that employs quantitative methods. As prerequisites, it is assumed that the student *may* have had a first course in differential equations and a first course in linear algebra or matrix analysis. These two subjects, however, are reviewed in Chapters 2 and 3, insofar as they are required for later developments.

The objective of the book, simply stated, is to help one develop the ability to analyze real dynamic phenomena and dynamic systems. This objective is pursued through the presentation of three important aspects of dynamic systems: (1) the *theory*, which explores properties of mathematical representations of dynamic systems, (2) example *models*, which demonstrate how concrete situations can be translated into appropriate mathematical representations, and (3) *applications*, which illustrate the kinds of questions that might be posed in a given situation, and how theory can help resolve these questions. Although the highest priority is, appropriately, given to the orderly presentation of the theory, significant samples of all three of these essential ingredients are contained in the book.

The organization of the book follows theoretical lines—as the chapter titles indicate. The particular theoretical approach, or style, however, is a blend of the traditional approach, as represented by many standard textbooks on differential equations, and the modern state-space approach, now commonly used as a setting for control theory. In part, this blend was selected so as to broaden the scope—to get the advantages of both approaches; and in part it was dictated by the requirements of the applications presented. It is recognized, however, that (as in every branch of mathematics) the root ideas of dynamic systems transcend any particular mathematical framework used to describe those ideas. Thus, although the theory in this book is presented within a certain framework, it is the intent that what is taught about dynamic systems is richer and less restrictive than the framework itself.

The content of the book is, of course, partly a reflection of personal taste, but in large portion it was selected to directly relate to the primary objective of developing the ability to analyze real systems, as stated earlier. The theoretical material in Chapters 2 through 5 is quite standard, although in addition to theory these chapters emphasize the relation between theory and analysis. Dominant eigenvector analysis is used as an extended illustration of this relationship. Chapter 6 extends the classical material of linear systems to the special and rich topic of positive systems. This chapter, perhaps more than any other, demonstrates the intimate relation between theory and intuition. The topic of Markov chains, in Chapter 7, has traditionally been treated most often as a distinct subject. Nevertheless, although it does have some unique features, a great deal of unity is achieved by regarding this topic as a branch of dynamic system theory. Chapter 8 outlines the concepts of system control—from both the traditional transform approach and the state-space approach. Chapters 9 and 10 treat nonlinear systems, with the Liapunov function concept serving to unify both the theory and a wide assortment of applications. Finally, Chapter 11 surveys the exciting topic of optimal control—which represents an important framework for problem formulation in many areas. Throughout all chapters there is an assortment of examples that not only illustrate the theory but have intrinsic value of their own. Although these models are abstractions of reality, many of these are "classic" models that have stood the test of time and have had great influence on scientific development. For developing effectiveness in analysis, the study of these examples is as valuable as the study of theory.

The book contains enough material for a full academic year course. There is room, however, for substantial flexibility in developing a plan of study. By omitting various sections, the book has been used at Stanford as the basis for a six-month course. The chapter dependency chart shown below can be used to plan suitable individual programs. As a further aid to this planning, difficult sections of the book that are somewhat tangential to the main development are designated by an asterisk*.

An important component of the book is the set of problems at the end of the chapters. Some of these problems are exercises, which are more or less straightforward applications of the techniques discussed in the chapter; some are extensions of the theory; and some introduce new application areas. A few

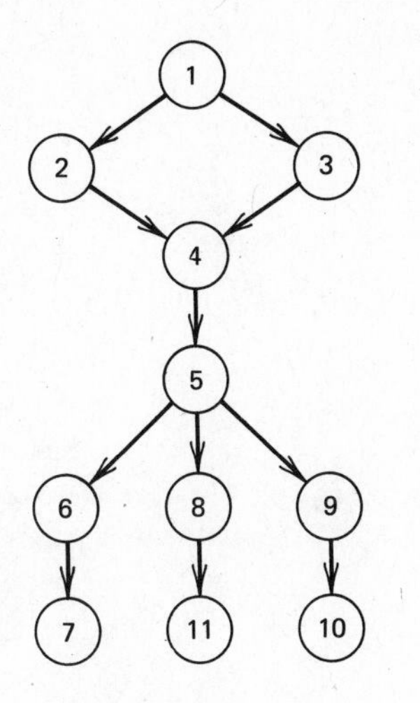

Chapter Dependency Chart (A chapter depends on all chapters leading to it in the chart.)

of each type should be attempted from each chapter. Especially difficult problems are marked with an asterisk*.

The preparation of this book has been a long task that could not have been completed without the help of many individuals. Many of the problems and examples in the book were developed jointly with teaching assistants and students. I wish to acknowledge the Department of Engineering-Economic Systems at Stanford which provided the atmosphere and resources to make this project possible. I wish to thank my family for their help, encouragement, and endurance. I wish to thank Lois Goularte who efficiently typed the several drafts and helped organize many aspects of the project. Finally, I wish to thank the scores of students, visitors, and colleagues who read primitive versions of the manuscript and made many valuable individual suggestions.

DAVID G. LUENBERGER
Stanford, California
January 1979

Contents

Introduction to Dynamic Systems

chapter 1.
Introduction

1.1 DYNAMIC PHENOMENA

The term *dynamic* refers to phenomena that produce time-changing patterns, the characteristics of the pattern at one time being interrelated with those at other times. The term is nearly synonymous with *time-evolution* or *pattern of change.* It refers to the unfolding of events in a continuing evolutionary process.

Nearly all observed phenomena in our daily lives or in scientific investigation have important dynamic aspects. Specific examples may arise in (a) a physical system, such as a traveling space vehicle, a home heating system, or in the mining of a mineral deposit; (b) a social system, such as the movement within an organizational hierarchy, the evolution of a tribal class system, or the behavior of an economic structure; or (c) a life system, such as that of genetic transference, ecological decay, or population growth. But while these examples illustrate the pervasiveness of dynamic situations and indicate the potential value of developing the facility for representing and analyzing dynamic behavior, it must be emphasized that the general concept of dynamics transcends the particular origin or setting of the process.

Many dynamic systems can be understood and analyzed intuitively, without resort to mathematics and without development of a general theory of dynamics. Indeed, we often deal quite effectively with many simple dynamic situations in our daily lives. However, in order to approach unfamiliar complex situations efficiently, it is necessary to proceed systematically. Mathematics can provide the required economy of language and conceptual framework.

With this view, the term *dynamics* soon takes on somewhat of a dual meaning. It is, first, as stated earlier, a term for the time-evolutionary phenomena in the world about us, and, second, it is a term for that part of mathematical science that is used for the representation and analysis of such phenomena. In the most profound sense the term refers simultaneously to both aspects: the real, the abstract, and the interplay between them.

Although there are endless examples of interesting dynamic situations arising in a spectrum of areas, the number of corresponding general forms for mathematical representation is relatively small. Most commonly, dynamic systems are represented mathematically in terms of either differential or difference equations. Indeed, this is so much the case that, in terms of pure mathematical content, at least the elementary study of dynamics is almost synonymous with the theory of differential and difference equations. It is these equations that provide the structure for representing time linkages among variables.

The use of either differential or difference equations to represent dynamic behavior corresponds, respectively, to whether the behavior is viewed as occurring in continuous or discrete time. Continuous time corresponds to our usual conception, where time is regarded as a continuous variable and is often viewed as flowing smoothly past us. In mathematical terms, continuous time of this sort is quantified in terms of the continuum of real numbers. An arbitrary value of this continuous time is usually denoted by the letter t. Dynamic behavior viewed in continuous time is usually described by differential equations, which relate the derivatives of a dynamic variable to its current value.

Discrete time consists of an ordered sequence of points rather than a continuum. In terms of applications, it is convenient to introduce this kind of time when events and consequences either occur or are accounted for only at discrete time periods, such as daily, monthly, or yearly. When developing a population model, for example, it may be convenient to work with yearly population changes rather than continuous time changes. Discrete time is usually labeled by simply indexing, in order, the discrete time points, starting at a convenient reference point. Thus time corresponds to integers 0, 1, 2, and so forth, and an arbitrary time point is usually denoted by the letter k. Accordingly, dynamic behavior viewed in discrete time is usually described by equations relating the value of a variable at one time to the values at adjacent times. Such equations are called *difference equations.*

1.2 MULTIVARIABLE SYSTEMS

The term *system*, as applied to general analysis, was originated as a recognition that meaningful investigation of a particular phenomenon can often only be achieved by explicitly accounting for its environment. The particular variables of interest are likely to represent simply one component of a complex, consisting of perhaps several other components. Meaningful analysis must consider the entire system and the relations among its components. Accordingly, mathematical models of systems are likely to involve a large number of interrelated variables—and this is emphasized by describing such situations as *multivariable systems.* Some examples illustrating the pervasiveness and importance of multivariable phenomena arise in consideration of (a) the migration patterns of population between various geographical areas, (b) the simultaneous interaction of various individuals in an economic system, or (c) the various age groups in a growing population.

The ability to deal effectively with large numbers of interrelated variables is one of the most important characteristics of mathematical system analysis. It is necessary therefore to develop facility with techniques that help one clearly think about and systematically manipulate large numbers of simultaneous relations. For one's own thinking purposes, in order to understand the essential elements of the situation, one must learn, first, to view the whole set of relations as a unit, suppressing the details; and, second, to see the important detailed interrelations when required. For purposes of manipulation, with the primary objective of computation rather than furthering insight, one requires a systematic and efficient representation.

There are two main methods for representing sets of interrelations. The first is vector notation, which provides an efficient representation both for computation and for theoretical development. By its very nature, vector notation suppresses detail but allows for its retrieval when required. It is therefore a convenient, effective, and practical language. Moreover, once a situation is cast in this form, the entire array of theoretical results from linear algebra is available for application. Thus, this language is also well matched to mathematical theory.

The second technique for representing interrelations between variables is by use of diagrams. In this approach the various components of a system are represented by points or blocks, with connecting lines representing relations between the corresponding components. This representation is exceedingly helpful for visualization of essential structure in many complex situations; however, it lacks the full analytical power of the vector method. It is for this reason that, although both methods are developed in this book, primary emphasis is placed on the vector approach.

Most situations that we investigate are both dynamic and multivariable. They are, accordingly, characterized by several variables, each changing with time and each linked through time to other variables. Indeed, this combination of multivariable and time-evolutionary structure characterizes the setting of the modern theory of dynamic systems.

That most dynamic systems are both time-evolutionary and multivariable implies something about the nature of the mathematics that forms the basis for their analysis. The mathematical tools are essentially a combination of differential (or difference) equations and vector algebra. The differential (or difference) equations provide the element of dynamics, and the vector algebra provides the notation for multivariable representation. The combination and interplay between these two branches of mathematics provides the basic foundation for all analysis in this book. It is for this reason that this introductory chapter is followed first by a chapter on differential and difference equations and then by a chapter on matrix algebra.

1.3 A CATALOG OF EXAMPLES

As in all areas of problem formulation and analysis, the process of passing from a "real world" dynamic situation to a suitable abstraction in terms of a mathematical model requires an expertise that is refined only through experience. In any given application there is generally no single "correct" model; rather, the degree of detail, the emphasis, and the choice of model form are subject to the discretionary choice of the analyst. There are, however, a number of models that are considered "classic" in that they are well-known and generally accepted. These classic models serve an important role, not only as models of the situation that they were originally intended to represent, but also as examples of the degree of clarity and reality one should strive to achieve in new situations. A proficient analyst usually possesses a large mental catalog of these classic models that serve as valuable reference points—as well-founded points of departure.

The examples in this section are in this sense all classic, and as such can form the beginnings of a catalog for the reader. The catalog expands as one works his way through succeeding chapters, and this growth of well-founded examples with known properties should be one of the most important objectives of one's study. A diverse catalog enriches the process of model development.

The first four examples are formulated in discrete time and are, accordingly, defined by difference equations. The last two are defined in continuous time and thus result in differential equations. It will be apparent from a study of the examples that the choice to develop a continuous-time or a discrete-time model of a specific phenomenon is somewhat arbitrary. The choice is usually resolved on the basis of data availability, analytical tractability, established convention in the application area, or simply personal preference.

Example 1 (Geometric Growth). A simple growth law, useful in a wide assortment of situations (such as describing the increase in human or other

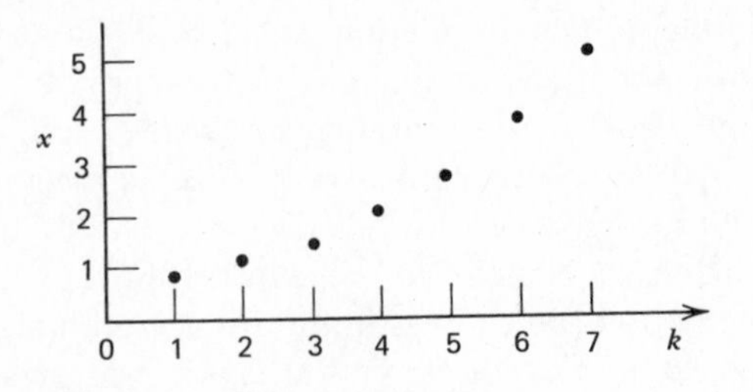

Figure 1.1. Geometric growth.

populations, the growth of vegetation, accumulated publications in a scientific field, consumption of raw materials, the accumulation of interest on a loan, etc.), is the linear law described by the difference equation

$$x(k+1) = ax(k)$$

The value $x(k)$ represents the magnitude of the variable (e.g., population) at time instant k. The parameter a is a constant that determines the rate of growth. For positive growth, the value of a must be greater than unity—then each successive magnitude is a fixed factor larger than its predecessor.

If an initial magnitude is given, say $x(0) = 1$, the successive values can be found recursively. In particular, it is easy to see that $x(1) = a$, $x(2) = a^2$, and, in general, $x(k) = a^k$ for $k = 0, 1, 2, \ldots$. A typical pattern of growth resulting from this model is shown in Fig. 1.1.

The growth pattern resulting from this simple linear model is referred to as *geometric growth* since the values grow as the terms of a geometric series. This form of growth pattern has been found to agree closely with empirical data in many situations, and there is often strong accompanying theoretical justification for the model, at least over a range of values.

Example 2 (Cohort Population Model). For many purposes (particularly in populations where the level of reproductive activity is nonuniform over a normal lifetime) the simple growth model given above is inadequate for comprehensive analysis of population change. More satisfactory models take account of the age distribution within the population. The classical model of this type is referred to as a *cohort population model.*

The population is divided into age groups (or cohorts) of equal age span, say five years. That is, the first group consists of all those members of the population between the ages of zero and five years, the second consists of those between five and ten years, and so forth. The cohort model itself is a discrete-time dynamic system with the duration of a single time period corresponding to the basic cohort span (five years in our example). By assuming that the male and female populations are identical in distribution, it is possible to simplify the model by considering only the female population. Let $x_i(k)$ be the (female) population of the ith age group at time period k. The groups are indexed sequentially from 0 through n, with 0 representing the lowest age group and n the largest. To describe system behavior, it is only necessary to describe how these numbers change during one time period.

First, aside from the possibility of death, which will be considered in a moment, it is clear that during one time period the cohorts in the ith age group simply move up to the $(i+1)$th age group. To account for the death rate of individuals within a given age group, this upward progression is attenuated by a survival factor. The net progression can be described by the simple equations

$$x_{i+1}(k+1)=\beta_i x_i(k), \qquad i=0,1,\ldots,n-1 \tag{1-1}$$

where β_i is the survival rate of the ith age group during one period. The factors β_i can be determined statistically from actuarial tables.

The only age group not determined by the equation above is $x_0(k+1)$, the group of individuals born during the last time period. They are offspring of the population that existed in the previous time period. The number in this group depends on the birth rate of each of the other cohort groups, and on how large each of these groups was during the previous period. Specifically,

$$x_0(k+1)=\alpha_0 x_0(k)+\alpha_1 x_1(k)+\alpha_2 x_2(k)+\cdots+\alpha_n x_n(k) \tag{1-2}$$

where α_i is the birth rate of the ith age group (expressed in number of female offspring per time period per member of age group i). The factor α_i also can be usually determined from statistical records.

Together Eqs. (1-1) and (1-2) define the system equations, determining how $x_i(k+1)$'s are found from $x_i(k)$'s. This is an excellent example of the combination of dynamics and multivariable system structure. The population system is most naturally visualized in terms of the variables representing the population levels of the various cohort groups, and thus it is a multivariable system. These variables are linked dynamically by simple difference equations, and thus the whole can be regarded as a composite of difference equations and multivariable structure.

Example 3 (National Economics). There are several simple models of national economic dynamics.* We present one formulated in discrete time, where the time between periods is usually taken as quarters of full years. At each time period there are four variables that define the model. They are

$Y(k)$ = National Income or National Product

$C(k)$ = Consumption

$I(k)$ = Investment

$G(k)$ = Government Expenditure

* See the notes and references for Sect. 4.8, at the end of Chapter 4.

The variable Y is defined to be the National Income: the total amount earned during a period by all individuals in the economy. Alternatively, but equivalently, Y can be defined as the National Product: the total value of goods and services produced in the economy during the period. Consumption C is the total amount spent by individuals for goods and services. It is the total of every individual's expenditure. The Investment I is the total amount invested in the period. Finally, G is the total amount spent by government during the period, which is equal to the government's current revenue. The basic national accounting equation is

$$Y(k)=C(k)+I(k)+G(k) \tag{1-3}$$

From an income viewpoint, the equation states that total individual income must be divided among consumption of goods and services, investment, or payments to the government. Alternatively, from a national product viewpoint, the total aggregate of goods and services produced must be divided among individual consumption, investment, or government consumption.

In addition to this basic definitional equation, two relationships are introduced that represent assumptions on the behavior of the economy. First, it is assumed that consumption is a fixed fraction of national income. Thus,

$$C(k)=mY(k) \tag{1-4}$$

for some m. The number m, which is restricted to the values $0<m<1$, is referred to as the *marginal propensity to consume.* This equation assumes that on the average individuals tend to consume a fixed portion of their income.

The second assumption concerning how the economy behaves relates to the influence of investment. The general effect of investment is to increase the productive capacity of the nation. Thus, present investment will increase national income (or national product) in future years. Specifically, it is assumed that the increase in national income is proportional to the level of investment. Or,

$$Y(k+1)-Y(k)=rI(k) \tag{1-5}$$

The constant r is the *growth factor*, and it is assumed that $r>0$.

The set of equations (1-3), (1-4), and (1-5) defines the operation of the economy. Of the three equations, only the last is dynamic. The first two, (1-3) and (1-4), are *static*, expressing relationships among the variables that hold at every k. These two static equations can be used to eliminate two variables from the model. Starting with

$$Y(k)=C(k)+I(k)+G(k)$$

substitution of (1-4) produces

$$Y(k)=mY(k)+I(k)+G(k)$$

Substitution of (1-5) then produces

$$Y(k)=mY(k)+\frac{Y(k+1)-Y(k)}{r}+G(k)$$

Rearrangement leads to the final result:

$$Y(k+1)=[1+r(1-m)]Y(k)-rG(k) \tag{1-6}$$

The quantity $G(k)$ appears as an input to the system. If $G(k)$ were held equal to zero, the model would be identical to the first-order (geometric) growth model discussed earlier.

Example 4 (Exponential Growth). The continuous-time version of the simple first-order growth model (the analog of geometric growth) is defined by the differential equation

$$\frac{dx(t)}{dt}=rx(t)$$

The growth parameter r can be any real value, but for (increasing) growth it must be greater than zero. The solution to the equation is found by writing it in the form

$$\frac{1}{x(t)}\frac{dx(t)}{dt}=r$$

Both sides can then be integrated with respect to t to produce

$$\log x(t)=rt+\log c=\log e^{rt}+\log c$$

where c is an arbitrary constant. Taking the antilog yields

$$x(t)=ce^{rt}$$

Finally, by setting $t=0$, it is seen that $x(0)=c$, so the solution can be written

$$x(t)=x(0)e^{rt}$$

This is the equation of *exponential growth.* The solution is sketched for various values of r in Fig. 1.2.

The pattern of solutions is similar to that of geometric growth shown in Fig. 1.1 in Sect. 1.6. Indeed, a series of values from the continuous-time solution at equally spaced time points make up a geometric growth pattern.

Example 5 (Newton's Laws). A wealth of dynamic system examples is found in mechanical systems governed by Newton's laws. In fact, many of the general techniques for dynamic system analysis were originally motivated by such applications. As a simple example, consider motion in a single dimension—of, say, a street car or cable car of mass M moving along a straight track. Suppose

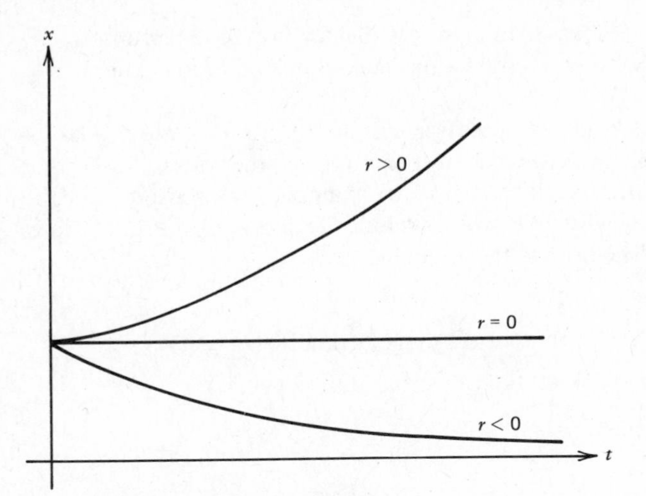

Figure 1.2. Exponential growth.

the position of the car along the track at time t is denoted by $y(t)$, and the force applied to the street car, parallel to the track, is denoted by $u(t)$. Newton's second law says that force is equal to mass times acceleration, or, mathematically,

$$u(t)=M\frac{d^2y}{dt^2}$$

Therefore, the motion is defined by a second-order differential equation.

A more detailed model would, of course, have many other variables and equations to account for spring action, rocking, and bouncing motion, and to account for the fact that forces are applied only indirectly to the main bulk through torque on the wheels or from a friction grip on a cable. The degree of detail constructed into the model would depend on the use to which the model were to be put.

Figure 1.3. Cable car.

Example 6 (Goats and Wolves). Imagine an island populated primarily by goats and wolves. The goats survive by eating the island vegetation. The wolves survive by eating the goats.

The modeling of this kind of population system, referred to as a predator–prey system, goes back to Volterra in response to the observation that populations of species often oscillated. In our example, goats would first be plentiful but wolves rare, and then wolves would be plentiful but goats rare. Volterra described the situation in the following way.

Let

$$N_1(t) = \text{number of goats at time } t$$

$$N_2(t) = \text{number of wolves at time } t$$

The proposed model is then

$$\frac{dN_1(t)}{dt} = aN_1(t) - bN_1(t)N_2(t)$$

$$\frac{dN_2(t)}{dt} = -cN_2(t) + dN_1(t)N_2(t)$$

where the constants a, b, c, and d are all positive.

This model, which is the archetype of predator–prey models, has a simple biological interpretation. In the absence of wolves $[N_2(t) = 0]$, the goat population is governed by simple exponential growth, with growth factor a. The goats thrive on the island vegetation. In the absence of goats $[N_1(t) = 0]$, on the other hand, the wolf population is governed by exponential decline, declining at a rate $-c$. This interpretation accounts for the first terms on the right-hand side of the differential equations.

When both goats and wolves are present on the island, there are encounters between the two groups. Under an assumption of random movement, the frequency of encounters is proportional to the product of the numbers in the two populations. Each encounter decreases the goat population and increases the wolf population. The effect of these encounters is accounted for by the second terms in the differential equations.

1.4 THE STAGES OF DYNAMIC SYSTEM ANALYSIS

The principal objectives of an analysis of a dynamic system are as varied as the range of possible application areas. Nevertheless, it is helpful to distinguish four (often overlapping) stages of dynamic analysis: representation of phenomena, generation of solutions, exploration of structural relations, and control or modification. Most analyses emphasize one or two of these stages, with the others having been completed previously or lying beyond the reach of current technique.

A recognition of these four stages helps motivate the assortment of theoretical principles associated with the mathematics of dynamic systems, for there is, naturally, great interplay between general theory and the analysis of given situations. On the one hand, the objectives for an analysis are strongly influenced by available theory, and, on the other hand, development of new theory is often motivated by the desire to conduct deeper analyses.

Representation

One of the primary objectives of the use of mathematics in complex dynamic systems is to obtain a mathematical representation of the system, and this is the first stage of analysis. The process of obtaining the representation is often referred to as *modeling*, and the final product a *model*. This stage is closely related to the sciences, for the development of a suitable model amounts to the employment or development of scientific theory. The theory employed in any given model may be well-founded and generally accepted, or it may be based only on one analyst's hypothesized relationships. A complex model will often have both strong and weak components. But in any case the model description is an encapsulation of a scientific theory.

Development of a meaningful representation of a complex system requires more than just scientific knowledge. The end product is likely to be most meaningful if one understands the theory of dynamic systems as well as the relevant scientific theory. Only then is it possible to assess, at least in qualitative terms, the dynamic significance of various assumptions, and thereby build a model that behaves in a manner consistent with intuitive expectations.

Generation of Solutions

The most direct use of a dynamic model is the generation of a specific solution to its describing equations. The resulting time pattern of the variables then can be studied for various purposes.

A specific solution can sometimes be found in analytical form, but more often it is necessary to generate specific solutions numerically by use of a calculator or digital computer—a process commonly referred to as *simulation*. As an example of this direct use of a model, a large cohort model of a nation's population growth can be solved numerically to generate predictions of future population levels, catalogued by age group, sex, and race. The results of such a simulation might be useful for various planning problems. Likewise, a model of the national economy can forecast future economic trends, thereby possibly suggesting the appropriateness of various corrective policies. Or, in the context of any situation, simulation might be used to test the reasonableness of a new model by verifying that a particular solution has the properties usually associated with the underlying phenomena.

It is of course rare that a single solution of a model is adequate for a meaningful analysis. Every model really represents a collection of solutions, each determined by different controlled inputs, different parameter values, and different starting conditions. In the population system, for example, the specific future population level is dependent on national immigration policy, on the birth rates in future years, and on the assumed level of current population. One may therefore find that it is necessary to generate solutions corresponding to various combinations of assumptions in order to conduct a meaningful analysis of probable future population.

As a general rule, the number of required solutions grows quickly with the number of different parameters and inputs that must be varied independently. Thus, although direct simulation is a flexible concept applicable to quite large and complex systems where analysis is difficult, it is somewhat limited in its capability to explore all ranges of input and parameter values.

Exploration of Structural Relations

Much of the theory of dynamic systems is motivated by a desire to go beyond the stage of simply computing particular solutions of a model to the point of establishing various structural relations as, say, between a certain parameter and its influence on the solution. Such relations are often obtained indirectly through the use of auxiliary concepts of analysis.

The payoff of this type of structural exploration manifests itself in two important and complementary ways. First, it develops intuitive insight into system behavior. With this insight, one is often able to determine the rough outlines of the solution to a complex system almost by inspection, and, more importantly, to foresee the nature of the effects of possible system modifications. But it is important to stress that the value of this insight goes well beyond the mere approximation of a solution. Insight into system behavior is reflected back, as an essential part of the creative process, to refinement of the formulation of the original model. A model will be finally accepted only when one is assured of its reasonableness—both in terms of its structure and in terms of the behavior patterns it generates.

The second payoff of structural exploration is that it often enables one to explicitly calculate relations that otherwise could be deduced only after examination of numerous particular solutions. For example, as is shown in Chapter 5, the natural rate of growth of a cohort population model can be determined directly from its various birth rate and survival rate coefficients, without generating even a single specific growth pattern. This leads, for example, to a specific relationship between changes in birth rates and changes in composite population growth. In a similar fashion, the stability of a complex economic process of price adjustment can often be inferred from its structural form, without generating solutions.

Most of the theoretical development in this book is aimed at revealing relationships of this kind between structure and behavior. By learning this theory we become more than just equation writers and equation solvers. Our analysis is not limited in its application to a particular problem with particular numerical constants, but instead is applicable to whole classes of models; and results from one situation can be readily transferred to another.

Control or Modification

Although study of a particular dynamic situation is sometimes motivated by the simple philosophic desire to understand the world and its phenomena, many analyses have the explicit motivation of devising effective means for changing a system so that its behavior pattern is in some way improved. The means for affecting behavior can be described as being either system modification or control. Modification refers to a change in the system, and hence in its describing equation. This might be a change in various parameter values or the introduction of new interconnective mechanisms. Examples of modification are: a change in the birth rates of a population system, a change of marriage rules in a class society, a change of forecasting procedure in an economic system, a change of promotion rate in an organizational hierarchy, and so forth. Control, on the other hand, generally implies a continuing activity executed throughout the operation of the system. The Federal Reserve Board controls the generation of new money in the economy on a continuing basis, a farmer controls the development of his herd of cattle by controlling the amount of grain they are fed, a pilot controls the behavior of his aircraft continuously, and so forth.

Determination of a suitable modification or control strategy for a system represents the fourth stage of analysis, and generally marks the conclusion of a complete analysis cycle. However, at the completion of the best analyses, the main outlines of the solution should be fairly intuitive—during the course of analysis the intuition should be heightened to a level sufficient to accept the conclusions. Mathematics serves as a language for organized thought, and thought development, not as a machine for generating complexity. The mathematics of dynamic systems is developed to expedite our requests for detail when required, and to enhance our insight into the behavior of dynamic phenomena we encounter in the world.

chapter 2.

Difference and Differential Equations

Ordinary difference and differential equations are versatile tools of analysis. They are excellent representations of many dynamic situations, and their associated theory is rich enough to provide substance to one's understanding. These equations are defined in terms of a single dynamic variable (that is, a single function of time) and therefore represent only a special case of more general dynamic models. However, ordinary difference and differential equations are quite adequate for the study of many problems, and the associated theory provides good background for more general multivariable theory. In other words, both with respect to problem formulation and theoretical development, difference and differential equations of a single variable provide an important first step in developing techniques for the mathematical analysis of dynamic phenomena.

2.1 DIFFERENCE EQUATIONS

Suppose there is defined a sequence of points, perhaps representing discrete equally spaced time points, indexed by k. Suppose also that there is a value $y(k)$ (a real number) associated with each of these points. A *difference equation* is an equation relating the value $y(k)$, at point k, to values at other (usually neighboring) points. A simple example is the equation

$$y(k+1) = ay(k) \qquad k = 0, 1, 2, \ldots \tag{2-1}$$

Difference equations may, however, be much more complicated than this. For example,

$$ky(k+2)y(k+1) = \tfrac{1}{2}\sqrt{y(k)y(k-1)} \qquad k = 0, 1, 2, \ldots \tag{2-2}$$

A difference equation is really a set of equations; one equation for each of the index points k. Therefore, part of the specification of a difference equation is the set of integers k for which it is to hold. In general, this set of integers must be a sequence of successive values, of either finite or infinite duration, such as $k=0, 1, 2, 3, \ldots, N$ or $k=0, 1, 2, 3, \ldots$. Often, if the sequence is not explicitly stated, it is to be understood that either it is arbitrary or that it is the most frequently used sequence $k=0, 1, 2, 3, \ldots$. In many cases the practical context of the equation makes the appropriate range clear. In any event, once the sequence is defined, the corresponding values of k can each be substituted into the difference equation to obtain an explicit equation relating various y's.

As an example, the simple difference equation (2-1) is equivalent to the following (infinite) set of equations:

$$\begin{aligned} y(1) &= ay(0) \\ y(2) &= ay(1) \\ y(3) &= ay(2) \\ &\vdots \end{aligned} \tag{2-3}$$

Difference equations, just as any set of equations, can be viewed in two ways. If the values $y(k)$ are known, or defined through some alternate description, the difference equation represents a relation among the different values. If, on the other hand, the values are not known, the difference equation is viewed as an equation that can be solved for the unknown $y(k)$ values. In either interpretation, it is often useful to regard $y(k)$ as a *function* on the index set. The difference equation then defines a relationship satisfied by the function.

The term *difference equation* is used in order to reflect the fact that the various time points in the equation slide along with the index k. That is, the terms involve the unknowns $y(k)$, $y(k+1)$, $y(k+2)$, $y(k-1)$, $y(k-2)$, and so forth, rather than a mixture of fixed and sliding indices, such as, say, $y(k)$, $y(k-1)$, $y(1)$, and $y(8)$. Indeed, since all indices slide along with k, it is possible by suitable (but generally tedious) manipulation to express a difference equation in terms of *differences* Δ^i of various orders, defined by $\Delta^0(k) = y(k)$, $\Delta^1(k) = \Delta^0(k+1) - \Delta^0(k)$, $\Delta^2(k) = \Delta^1(k+1) - \Delta^1(k)$, and so forth. This difference formulation arises naturally when a difference equation is defined as an approximation to a differential equation, but in most cases the more direct form is both more natural and easier to work with.

The *order* of a difference equation is the difference between the highest and lowest indices that appear in the equation. Thus (2-1) is first-order, and (2-2) is third-order.

A difference equation is said to be *linear* if it has the form

$$a_n(k)y(k+n)+a_{n-1}(k)y(k+n-1)+\cdots+a_1(k)y(k+1)+a_0(k)y(k)=g(k) \tag{2-4}$$

for some given functions $g(k)$ and $a_i(k)$, $i=0,1,2,\ldots,n$. The unknown function y appears linearly in the equation. The $a_i(k)$'s in these equations are referred to as *coefficients* of the linear equation. If these coefficients do not depend on k, the equation is said to have *constant coefficients* or to be *time-invariant.* The function $g(k)$ is variously called the *forcing term*, the *driving term*, or simply the *right-hand side.*

Solutions

A *solution* of a difference equation is a function $y(k)$ that reduces the equation to an identity. For example, corresponding to the first-order equation

$$y(k+1)=ay(k)$$

the function $y(k)=a^k$ reduces the equation to an identity, since $y(k+1)=a^{k+1}=aa^k=ay(k)$.

A solution to a difference equation can alternatively be viewed as a *sequence* of numbers. Thus, for the equation above with $a=1/2$ a solution is represented by the sequence 1, 1/2, 1/4, 1/8, ... The solution is easily expressed in this case as $(1/2)^k$. In general, however, there may not be a simple representation, and it is therefore often preferable, in order to simplify conceptualization, to view a solution as a sequence—stepping along with the time index k. The two viewpoints of a solution as some (perhaps complicated) function of k and as a sequence of numbers are, of course, equivalent.

Example 1. Consider the linear difference equation

$$(k+1)y(k+1)-ky(k)=1$$

for $k=1,2,\ldots$ A solution is

$$y(k)=1-1/k$$

To check this we note that $y(k)=(k-1)/k$, $y(k+1)=k/(k+1)$, and thus $(k+1)y(k+1)-ky(k)=k-(k-1)=1$.

There are other solutions as well. Indeed, it is easily seen that

$$y(k)=1-A/k$$

is a solution for any constant A.

Example 2. A nonlinear difference equation that arises in genetics (see Chapter 10) is

$$y(k+1)=\frac{y(k)}{1+y(k)}, \qquad k=0,1,2,\ldots$$

It has the solution

$$y(k)=\frac{A}{1+Ak}$$

where A is an arbitrary constant.

Example 3. Consider the nonlinear difference equation

$$y(k+1)^2+y(k)^2=-1$$

Since $y(k)$ is defined as a real-valued function, the left-hand side can never be less than zero; hence no solution can exist.

2.2 EXISTENCE AND UNIQUENESS OF SOLUTIONS

As with any set of equations, a difference equation need not necessarily possess a solution, and if it does have a solution, the solution may not be unique. These facts are illustrated by the examples in Sect. 2.1. We now turn to a general examination of the existence and uniqueness questions.

Initial Conditions

One characteristic and essential feature of a difference equation is that, over a finite interval of time, as indexed by k, there are more unknowns than equations. For example, the first-order difference equation $y(k+1)=2y(k)$ when enumerated for two time periods $k=0, 1$ becomes

$$y(1)=2y(0)$$
$$y(2)=2y(1)$$

which is a system of *two* equations and *three* unknowns. Therefore, from the elementary theory of equations, we expect that it may be necessary to assign a value to one of the unknown variables in order to specify a unique solution. If the difference equations were applied to a longer sequence of index values, each new equation would add both one new equation and one new unknown. Therefore, no matter how long the sequence, there would always be one more unknown than equations.

In the more general situation where the difference equation for each fixed k involves the value of $y(k)$ at $n+1$ successive points, there are n more unknowns than equations in any finite set. This can be seen from the fact that the first equation involves $n+1$ unknown variables, and again each additional equation adds both one more unknown and one more equation—keeping the surplus constant at n. This surplus allows the values of n variables to be specified arbitrarily, and accordingly, there are n degrees of freedom in the solution of a difference equation. These degrees of freedom show up in the form of arbitrary constants in the expression for the general solution of the equation.

In principle, the n arbitrary components of the solution can be specified in various ways. However, it is most common, particularly in the context of dynamic systems evolving forward in time, to specify the first n values of $y(k)$; that is, the values $y(0), y(1), \ldots, y(n-1)$. The corresponding specified values are referred to as *initial conditions.* For many difference equations, specification of a set of values for initial conditions leads directly to a corresponding unique solution of the equation.

Example 1. The first-order difference equation

$$y(k+1) = ay(k)$$

corresponding to geometric growth, has the general solution $y(k) = Ca^k$. Substituting $k=0$, we see that $y(0) = C$, and the solution can be written in terms of the initial condition as $y(k) = y(0)a^k$.

Example 2. Consider the second-order difference equation

$$y(k+2) = y(k)$$

This equation can be regarded as applying separately to the even and the odd indices k. Once $y(0)$ is specified, the equation implies the same value of $y(k)$ for all even k's, but the single value of $y(k)$ for all odd k's remains arbitrary. Once $y(1)$ is also specified, the entire sequence is determined. Thus, specification of $y(0)$ and $y(1)$ determine a unique solution. The solution can be written as

$$y(k) = \left[\frac{y(0)+y(1)}{2}\right] + (-1)^k \left[\frac{y(0)-y(1)}{2}\right]$$

Existence and Uniqueness Theorem

Although, in general, difference equations may not possess solutions, most difference equations encountered in applications do. Moreover, it is usually not necessary to exhibit a solution in order to be assured of its existence, for the very structure of the most common difference equations implies that a solution exists.

As indicated above, even if existence is guaranteed, we do not expect that the solution to a difference equation will be unique. The solution must be restricted further by specifying a set of initial conditions. The theorem proved below is a formal statement of this fact. The assumption of suitable structure, together with appropriately specified initial conditions, guarantees existence of a unique solution.

The essential idea of the theorem is quite simple. It imposes a rather modest assumption that allows the solution of a difference equation to be computed forward recursively, starting with the given set of initial conditions and successively determining the values of the other unknowns. Stated another way, the theorem imposes assumptions guaranteeing that the difference equation represents a truly dynamic system, which evolves forward in time.

Existence and Uniqueness Theorem. *Let a difference equation of the form*

$$y(k+n)+f[y(k+n-1), y(k+n-2), \ldots, y(k), k]=0 \tag{2-5}$$

where f is an arbitrary real-valued function, be defined over a finite or infinite sequence of consecutive integer values of k ($k = k_0, k_0+1, k_0+2, \ldots$). The equation has one and only one solution corresponding to each arbitrary specification of the n initial values $y(k_0)$, $y(k_0+1), \ldots, y(k_0+n-1)$.

Proof. Suppose the values $y(k_0), y(k_0+1), \ldots, y(k_0+n-1)$ are specified. Then the difference equation (2-5), with $k = k_0$, can be solved uniquely for $y(k_0+n)$ simply by evaluating the function f. Then, once $y(k_0+n)$ is known, the difference equation (2-5) with $k = k_0+1$ can be solved for $y(k_0+n+1)$, and so forth for all consecutive values of k. ∎

It should be noted that no restrictions are placed on the real-valued function f. The function can be highly nonlinear. The essential ingredient of the result is that the y of leading index value can be determined from previous values, and this leading index increases stepwise. A special class of difference equations which satisfies the theorem's requirements is the nth-order linear difference equation

$$y(k+n)+a_{n-1}(k)y(k+n-1)+\cdots+a_0(k)y(k)=g(k)$$

This equation conforms to (2-5), with the function f being just a sum of terms.

2.3 A FIRST-ORDER EQUATION

The first-order difference equation

$$y(k+1)=ay(k)+b \tag{2-6}$$

arises in many important applications, and its analysis motivates much of the general theory of difference equations. The equation is linear, has a constant coefficient a, and a constant forcing term b.

The general solution to this equation is easily deduced. The most straightforward solution procedure is to determine successive values recursively, as outlined in the previous section. Thus, we arbitrarily specify the value of y at an initial point k_0, say $k_0 = 0$, and specify $y(0) = C$. This leads immediately to the following successive values:

$$\begin{aligned} y(0) &= C \\ y(1) &= ay(0) + b = aC + b \\ y(2) &= ay(1) + b = a^2C + ab + b \\ y(3) &= a^3C + a^2b + ab + b \end{aligned}$$

The general term is

$$y(k) = a^kC + (a^{k-1} + a^{k-2} + \cdots + a + 1)b \tag{2-7}$$

For $a = 1$, the expression reduces simply to

$$y(k) = C + kb$$

For $a \neq 1$, the expression can be somewhat simplified by collapsing the geometric series, using

$$1 + a + a^2 + \cdots + a^{k-1} = \frac{1 - a^k}{1 - a}$$

Therefore, the desired solution in closed-form is

$$y(k) = \begin{cases} C + kb, & a = 1 \\ a^kC + \dfrac{1 - a^k}{1 - a}b, & a \neq 1 \end{cases} \tag{2-8}$$

This solution can be checked by substituting it into the original difference equation (2-6).

When $a \neq 1$ another way of displaying the general solution (2-8) is sometimes more convenient:

$$y(k) = Da^k + \frac{b}{1 - a}$$

where D is an arbitrary constant. Clearly this new constant D is related to the earlier constant C by $D = C - [b/(1 - a)]$. In this form, it is apparent that the solution function is the sum of two elementary functions: the constant function $b/(1 - a)$ and the geometric sequence Da^k.

In addition to acquiring familiarity with the analytic solutions to simple difference equations, it is desirable that one be able to infer these solutions intuitively. To begin developing this ability, consider the special case corresponding to $a = 1$ in (2-6). For this case, the equation states that the new value of y equals the old value plus the constant b. Therefore, successive y's merely accumulate successive additions of the constant b. The general solution is clearly $y(k)=C+kb$, where C is the initial value, $y(0)$.

If $a\neq 1$ the difference equation multiplies the old value by the factor a each period and adds the constant b. It is like storing up value and either paying interest (if $a>1$) or deducting a tax (if $0<a<1$). Clearly, an initial amount C will, after k periods of such a process, be transformed to a^kC. The term b in the equation acts like additional deposits made each period. This leads immediately to (2-7), and then by manipulation to (2-8). This interpretation of the equation is explored formally in Sect. 2.4 where the classical amortization formula is derived.

2.4 CHAIN LETTERS AND AMORTIZATION

The examples presented here and in Sect. 2.5 illustrate how first-order difference equations arise in various situations, and how the general solution formula provides a basis for analysis. Although the three examples all lead to the same form of difference equation, they have three different analysis objectives. The chain letter problem is simply one of computing the solution for a particular value of k. The amortization problem is one of determining an appropriate repayment level; it requires full use of the solution formula. The cobweb model of economic interaction leads to an analysis of stability, relating solution behavior to the model parameters.

Example 1 (The Chain Letter). Suppose you receive a chain letter that lists six names and addresses. The letter asks you to send 10¢ to the first person on the list. You are then to make up a new letter with the first name deleted and your name added to the bottom of the list. You are instructed to send a copy of this new letter to each of five friends. You are promised that within a few weeks you will receive up to $1562.50.

Although chain letters are illegal, you might find it amusing to verify the letter's promise under the hypothesis that you and everyone else were to follow the instructions, thus not "breaking the chain." The spreading of these letters can be formulated in terms of a difference equation.

Let us follow only those letters that derive from the letter you receive. Let $y(k)$ denote the number of letters in the kth generation, with the letter you receive corresponding to $y(0)=1$, the letters written by you corresponding to $y(1)$, the letters written by those you contact as $y(2)$, and so forth. Each letter written induces five letters in the next generation. Thus, the appropriate relation between successive generations is

$$y(k+1)=5y(k)$$

With the initial condition $y(0)=1$, the solution is

$$y(k)=5^k$$

According to the letter's instructions, all recipients of sixth generation letters should send you 10¢. This would be $5^6=15{,}625$ letters and \$1562.50.

Example 2 (Interest and Amortization). As mentioned earlier, the accumulation of bank deposits can be described by a first-order difference equation. Suppose deposits are made at the end of each year and let $y(k)$ denote the amount in the account at the beginning of year k. If the bank pays no interest, then the account is simply a storage mechanism governed by the equation

$$y(k+1)=y(k)+b(k)$$

where $b(k)$ is the amount of deposit at the beginning of year k. If equal deposits of amount b are made each year, the balance in the account will grow linearly.

If the bank pays interest i, compounded annually, the account balance is governed by

$$y(k+1)=(1+i)y(k)+b$$

since in addition to the simple holding action the bank pays $iy(k)$ at the end of the year as interest. If equal deposits are made, the account will grow according to the solution of the first-order equation.

A similar structure arises when one borrows money at an interest rate i. The total debt increases just as would the balance in an account paying that interest. *Amortization* is a method for repaying an initial debt, including the interest and original principal, by a series of payments (usually at equal intervals and of equal magnitude). If a payment B is made at the end of each year the total debt will satisfy the equation

$$d(k+1)=(1+i)d(k)-B$$

where $d(0)=D$, the initial debt. If it is desired to amortize the debt so that it is paid off at the end of n years, it is necessary to select B so that $d(n)=0$.

The general solution developed in Sect. 2.3, implies

$$d(n)=D(1+i)^n-\frac{1-(1+i)^n}{1-(1+i)}B$$

Setting $d(n)=0$ yields

$$B\frac{1-(1+i)^n}{-i}=D(1+i)^n$$

which simplifies to the standard amortization formula

$$B = \frac{iD}{1-(1+i)^{-n}}$$

2.5 THE COBWEB MODEL

A classic dynamic model of supply and demand interaction is described by a first-order dynamic equation. The model's behavior can be conveniently displayed using the graph of supply and demand curves. The diagram resulting from this analysis resembles a cobweb, and this is responsible for the model's name.

The cobweb model is concerned with a single commodity, say corn. The demand d for the commodity depends on its price p through a function $d(p)$. Since the amount that consumers buy decreases as the price increases, $d(p)$ decreases as p increases. For purposes of this example, we assume that the demand function is linear

$$d(p) = d_0 - ap$$

where d_0 and a are positive constants.

Likewise, the amount s of the commodity that will be supplied by producers also depends on the price p through a function $s(p)$. Usually, $s(p)$ increases as p increases. (For instance, a high price will induce farmers to plant more corn.) We assume the specific linear form

$$s(p) = s_0 + bp$$

where b is positive (s_0 may have any value, but is usually negative).

The two curves are shown together in Fig. 2.1*a*. In equilibrium the demand must equal supply, which corresponds to the point where the two curves intersect. The equilibrium price is attained, however, only after a series of adjustments by consumers and producers. Each adjustment corresponds to movement along the appropriate demand or supply curve. It is the dynamics of this adjustment process that we wish to describe.

We assume that at period k there is a prevailing price $p(k)$. The producer bases his production in period k on this price. However, due to the time lag in the production process (growing corn in our example) the resulting supply is not available until the next period, $k+1$. When that supply is available, its price will be determined by the demand function—the price will adjust so that all of the available supply will be sold. This new price at $k+1$ is observed by the producers who then, accordingly, initiate production for the next period, and a new cycle begins.

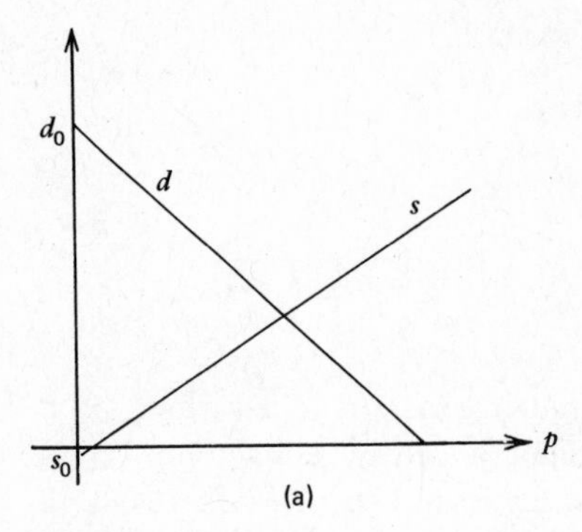

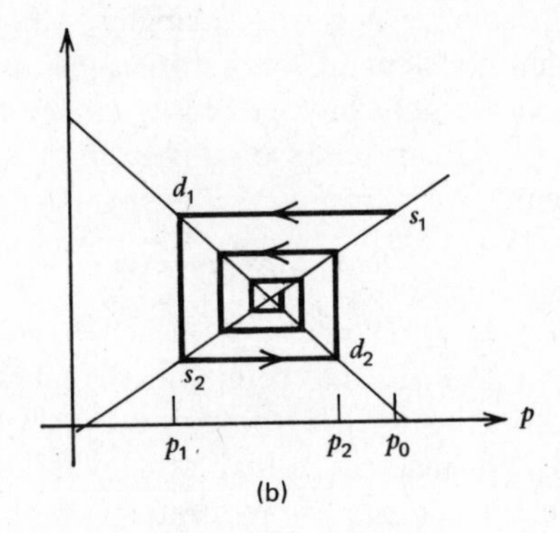

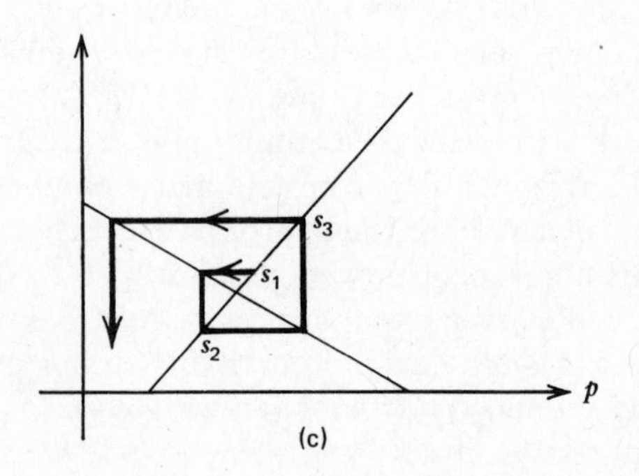

Figure 2.1. The cobweb model.

The equation

$$s(k+1) = s_0 + bp(k)$$

expresses the fact that the supply at period $k+1$ is determined by the price at period k through the supply function. Also,

$$d(k+1) = d_0 - ap(k+1)$$

formally states the fact that demand at period $k+1$ is determined by the price at period $k+1$ through the demand function. Finally, imposing the condition that at equilibrium supply must equal demand leads to the dynamic equation

$$s_0 + bp(k) = d_0 - ap(k+1)$$

which can be restated in the standard form for difference equations

$$p(k+1) = -\frac{b}{a}\,p(k) + \frac{d_0 - s_0}{a}$$

The equilibrium price can be found by setting $p(k) = p(k+1)$ leading to

$$p = \frac{d_0 - s_0}{a+b}$$

This price, if once established, would persist indefinitely. The question arises, however, as to whether this price will ever be established or even whether over successive periods the price will tend toward this equilibrium price, rather than diverging away from it.

Using the general solution of the first-order equation we find

$$p(k) = (-b/a)^k p(0) + \frac{1 - (-b/a)^k}{a+b}(d_0 - s_0)$$

If $b < a$, it follows that as $k \to \infty$ the solution will tend toward the equilibrium value since the $(-b/a)^k$ terms all go to zero. Equilibrium will be attained (at least in the limit) no matter what the initial price. Obviously $b < a$ is both necessary and sufficient for this convergence property to hold.

Now for the cobweb interpretation. Let us trace the path of supply and demand over successive periods on the graphs in Fig. 2.1*a*. The results are shown in Figs. 2.1*b* and 2.1*c*, which represent, respectively, a converging and a diverging situation. The price $p(0)$ determines the supply s_1 that will be available in the next period. This supply determines the demand d_1 and hence the price p_1, and so forth. By this reasoning we are led to trace out a rectangular spiral. If $b < a$, the spiral will converge inward. If $b > a$, it will diverge outward.

From this analysis we have deduced an important general conclusion for the cobweb model of economic interaction. For equilibrium to be approached, the slope b of the supply curve must be less than the slope a of the demand curve. Said another way, for equilibrium to be attained, the producers must be less sensitive to price changes than the consumers.

2.6 LINEAR DIFFERENCE EQUATIONS

As defined earlier, a difference equation is *linear* if it has the form

$$a_n(k)y(k+n)+a_{n-1}(k)y(k+n-1)+\cdots+a_1(k)y(k+1) + a_0(k)y(k)=g(k) \quad (2\text{-}9)$$

Linearity makes it possible to explicitly examine the relationship between various solutions to the difference equation, leading to a rather complete theory. In some cases this theory actually leads to analytical expressions for solutions, but even when such expressions are not obtainable, the theory of linear equations provides important structural information.

The Homogeneous Equation

A basic concept underlying the general theory is that of a homogeneous linear difference equation. The linear difference equation (2-9) is said to be *homogeneous* if $g(k)=0$ for all k in the set over which the equation is defined. That is, a linear equation is homogeneous if its forcing term (or right-hand side) is zero. Equation (2-9) is said to be *nonhomogeneous* if $g(k)\neq 0$ for some k. Given a general nonhomogeneous linear difference equation, one associates with it the corresponding homogeneous equation (2-10) obtained by setting $g(k)$ to be zero for all k:

$$a_n(k)y(k+n)+a_{n-1}(k)y(k+n-1)+\cdots+a_1(k)y(k+1)+a_0(k)y(k)=0 \quad (2\text{-}10)$$

This corresponding homogeneous equation plays a central role in specifying solutions to the original nonhomogeneous equation.

Two observations indicate the importance of homogeneous equations for the theory of linear equations. First, the difference between two solutions of the nonhomogeneous equation (2-9) must satisfy the homogeneous equation (2-10). Second, if a solution to the homogeneous equation is added to a solution to the nonhomogeneous equation, the result is also a solution to the nonhomogeneous equation. These two observations (which are carefully justified in the proof below) are converses, and together they imply the following theorem characterizing the structure of the solution set of a nonhomogeneous equation.

Theorem 1. *Let $\bar{y}(k)$ be a given solution to the linear difference equation* (2-9). *Then the collection of all solutions to this equation is the collection of all functions of the form $y(k)=\bar{y}(k)+z(k)$, where $z(k)$ is a solution of the corresponding homogeneous equation* (2-10).

Proof. Two things must be proved. First, it must be shown that if $y(k)$ and $\bar{y}(k)$ are both solutions to the nonhomogeneous equation, then the difference $z(k)=y(k)-\bar{y}(k)$ is a solution to the homogeneous equation. To prove this note that if $y(k)$ and $\bar{y}(k)$ are both solutions to (2-9), then

$$a_n(k)y(k+n)+\cdots+a_0(k)y(k)=a_n(k)\bar{y}(k+n)+\cdots+a_0(k)\bar{y}(k)$$

Corresponding terms can be combined to yield

$$a_n(k)[y(k+n)-\bar{y}(k+n)]+\cdots+a_0(k)[y(k)-\bar{y}(k)]=0$$

Therefore, the difference $z(k)=y(k)-\bar{y}(k)$ satisfies the homogeneous equation.

Second, it must be shown that (conversely) when any solution of the homogeneous equation is added to a solution of the nonhomogeneous equation, the sum is also a solution of the nonhomogeneous equation. To prove this, let $\bar{y}(k)$ and $z(k)$ be solutions of the nonhomogeneous and homogeneous equations, respectively. Let $y(k)=\bar{y}(k)+z(k)$. It then follows that

$$\begin{aligned}
a_n(k)[\bar{y}(k+n)+z(k+n)]&+a_{n-1}(k)[\bar{y}(k+n-1)+z(k+n-1)]\\
&+\cdots+a_0(k)[\bar{y}(k)+z(k)]\\
&=a_n(k)\bar{y}(k+n)+a_{n-1}(k)\bar{y}(k+n-1)+\cdots+a_0(k)\bar{y}(k)\\
&\quad+a_n(k)z(k+n)+a_{n-1}(k)z(k+n-1)+\cdots+a_0(k)z(k)\\
&=g(k)+0\\
&=g(k)
\end{aligned}$$

Therefore, $y(k)=\bar{y}(k)+z(k)$ is a solution of the nonhomogeneous equation. ∎

Theorem 1 reveals the importance of homogeneous equations in defining general solutions to nonhomogeneous equations. This result is useful for investigating some simple linear difference equations.

Example 1. Consider again (see Sect. 2.1) the linear difference equation

$$(k+1)y(k+1)-ky(k)=1$$

defined for $k\geq 1$. By inspection, one solution is $\bar{y}(k)=1$. From Theorem 1 we know that the general solution is the sum of this particular solution and the general solution of the associated homogeneous equation

$$(k+1)z(k+1)-kz(k)=0$$

The general solution to this homogeneous equation is $z(k) = A/k$, where A is an arbitrary constant. Thus by Theorem 1, the general solution to the original nonhomogeneous equation is

$$y(k) = 1 + A/k$$

One interpretation of Theorem 1 is that the solutions of homogeneous equations provide the flexibility, or the degrees of freedom, in solutions to linear difference equations. From this view it is natural to seek a characterization of the solution set of a homogeneous equation. The remainder of this section addresses this subject. The next theorem, a first result in this direction, establishes the linearity of the solution set.

Theorem 2. *If* $z_1(k), z_2(k), \ldots, z_m(k)$ *are all solutions to the homogeneous equation* (2-10), *then any linear combination of these* m *solutions*

$$z(k) = c_1 z_1(k) + c_2 z_2(k) + \cdots + c_m z_m(k)$$

where $c_1, c_2, \ldots, c_m$ *are arbitrary constants, is also a solution of* (2-10).

Proof. For notational convenience let us consider the case $m = 2$. The general case can be proved in the same way.

If $z(k) = c_1 z_1(k) + c_2 z_2(k)$ is substituted into the left-hand side of (2-10), one obtains

$$a_n(k)[c_1 z_1(k+n) + c_2 z_2(k+n)] + a_{n-1}(k)[c_1 z_1(k+n-1) + c_2 z_2(k+n-1)] + \cdots + a_0(k)[c_1 z_1(k) + c_2 z_2(k)]$$

This can be rewritten as

$$c_1\{a_n(k) z_1(k+n) + a_{n-1}(k) z_1(k+n-1) + \cdots + a_0(k) z_1(k)\} + c_2\{a_n(k) z_2(k+n) + a_{n-1}(k) z_2(k+n-1) + \cdots + a_0(k) z_2(k)\}$$

The expression is zero, since each of the two bracketed expressions is zero. ▮

Theorem 2 shows that a large collection of solutions to the homogeneous equation can be derived from a few known solutions. This raises the rather obvious question as to whether it is possible to find a special finite number of solutions that can be used to generate all other solutions. This is the question to which we now turn.

A Fundamental Set of Solutions

We focus on the important special case where $a_n(k) \neq 0$. In that case, Eq. (2-10) can be divided by $a_n(k)$ and, without loss of generality, it can be assumed that the homogeneous equation under investigation is

$$z(k+n) + a_{n-1}(k) z(k+n-1) + \cdots + a_0(k) z(k) = 0 \qquad (2\text{-}11)$$

for $k=0, 1, 2, \ldots$. In this form the basic existence and uniqueness theorem of Sect. 2.2 is applicable. There is, therefore, a natural correspondence between a set of initial conditions and a solution.

Let us construct a special set of n different solutions $\bar{z}_1(k)$, $\bar{z}_2(k), \ldots, \bar{z}_n(k)$. Define $\bar{z}_1(k)$ to be the solution corresponding to the initial conditions $\bar{z}_1(0)=1$, $\bar{z}_1(1)=0$, $\bar{z}_1(2)=0, \ldots, z_1(n-1)=0$. And, in general, let $\bar{z}_i(k)$ be the solution corresponding to initial conditions that are all zero except the $(i-1)$th, which equals one. This set of n distinct solutions is called a *fundamental set.** It can be generated by solving the difference equation n times, once for each of the special sets of initial conditions. These n special solutions can be used, as described in the following theorem, to construct all solutions to the homogeneous equation (2-11).

Theorem 3. *If $z(k)$ is any solution to the homogeneous equation* (2-11), *then $z(k)$ can be expressed in terms of the n fundamental solutions in the form*

$$z(k)=c_1\bar{z}_1(k)+c_2\bar{z}_2(k)+\cdots+c_n\bar{z}_n(k)$$

for some constants $c_1, c_2, \ldots, c_n$.

Proof. Let $z(k)$ be an arbitrary solution to (2-11). Corresponding to its initial values define

$$c_i=z(i-1) \qquad i=1, 2, \ldots, n$$

Now consider the special solution $y(k)$ defined by

$$y(k)=c_1\bar{z}_1(k)+c_2\bar{z}_2(k)+\cdots+c_n\bar{z}_n(k)$$

It has the same n initial conditions as the original solution $z(k)$, and therefore it follows by the existence and uniqueness theorem of Sect. 2.2 that $y(k)= z(k)$. ∎

At this point it is perhaps useful to point out that the approach presented in this section represents the classical theory of difference equations. In this approach it is recognized that the solution to a linear difference equation is in general not unique. An nth-order equation has n degrees of freedom expressed earlier by the fact that n arbitrary initial conditions can be specified. Theorems 1 and 2 of this section provide an alternative characterization of this nonuniqueness in terms of solutions to the homogeneous equation, which themselves can be combined in arbitrary combinations. This classical approach, focusing on the solution function as a unit, is rather algebraic in its viewpoint, and somewhat suppresses the inherently dynamic character of difference equations. Essentially, the classic approach exploits linearity more than dynamic structure.

* This is not the only fundamental set of solutions, but it is the most convenient. In general, as explained more fully in Chapter 4, a fundamental set is any set of n linearly independent solutions.

Theorem 3 bridges the gap between purely algebraic and dynamic viewpoints. Each of the n free initial conditions defines a single degree of freedom in the dynamic viewpoint and corresponds directly to a fundamental solution of the homogeneous equation. Therefore, an arbitrary solution can be specified either by assigning its n initial conditions or, equivalently, by assigning weights to its component fundamental solutions.

Example 2. Consider the linear homogeneous difference equation

$$z(k+2)-2z(k+1)+z(k)=0$$

Since it is second order, we know that there will be two degrees of freedom in its general solution. This freedom will be manifested by the fact that two initial conditions can be specified, or by the fact that two fundamental solutions can be found.

By inspection it is clear that the two functions

$$z_1(k)=1 \qquad z_2(k)=k$$

are both solutions. The two fundamental solutions can be easily found to be

$$\bar{z}_1(k)=1-k$$
$$\bar{z}_2(k)=k$$

An arbitrary solution, therefore, has the form

$$\begin{aligned} z(k) &= c_1\bar{z}_1(k)+c_2\bar{z}_2(k)=c_1(1-k)+c_2k \\ &= c_1+(c_2-c_1)k \end{aligned}$$

or, since both c's are arbitrary,

$$z(k)=c+dk$$

for arbitrary c and d.

Linear Independence

Given a finite set of functions $z_1(k)$, $z_2(k), \ldots, z_m(k)$ defined for a set of integers, say $k=0, 1, 2, \ldots, N$, we say that these functions are *linearly independent* if it is impossible to find a relation of the form

$$c_1z_1(k)+c_2z_2(k)+\cdots+c_mz_m(k)=0$$

valid for all $k=0, 1, 2, \ldots, N$, except by setting $c_1=c_2=c_3=\cdots=c_m=0$. If it *is* possible to find such a relation, the set of functions is said to be *linearly dependent.*

An example of a linearly *dependent* set of functions for $k=0, 1, 2, \ldots$ is the set $z_1(k)=1$, $z_2(k)=2^k$, $z_3(k)=2^{k+1}-3$ because $3z_1(k)-2z_2(k)+z_3(k)=$

0. An example of a linearly independent set of functions is the set $z_1(k)=1$, $z_2(k)=2^k$, $z_3(k)=3^k$ because if not all coefficients are zero, there is no linear combination of these that is identically equal to zero.

The fundamental set of solutions $\bar{z}_1(k)$, $\bar{z}_2(k), \ldots, \bar{z}_n(k)$ corresponding to the homogeneous difference equation (2-11) is a linearly independent set. This is easy to see. Since $\bar{z}_i(k)$ is the only one in the set that is nonzero for $k=i-1$, the only way to get $c_1\bar{z}_1(k)+c_2\bar{z}_2(k)+\cdots+c_n\bar{z}_n(k)=0$ for $k=i-1$ is for $c_i=0$. This argument is valid for all $i=1,2,\ldots,n$. Therefore, the only linear combination that is identically zero is the one having all zero coefficients.

An extension of Theorem 3 is that *any* set of n linearly independent solutions of the homogeneous equation can play the role of the fundamental solutions. This theorem is the final characterization of the solution set. The details of the proof are not given here, since a more general version is established in Chapter 4.

Theorem 4. *Suppose $z_1(k), z_2(k), \ldots, z_n(k)$ is a linearly independent set of solutions to the homogeneous equation* (2-11). *Then any solution $z(k)$ to* (2.11) *can be expressed as a linear combination*

$$z(k)=c_1z_1(k)+c_2z_2(k)+\cdots+c_nz_n(k)$$

for some constants $c_1, c_2, \ldots, c_n$.

Solution of Nonhomogeneous Equation

The theory of this section leads to a general method for finding a solution to a nonhomogeneous equation of the form

$$y(k+n)+a_{n-1}(k)y(k+n-1)+\cdots+a_0(k)y(k)=g(k) \tag{2-12}$$

which satisfies a given set of initial conditions. The procedure is to find (a) a set of n linearly independent solutions to the corresponding homogeneous equation, and (b) a particular solution to the nonhomogeneous equation that does not necessarily satisfy the given conditions. The solution to the nonhomogeneous equation is then modified by the addition of suitable linear combinations of solutions to the homogeneous equation so that the initial conditions are satisfied.

If $\bar{y}(k)$ is a particular solution and $z_1(k)$, $z_2(k), \ldots, z_n(k)$ are linearly independent solutions to the corresponding homogeneous equation, then the general solution of (2-12) is

$$y(k)=\bar{y}(k)+c_1z_1(k)+c_2z_2(k)+\cdots+c_nz_n(k)$$

If a different particular solution $\bar{y}(k)$ were used, it would simply change the values of the c_i's in the general solution.

Forward recursion methods can always be used to find a particular solution to the nonhomogeneous equation and the linearly independent solutions to the homogeneous equations. Analytical methods of finding these solutions are available only for special cases.

2.7 LINEAR EQUATIONS WITH CONSTANT COEFFICIENTS

In the important case of linear difference equations with constant coefficients, it is possible to find all solutions to the homogeneous equation. As shown in Sect. 2.6, these provide the means for calculating general solutions to an equation with constant coefficients, once a particular solution is known.

The key result is that corresponding to every linear homogeneous equation with constant coefficients, there is a geometric sequence that is a solution; that is, there is a solution of the form $z(k)=\lambda^k$ for some suitable constant λ. Because of this fact, geometric sequences play a major role in the theory of linear homogeneous equations with constant coefficients.

The Characteristic Equation

Consider the linear difference equation with constant coefficients

$$z(k+n)+a_{n-1}z(k+n-1)+\cdots+a_0z(k)=0 \tag{2-13}$$

We hypothesize a solution of the form

$$z(k)=\lambda^k \tag{2-14}$$

where λ is a constant (not yet specified). Substituting this trial solution into (2-13) yields

$$\lambda^{k+n}+a_{n-1}\lambda^{k+n-1}+\cdots+a_0\lambda^k=0 \tag{2-15}$$

and multiplying this by λ^{-k} yields

$$\lambda^n+a_{n-1}\lambda^{n-1}+\cdots+a_1\lambda+a_0=0 \tag{2-16}$$

which depends on λ, but not on k. This last equation is called the *characteristic equation* of the difference equation (2-13). It is clear from the above argument that λ must satisfy the characteristic equation if $z(k)=\lambda^k$ is to be a solution to the difference equation (2-13). Conversely, since the above steps are reversible, any λ satisfying the characteristic equation provides a solution of the form (2-14) to the difference equation. Accordingly, the role of the characteristic equation is summarized by the following statement.

Theorem. *A necessary and sufficient condition for the geometric sequence $z(k)=\lambda^k$ to be a solution to* (2-13) *is that the constant λ satisfy the characteristic equation* (2-16).

The left-hand side of the characteristic equation is a polynomial of degree n—generally referred to as the *characteristic polynomial.* A root of this polynomial is called a *characteristic value.* By the fundamental theorem of algebra, it is known that such a polynomial can be factored into n first-degree terms so that the polynomial has n roots (although the roots may not be distinct, and some may be complex numbers). Therefore, there is always at least one solution to the characteristic equation, and, accordingly, there is always a geometric sequence that is a solution to the homogeneous difference equation.

If there are n distinct solutions to the characteristic equation, each of them provides a distinct geometric sequence that is a solution to the difference equation. Moreover, it can be easily shown that these n solutions are linearly independent; hence by linear combination they can be used to generate *all* solutions to the homogeneous equation. Thus, for this case, the n distinct roots of the characteristic polynomial, when translated to geometric sequences, provide a complete resolution to the problem of determining solutions to the homogeneous equation.

In some cases the characteristic equation will have complex roots. However, because the coefficients of the characteristic polynomial are all real, complex roots must occur in complex conjugate pairs. That is, if $\lambda_1 = a + ib$ is a root, then so is $\lambda_2 = a - ib$. The expression $c_1\lambda_1^k + c_2\lambda_2^k$ in the general solution will be real-valued if c_1 and c_2 are selected as complex conjugates. Thus, even though we are interested exclusively in real solutions to difference equations, complex roots often are used in the construction of such solutions.

Example 1 (First-Order Equation). Consider the familiar first-order equation

$$y(k+1) = ay(k)$$

The characteristic equation corresponding to this (homogeneous) difference equation is

$$\lambda - a = 0$$

which has the single solution $\lambda = a$. Therefore, we expect solutions of the form

$$y(k) = Ca^k$$

which we know to be correct from our earlier discussion of this equation.

Now consider the nonhomogeneous equation

$$y(k+1) = ay(k) + b$$

with $a \neq 1$. As a trial solution let us set $y(k) = c$ for some constant c. If this is to be a solution, the equation

$$c = ac + b$$

must hold. Thus,

$$c = \frac{b}{1-a}$$

corresponds to a particular solution. The general solutions are the sum of this particular solution and solutions to the homogeneous equation. Therefore, the general solution is

$$y(k) = Ca^k + \frac{b}{1-a}$$

which agrees with what was deduced by forward recursion in Sect. 2.3.

Example 2 (Second-Order Equation). Consider the difference equation

$$y(k+2) - 3y(k+1) + 2y(k) = 3^k$$

As a particular solution let us try $y(k) = C3^k$. Substitution of this into the equation yields

$$C(3^{k+2} - 3 \cdot 3^{k+1} + 2 \cdot 3^k) = 3^k$$

Or, multiplying by 3^{-k}, $(9-9+2)C = 1$. Thus, this form of solution is suitable provided

$$C = \tfrac{1}{2}$$

The corresponding characteristic equation is

$$\lambda^2 - 3\lambda + 2 = 0$$

which can be factored to yield

$$(\lambda - 2)(\lambda - 1) = 0$$

The two roots, which are distinct, are $\lambda = 1$ and $\lambda = 2$. Therefore, the general solution to the original nonhomogeneous equation is

$$y(k) = \tfrac{1}{2}\, 3^k + C_1 + C_2 2^k$$

Example 3 (Fibonacci Sequence). The series of numbers

$$1, 1, 2, 3, 5, 8, 13, 21, \ldots$$

is called the Fibonacci sequence. Its terms are generated by the difference equation

$$y(k+2) = y(k+1) + y(k)$$

together with the initial conditions $y(1) = y(2) = 1$. The direct way to calculate the members of the sequence is recursively, summing the last two to get the next. Alternatively, an analytical expression for the general term can be found once the characteristic equation is solved.

The characteristic equation corresponding to the Fibonacci sequence is

$$\lambda^2 - \lambda - 1 = 0$$

Its solutions are

$$\lambda = \frac{1 \pm \sqrt{1+4}}{2}$$

Thus, the two values are

$$\lambda_1 = \frac{1+\sqrt{5}}{2} \simeq 1.618$$

$$\lambda_2 = \frac{1-\sqrt{5}}{2} \simeq -.618$$

The number λ_1 is known as the *golden section* ratio and was considered by early Greeks to be the most aesthetic value for the ratio of two adjacent sides of a rectangle.

In terms of these values, the solution to the Fibonacci difference equation is

$$y(k) = A\lambda_1^k + B\lambda_2^k$$

for some constants A and B. Substitution of the initial conditions for $k = 1$ and $k = 2$ yield, respectively, the equations

$$1 = A\lambda_1 + B\lambda_2$$

$$1 = A\lambda_1^2 + B\lambda_2^2$$

After a fair amount of algebra the solutions can be found to be

$$A = 1/\sqrt{5}$$

$$B = -1/\sqrt{5}$$

Therefore, the expression for the general term is

$$y(k) = \left\{\left(\frac{1+\sqrt{5}}{2}\right)^k - \left(\frac{1-\sqrt{5}}{2}\right)^k\right\}\frac{1}{\sqrt{5}}$$

It might be surprising that this expression, involving as it does several appearances of $\sqrt{5}$, generates a sequence composed entirely of integers. Nevertheless, this *is* the solution.

Example 4 (An Imaginary Root Example). Consider the second-order equation

$$y(k+2)+y(k)=0$$

with initial conditions $y(0)=1$, $y(1)=0$. The characteristic equation is

$$\lambda^2+1=0$$

which has the roots $\lambda=\pm i$ (where $i=\sqrt{-1}$). The general solution is therefore

$$y(k)=c_1(i)^k+c_2(-i)^k$$

Substitution of the given initial conditions yields the equations

$$c_1+c_2=1$$

$$c_1(i)+c_2(-i)=0$$

Thus, $c_1=c_2=\frac{1}{2}$. The desired solution is

$$y(k)=\tfrac{1}{2}(i)^k+\tfrac{1}{2}(-i)^k$$

Although this solution involves imaginary numbers, the solution is actually real for all values of k—the imaginary values all cancel. Indeed the solution is the sequence $1, 0, -1, 0, 1, 0, \ldots$.

Example 5 (Gambler's Ruin). Consider a gambling situation involving two players A and B. An example is roulette where, say, player A is a "guest" and player B is the "house." During any one play of the game there is a probability p, $0<p<1$, that player A wins a chip (or coin) from player B, and a probability $q=1-p$ that player B wins a chip from player A. The players begin with initial holdings of a and b chips, respectively. A player wins overall if he obtains all the chips. What is the probability that player A wins?

To solve this classic problem, consider the general situation where A has k chips, $0\le k\le a+b$, and B has $a+b-k$ chips. Denote the probability under these circumstances that player A eventually wins by $u(k)$. We can deduce a difference equation for $u(k)$.

Assuming player A has k chips, at the conclusion of the next play he will have either $k+1$ or $k-1$ chips, depending on whether he wins or loses that play. The probabilities of eventually winning must therefore satisfy the difference equation

$$u(k)=pu(k+1)+qu(k-1)$$

In addition we have the two auxiliary conditions

$$u(0)=0 \qquad u(a+b)=1$$

This difference equation for $u(k)$ is linear, homogeneous, and has constant

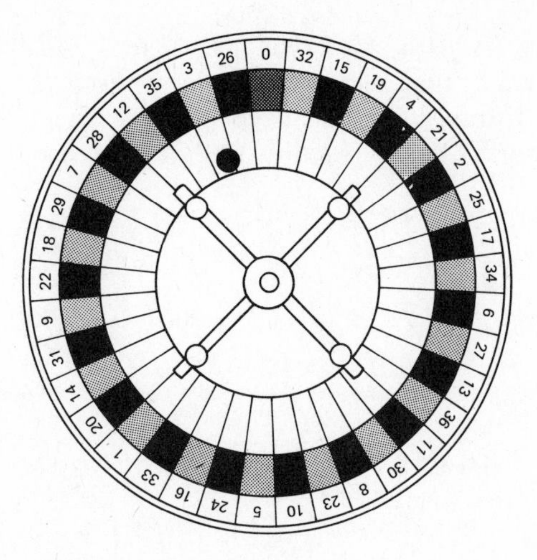

Figure 2.2. Roulette wheel.

coefficients. Its characteristic equation is

$$-p\lambda^2+\lambda-q=0$$

The corresponding roots are $\lambda=1$, $\lambda=q/p$. Accordingly, the general solution (assuming $q\neq p$) is

$$u(k)=c_1+c_2(q/p)^k$$

The two auxiliary conditions give the equations

$$0=c_1+c_2$$

$$1=c_1+c_2(q/p)^{a+b}$$

These can be solved for c_1 and c_2 and the result substituted into the general solution. This leads to

$$u(k)=\frac{1-(q/p)^k}{1-(q/p)^{a+b}}$$

Finally, at the original position where player A has a chips, the corresponding probability of winning is

$$u(a)=\frac{1-(q/p)^a}{1-(q/p)^{a+b}}$$

As a specific example, suppose you play a roulette wheel that has 37 divisions: 18 are red, 18 are black, and one (number 0) is green (see Fig. 2.2).

If you bet on either red or black you win a sum equal to your bet if the outcome is a division of that color. Otherwise you lose your bet.* If the house bank has 1,000,000 francs and you have 100,000 francs, what is the chance that you can "break the bank," betting 1000 francs on red or black each spin of the wheel?

In this case

$$p=\frac{18}{37} \quad q=\frac{19}{37} \qquad a=100 \quad b=1{,}000$$

Thus,

$$u(100)=\frac{1-(19/18)^{100}}{1-(19/18)^{1100}}=3.29\cdot 10^{-24}$$

2.8 DIFFERENTIAL EQUATIONS

Suppose there is an interval, say $t_0 \leq t \leq t_1$, representing, perhaps, an interval of continuous time. Suppose also that there is a value $y(t)$ associated with each point t in that interval. Then $y(t)$ is a function defined on the interval. A differential equation is an equation connecting such a function and some of its derivatives. A simple example is the equation

$$\frac{dy}{dt}=ay$$

and a more complicated example is the equation

$$\frac{d^2y}{dt^2}+[\sin y]\frac{dy}{dt}=\cos t$$

The *order* of a differential equation is the order of the highest derivative that appears in the equation. Thus, the first example above is first order and the second is second order.

As should be reasonably obvious, the mathematics of differential equations is in many respects analogous to that of difference equations. As a general rule any *concept* for one of these mathematical structures has a direct analog for the other, although in some cases the mechanics or the sharpness of theoretical statements may vary when the concept is implemented in the two structures. Thus, just as for difference equations, the concepts of initial conditions, linearity, constant coefficients, and homogeneous linear equations

* In many European casinos the actual procedure following an outcome of green is quite different, yielding more favorable odds to the players. See Problem 12.

all are important for the study of differential equations. These are outlined in the remainder of this chapter.

Quite analogously to the notions associated with difference equations, a differential equation is said to be *linear* if it has the form

$$a_n(t)\frac{d^n y}{dt^n}+a_{n-1}(t)\frac{d^{n-1}y}{dt^{n-1}}+\cdots+a_1(t)\frac{dy}{dt}+a_0(t)y=g(t)$$

for some functions $a_i(t)$, $i=0,1,2,\ldots,n$ and $g(t)$ defined on the given interval. Again the $a_i(t)$'s are referred to as *coefficients* and $g(t)$ as the *driving term* or *right-hand side.*

Initial Conditions

It is usually necessary, just as for difference equations, to specify a set of auxiliary conditions in order to completely specify a unique solution to a differential equation. For example, the first-order equation

$$\frac{dy}{dt}=ay$$

has solutions of the form

$$y(t)=Ce^{at}$$

where C is an arbitrary constant. In order to specify a unique solution, the value of this constant must be pinned down. One way to do this is to specify the initial value $y(0)$, which then determines C by $C=y(0)$.

In general, higher-order differential equations require that additional auxiliary conditions be specified. These additional conditions often are specified by assigning initial conditions to the derivatives of the function as well as assigning its initial value.

Example. Consider the second-order differential equation

$$\frac{d^2y}{dt^2}=0$$

This has a general solution of the form

$$y(t)=A+Bt$$

where A and B are arbitrary constants. To specify a unique solution, two auxiliary conditions must be given. If in this case $y(0)$ and $dy(0)/dt$ are specified, then the constants A and B are determined by

$$A=y(0)$$

$$B=\frac{dy(0)}{dt}$$

Existence and Uniqueness Theorem

It is considerably more difficult to establish an existence proof for differential equations than for difference equations. This is because, although the *concept* of computing a solution by moving forward from a set of initial conditions is usually still valid, the mechanics cannot be reduced to a finite recursion. In order to avoid difficult mathematical developments, which are somewhat tangential to our primary objectives, we simply state an existence result for linear differential equations that is adequate for most of our needs.

Existence and Uniqueness Theorem. *Suppose the coefficients $a_i(t)$, $i = 0, 1, 2, \ldots, n-1$ and the function $g(t)$ are continuous on an interval $0 \leq t \leq T$. Then for any set of values b_i, $i = 0, 1, 2, \ldots, n-1$, there is a unique solution to the linear differential equation*

$$\frac{d^n y}{dt^n} + a_{n-1}(t)\frac{d^{n-1}y}{dt^{n-1}} + \cdots + a_0(t)y = g(t)$$

satisfying the initial conditions

$$y(0) = b_0$$

$$\frac{dy(0)}{dt} = b_1$$

$$\vdots$$

$$\frac{d^{n-1}y(0)}{dt^{n-1}} = b_{n-1}$$

This theorem allows us to think in terms of solving a differential equation by moving forward in t. Once a set of initial conditions is specified, one can imagine moving forward along the resulting solution to a value, say, $t_1 > 0$. At this point, even if the original initial conditions are forgotten, the n corresponding conditions at t_1 serve to specify the unique solution for all $t > t_1$. In general, the conditions at any point determine the entire future behavior of the solution. Therefore, one can consider the solution to be generated by moving forward in time, forgetting the past, but keeping track of the current derivatives that serve as the newest initial conditions for the generation process.

2.9 LINEAR DIFFERENTIAL EQUATIONS

Linear differential equations have an associated theory that is parallel to that of linear difference equations. Again homogeneous equations play a central role.

Homogeneous Equations

A linear differential equation

$$\frac{d^n y}{dt^n} + a_{n-1}(t)\frac{d^{n-1}y}{dt^{n-1}} + \cdots + a_0(t)y = g(t) \tag{2-17}$$

is said to be *homogeneous* if $g(t) = 0$, otherwise it is *nonhomogeneous.* Associated with a general linear equation (2-17) is the corresponding homogeneous equation obtained by setting $g(t) = 0$. It is quite easy to establish the following basic results.

Theorem 1. *Let $\bar{y}(t)$ be a given solution to the linear differential equation* (2-17). *Then the collection of all solutions to this equation is the collection of all functions of the form $y(t) = \bar{y}(t) + z(t)$, where $z(t)$ is a solution to the corresponding homogeneous equation.*

Theorem 2. *If $z_1(t)$, $z_2(t), \ldots, z_m(t)$ are all solutions to a linear homogeneous differential equation, then any linear combination of these m solutions*

$$z(t) = c_1 z_1(t) + c_2 z_2(t) + \cdots + c_m z_m(t)$$

where $c_1, c_2, \ldots, c_m$ are arbitrary constants, is also a solution.

The interpretations and proofs of these results are virtually identical to those for difference equations. The flexibility of solution to a differential equation, as characterized earlier in terms of a set of arbitrary initial conditions, can be interpreted in terms of the addition of arbitrary combinations of solutions to the homogeneous equation. Again, this is the classical viewpoint.

Example 1 (First-Order Equation). Consider the first-order, constant coefficient equation

$$\frac{dy}{dt} = ay + b$$

This equation arises in many applications, and serves as a building block for more complex equations.

It can be seen by inspection that one solution is the constant function

$$\bar{y}(t) = -b/a$$

This solution is regarded as a *particular solution* and all other solutions can be written as the sum of this solution and a solution to the corresponding homogeneous equation

$$\frac{dz}{dt} = az$$

This homogeneous equation has solutions of the form

$$z(t) = Ce^{at}$$

where C is an arbitrary constant. Therefore, the general solution to the original nonhomogeneous equation is

$$y(t) = Ce^{at} - b/a$$

Fundamental Solutions

Corresponding to the homogeneous equation

$$\frac{d^n z}{dt^n} + a_{n-1}(t)\frac{d^{n-1}z}{dt^{n-1}} + \cdots + a_0(t)z = 0 \tag{2-18}$$

it is natural to define the set of n fundamental solutions $\bar{z}_1(t), \bar{z}_2(t), \ldots, \bar{z}_n(t)$ by assigning the special initial conditions

$$\frac{d^k \bar{z}_i(0)}{dt^k} = \begin{cases} 1 & k = i-1 \\ 0 & \text{otherwise} \end{cases}$$

(the zero-th derivative is defined to be the function itself). Thus, each of these solutions has only one nonzero initial condition. It is then easy to prove the analog of the earlier result for fundamental solutions to difference equations.

Theorem 3. *If $z(t)$ is any solution to the homogeneous equation* (2-18), *then $z(t)$ can be expressed in terms of the n fundamental solutions in the form*

$$z(t) = c_1\bar{z}_1(t) + c_2\bar{z}_2(t) + \cdots + c_n\bar{z}_n(t)$$

for some constants $c_1, c_2, \ldots, c_n$.

The concept of linear independence of functions extends to continuous as well as discrete time. The functions $y_1(t), y_2(t), \ldots, y_m(t)$ are *linearly independent* on an interval $t_0 \le t \le t_1$ if there is no set of constants, $c_1, c_2, \ldots, c_m$, at least one of which is nonzero, for which $c_1y_1(t) + c_2y_2(t) + \cdots + c_my_m(t) = 0$ for all t, $t_0 \le t \le t_1$.

Just as for difference equations, it can be shown that the n fundamental solutions are linearly independent. Accordingly, the result of Theorem 3 can be extended to an arbitrary set of n linearly independent solutions.

Theorem 4. *Suppose $z_1(t), z_2(t), \ldots, z_n(t)$ is a linearly independent set of solutions to the homogeneous equation* (2-18). *Then any solution $z(t)$ can be expressed as a linear combination*

$$z(t) = c_1z_1(t) + c_2z_2(t) + \cdots + c_nz_n(t)$$

for some constants $c_1, c_2, \ldots, c_n$.

The Characteristic Equation

For linear differential equations with constant coefficients, the corresponding homogeneous equation can be solved by consideration of an associated characteristic equation. This method is based on assuming that solutions of the form $z(t)=e^{\lambda t}$ exist for some constant λ. Substituting this expression into the equation yields a polynomial equation for λ. Thus, just as for difference equations, the assumptions of linearity and time-invariance are jointly sufficient to simplify greatly the solution of homogeneous differential equations.

To be specific, consider the homogeneous differential equation

$$\frac{d^n z}{dt^n}+a_{n-1}\frac{d^{n-1}z}{dt^{n-1}}+\cdots+a_0 z=0 \tag{2-19}$$

Suppose there exists a solution of the form $z(t)=e^{\lambda t}$ for some constant λ. Substituting this into the equation yields

$$\lambda^n e^{\lambda t}+a_{n-1}\lambda^{n-1}e^{\lambda t}+\cdots+a_0 e^{\lambda t}=0$$

Cancelling $e^{\lambda t}$ (since it is never zero) leads to the equation

$$\lambda^n+a_{n-1}\lambda^{n-1}+\cdots+a_0=0 \tag{2-20}$$

This is the *characteristic equation.* The left-hand side is the *characteristic polynomial,* and any root of this polynomial is a *characteristic value.*

It is clear that if $z(t)=e^{\lambda t}$ is a solution to the homogeneous equation (2-19), then λ must satisfy the characteristic equation. Conversely, if λ is a value satisfying the characteristic equation, then one can trace backward through the above argument to conclude that $z(t)=e^{\lambda t}$ is a solution to the differential equation. Therefore, if the roots of the characteristic polynomial are distinct, n different solutions are obtained in this way, corresponding to the n degrees of freedom inherent in the original equation.

Example 2 (First-order Equation). The first-order differential equation

$$\frac{dy}{dt}=ay$$

has the characteristic equation

$$\lambda-a=0$$

There is only one root, $\lambda=a$, leading to the solution function

$$y(t)=Ce^{at}$$

Example 3 (A Second-order Equation). Consider the homogeneous differential equation

$$\frac{d^2y}{dt^2}-5\frac{dy}{dt}+6y=0$$

The corresponding characteristic equation is

$$\lambda^2 - 5\lambda + 6 = 0$$

with roots $\lambda = 2, 3$. Therefore, two solutions to the differential equation are e^{2t} and e^{3t}. The general solution to the homogeneous differential equation is, accordingly,

$$y(t) = c_1 e^{2t} + c_2 e^{3t}$$

where c_1 and c_2 are arbitrary constants.

2.10 HARMONIC MOTION AND BEATS

To illustrate the analysis of linear time-invariant differential equations, let us consider the important topic of pure oscillatory motion, referred to as *harmonic motion.* This arises in many simple physical phenomena, including the motion of a mass bouncing on a spring, small oscillations of a pendulum, small vibrations of a violin string, oscillations of a tuned electric circuit, and some atomic phenomena.

Such motion is defined by the second-order homogeneous differential equation

$$\frac{d^2y}{dt^2} + \omega^2 y = 0 \tag{2-21}$$

where ω is a fixed positive constant. Figure 2.3 illustrates this for a mass on a spring. If the spring exerts a force that is proportional to its displacement from equilibrium, the displacement y will be governed by this equation. If the force is $-ky$ and the mass is m, then equating mass times acceleration to force yields

$$m\frac{d^2y}{dt^2} = -ky$$

Thus, in this case $\omega^2 = k/m$.

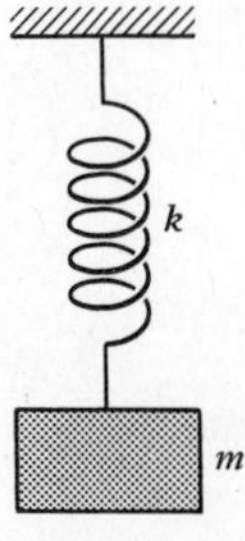

Figure 2.3. Mass and spring.

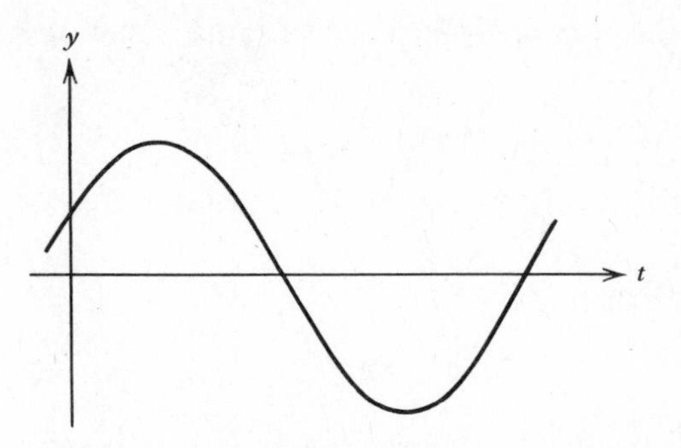

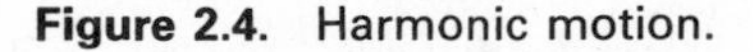
Figure 2.4. Harmonic motion.

The corresponding characteristic equation is

$$\lambda^2 + \omega^2 = 0 \tag{2-22}$$

This has solutions $\lambda = \pm i\omega$, where $i = \sqrt{-1}$. Thus, the roots of the characteristic polynomial are imaginary.

It follows that the general solution to (2-21) is

$$y(t) = c_1 e^{i\omega t} + c_2 e^{-i\omega t} \tag{2-23}$$

which in general is a complex value for each value of t. If, however, attention is restricted to real-valued solutions, the values of c_1 and c_2 must be restricted so that $c_1 + c_2$ is real, and $c_1 - c_2$ is imaginary. In this case the solution can be expressed equivalently as

$$y(t) = A \sin \omega t + B \cos \omega t$$

where A and B are arbitrary real constants. Indeed, the functions $\cos \omega t$ and $\sin \omega t$ form a fundamental set of solutions to the homogeneous equation.

The pattern of solution is the pure harmonic motion illustrated in Fig. 2.4. It consists of a pure sine or cosine wave. Variation of A and B acts only to change the height of oscillations and the displacement of phase. It should be noted that rather than specifying A and B it is possible to specify the initial conditions $y(0)$ and $(dy/dt)(0)$ to determine a particular solution.

Beats

Suppose now that an oscillatory system is subjected to an additional force, which itself varies harmonically but at a different frequency. This corresponds (roughly) to motion of a child's swing being pushed at other than the natural rate, or to one violin string being subjected to the force of air motion generated by the vigorous vibrations of a nearby string. We seek to characterize the general form of the induced vibration.

The new equation of motion is the nonhomogeneous differential equation

$$\frac{d^2y}{dt^2}+\omega^2 y=\sin \omega_0 t \tag{2-24}$$

where $\omega \neq \omega_0$. The magnitude of the external forcing term is set arbitrarily to one.

As a trial solution it seems reasonable to set

$$y(t)=C \sin \omega_0 t$$

Indeed, substituting this into the equation yields

$$-C\omega_0^2 \sin \omega_0 t+C\omega^2 \sin \omega_0 t=\sin \omega_0 t$$

which is satisfied if

$$C=\frac{1}{\omega^2-\omega_0^2}$$

Therefore, a particular solution is

$$y(t)=\frac{1}{\omega^2-\omega_0^2} \sin \omega_0 t \tag{2-25}$$

The general solution is found simply by adding to this solution the general solution of the homogeneous equation. Therefore, the general solution to the whole equation is

$$y(t)=A \sin \omega t+B \cos \omega t+\frac{1}{\omega^2-\omega_0^2} \sin \omega_0 t \tag{2-26}$$

If, for example, the system were known to be initially at rest, it is possible to find A and B explicitly. Evaluating (2-26) at $t=0$ leads to $B=0$. Evaluating the derivative at $t=0$ leads to

$$A=\frac{-\omega_0/\omega}{\omega^2-\omega_0^2} \tag{2-27}$$

Therefore, the solution corresponding to zero initial conditions is

$$y(t)=\frac{-\omega_0/\omega}{\omega^2-\omega_0^2} \sin \omega t+\frac{1}{\omega^2-\omega_0^2} \sin \omega_0 t \tag{2-28}$$

If it is assumed that ω_0 is very close to ω, the solution can be approximated as

$$y(t)=\frac{1}{\omega^2-\omega_0^2} [\sin \omega_0 t-\sin \omega t]$$

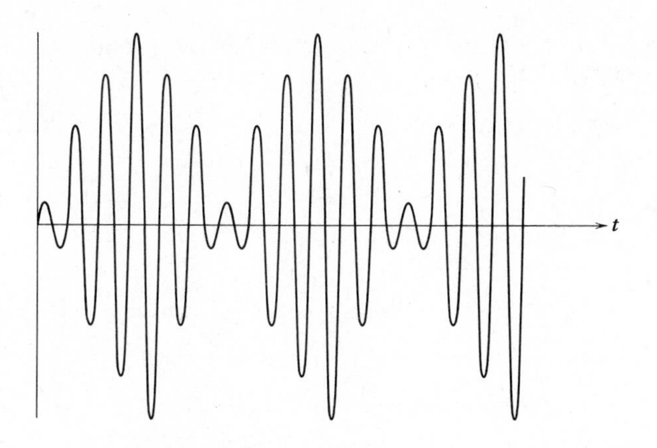

Figure 2.5. Beats

which, using a standard trigonometric identity, can be expressed as

$$y(t) = \frac{2}{\omega^2 - \omega_0^2} [\sin \tfrac{1}{2}(\omega_0 - \omega)t][\cos \tfrac{1}{2}(\omega_0 + \omega)t]$$

This solution is sketched in Fig. 2.5. It consists of an oscillation at nearly the driving frequency ω_0, but modified by a "beat" frequency $\frac{1}{2}(\omega_0 - \omega)$.

2.11 PROBLEMS

1. Solve the nonlinear difference equation

$$y(k+1) = \frac{y(k)}{b + y(k)}$$

by finding a change of variable that converts it to a linear difference equation.

2. A bank offers 7% annual interest. What would be the overall annual rate if the 7% interest were compounded quarterly?

3. Assuming $y(0) = y(1) = y(2) = 0$, find, by direct numerical recursion, the values of $y(k)$, $k = 3, 4, 5$, satisfying the difference equation

$$y(k+3) - y(k+2) + [y(k+1) + y(k)]^2 = 1$$

4. Consider the difference equation

$$(k+1)^2 y(k+1) - k^2 y(k) = 1$$

for $k = 1, 2, \ldots$. Find the general solution.

5. *Intelligence Test Sequence.* Find the second-order linear homogeneous difference equation which generates the sequence

$$1, 2, 5, 12, 29, 70, 169$$

What is the limiting ratio of consecutive terms?

6. *Binomial Coefficients.* The sequence

$$0, 1, 3, 6, 10, 15, 21, 28, 36, \ldots$$

is a sequence of binomial coefficients. Show (assuming the first term corresponds to $k = 1$) that this sequence can be generated in any of the following ways:

(a) $y(k) = \dfrac{k!}{2!\,(k-2)!}$ — 0th order varying coefficient equation

(b) $y(k+1) = \left(\dfrac{k+1}{k-1}\right) y(k)$ — 1st order, varying coefficient, homogeneous

(c) $y(k+1) = y(k) + k$ — 1st order, constant-coefficient, with varying input term

(d) $y(k+1) = 2y(k) - y(k-1) + 1$ — 2nd order, constant-coefficient, with constant input term

(e) $y(k+1) = 3y(k) - 3y(k-1) + y(k-2)$ — 3rd order homogeneous, constant-coefficient

Find the roots of the characteristic polynomial corresponding to part (e).

*7. The Fibonacci numbers are $F_1 = 1$, $F_2 = 1$, $F_3 = 2$, $F_4 = 3$, and so forth. Given that $F_{48} = 4{,}807{,}526{,}976$ and $F_{49} = 7{,}778{,}742{,}049$, what is the sum of the *squares* of the first forty-eight Fibonacci numbers?

8. *Supply and Demand Equilibrium—The Effect of Price Prediction.* Assume that the demand for a product at time k is given by

$$d(k) = d_0 - ap(k)$$

where d_0 and a are positive constants and $p(k)$ is the price at time k. In the simple classical cobweb analysis the supply is assumed to be governed by the equation

$$s(k) = s_0 + bp(k-1)$$

where s_0 and b are positive constants. Equating supply and demand leads to the dynamic equation for $p(k)$, which is convergent if $b < a$.

The $p(k-1)$ that appears in the supply equation can be considered to be an estimate of the future price. In other words, when planning at time $k-1$ how much to supply at time k, suppliers would really like to know what $p(k)$ will be. Since they cannot observe the actual price in advance, they do their planning on the basis of $p(k-1)$, using it as an estimate of $p(k)$. It is possible, however, to conceive more

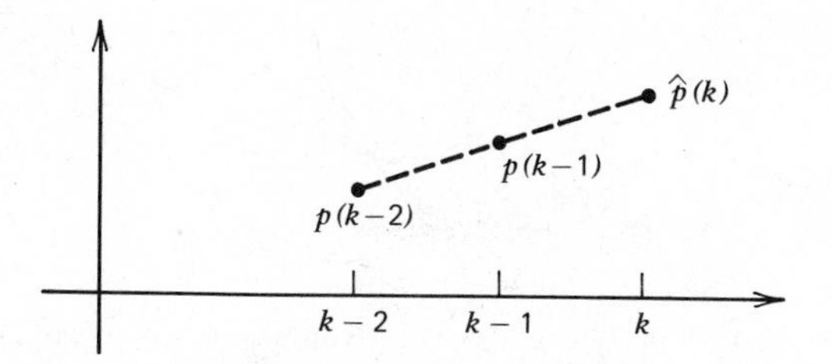

Figure 2.6. Price extrapolation.

complex procedures for estimating the future price. One such procedure is to use linear extrapolation, illustrated in Fig. 2.6. This yields $\hat{p}(k) = 2p(k-1) - p(k-2)$, where $\hat{p}(k)$ denotes the estimate of $p(k)$ based on previous prices. On the surface, it would seem that this "more sophisticated" estimation scheme might be better.

(a) Using $\hat{p}(k)$ in place of $p(k-1)$ in the supply equation, equate supply and demand to find a dynamic equation for $p(k)$. (Answer should be in the form of a difference equation.)

(b) For notational simplicity let $c = b/a$. What are the characteristic values of the equation?

9. *Information Theory.* Imagine an information transmission system that uses an alphabet consisting of just two symbols—dot and dash. Messages are transmitted by first encoding them into a string of these symbols. Each symbol requires some length of time for its transmission. Therefore, for a fixed total time duration only a finite number of different message strings is possible. Let N_t denote the number of different message strings of duration t. Shannon defined the *capacity* of the transmission system (measured in bits per time unit) to be

$$C = \lim_{t \to \infty} \frac{\log_2 N_t}{t}$$

If, for example, the dot and the dash each required one unit of time for transmission, it would follow that $N_t = 2^t$ and, accordingly,

$$C = \lim_{t \to \infty} \frac{\log_2 2^t}{t} = 1 \text{ bit per time unit}$$

Suppose now that the dot requires one unit of time for transmission while the dash requires two units.

(a) What are the values for N_1 and N_2? (Note: Dot and dash are the only two allowed symbols. A blank space is *not* allowed.)

(b) Justify the second-order difference equation $N_t = N_{t-1} + N_{t-2}$.

(c) Find the capacity C of this transmission system.

10. *Repeated Roots.* Consider the second-order difference equation

$$y(k+2) - 2ay(k+1) + a^2 y(k) = 0$$

Its characteristic polynomial has both roots equal to $\lambda = a$.

(a) Show that both

$$y(k)=a^k \quad \text{and} \quad y(k)=ka^k$$

are solutions.

(b) Find the solution to this equation that satisfies the auxiliary conditions $y(0)=1$ and $y(1)=0$.

11. Find the solution to the Gambler's ruin problem in the important special case where $p=q=\frac{1}{2}$. [*Hint:* Use the result of problem 10.]

12. *Monte Carlo Roulette.* In many European casinos, including Monte Carlo, a bet on red or black is not lost outright if the outcome is green. (See Example 5, Sect. 2.7.) Instead, the bet is "imprisoned" and play continues until the ball lands on either red or black. At that point the original bet is either returned to the player or lost, depending on whether the outcome matches the color originally selected.

(a) Argue that an appropriate difference equation for a gambler's ruin problem in this case is

$$u(k)=\tfrac{18}{37}u(k-1)+\tfrac{1}{74}u(k)+\tfrac{1}{74}u(k-1)+\tfrac{18}{37}u(k+1)$$

Find the probability that you can "break the bank at Monte Carlo" under the conditions of Example 5, with this set of rules.

(b) Note that if the outcome is green, there is a probability of $\frac{1}{2}$ that the bet will be returned. This is, on the average, equivalent to a probability of $\frac{1}{4}$ that twice the bet will be returned. Therefore, an "equivalent" game, with the same odds but having the standard form of Example 5, is obtained by setting

$$p=\tfrac{18}{37}+\tfrac{1}{4}\cdot\tfrac{1}{37}=\tfrac{73}{148} \qquad q=1-p$$

Show that although this equivalent game exactly matches the odds of Monte Carlo roulette, its ruin probabilities are not exactly the same as those found in part (a).

*13. *Discrete Queue.* A small business receives orders for work, and services those orders on a first-come, first-served basis. In any given hour of the day there is a probability p (very small) that the business will receive an order. It almost never receives two orders in one hour. If the company has at least one order in hand, there is a probability q (also small, but $q>p$) that it will complete service on the order within an hour. It never completes two orders in an hour. On the average, how many orders will there be waiting for service to be completed? [*Hint*: Let $u(n)$ be the probability at any one time that the length of the line is n. Then neglecting pq terms, argue that

$$u(n)=u(n-1)p+u(n+1)q+u(n)(1-p-q), \qquad n>0$$

$$u(0)=u(1)q+u(0)(1-p), \qquad n=0$$

Use the subsidiary condition $\sum u(n)=1$ to find the $u(n)$'s. The average number of waiting orders is $\sum nu(n)$.]

14. *Geometric Forcing Term.* Consider a nonhomogeneous difference equation of the form

$$y(k+n)+a_{n-1}y(k+n-1)+\cdots+a_0y(k)=br^k$$

for $k=0,1,2,\ldots$. Suppose that the roots $\lambda_1, \lambda_2, \ldots, \lambda_n$ of the characteristic polynomial are distinct and that $r\neq\lambda_i$, $i=1,2,\ldots,n$. Show that the general solution to this difference equation is

$$y(k)=c_0r^k+c_1\lambda_1^k+c_2\lambda_2^k+\cdots+c_n\lambda_n^k$$

Show how to determine the constant c_0. (The term br^k can be regarded as an input or forcing term applied to the homogeneous difference equation. From this viewpoint, the above result shows that the response of a constant coefficient linear difference equation to a geometric series is a multiple of that series plus other geometric series defined by the homogeneous equation.)

15. *Numerical Solution of Differential Equations.* Differential equations are often solved numerically by a discrete forward recursion method. Consider the equation

$$\frac{dx}{dt}=f(x) \tag{2-29}$$

where x is scalar-valued. To solve this equation numerically one considers the sequence of discrete points $0, s, 2s, 3s, \ldots$, where s is a positive "step length."

The simplest solution technique is the *Euler method,* which calculates a sequence of values according to the recursion

$$x(k+1)=x(k)+sf[x(k)] \tag{2-30}$$

This procedure can be viewed as simple linear extrapolation, over the step length s, on the basis of x at $t=ks$ and its derivative. See Fig. 2.7.

(a) Assuming $f(x)=ax$, for what range of values of the constant a does the solution of (2-29) converge to zero, as $t\to\infty$?

(b) For what range of values of the constant a does the solution of (2-30) converge to zero?

(c) For a fixed $a<0$, what is the largest steplength that guarantees convergence in Euler's method?

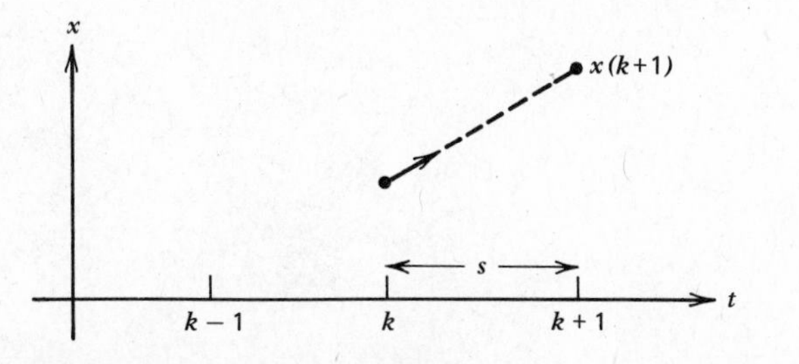

Figure 2.7. Euler's method.

16. *Alternate Method.* An alternate method for solving a differential equation (see previous problem) is to calculate $x(k+1)$ on the basis of a higher-order extrapolation. If $x(k-1)$ and $x(k)$ are known, then the function

$$g(t)=\frac{\{x(k-1)-x(k)+f[x(k)]s\}}{s^2}t^2+f[x(k)]t+x(k)$$

is the quadratic function that satisfies

$$g(0)=x(k)$$
$$g'(0)=f[x(k)]$$
$$g(-s)=x(k-1)$$

The value $g(s)$ would seem to be a good choice for $x(k+1)$. See Fig. 2.8.

(a) Again assuming $f(x)=ax$ for a fixed $a<0$, what condition must be satisfied by s for this new method to be convergent?

(b) Show that no value of $s>0$ satisfies these conditions.

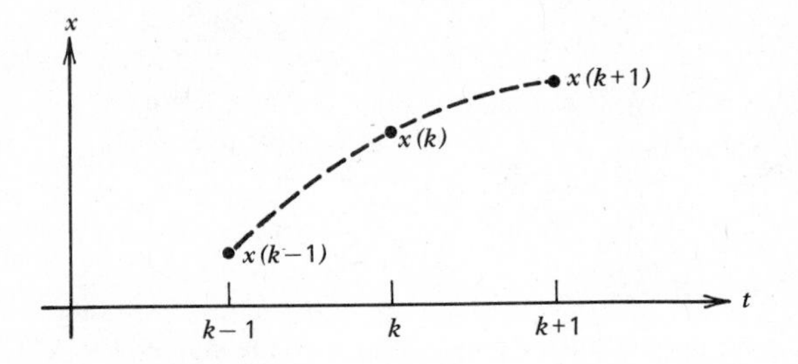

Figure 2.8. Alternate method.

17. Find the solutions to the following difference equations, for $k=0, 1, 2, \cdots$

(a) $(k+2)^2y(k+1)-(k+1)^2y(k)=2k+3$
$y(0)=0.$

(b) $y(k+2)-5y(k+1)+6y(k)=0$
$y(0)=y(1)=1.$

(c) $y(k+2)+y(k+1)+y(k)=0$
$y(0)=0, y(1)=1.$

(d) $y(k+2)-2y(k+1)-4y(k)=0$
$y(0)=1, y(1)=0.$

(e) $y(k+2)-3y(k+1)+2y(k)=1$
$y(0)=2, y(1)=2.$

18. *Radioactive Dating.* Normal carbon has an atomic weight of 12. The radioisotope C^{14}, with an atomic weight of 14, is produced continuously by cosmic radiation and is distributed throughout the earth's atmosphere in a form of carbon dioxide. Carbon dioxide is absorbed by plants, these plants are eaten by animals, and, consequently, all living matter contains radioactive carbon. The isotope C^{14} is unstable—by emitting an electron, it eventually disintegrates to nitrogen. Since at death the carbon in plant and animal tissue is no longer replenished, the percentage of C^{14} in such tissue begins to decrease. It decreases exponentially with a half-life of 5685 years (that is, after 5685 years one half of the C^{14} atoms will have disintegrated).

    Suppose charcoal from an ancient ruin produced a count of 1 disintegration/min/g on a geiger counter while living wood gave a count of 7. Estimate the age of the ruins.

19. *Newton Cooling.* According to Newton's law of cooling, an object of higher temperature than its environment cools at a rate that is proportional to the difference in temperature.
    (a) A thermometer reading 70°F, which has been inside a house for a long time, is taken outside. After one minute the thermometer reads 60°F; after two minutes it reads 53°F. What is the outside temperature?
    (b) Suppose you are served a hot cup of coffee and a small pitcher of cream (which is cold). You want to drink the coffee only after it cools to your favorite temperature. If you wish to get the coffee to proper temperature as quickly as possible, should you add the cream immediately or should you wait awhile?

20. The equation

$$\frac{d^2y}{dt^2}+\frac{4}{t}\frac{dy}{dt}+\frac{2}{t^2}y=0$$

    is an example of an *equi-dimensional differential equation.* Find a set of linearly independent solutions. [*Hint*: Try $y(t)=t^p$.]

21. *An Elementary Seismograph.* A seismograph is an instrument that records sudden ground movements. The simplest kind of seismograph, measuring horizontal displacement, consists of a mass attached to the instrument frame by a spring. The frame moves when hit by a seismic wave, whereas the mass, isolated by the spring, initially tends to remain still. A recording pen, attached to the mass, traces a displacement in a direction opposite to the displacement of the frame. The mass will of course soon begin to oscillate. In order to be able to faithfully record additional seismic waves, it is therefore desirable to suppress the oscillation of the mass by the addition of a damper (often consisting of a plunger in a viscous fluid). To be most effective the seismograph must have a proper combination of mass, spring, and damper. (See Fig. 2.9.) If the force exerted by the spring on the mass is proportional to displacement x and in an opposite direction; the force exerted by the damper is proportional to the velocity dx/dt and in an opposite direction; and the total force is

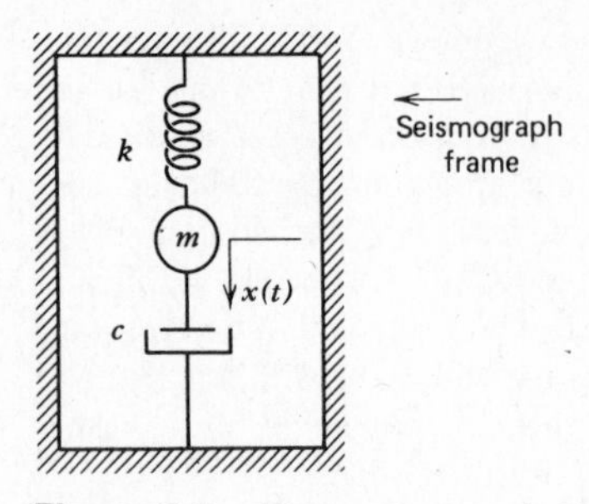

Figure 2.9. Elementary seismograph.

equal to mass times acceleration, the equation that describes the motion is

$$m\frac{d^2x}{dt^2}+c\frac{dx}{dt}+kx=0$$

(a) Find the roots of the characteristic equation in terms of c, m, and k.
(b) Distinguish three cases, overdamping, underdamping, and critical damping based on the relationship among c, m, and k as implied by the solution of the characteristic equation. Find the general solutions for $x(t)$ for all three cases.
(c) Which case is best for a seismograph as described in this problem? Why?

22. Prove Theorems 1, 2, and 3 of Sect. 2.9.

NOTES AND REFERENCES

General. The elementary theories of difference and differential equations are so similar that mastery of one essentially implies mastery of the other. However, because there are many more texts on differential equations than difference equations, the reader interested in supplemental material may find it most convenient to study differential equations. Some excellent popular general texts are Rainville and Bedient [R1], Coddington [C5], Braun [B11], and Martin and Reissner [M2]. An excellent text on difference equations, which includes many examples, is Goldberg [G8]. See also Miller [M5].

Section 2.5. The cobweb model is an important classic model. For further discussion see Henderson and Quandt [H2].

Section 2.7. The Gambler's ruin problem (Example 5) is treated extensively in Feller [F1]. It is also discussed further in Chapter 7 of this book.

Section 2.11. Information theory, as discussed briefly in Problem 9, is due to Shannon. See Shannon and Weaver [S4].

chapter 3.

Linear Algebra

Linear algebra is a nearly indispensable tool for modern analysis. It provides both a streamlined notation for problems with many variables and a powerful format for the rich theory of linear analysis. This chapter is an introductory account of that portion of linear algebra that is needed for a basic study of dynamic systems. In particular, the first three sections of the chapter are essential prerequisites for the next chapter, and the remaining sections are prerequisites for later chapters. Other results from linear algebra that are important in the analysis of dynamic systems are discussed in individual sections in later portions of the text.

In some respects this chapter can be regarded as a kind of appendix on linear algebra. As such it is suggested that the reader may wish to skim much of the material, briefly reviewing that part which is familiar, and spending at least some preliminary effort on the parts that are unfamiliar. Many of the concepts presented here strictly from the viewpoint of linear algebra, particularly those related to eigenvectors, are reintroduced and elaborated on with applications in Chapter 5 in the context of dynamic systems. Accordingly, many readers will find it advantageous to study this material by referring back and forth between the two chapters.

ALGEBRAIC PROPERTIES

3.1 FUNDAMENTALS

Much of linear algebra is motivated by consideration of the general system of m linear algebraic equations in n unknowns:

$$\begin{aligned}
a_{11}x_1+a_{12}x_2+\cdots+a_{1n}x_n &= y_1\\
a_{21}x_1+a_{22}x_2+\cdots+a_{2n}x_n &= y_2\\
\vdots \qquad\qquad\qquad\qquad & \quad \vdots\\
a_{m1}x_1+a_{m2}x_2+\cdots+a_{mn}x_n &= y_m
\end{aligned} \tag{3-1}$$

where the x_j, $j=1, 2, \ldots, n$ are the *dependent variables*, the y_i, $i=1, 2, \ldots, m$ are the *independent variables*, and the a_{ij}, $i=1, 2, \ldots, m$, $j=1, 2, \ldots, n$ are constant *coefficients*. The values of the y_i's are generally considered to be known (or given) and the x_j's are considered unknown.

Rather than write out this set in full detail, it is often convenient to use the alternative summation representation

$$\sum_{j=1}^{n} a_{ij}x_j = y_i \qquad i=1, 2, \ldots, m \tag{3-2}$$

This simplifies the notation somewhat, but even it is a bit cumbersome.

A representation that is even more compact but still highly suggestive of the original detailed form is the matrix notation

$$\mathbf{Ax}=\mathbf{y} \tag{3-3}$$

For this simple notation to be meaningful, however, an associated machinery of auxiliary definitions must be carefully developed.

Matrices and Vectors

In general a *matrix* is a rectangular array of elements. If the array has m rows and n columns it is said to be an $m\times n$ (read m by n) matrix, or, equivalently, the matrix is said to $m\times n$ *dimensional*. Matrices are generally denoted by boldface capital letters, such as **A**. Elements of the matrix are denoted, correspondingly, by lower case letters with subscripts to indicate the position of the element. Thus, the element in the ith row and jth column of the matrix **A** is denoted a_{ij}. To highlight this correspondence, the matrix is sometimes written $\mathbf{A}=[a_{ij}]$.

A special class of matrices are those having $m=1$ or $n=1$, corresponding to a matrix having either a single row or a single column. In either case, the corresponding matrix is said to be a *vector.* Vectors are usually denoted by lower case boldface letters, and their elements have but a single subscript. A vector of the form (with $m=1$)

$$\mathbf{a}=[a_1, a_2, \ldots, a_n]$$

is a *row vector,* while a vector of the form (with $n=1$)

$$\mathbf{b}=\begin{bmatrix} b_1 \\ b_2 \\ \cdot \\ \cdot \\ \cdot \\ b_m \end{bmatrix}$$

is a *column vector.* Column vectors are used for most purposes, particularly in systems of equations, but row vectors also arise naturally.

Special Matrices

For any dimension, one special matrix is the matrix whose elements are all zero. Such a matrix is denoted by $\mathbf{0}$, and is called the *zero matrix.*

A matrix that has the same number of rows as columns ($m=n$) is said to be *square.* Corresponding to a square $n \times n$ matrix $\mathbf{A}$, the elements a_{ii}, $i=1, 2, \ldots, n$ are referred to as the *diagonal elements* of $\mathbf{A}$. If all elements except possibly the diagonal elements are zero, the square matrix $\mathbf{A}$ is said to be *diagonal.* A very special case of a diagonal matrix is the $n \times n$ square matrix whose elements are zero, except on the diagonal where they are equal to one. This matrix (for any dimension n) is denoted $\mathbf{I}$, and called the *identity matrix.* Thus,

$$\mathbf{I}=\begin{bmatrix} 1 & 0 & 0 & \cdots & 0 \\ 0 & 1 & 0 & \cdots & 0 \\ \cdot & & & & \cdot \\ \cdot & & & & \cdot \\ \cdot & & & & \cdot \\ 0 & & & & 0 \\ 0 & \cdots & & 0 & 1 \end{bmatrix}$$

Elementary Operations

Addition of Matrices. If two matrices $\mathbf{A}$ and $\mathbf{B}$ are of the same dimension, then their sum can be defined and is a matrix $\mathbf{C}$, also of the same dimension. If $\mathbf{A}=[a_{ij}]$, $\mathbf{B}=[b_{ij}]$, and $\mathbf{C}=[c_{ij}]$, where $\mathbf{C}=\mathbf{A}+\mathbf{B}$, then the elements of $\mathbf{C}$ are defined by $c_{ij} = a_{ij} + b_{ij}$. In other words, the addition is carried out element by element.

Example 1. Suppose $\mathbf{A}$ and $\mathbf{B}$ are defined as the 2×3 matrices

$$\mathbf{A} = \begin{bmatrix} 1 & 2 & 3 \\ 0 & 4 & 2 \end{bmatrix} \qquad \mathbf{B} = \begin{bmatrix} 2 & 3 & 4 \\ 0 & -1 & -4 \end{bmatrix}$$

The sum $\mathbf{C} = \mathbf{A} + \mathbf{B}$ is the matrix

$$\mathbf{C} = \begin{bmatrix} 3 & 5 & 7 \\ 0 & 3 & -2 \end{bmatrix}$$

It is easily verified that matrix addition satisfies the following two laws:

$$\text{(i)}\ \mathbf{A}+\mathbf{B} = \mathbf{B}+\mathbf{A} \quad \text{(commutative law)}$$
$$\text{(ii)}\ \mathbf{A}+(\mathbf{B}+\mathbf{C}) = (\mathbf{A}+\mathbf{B})+\mathbf{C} \quad \text{(associative law)}$$

Therefore, addition of matrices acts much like addition of numbers.

Scalar Multiplication. For any matrix $\mathbf{A}$ and any scalar (real or complex number) α, the product $\alpha\mathbf{A}$ is the matrix obtained by multiplying every element of the matrix $\mathbf{A}$ by the factor α. In other words, if $\mathbf{A} = [a_{ij}]$, then $\alpha\mathbf{A} = [\alpha a_{ij}]$.

Example 2. If $\mathbf{A}$ is the matrix

$$\mathbf{A} = \begin{bmatrix} 2 & 1 & 0 \\ 1 & 4 & -1 \end{bmatrix}$$

Then (using $\alpha = 2$)

$$2\mathbf{A} = \begin{bmatrix} 4 & 2 & 0 \\ 2 & 8 & -2 \end{bmatrix}$$

Matrix Multiplication. Multiplication of two matrices to obtain a third is perhaps the most important of the elementary operations. This is the operation that neatly packages the bulky individual operations associated with defining and manipulating systems of linear algebraic equations.

If $\mathbf{A}$ is an $m\times n$ matrix and $\mathbf{B}$ is an $n\times p$ matrix, the matrix $\mathbf{C} = \mathbf{AB}$ is defined as the $m\times p$ matrix with elements

$$c_{ik} = \sum_{j=1}^{n} a_{ij} b_{jk} \tag{3-4}$$

This definition of matrix multiplication has several important interpretations. First, it should be noted that it is consistent with the matrix notation for a system of linear equations, as described by (3-1) and (3-3). Thus, for an $m\times n$ matrix $\mathbf{A}$ and an $n\times 1$ matrix $\mathbf{x}$ (a column vector) the product $\mathbf{Ax}$ is the $m\times 1$

matrix (another column vector) $\mathbf{y}$ with ith element equal to

$$y_i = \sum_{j=1}^{n} a_{ij} x_j \tag{3-5}$$

Second, the product $\mathbf{C} = \mathbf{AB}$ when $\mathbf{B}$ has p columns can be viewed as $\mathbf{A}$ multiplying each of these p columns separately. That is, the first column of $\mathbf{C}$ is $\mathbf{A}$ times the first column of $\mathbf{B}$, the second column of $\mathbf{C}$ is $\mathbf{A}$ times the second column of $\mathbf{B}$, and so forth. Thus, $\mathbf{C} = \mathbf{AB}$ can be regarded as p separate column relations.

Matrix multiplication satisfies

$$\mathbf{A}(\mathbf{BC}) = (\mathbf{AB})\mathbf{C} \quad \text{(associative law)}$$

However, it is *not* commutative. Thus, in general,

$$\mathbf{AB} \neq \mathbf{BA}$$

even if both products are defined.

Finally, it should be noted that if $\mathbf{A}$ is an arbitrary $m \times n$ matrix and $\mathbf{I}$ is the $m \times m$ identity matrix, then $\mathbf{IA} = \mathbf{A}$.

Example 3. Suppose $\mathbf{A}$ and $\mathbf{B}$ are defined as

$$\mathbf{A} = \begin{bmatrix} 1 & -2 & 0 \\ 2 & 1 & 3 \end{bmatrix} \qquad \mathbf{B} = \begin{bmatrix} 1 & -3 & 3 & 0 \\ 1 & 4 & -1 & 0 \\ 2 & 1 & 2 & 1 \end{bmatrix}$$

Then the product $\mathbf{C} = \mathbf{AB}$ is

$$\mathbf{C} = \begin{bmatrix} -1 & -11 & 5 & 0 \\ 9 & 1 & 11 & 3 \end{bmatrix}$$

Example 4 (Inner Product). A special case of matrix multiplication is the *dot* or *inner product* of two vectors. This is just the product of an n-dimensional row vector, say $\mathbf{r}$, and an n-dimensional column vector, say $\mathbf{c}$. The product, according to the general definition (3-4), is

$$\mathbf{rc} = \sum_{i=1}^{n} r_i c_i \tag{3-6}$$

which is 1×1; that is, it is simply a scalar.

One common way that the inner product arises is when one vector represents quantities and another represents corresponding unit prices. Thus, grocery purchases of sugar, flour, and potatoes might be represented by the vectors

$$\mathbf{x} = \begin{bmatrix} x_1 \\ x_2 \\ x_3 \end{bmatrix} \qquad \mathbf{p} = [p_1 \quad p_2 \quad p_3]$$

where x_1, x_2, x_3 are, respectively, the amounts of the three commodities purchased, and p_1, p_2, p_3 are their respective unit prices. Then the product

$$\mathbf{px} = \sum_{i=1}^{3} p_i x_i$$

is the total purchase price of the groceries.

Transpose. Corresponding to an $m \times n$ matrix $\mathbf{A} = [a_{ij}]$, the *transpose* of $\mathbf{A}$, denoted $\mathbf{A}^T$, is defined as the $n \times m$ matrix $\mathbf{A}^T = [a_{ij}^T]$ with $a_{ij}^T = a_{ji}$. This means that $\mathbf{A}^T$ is defined by interchanging rows and columns in $\mathbf{A}$. As an example,

$$\begin{bmatrix} 1 & 2 & 3 \\ 0 & 4 & 5 \end{bmatrix}^T = \begin{bmatrix} 1 & 0 \\ 2 & 4 \\ 3 & 5 \end{bmatrix}$$

An important property of transposes is the way they are transformed in matrix multiplication. The reader can verify the rule $(\mathbf{AB})^T = \mathbf{B}^T\mathbf{A}^T$. Thus, *the transpose of a product is equal to the product of the transposes in the reverse order.*

Differentiation. If the elements of a matrix depend on a variable t, making the elements functions rather than constants, it is possible to consider differentiation of the matrix. Differentiation is simply defined by differentiating each element of the matrix individually. Thus, if

$$\mathbf{A}(t) = \begin{bmatrix} a_{11}(t) & a_{12}(t) & \cdots & a_{1n}(t) \\ a_{21}(t) & a_{22}(t) & \cdots & a_{2n}(t) \\ \vdots & & & \vdots \\ a_{m1}(t) & a_{m2}(t) & \cdots & a_{mn}(t) \end{bmatrix} \tag{3-7}$$

then

$$\frac{d\mathbf{A}(t)}{dt} \equiv \dot{\mathbf{A}}(t) = \begin{bmatrix} \dot{a}_{11}(t) & \dot{a}_{12}(t) & \cdots & \dot{a}_{1n}(t) \\ \dot{a}_{21}(t) & \dot{a}_{22}(t) & \cdots & \dot{a}_{2n}(t) \\ \vdots & & & \vdots \\ \dot{a}_{m1}(t) & \dot{a}_{m2}(t) & \cdots & \dot{a}_{mn}(t) \end{bmatrix} \tag{3-8}$$

Integration. In a manner analogous to differentiation, integration of a matrix whose elements depend on a variable t is defined in terms of the integrals of the individual elements. Thus for $\mathbf{A}(t)$ as in (3-7), there is defined

$$\int \mathbf{A}(t)\,dt = \begin{bmatrix} \int a_{11}(t)\,dt & \int a_{12}(t)\,dt & \cdots & \int a_{1n}(t)\,dt \\ \int a_{21}(t)\,dt & \int a_{22}(t)\,dt & \cdots & \int a_{2n}(t)\,dt \\ \cdot & & & \cdot \\ \cdot & & & \cdot \\ \cdot & & & \cdot \\ \int a_{m1}(t)\,dt & \int a_{m2}(t)\,dt & \cdots & \int a_{mn}(t)\,dt \end{bmatrix} \tag{3-9}$$

An Example: The Leontief Economic Model

To illustrate the value of matrix notation in the description of complex systems, consider an economy consisting of n industries, each of which produces a single product. In order to produce its product, each industry must have on hand various amounts of the products of other industries (and perhaps some of its own). For example, the automotive industry purchases steel from the steel industry and tires from the rubber industry, while the agriculture industry purchases tractors from the automotive industry and fertilizers from the chemical industry.

Assume that the basic production cycle is one year in duration, and that for each unit of output from industry j, a_{ij} units of the product of industry i are required. The constants a_{ij} are called *technical coefficients.* Denote by $x_1, x_2, \ldots, x_n$ the amounts of the products produced in the n industries. Then the amount of product i required for this pattern of production is

$$a_{i1}x_1 + a_{i2}x_2 + \cdots + a_{in}x_n$$

The total amount of product i produced goes in part to help produce other products as described above, and in part to consumers to meet their demand. Therefore,

$$x_i = a_{i1}x_1 + a_{i2}x_2 + \cdots + a_{in}x_n + d_i, \qquad i = 1, 2, \ldots, n$$

where d_i is the demand for product i. Thus, total production of a product exceeds the actual consumer demand because of the use of the product in various production processes.

Introducing the matrix $\mathbf{A} = [a_{ij}]$ and the column vectors $\mathbf{x}$ and $\mathbf{d}$ with components x_i, d_i, $i = 1, 2, \ldots, n$, respectively, these equations can be written as

$$\mathbf{x} = \mathbf{A}\mathbf{x} + \mathbf{d}$$

or, equivalently, $[\mathbf{I}-\mathbf{A}]\mathbf{x}=\mathbf{d}$. This is a compact representation of the complex interrelations among industries. The coefficient matrix is the sum of the identity $\mathbf{I}$ and $(-1)\mathbf{A}$. If a given set of consumer demands is specified (as for example by a yearly forecast of demand) the required total level of production in each of the industries can be found by solving for $\mathbf{x}$.

3.2 DETERMINANTS

The *determinant* of a square matrix is a scalar value that arises naturally in the solution of sets of linear equations. The determinant of the matrix

$$\mathbf{A}=\begin{bmatrix} a_{11} & a_{12} & \cdots & a_{1n} \\ a_{21} & a_{22} & \cdots & a_{2n} \\ \cdot & & & \cdot \\ \cdot & & & \cdot \\ \cdot & & & \cdot \\ a_{n1} & a_{n2} & \cdots & a_{nn} \end{bmatrix}$$

is denoted $|\mathbf{A}|$, det $\mathbf{A}$, or by simply enclosing the corresponding array with two vertical lines, as

$$\begin{vmatrix} a_{11} & a_{12} & \cdots & a_{1n} \\ a_{21} & a_{22} & \cdots & a_{2n} \\ \cdot & & & \cdot \\ \cdot & & & \cdot \\ \cdot & & & \cdot \\ a_{n1} & a_{n2} & \cdots & a_{nn} \end{vmatrix}$$

The determinant of a simple 1×1 matrix $A=[a]$ is defined to be $|A|=a$. The determinant of the general 2×2 matrix is given by the formula

$$\begin{vmatrix} a_{11} & a_{12} \\ a_{21} & a_{22} \end{vmatrix} = a_{11}a_{22}-a_{12}a_{21} \tag{3-10}$$

Laplace's Expansion

The value of the determinant corresponding to a general $n\times n$ matrix can be found in terms of lower-order determinants through use of Laplace's expansion. This expansion is defined in terms of *minors* or *cofactors* of elements of the matrix.

The *minor* M_{ij} of the element a_{ij} in a matrix is the determinant of the array formed by deleting the ith row and the jth column from the original matrix. Thus, if $\mathbf{A}$ is an $n\times n$ matrix, each minor is an $(n-1)\times(n-1)$ determinant.

The *cofactor* C_{ij} corresponding to the element a_{ij} of $\mathbf{A}$ is $(-1)^{i+j}M_{ij}$. Thus, the cofactors are identical to the minors, except for a possible change in sign.

In terms of Laplace's expansion, the determinant of a matrix $\mathbf{A}$ is

$$\det \mathbf{A} = \sum_{j=1}^{n} a_{ij}C_{ij} \tag{3-11}$$

for any i. Or, equivalently,

$$\det \mathbf{A} = \sum_{i=1}^{n} a_{ij}C_{ij} \tag{3-12}$$

for any j. The first of these is called an expansion along the ith row, while the second is an expansion along the jth column. All such expansions yield identical values.

A Laplace expansion expresses an nth-order determinant as a combination of $(n-1)$th-order determinants. Each of the required $(n-1)$th-order determinants can itself be expressed, by a Laplace expansion, in terms of $(n-2)$th-order determinants, and so on, all the way down to first order if necessary. Therefore, this expansion together with the definition of the determinant for 1×1 matrices is sufficient to determine the value of any determinant.

Example 1. Let us evaluate the fourth-order determinant

$$\det \mathbf{A} = \begin{vmatrix} 3 & 2 & 1 & 0 \\ 1 & 0 & 0 & 2 \\ 1 & 0 & 2 & 1 \\ 4 & 2 & 5 & 0 \end{vmatrix}$$

Since the second row has two zeros, it is convenient to expand along that row. Thus,

$$\det \mathbf{A} = (-1)\begin{vmatrix} 2 & 1 & 0 \\ 0 & 2 & 1 \\ 2 & 5 & 0 \end{vmatrix} + 2\begin{vmatrix} 3 & 2 & 1 \\ 1 & 0 & 2 \\ 4 & 2 & 5 \end{vmatrix}$$

The first third-order determinant in this expression can be expanded along the third column, and the second determinant can be expanded along the second row, yielding

$$\det \mathbf{A} = (-1)(-1)\begin{vmatrix} 2 & 1 \\ 2 & 5 \end{vmatrix} + (2)(-1)\begin{vmatrix} 2 & 1 \\ 2 & 5 \end{vmatrix} + (2)(-2)\begin{vmatrix} 3 & 2 \\ 4 & 2 \end{vmatrix}$$

All these second-order determinants can be evaluated by use of (3-10), resulting in

$$\det \mathbf{A} = (1)(8) + (-2)(8) + (-4)(-2)$$
$$\det \mathbf{A} = 0$$

Determinants of Triangular Matrices

A matrix **A** is said to be either *upper* or *lower triangular* if it has the form

$$\mathbf{A} = \begin{bmatrix} a_{11} & a_{12} & \cdot & \cdot & \cdot & a_{1n} \\ 0 & a_{22} & \cdot & \cdot & \cdot & a_{2n} \\ 0 & 0 & a_{33} & & & a_{3n} \\ \vdots & & & & & \vdots \\ 0 & 0 & \cdot & \cdot & \cdot\ 0 & a_{nn} \end{bmatrix}$$

or

$$\mathbf{A} = \begin{bmatrix} a_{11} & 0 & 0 \cdot & \cdot & \cdot\ 0 \\ a_{21} & a_{22} & 0 \cdot & \cdot & \cdot\ 0 \\ a_{31} & a_{32} & a_{33} & \cdot & \cdot\ 0 \\ \vdots & & & & \vdots \\ a_{n1} & a_{n2} & \cdot & \cdot & \cdot\ a_{nn} \end{bmatrix}$$

respectively. The determinant of a triangular matrix is equal to the product of its diagonal elements. We can prove this easily using induction on the dimension n together with Laplace's expansion. It is certainly true for $n = 1$. Suppose then that it is true for $n-1$. Then, for the upper triangular case, expansion down the first column yields $\det \mathbf{A} = a_{11}M_{11}$. (For the lower triangular case, we would expand along the first row.) Using the induction hypothesis M_{11} is the product of its diagonal elements, and therefore,

$$\det \mathbf{A} = a_{11}a_{22}a_{33} \cdots a_{nn} \tag{3-13}$$

This simple result is useful in numerous applications.

Products and Transposes

Two important properties of determinants are the product formula

$$\det(\mathbf{AB}) = (\det \mathbf{A})(\det \mathbf{B}) \tag{3-14}$$

where **A** and **B** are both $n \times n$ square matrices, and the transpose rule

$$\det(\mathbf{A}^T) = \det(\mathbf{A}) \tag{3-15}$$

Linear Combinations

Determinants can sometimes be evaluated easily by transforming them to equivalent but simpler forms. This is accomplished by use of rules governing the change in the value of a determinant when rows or columns of its array are linearly combined. There are three basic row operations, and associated rules, from which the effect of any linear combination of rows on the value of a determinant can be deduced:

(a) If all elements in one row are multiplied by a constant c, the value of the corresponding new determinant is c times the original value.
(b) If two rows are interchanged, the value of the corresponding new determinant is the negative of the original value.
(c) If any multiple of one row is added to another row, element by element, the value of the determinant is unchanged.

Each of these rules can be easily deduced from Laplace's expansion. Moreover, since the determinant of the transpose of a matrix is equal to the determinant of the matrix itself, as given by (3-15), three identical rules hold for column operations.

Example 2. Using the above rules, the determinant of the matrix below is manipulated step by step to triangular form, from which the value is easily determined:

$$\mathbf{A} = \begin{bmatrix} 4 & -2 & 0 \\ 0 & 4 & 1 \\ 2 & 1 & 4 \end{bmatrix}$$

Multiply the first row by $\frac{1}{2}$, yielding

$$\det \mathbf{A} = 2 \begin{vmatrix} 2 & -1 & 0 \\ 0 & 4 & 1 \\ 2 & 1 & 4 \end{vmatrix}$$

Interchange the second and third rows:

$$\det \mathbf{A} = (-2) \begin{vmatrix} 2 & -1 & 0 \\ 2 & 1 & 4 \\ 0 & 4 & 1 \end{vmatrix}$$

Subtract the first row from the second

$$\det \mathbf{A} = (-2) \begin{vmatrix} 2 & -1 & 0 \\ 0 & 2 & 4 \\ 0 & 4 & 1 \end{vmatrix}$$

Subtract twice the second row from the third

$$\det \mathbf{A} = (-2)\begin{vmatrix} 2 & -1 & 0 \\ 0 & 2 & 4 \\ 0 & 0 & -7 \end{vmatrix}$$

Therefore, $\det \mathbf{A} = (-2)\cdot 2 \cdot 2 \cdot (-7) = 56$.

3.3 INVERSES AND THE FUNDAMENTAL LEMMA

Consider a square $n \times n$ matrix $\mathbf{A}$. An $n \times n$ matrix $\mathbf{A}^{-1}$ is said to be the *inverse* of $\mathbf{A}$ if $\mathbf{A}^{-1}\mathbf{A} = \mathbf{I}$. That is, the product of $\mathbf{A}^{-1}$ and $\mathbf{A}$ is the identity matrix.

Not every square matrix has an inverse. Indeed, as discussed below, a square matrix has an inverse if and only if its determinant is nonzero. If the determinant is zero the matrix is said to be *singular*, and no inverse exists.

Cofactor Formula for Inverses

Perhaps the simplest way to prove that an inverse exists if the determinant is not zero is to display an explicit formula for the inverse. There is a simple formula deriving from Cramer's rule for solving sets of linear equations, which is expressed in terms of the cofactors of the matrix. Denoting the elements of $\mathbf{A}^{-1}$ by a_{ij}^{-1}, that is, $\mathbf{A}^{-1} = [a_{ij}^{-1}]$, the formula is

$$\mathbf{A}^{-1} = [a_{ij}^{-1}] = \frac{1}{\Delta}[C_{ji}] \tag{3-16}$$

where Δ is the determinant of $\mathbf{A}$. This formula can be verified using Laplace's expansion as follows. The ikth element of $\mathbf{B} = \mathbf{A}^{-1}\mathbf{A}$ is

$$b_{ik} = \sum_{j=1}^{n} \frac{C_{ji}}{\Delta} a_{jk}$$

For $i = k$ we obtain from Laplace's expansion [Eq. (3-12)] that $b_{ii} = 1$, and hence the diagonal elements are all unity. To verify that the off-diagonal elements are zero, for a given i and k, $i \neq k$, consider the matrix obtained from $\mathbf{A}$ by setting the elements in the ith column all zero. Clearly the determinant of this new matrix is zero. The value of this determinant is unchanged if we now add the kth column to the ith column, forming the matrix $\bar{\mathbf{A}}$. However,

$$0 = \det \bar{\mathbf{A}} = \sum_{j=1}^{n} \bar{a}_{ji} C_{ji} = \sum_{j=1}^{n} a_{jk} C_{ji}$$

This shows that $b_{ik} = 0$. Thus, $\mathbf{A}^{-1}\mathbf{A} = \mathbf{I}$.

Example. Let us compute the inverse of the 3×3 matrix

$$\mathbf{A}=\begin{bmatrix}1 & 0 & 2\\ 3 & 1 & 0\\ 0 & 1 & 4\end{bmatrix}$$

We find

$$\Delta = 10$$

$$C_{11}=(1)\begin{vmatrix}1 & 0\\ 1 & 4\end{vmatrix}=4$$

$$C_{12}=(-1)\begin{vmatrix}3 & 0\\ 0 & 4\end{vmatrix}=-12$$

$$C_{13}=(1)\begin{vmatrix}3 & 1\\ 0 & 1\end{vmatrix}=3$$

$$C_{21}=(-1)\begin{vmatrix}0 & 2\\ 1 & 4\end{vmatrix}=2$$

$$C_{22}=4,\quad C_{23}=-1$$

$$C_{31}=-2,\quad C_{32}=6,\quad C_{33}=1$$

Therefore,

$$\mathbf{A}^{-1}=\frac{1}{10}\begin{bmatrix}4 & 2 & -2\\ -12 & 4 & 6\\ 3 & -1 & 1\end{bmatrix}$$

Properties of Inverses

If $\mathbf{A}$ is a square nonsingular matrix and $\mathbf{A}^{-1}$ is its inverse, then by definition

$$\mathbf{A}^{-1}\mathbf{A}=\mathbf{I}$$

It also can be verified that $\mathbf{A}$ acts as the inverse of $\mathbf{A}^{-1}$. That is,

$$\mathbf{A}\mathbf{A}^{-1}=\mathbf{I}$$

Finally, suppose $\mathbf{A}$ and $\mathbf{B}$ are nonsingular $n\times n$ matrices. Let us compute $(\mathbf{AB})^{-1}$ in terms of the inverses of the individual matrices. We write

$$(\mathbf{AB})^{-1}=\mathbf{C}$$

Then

$$\mathbf{I}=\mathbf{ABC}$$

$$\mathbf{A}^{-1}=\mathbf{BC}$$

$$\mathbf{B}^{-1}\mathbf{A}^{-1}=\mathbf{C}$$

Therefore,

$$(\mathbf{AB})^{-1} = \mathbf{B}^{-1}\mathbf{A}^{-1} \tag{3-17}$$

The general rule is: *The inverse of a product of square matrices is equal to the product of the inverses in the reverse order.*

Homogeneous Linear Equations

One of the most fundamental results of linear algebra is concerned with the existence of nonzero solutions to a set of linear homogeneous equations. Because of its importance, we display this result as a formal lemma, and give a complete proof.

Fundamental Lemma. *Let* $\mathbf{A}$ *be an* $n \times n$ *matrix. Then the homogeneous equation*

$$\mathbf{Ax} = \mathbf{0} \tag{3-18}$$

has a nonzero solution (a vector $\mathbf{x}$ *whose components are not all zero) if and only if the matrix* $\mathbf{A}$ *is singular.*

Proof. The "only if" portion is quite simple. To see this suppose there is a nonzero solution. If $\mathbf{A}$ were nonsingular, the equation could be multiplied through by $\mathbf{A}^{-1}$ yielding

$$\mathbf{A}^{-1}\mathbf{Ax} = \mathbf{0}$$

or, equivalently, $\mathbf{x} = \mathbf{0}$, which is a contradiction. Therefore, there can be a nonzero solution only if $\mathbf{A}$ is singular.

The "if" portion is proved by induction on the dimension n. Certainly the statement is true for $n = 1$. Suppose that it is true for $n - 1$. When written out in detail the set of equations has the form

$$\begin{aligned}
a_{11}x_1 + a_{12}x_2 + \cdots + a_{1n}x_n &= 0 \\
a_{21}x_1 + a_{22}x_2 + \cdots + a_{2n}x_n &= 0 \\
\vdots \qquad\qquad\qquad & \vdots \\
a_{n1}x_1 + a_{n2}x_2 + \cdots + a_{nn}x_n &= 0
\end{aligned}$$

Suppose the corresponding matrix $\mathbf{A}$ is singular. We must construct a nonzero solution.

In this set of equations, if all the coefficients in the first column (the coefficients of the form a_{i1}) are all zero, then the solution $x_1 = 1$, $x_i = 0$, $i > 1$, satisfies the conditions and the conclusion would follow. Otherwise at least one such coefficient must be nonzero, and without loss of generality it may be assumed that $a_{11} \neq 0$.

By subtracting appropriate multiples of the first equation from the remaining equations, one obtains the equivalent set of equations:

$$\begin{aligned}
&a_{11}x_1 \quad + a_{12}x_2 \quad + a_{13}x_3 + \cdots \quad + a_{1n}x_n = 0\\
&\left(a_{22} - a_{12}\frac{a_{21}}{a_{11}}\right)x_2 + \left(a_{23} - a_{13}\frac{a_{21}}{a_{11}}\right)x_3 + \cdots \left(a_{2n} - a_{1n}\frac{a_{21}}{a_{11}}\right)x_n = 0\\
&\left(a_{32} - a_{12}\frac{a_{31}}{a_{11}}\right)x_2 + \left(a_{33} - a_{13}\frac{a_{31}}{a_{11}}\right)x_3 + \cdots \left(a_{3n} - a_{1n}\frac{a_{31}}{a_{11}}\right)x_n = 0\\
&\qquad \vdots \qquad\qquad\qquad\qquad\qquad\qquad \vdots\\
&\left(a_{n2} - a_{12}\frac{a_{n1}}{a_{11}}\right)x_2 + \left(a_{n3} - a_{13}\frac{a_{n1}}{a_{11}}\right)x_3 + \cdots \left(a_{nn} - a_{1n}\frac{a_{n1}}{a_{11}}\right)x_n = 0
\end{aligned}$$

In this form the system can be regarded as consisting of a first equation and a $(n-1)$-dimensional system in the variables $x_2, x_3, \ldots, x_n$.

The determinant of the entire transformed n-dimensional set is exactly equal to the determinant of the original set, since the transformed set was obtained by subtracting multiples of the first row. Laplace's expansion down the first column, however, shows that the value of the determinant of the transformed set is just a_{11} times the determinant of the $n-1$ dimensional system. Since the $n \times n$ original determinant is assumed to be zero and $a_{11} \neq 0$, it follows that the determinant of the $(n-1)$-dimensional system is zero. By the induction hypothesis this smaller system has a nonzero solution $x_2, x_3, \ldots, x_n$. If this solution is substituted into the very first equation, a corresponding value for x_1 can be found. The resulting set of n values $x_1, x_2, \ldots, x_n$ then comprises a nonzero solution to the complete n-dimensional system. ∎

GEOMETRIC PROPERTIES

3.4 VECTOR SPACE

For purposes of manipulation, the formalism of matrix algebra, as outlined in the first three sections of this chapter, is extremely valuable. It simultaneously provides both a compact notational framework and a set of systematic procedures for what might otherwise be complicated operations.

For purposes of conceptualization, however, to most effectively explore new ideas related to multivariable systems, it is useful to take yet another step away from detail. The appropriate step is to introduce the concept of *vector space* where vectors are regarded simply as elements in a space, rather than as special one-dimensional arrays of coefficients.

Define the space E^n as the set of all vectors of the form

$$\mathbf{x} = \begin{bmatrix} x_1 \\ x_2 \\ \cdot \\ \cdot \\ \cdot \\ x_n \end{bmatrix}$$

where each x_i is a scalar (real or complex number). Vectors of this form can be visualized as points in n-dimensional space or as directed lines emanating from the origin, and indeed this vector space is equal to what is generally referred to as (complex) n-dimensional space.

If the *coordinate* or *basis vectors*

$$\mathbf{u}_1 = \begin{bmatrix} 1 \\ 0 \\ 0 \\ \cdot \\ \cdot \\ \cdot \\ 0 \end{bmatrix} \quad \mathbf{u}_2 = \begin{bmatrix} 0 \\ 1 \\ 0 \\ \cdot \\ \cdot \\ \cdot \\ 0 \end{bmatrix} \quad \cdots \quad \mathbf{u}_n = \begin{bmatrix} 0 \\ 0 \\ 0 \\ \cdot \\ \cdot \\ \cdot \\ 1 \end{bmatrix}$$

are defined, a given vector $\mathbf{x}$ can be thought of as being constructed from these vectors. The components of $\mathbf{x}$ are the amounts of the various n coordinate vectors that comprise $\mathbf{x}$. This is illustrated in Fig. 3.1*a*. For purposes of discussion and conceptualization, however, it is not really necessary to continually think about the coordinates and the components, for they clutter up our visualization. Instead, one imagines the vector simply as an element in the space, as illustrated in Fig. 3.1*b*. Furthermore, vectors can be added together or multiplied by a constant without explicit reference to the components, as

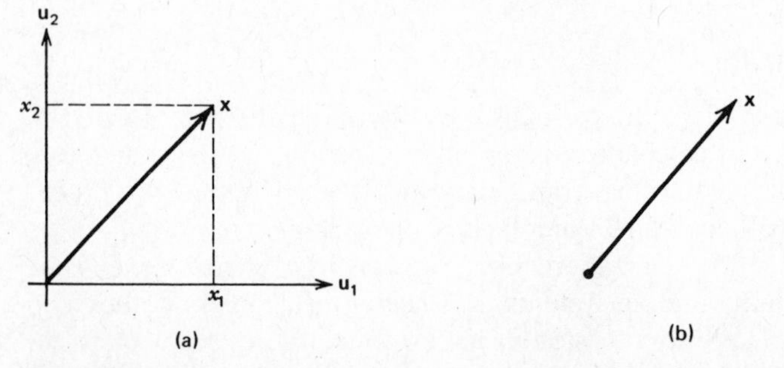

Figure 3.1. (a) Coordinate representation, (b) Vector space representation.

Figure 3.2. (a) Vector addition. (b) Scalar multiplication.

illustrated in Figs. 3.2*a* and 3.2*b*. In this view, a vector has a meaning, and can be conceptually manipulated, quite apart from its representation in terms of the coordinate system.

Linear Independence

A set of vectors $\mathbf{a}_1, \mathbf{a}_2, \ldots, \mathbf{a}_m$ is said to be *linearly dependent* if there is a set of numbers $\alpha_1, \alpha_2, \alpha_3, \ldots, \alpha_m$, not all zero, such that

$$\alpha_1\mathbf{a}_1 + \alpha_2\mathbf{a}_2 + \cdots + \alpha_m\mathbf{a}_m = \mathbf{0}$$

Visually, this means that two vectors are linearly dependent if they point in the same direction (or in directly opposite directions), three vectors are linearly dependent if they lie in a common plane passing through the origin. A set of vectors is *linearly independent* if it is not linearly dependent. In general, to be linearly independent m vectors must "fill out" m dimensions.

In E^n there is a simple test based on evaluating a determinant to check whether n given vectors are linearly independent. The validity of the test rests on the Fundamental Lemma for linear homogeneous equations.

Suppose

$$\mathbf{a}_1 = \begin{bmatrix} a_{11} \\ a_{21} \\ \cdot \\ \cdot \\ \cdot \\ a_{n1} \end{bmatrix} \qquad \mathbf{a}_2 = \begin{bmatrix} a_{12} \\ a_{22} \\ \cdot \\ \cdot \\ \cdot \\ a_{n2} \end{bmatrix} \qquad \cdots \qquad \mathbf{a}_n = \begin{bmatrix} a_{1n} \\ a_{2n} \\ \cdot \\ \cdot \\ \cdot \\ a_{nn} \end{bmatrix}$$

are n given vectors. Stacking them side by side, one can form an $n \times n$ matrix $\mathbf{A}$. To test the linear independence of the vectors $\mathbf{a}_i$, $i = 1, 2, \ldots, n$, one evaluates the determinant of $\mathbf{A}$, as spelled out below.

Theorem. *The vectors $\mathbf{a}_1, \mathbf{a}_2, \ldots, \mathbf{a}_n$ comprising the columns of the $n \times n$ matrix $\mathbf{A}$ are linearly independent if and only if the matrix $\mathbf{A}$ is nonsingular.*

Proof. A linear combination of the vectors $\mathbf{a}_1, \mathbf{a}_2, \ldots, \mathbf{a}_n$ with respective weights $x_1, x_2, \ldots, x_n$ can be represented as $\mathbf{Ax}$. By the Fundamental Lemma, Sect. 3.3, there is a nonzero solution to $\mathbf{Ax}=\mathbf{0}$ if and only if $\mathbf{A}$ is singular. ∎

Rank

Suppose now that $\mathbf{A}$ is an arbitrary $m \times n$ matrix. The *rank* of $\mathbf{A}$ is the number of linearly independent columns in $\mathbf{A}$.

An important result (which we do not prove) is that the rank of $\mathbf{A}^T$ is equal to the rank of $\mathbf{A}$. That means that the number of linearly independent rows of $\mathbf{A}$ is the same as the number of linearly independent columns. It is therefore apparent that the rank of an $m \times n$ matrix $\mathbf{A}$ can be at most equal to the smaller of the two integers m and n. Thus, a matrix with two rows can have rank at most equal to 2, no matter how many columns it has.

Basis

A *basis* for E^n is any set of n linearly independent vectors. The *standard basis* is the set of vectors $\mathbf{u}_1, \mathbf{u}_2, \ldots, \mathbf{u}_n$ defined earlier. An arbitrary vector can be represented as a linear combination of basis vectors. In particular, as is familiar, a vector

$$\mathbf{x} = \begin{bmatrix} x_1 \\ x_2 \\ \cdot \\ \cdot \\ \cdot \\ x_n \end{bmatrix}$$

can be expressed in terms of the standard basis as

$$\mathbf{x} = x_1\mathbf{u}_1 + x_2\mathbf{u}_2 + \cdots + x_n\mathbf{u}_n \tag{3-19}$$

The elements x_i are referred to as the *components* of $\mathbf{x}$ with respect to the standard basis.

Suppose now that a new basis is introduced. This basis consists of a set of n linearly independent vectors, say $\mathbf{p}_1, \mathbf{p}_2, \ldots, \mathbf{p}_n$. The vector $\mathbf{x}$ will have a representation as a linear combination of these vectors in the form

$$\mathbf{x} = z_1\mathbf{p}_1 + z_2\mathbf{p}_2 + \cdots + z_n\mathbf{p}_n \tag{3-20}$$

where now the z_i's are the components of $\mathbf{x}$ with respect to this new basis. Stacking the n vectors $\mathbf{p}_i$, $i = 1, 2, \ldots, n$ into a matrix $\mathbf{P}$, the above can be written as

$$\mathbf{x} = \mathbf{Pz} \tag{3-21}$$

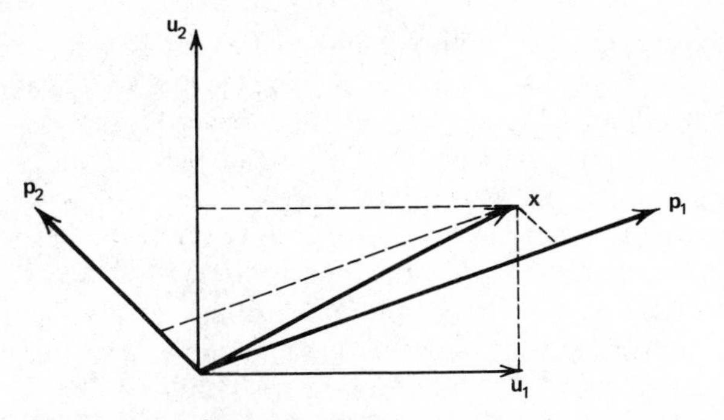

Figure 3.3. Change of basis.

where $\mathbf{z}$ is the column vector with elements z_i. Thus, since we are assured that $\mathbf{P}$ is nonsingular because the $\mathbf{p}_i$'s are linearly independent, we can write

$$\mathbf{z} = \mathbf{P}^{-1}\mathbf{x} \tag{3-22}$$

This equation gives the new components in terms of the old. Both sets of components represent the same point in the vector space—they just define that point in terms of different bases.

This process of changing basis is illustrated in Fig. 3.3. The vector $\mathbf{x}$ is shown as being defined both in terms of the standard basis and in terms of a new basis consisting of the two vectors $\mathbf{p}_1$ and $\mathbf{p}_2$.

Example. As a specific example suppose that, in terms of the standard basis, we have

$$\mathbf{u}_1 = \begin{bmatrix} 1 \\ 0 \end{bmatrix} \qquad \mathbf{u}_2 = \begin{bmatrix} 0 \\ 1 \end{bmatrix}$$

$$\mathbf{x} = \begin{bmatrix} 2 \\ 1 \end{bmatrix} \qquad \mathbf{p}_1 = \begin{bmatrix} 3 \\ 1 \end{bmatrix} \qquad \mathbf{p}_2 = \begin{bmatrix} -1 \\ 1 \end{bmatrix}$$

Then

$$\mathbf{P} = \begin{bmatrix} 3 & -1 \\ 1 & 1 \end{bmatrix} \qquad \mathbf{P}^{-1} = \frac{1}{4}\begin{bmatrix} 1 & 1 \\ -1 & 3 \end{bmatrix}$$

and therefore

$$\mathbf{z} = \mathbf{P}^{-1}\mathbf{x} = \begin{bmatrix} \frac{3}{4} \\ \frac{1}{4} \end{bmatrix}$$

3.5 TRANSFORMATIONS

Once vector space is introduced from a geometric viewpoint, it is possible to introduce the concept of a transformation which is also geometrically based. A *transformation* on a vector space is a function taking vectors into vectors. Geometrically, if one visualizes a vector as a point in n-dimensional space, a transformation associates a new point with each point in the space. A simple example of a transformation is a rotation of, say, 90° counterclockwise about a given axis. Another is an elongation where every vector is multiplied by a constant, such as 3, and thus moves further away from the zero vector. In general, a transformation is defined on the vector space itself and has a meaning that is independent of the method used for representing vectors.

An $n \times n$ matrix $\mathbf{A}$ (together with a specified basis) defines a *linear transformation.* If a vector is represented (in the standard basis) by

$$\mathbf{x} = \begin{bmatrix} x_1 \\ x_2 \\ \cdot \\ \cdot \\ \cdot \\ x_n \end{bmatrix}$$

then a new vector $\mathbf{y}$ with components

$$\mathbf{y} = \mathbf{A}\mathbf{x} \tag{3-23}$$

defines the result of the transformation. Thus, a matrix transforms vectors into vectors.

Example 1 (Book Rotations). Let us think of x_1, x_2, x_3 as coordinates of a point in three-dimensional space. The matrix

$$\mathbf{A} = \begin{bmatrix} 0 & -1 & 0 \\ 1 & 0 & 0 \\ 0 & 0 & 1 \end{bmatrix}$$

can then be visualized as corresponding to a counterclockwise rotation of 90° about the x_3 axis. As a concrete visualization, one can hold a book vertically, and face the front cover. The x_1 direction is to the right, the x_2 direction is upward, and x_3 is a ray coming out toward the viewer. Rotation of the book 90° counterclockwise corresponds to $\mathbf{A}$. To verify that, we note that the vector $\mathbf{u}_1$ corresponding to the center of the right edge of the book is transformed to $\mathbf{u}_2$. Likewise, $\mathbf{u}_2$ is transformed to $-\mathbf{u}_1$, and so forth.

In a similar way the matrix

$$\mathbf{B} = \begin{bmatrix} 0 & 0 & -1 \\ 0 & 1 & 0 \\ 1 & 0 & 0 \end{bmatrix}$$

corresponds to a 90° clockwise (as viewed from above) rotation about the vertical axis. If one holds the book and carries out these two rotations successively, it will be found that the result is the rotation **BA**. If these two rotations are carried out in the opposite order the result is the rotation **AB**, and it is easily verified that $\mathbf{AB} \neq \mathbf{BA}$. In general, one linear transformation followed by another corresponds to multiplication of the two associated matrices; and since matrix multiplication is not commutative, the order of the transformations is important.

Change of Basis

Suppose now that starting with the standard basis $\mathbf{u}_1, \mathbf{u}_2, \ldots, \mathbf{u}_n$ there is a given $n \times n$ matrix **A** defining a transformation. Let us consider the effect of a change of the basis on the representation of the transformation. The new basis introduces a new representation of vectors in terms of new components. We want to construct a matrix that in terms of this basis has the same effect on vectors as the original matrix.

Suppose the new basis consists of the columns of an $n \times n$ matrix **P**. Then a vector **x** having components $x_1, x_2, \ldots, x_n$ with respect to the original standard basis will be represented by the components z_i in the new basis. The two sets of components are related, as shown in Sect. 3.4, by the equation

$$\mathbf{x} = \mathbf{P}\mathbf{z} \tag{3-24}$$

The vector that is represented by $\mathbf{y} = \mathbf{A}\mathbf{x}$ in the standard basis will be represented by $\mathbf{w} = \mathbf{P}^{-1}\mathbf{y}$ in the new basis. Therefore, we have $\mathbf{w} = \mathbf{P}^{-1}\mathbf{A}\mathbf{x}$, or equivalently, $\mathbf{w} = \mathbf{P}^{-1}\mathbf{A}\mathbf{P}\mathbf{z}$. Thus, in terms of the new basis, the matrix $\mathbf{P}^{-1}\mathbf{A}\mathbf{P}$ transforms the point represented by **z** into the point represented by **w**. For reference, we write

$$\mathbf{A} \leftrightarrow \mathbf{P}^{-1}\mathbf{A}\mathbf{P} \tag{3-25}$$

to indicate how a transformation represented by the matrix **A** in the standard basis is represented in a new basis.

Let us review this important argument. Given the standard basis, vectors are defined as an array of n components. A given matrix **A** acts on these components yielding a new array of n components, and correspondingly a new vector defined by these components. The result of this action of **A** defines a transformation on the vector space—transforming vectors into new vectors.

If a basis other than the standard basis is introduced, vectors will have new representations; that is, new components. It is to be expected that to define the same transformation as before, transforming vectors just as before, a new matrix must be derived. This matrix must transform components with respect to the new basis so that the corresponding action is geometrically equivalent to the way the original matrix transforms components with respect to the standard basis. The appropriate new matrix is $\mathbf{P}^{-1}\mathbf{A}\mathbf{P}$.

Changes of basis are used frequently in connection with eigenvector analysis, as discussed in the remaining sections of this chapter. The new basis is selected so as to simplify the representation of a transformation.

Example 2. Consider the 2×2 matrix

$$\mathbf{A}=\begin{bmatrix}0 & -1\\ 1 & 0\end{bmatrix}$$

which, with respect to the standard basis, represents a counterclockwise rotation of 90°. Let us introduce the new basis defined by

$$\mathbf{P}=\begin{bmatrix}3 & -1\\ 1 & 1\end{bmatrix}$$

used in the example of the last section. According to the above result, the rotation transformation is represented by the matrix

$$\mathbf{B}=\mathbf{P}^{-1}\mathbf{A}\mathbf{P}$$

with respect to the new basis. This works out to be

$$\mathbf{B}=\begin{bmatrix}\frac{1}{2} & -\frac{1}{2}\\ \frac{5}{2} & -\frac{1}{2}\end{bmatrix}$$

which in this case is somewhat more complicated than the original representation.

We can check that the new matrix is consistent for the vector represented by

$$\mathbf{x}=\begin{bmatrix}2\\ 1\end{bmatrix}$$

in the standard basis. We easily calculate

$$\mathbf{y}=\mathbf{A}\mathbf{x}=\begin{bmatrix}-1\\ 2\end{bmatrix}$$

On the other hand, the original vector is represented by

$$\mathbf{z}=\begin{bmatrix}\frac{3}{4}\\ \frac{1}{4}\end{bmatrix}$$

in the new basis. Then

$$\mathbf{w}=\mathbf{B}\mathbf{z}=\begin{bmatrix}\frac{1}{4}\\ \frac{7}{4}\end{bmatrix}$$

This corresponds to $\mathbf{y}$, since it is easily verified that $\mathbf{y}=\mathbf{P}\mathbf{w}$.

3.6 EIGENVECTORS

The remainder of this chapter deals with a structural analysis of linear transformations based on eigenvectors. In essence, the objective of this study is to find, for a given transformation, a new basis in which the transformation has a simple representation—perhaps as a diagonal matrix. This topic forms the framework for much of the study of linear time-invariant systems that is a central subject of later chapters.

Definition. A number λ is an *eigenvalue* of an $n \times n$ matrix $\mathbf{A}$ if there is a nonzero n vector $\mathbf{x}$ such that

$$\mathbf{A}\mathbf{x} = \lambda \mathbf{x}$$

The corresponding vector $\mathbf{x}$ is said to be an *eigenvector* of the matrix $\mathbf{A}$.

The terms *characteristic value* and *characteristic vector* are sometimes used for eigenvalue and eigenvector. The geometric interpretation of an eigenvector is that operation by $\mathbf{A}$ on the vector merely changes the length (and perhaps the sign) of the vector. It does not rotate the vector to a new position.

The Characteristic Polynomial

For a given value of λ, the eigenvector equation

$$\mathbf{A}\mathbf{x} = \lambda \mathbf{x}$$

is equivalent to the linear homogeneous equation

$$[\mathbf{A} - \lambda \mathbf{I}]\mathbf{x} = \mathbf{0} \tag{3-26}$$

From the Fundamental Lemma (Sect. 3.3) it is known that such an equation possesses a nonzero solution if and only if the determinant of the coefficient matrix vanishes. Therefore, a necessary and sufficient condition for a value λ to be an eigenvalue of the matrix $\mathbf{A}$ is that

$$\det[\mathbf{A} - \lambda \mathbf{I}] = 0 \tag{3-27}$$

This equation is called the *characteristic equation* of $\mathbf{A}$.

The value of $\det[\mathbf{A} - \lambda \mathbf{I}]$ is a function of the variable λ. Indeed, it can be seen that $\det[\mathbf{A} - \lambda \mathbf{I}]$, when expanded out, is a polynomial of degree n in the variable λ with the coefficient of λ^n being $(-1)^n$. (See Problem 13.) This polynomial $p(\lambda)$ is called the *characteristic polynomial* of the matrix $\mathbf{A}$. From the discussion above, it is clear that there is a direct correspondence between roots of the characteristic polynomial and eigenvalues of the matrix $\mathbf{A}$.

From the Fundamental Theorem of algebra it is known that every polynomial of degree $n \geq 1$ has at least one (possibly complex) root, and can be decomposed into first-degree factors. The characteristic polynomial can be written in factored form as

$$p(\lambda) = (\lambda_1 - \lambda)(\lambda_2 - \lambda) \cdots (\lambda_n - \lambda)$$

The λ_i are the (not necessarily distinct) roots of the polynomial. It follows that there is always at least one solution to the characteristic equation, and hence always at least one eigenvalue. To summarize:

Theorem. *Every $n \times n$ matrix $\mathbf{A}$ possesses at least one eigenvalue and a corresponding (nonzero) eigenvector.*

Example 1. Let

$$\mathbf{A}=\begin{bmatrix} 2 & 1 \\ 2 & 3 \end{bmatrix}$$

The characteristic polynomial is

$$\begin{vmatrix} 2-\lambda & 1 \\ 2 & 3-\lambda \end{vmatrix} = (2-\lambda)(3-\lambda)-2$$

$$= \lambda^2 - 5\lambda + 4$$

The characteristic polynomial can be factored as $\lambda^2-5\lambda+4=(\lambda-1)(\lambda-4)$. Therefore, the polynomial has the two roots: $\lambda=1$, $\lambda=4$. These are the eigenvalues of the matrix.

To find the corresponding eigenvectors, we first set $\lambda=1$ in the homogeneous equation $[\mathbf{A}-\lambda\mathbf{I}]\mathbf{x}=\mathbf{0}$. This leads to

$$\begin{bmatrix} 1 & 1 \\ 2 & 2 \end{bmatrix}\begin{bmatrix} x_1 \\ x_2 \end{bmatrix}=\begin{bmatrix} 0 \\ 0 \end{bmatrix}$$

The two scalar equations defined by this set are equivalent to $x_1=-x_2$. Thus, one solution is

$$\mathbf{x}=\begin{bmatrix} 1 \\ -1 \end{bmatrix}$$

and the general solution is

$$\mathbf{x}=\begin{bmatrix} a \\ -a \end{bmatrix}$$

for $a \neq 0$. These vectors are the eigenvectors corresponding to the eigenvalue $\lambda=1$.

For $\lambda=4$ we are led to

$$\begin{bmatrix} -2 & 1 \\ 2 & -1 \end{bmatrix}\begin{bmatrix} x_1 \\ x_2 \end{bmatrix}=\begin{bmatrix} 0 \\ 0 \end{bmatrix}$$

Thus, one corresponding eigenvector is

$$\mathbf{x}=\begin{bmatrix} 1 \\ 2 \end{bmatrix}$$

and the general solution is

$$\mathbf{x}=\begin{bmatrix} b \\ 2b \end{bmatrix}$$

for $b \neq 0$.

It is a general property that eigenvectors are defined only to within a scalar multiple. If $\mathbf{x}$ is an eigenvector, then so is $\alpha\mathbf{x}$ for any nonzero scalar α.

Example 2 (Complex Eigenvalues). Let

$$\mathbf{A}=\begin{bmatrix} 3 & 2 \\ -1 & 1 \end{bmatrix}$$

The corresponding characteristic equation is

$$\begin{vmatrix} 3-\lambda & 2 \\ -1 & 1-\lambda \end{vmatrix}=0$$

or

$$(3-\lambda)(1-\lambda)+2=0$$

Equivalently,

$$\lambda^2-4\lambda+5=0$$

There are two complex roots:

$$\lambda=2+i$$
$$\lambda=2-i$$

which, as is always the case for real matrices, are complex conjugate pairs.

Corresponding to $\lambda=2+i$, one can find the eigenvector

$$\mathbf{e}_1=\begin{bmatrix} 2 \\ -1+i \end{bmatrix}$$

Likewise, corresponding to the eigenvalue $\lambda=2-i$, there is the eigenvector

$$\mathbf{e}_2=\begin{bmatrix} 2 \\ -1-i \end{bmatrix}$$

Linear Independence of Eigenvectors

Each distinct root of the characteristic polynomial defines an eigenvalue of the matrix $\mathbf{A}$. Associated with each of these distinct eigenvalues there is at least one eigenvector. As stated below, a set of such eigenvectors, each corresponding to a different eigenvalue, is always a linearly independent set.

Proposition. *Let $\lambda_1, \lambda_2, \ldots, \lambda_m$ be distinct eigenvalues of the matrix $\mathbf{A}$. Then any set $\mathbf{e}_1, \mathbf{e}_2, \ldots, \mathbf{e}_m$ of corresponding eigenvectors is linearly independent.*

Proof. Suppose that the eigenvectors were linearly dependent. Then there would be a nonzero linear combination of these vectors that was equal to zero. From the possible such linear combinations, select one which has the minimum number of nonzero coefficients. Without loss of generality it can be assumed that these coefficients correspond to the first k eigenvectors, and that the first coefficient is unity. That is, the relation is of the form

$$\mathbf{e}_1 + \sum_{i=2}^{k} \alpha_i \mathbf{e}_i = \mathbf{0} \tag{3-28}$$

for some set of α_i's, $i = 2, 3, \ldots, k$, $\alpha_i \neq 0$.

Multiplication of this equation by the matrix $\mathbf{A}$ gives

$$\mathbf{A}\mathbf{e}_1 + \sum_{i=2}^{k} \alpha_i \mathbf{A}\mathbf{e}_i = \mathbf{0} \tag{3-29}$$

Using the fact that the $\mathbf{e}_i$'s are eigenvectors, this last equation is equivalent to

$$\lambda_1 \mathbf{e}_1 + \sum_{i=2}^{k} \alpha_i \lambda_i \mathbf{e}_i = \mathbf{0} \tag{3-30}$$

Multiplying (3-28) by λ_1 and subtracting it from (3-30) yields

$$\sum_{i=2}^{k} \alpha_i (\lambda_i - \lambda_1) \mathbf{e}_i = \mathbf{0}$$

This, however, is a linear combination of only $k-1$ terms, contradicting the definition of k as the minimum possible value. ∎

It is important to note that this result on linear independence is true even if the eigenvalues of $\mathbf{A}$ are not all distinct. Any set of eigenvectors, one for each of the distinct eigenvalues, will be an independent set.

3.7 DISTINCT EIGENVALUES

An important special situation is where the n eigenvalues determined from the characteristic polynomial of an $n \times n$ matrix $\mathbf{A}$ are all distinct. In that case, as is shown in this section, the corresponding n eigenvectors serve as a convenient new set of basis vectors, and with respect to this basis the original transformation is represented by a diagonal matrix.

Suppose the $n \times n$ matrix $\mathbf{A}$ has the n (distinct) eigenvalues $\lambda_1, \lambda_2, \ldots, \lambda_n$, and corresponding eigenvectors $\mathbf{e}_1, \mathbf{e}_2, \ldots, \mathbf{e}_n$. According to the Proposition of the last section, the set of eigenvectors is in this case linearly independent. Therefore, the n eigenvectors can serve as a basis for the vector space E^n. In particular, any vector $\mathbf{x}$ can be expressed as a linear combination of these basis vectors in the form

$$\mathbf{x} = z_1\mathbf{e}_1 + z_2\mathbf{e}_2 + \cdots + z_n\mathbf{e}_n \tag{3-31}$$

for some constants z_i, $i = 1, 2, \ldots, n$. Expressed in this form, it is quite easy to find the corresponding representation for $\mathbf{Ax}$. Indeed, it follows immediately that

$$\mathbf{Ax} = \lambda_1 z_1\mathbf{e}_1 + \lambda_2 z_2\mathbf{e}_2 + \cdots + \lambda_n z_n\mathbf{e}_n \tag{3-32}$$

Thus, the new coefficients of the basis vectors are just multiples of the old coefficients. There is no mixing among coefficients as there would be in an arbitrary basis.

This simple but valuable idea can be translated into the mechanics of matrix manipulation, where it takes on a form directly suitable for computation. Define the *modal matrix* of $\mathbf{A}$ to be the $n \times n$ matrix

$$\mathbf{M} = [\mathbf{e}_1\ \mathbf{e}_2 \cdots \mathbf{e}_n] \tag{3-33}$$

That is, $\mathbf{M}$ has the eigenvectors as its n columns. The vector $\mathbf{x}$ and its representation in the new basis with components z_i, $i = 1, 2, \ldots, n$ are then related by

$$\mathbf{x} = \mathbf{Mz} \tag{3-34}$$

In the new set of coordinates, using the new basis, the matrix $\mathbf{A}$, as derived in Sect. 3.5, will be represented as

$$\mathbf{\Lambda} = \mathbf{M}^{-1}\mathbf{AM} \tag{3-35}$$

However, from (3-32) it is known that in the new basis the matrix is represented by a diagonal matrix, for action by $\mathbf{A}$ simply multiplies the ith component value by λ_i. Thus, the matrix $\mathbf{\Lambda}$ is the diagonal matrix

$$\mathbf{\Lambda} = \begin{bmatrix} \lambda_1 & 0 & \cdots & 0 \\ 0 & \lambda_2 & & \cdot \\ \cdot & & \cdot & \cdot \\ \cdot & & & \cdot \\ 0 & \cdot & \cdot & \lambda_n \end{bmatrix} \tag{3-36}$$

Thus, we may state the following very useful result.

Theorem. *Any square matrix with distinct eigenvalues can be put in diagonal form by a change of basis. Specifically, corresponding to an $n \times n$ matrix $\mathbf{A}$ with distinct eigenvalues, there holds*

$$\mathbf{\Lambda} = \mathbf{M}^{-1}\mathbf{AM}$$

where $\mathbf{\Lambda}$ is defined by (3-36) and $\mathbf{M}$ is the modal matrix of $\mathbf{A}$.

Equation (3-34) is frequently used in the reverse direction as

$$\mathbf{A} = \mathbf{M}\mathbf{\Lambda}\mathbf{M}^{-1} \tag{3-37}$$

which gives a representation for the matrix **A** in terms of its eigenvectors and eigenvalues.

Another way to write this relation is

$$\mathbf{A}\mathbf{M} = \mathbf{M}\mathbf{\Lambda} \tag{3-38}$$

which is a form that is directly equivalent to the original definition of the eigenvectors of **A**. This is seen by viewing the matrix equation one column at a time. For example, the first column on the left-hand side of the equation is **A** times the first column in **M**; that is, **A** times the first eigenvector. Correspondingly, the first column on the right-hand side of the equation is just λ_1 times the first eigenvector. Thus, the correspondence of the first columns is equivalent to the equation $\mathbf{A}\mathbf{e}_1 = \lambda_1 \mathbf{e}_1$. Identical interpretations apply to the other columns.

Example. Consider again the matrix

$$\mathbf{A} = \begin{bmatrix} 2 & 1 \\ 2 & 3 \end{bmatrix}$$

It was found, in Example 1, Sect. 3.6, that the eigenvalues and corresponding eigenvectors of **A** are $\lambda_1 = 1$, $\lambda_2 = 4$:

$$\mathbf{e}_1 = \begin{bmatrix} 1 \\ -1 \end{bmatrix}, \qquad \mathbf{e}_2 = \begin{bmatrix} 1 \\ 2 \end{bmatrix}$$

The modal matrix of **A** is therefore

$$\mathbf{M} = \begin{bmatrix} 1 & 1 \\ -1 & 2 \end{bmatrix}$$

and it is readily computed that

$$\mathbf{M}^{-1} = \frac{1}{3} \begin{bmatrix} 2 & -1 \\ 1 & 1 \end{bmatrix}$$

Then

$$\mathbf{A}\mathbf{M} = \begin{bmatrix} 1 & 4 \\ -1 & 8 \end{bmatrix}$$

and, finally,

$$\mathbf{M}^{-1}\mathbf{A}\mathbf{M} = \begin{bmatrix} 1 & 0 \\ 0 & 4 \end{bmatrix} = \mathbf{\Lambda}$$

3.8 RIGHT AND LEFT EIGENVECTORS

As defined to this point, eigenvectors are *right eigenvectors* in the sense that they appear as columns on the right-hand side of the $n \times n$ matrix $\mathbf{A}$ in the equation

$$\mathbf{A}\mathbf{e}_i = \lambda_i \mathbf{e}_i \tag{3-39}$$

It is also possible to consider *left eigenvectors* that are multiplied as rows on the left-hand side of $\mathbf{A}$ in the form

$$\mathbf{f}_i^T \mathbf{A} = \lambda_i \mathbf{f}_i^T \tag{3-40}$$

The vector $\mathbf{f}_i$ is an n-dimensional column, and thus $\mathbf{f}_i^T$ is an n-dimensional row vector.

Equation (3-40) can be rewritten in column form by taking the transpose of both sides, yielding

$$\mathbf{A}^T \mathbf{f}_i = \lambda_i \mathbf{f}_i \tag{3-41}$$

Therefore, a left eigenvector of $\mathbf{A}$ is really the same thing as an ordinary right eigenvector of $\mathbf{A}^T$. For most purposes, however, it is more convenient to work with left and right eigenvectors than with transposes.

The characteristic polynomial of $\mathbf{A}^T$ is $\det[\mathbf{A}^T - \lambda \mathbf{I}]$, which, since the determinants of a matrix and its transpose are equal, is identical to the characteristic polynomial of $\mathbf{A}$. Therefore, the right and left *eigenvalues* (not eigenvectors) are identical.

Example. For the matrix

$$\mathbf{A} = \begin{bmatrix} 2 & 1 \\ 2 & 3 \end{bmatrix}$$

it has been shown that $\lambda_1 = 1$, $\lambda_2 = 4$ with corresponding right eigenvectors

$$\mathbf{e}_1 = \begin{bmatrix} 1 \\ -1 \end{bmatrix} \qquad \mathbf{e}_2 = \begin{bmatrix} 1 \\ 2 \end{bmatrix}$$

Let us find the corresponding left eigenvectors. First, for $\lambda_1 = 1$ we must solve

$$[y_1 \quad y_2] \begin{bmatrix} 1 & 1 \\ 2 & 2 \end{bmatrix} = [0 \quad 0]$$

A solution is $y_1 = 2$, $y_2 = -1$, giving the left eigenvector

$$\mathbf{f}_1^T = [2 \quad -1]$$

For $\lambda_2 = 4$, we solve

$$[y_1 \quad y_2] \begin{bmatrix} -2 & 1 \\ 2 & -1 \end{bmatrix} = [0 \quad 0]$$

A solution is $y_1=1$, $y_2=1$, giving the left eigenvector

$$\mathbf{f}_2^T=[1 \quad 1]$$

Orthogonality

There is an important relation between right and left eigenvectors. Suppose λ_i and λ_j are any two (distinct) eigenvalues of the matrix $\mathbf{A}$. Let $\mathbf{e}_i$ be a *right* eigenvector corresponding to λ_i and let $\mathbf{f}_j$ be a *left* eigenvector corresponding to λ_j. Then

$$\mathbf{A}\mathbf{e}_i=\lambda_i\mathbf{e}_i$$

$$\mathbf{f}_j^T\mathbf{A}=\lambda_j\mathbf{f}_j^T$$

Multiplying the first of these equations by $\mathbf{f}_j^T$ on the left, and the second by $\mathbf{e}_i$ on the right, yields the two equations

$$\mathbf{f}_j^T\mathbf{A}\mathbf{e}_i=\lambda_i\mathbf{f}_j^T\mathbf{e}_i$$

$$\mathbf{f}_j^T\mathbf{A}\mathbf{e}_i=\lambda_j\mathbf{f}_j^T\mathbf{e}_i$$

Subtracting we obtain

$$0=(\lambda_i-\lambda_j)\mathbf{f}_j^T\mathbf{e}_i$$

Since $\lambda_i \neq \lambda_j$ it follows that

$$\mathbf{f}_j^T\mathbf{e}_i=0$$

This relation is referred to as an *orthogonality* relation. It says that the inner product (or the dot product) of the vectors $\mathbf{f}_j$ and $\mathbf{e}_i$ is zero. (The reader may wish to check this relation on the example above.) As a formal statement this result is expressed by the following theorem.

Theorem. *For any two distinct eigenvalues of a matrix, the left eigenvector of one eigenvalue is orthogonal to the right eigenvector of the other.*

3.9 MULTIPLE EIGENVALUES

If an $n\times n$ matrix has nondistinct eigenvalues (that is, repeated or multiple roots to its characteristic equation) a more involved analysis may be required. For some matrices with multiple roots it may still be possible to find n linearly independent eigenvectors and use these as a new basis, leading to a diagonal representation. The simplest example is the identity matrix $\mathbf{I}$ that has 1 as an eigenvalue repeated n times. This matrix is, of course, already diagonal. In general, however, matrices with multiple roots may or may not be diagonalizable by a change of basis.

Two important concepts for matrices with multiple roots, which help characterize the complexity of a given matrix, are the notions of algebraic and geometric multiplicity. The *algebraic multiplicity* of an eigenvalue λ_i is the multiplicity determined by the characteristic polynomial. It is the integer α_i associated with $(\lambda-\lambda_i)^{\alpha_i}$ as it appears when the polynomial is factored into distinct factors. If the algebraic multiplicity is one, the eigenvalue is said to be *simple.*

The *geometric multiplicity* of λ_i is the number of linearly independent eigenvectors that can be associated with λ_i. For any eigenvalue, the geometric multiplicity is always at least unity. Also, the geometric multiplicity never exceeds the algebraic multiplicity.

As an example consider the 2×2 matrix

$$\mathbf{A}=\begin{bmatrix}5 & 1\\ 0 & 5\end{bmatrix}$$

It has characteristic polynomial $(5-\lambda)^2$, and hence the only eigenvalue is 5, with algebraic multiplicity of two. A corresponding eigenvector must satisfy the equation

$$\begin{bmatrix}0 & 1\\ 0 & 0\end{bmatrix}\begin{bmatrix}x_1\\ x_2\end{bmatrix}=\begin{bmatrix}0\\ 0\end{bmatrix}$$

The only nonzero solutions to this set are of the form $x_1=\alpha$, $x_2=0$ for some $\alpha\neq 0$. Thus, there is only one linearly independent eigenvector, which can be taken to be

$$\mathbf{x}=\begin{bmatrix}1\\ 0\end{bmatrix}$$

Thus, the geometric multiplicity of λ is one.

Jordan Canonical Form

In the general case, when there is not a full set of eigenvectors, a matrix cannot be transformed to diagonal form by a change of basis. It is, however, always possible to find a basis in which the matrix is nearly diagonal, as defined below. The resulting matrix is referred to as the *Jordan Canonical Form* of the matrix. Since derivation of the general result is quite complex and because the Jordan form is only of modest importance for the development in other chapters, we state the result without proof.

Theorem (Jordan Canonical Form). *Denote by* $\mathbf{L}_k(\lambda)$ *the* $k \times k$ *matrix*

$$\mathbf{L}_k(\lambda) = \begin{bmatrix} \lambda & 1 & 0 & \cdots & & 0 \\ 0 & \lambda & 1 & & & \\ \cdot & & \lambda & 1 & & \\ \cdot & & & \ddots & \ddots & \\ \cdot & & & & \ddots & 1 \\ 0 & 0 & \cdots & & & \lambda \end{bmatrix}$$

Then for any $n \times n$ *matrix* $\mathbf{A}$ *there exists a nonsingular matrix* $\mathbf{T}$ *such that*

$$\mathbf{T}^{-1}\mathbf{A}\mathbf{T} = \begin{bmatrix} \mathbf{L}_{k_1}(\lambda_1) & & & & \\ & \mathbf{L}_{k_2}(\lambda_2) & & & \\ & & \cdot & & \\ & & & \cdot & \\ & & & & \mathbf{L}_{k_r}(\lambda_r) \end{bmatrix}$$

where $k_1 + k_2 + \cdots + k_r = n$, *and where the* λ_i, $i = 1, 2, \ldots, r$ *are the (not necessarily distinct) eigenvalues of* $\mathbf{A}$.

3.10 PROBLEMS

1. Prove that matrix multiplication is associative, and construct an example showing that it is not commutative.

2. *Differentiation Formulas.* (a) Suppose $\mathbf{A}(t)$ and $\mathbf{B}(t)$ are $m \times n$ and $n \times p$ matrices, respectively. Find a formula for

$$\frac{d}{dt}[\mathbf{A}(t)\mathbf{B}(t)]$$

in terms of the derivatives of the individual matrices.
(b) If $\mathbf{A}(t)$ is $n \times n$ and invertible, find a formula for

$$\frac{d}{dt}[\mathbf{A}(t)]^{-1}$$

3. Show that for any n, the $n \times n$ identity matrix $\mathbf{I}$ has determinant equal to unity.

4. Using Laplace's expansion evaluate the determinants of the matrices below:

$$\begin{bmatrix} 3 & 0 & 1 \\ 2 & 4 & 3 \\ 1 & 1 & 2 \end{bmatrix} \qquad \begin{bmatrix} 1 & 2 & 3 & 4 \\ 2 & 3 & 4 & 5 \\ 3 & 4 & 5 & 6 \\ 4 & 5 & 6 & 7 \end{bmatrix}$$

5. Prove that $\det(\mathbf{A}^T)=\det(\mathbf{A})$.

6. Using Laplace's expansion, prove the linear combination properties of determinants.

7. Evaluate the determinants below using the rules for row and column operations:

$$\begin{vmatrix} 3 & 2 & 0 & 4 \\ 1 & 1 & 2 & 3 \\ -1 & 1 & 1 & 2 \\ 0 & 0 & 4 & 3 \end{vmatrix} \quad \begin{vmatrix} 5 & 4 & 6 & 2 \\ 3 & 4 & 5 & 3 \\ 1 & 1 & 3 & 1 \\ 2 & 2 & 2 & 2 \end{vmatrix}$$

8. Find the inverses of the matrices of Problem 4.

9. Prove Theorem 4, Sect. 2.6.

10. Consider the two systems of linear equations

$$\mathbf{Ax}=\mathbf{y}$$

$$\mathbf{Bz}=\mathbf{x}$$

where $\mathbf{x}$ is $n\times 1$, $\mathbf{y}$ is $m\times 1$, $\mathbf{z}$ is $p\times 1$, $\mathbf{A}$ is $m\times n$, and $\mathbf{B}$ is $n\times p$. Show in detail that if the x_j variables are eliminated the resulting system can be expressed as

$$\mathbf{ABz}=\mathbf{y}$$

11. Let $\mathbf{p}_1, \mathbf{p}_2, \ldots, \mathbf{p}_n$ be a basis for E^n and let $\mathbf{x}$ be a given vector in E^n. Show that the representation $\mathbf{x}=\alpha_1\mathbf{p}_1+\alpha_2\mathbf{p}_2+\cdots+\alpha_n\mathbf{p}_n$ is unique. That is, show that the α_i's are unique.

12. Consider the basis for E^3 consisting of the columns of the matrix

$$\mathbf{P}=\begin{bmatrix} 2 & 3 & 1 \\ 1 & 2 & 1 \\ 1 & 1 & 1 \end{bmatrix}$$

Find $\mathbf{P}^{-1}$. Suppose that in the standard basis a vector $\mathbf{x}$ is given by

$$\mathbf{x}=\begin{bmatrix} 2 \\ 1 \\ 4 \end{bmatrix}$$

Find the representation of $\mathbf{x}$ with respect to the basis defined by $\mathbf{P}$.

13. Prove, by induction on the dimension n, that $\det[\mathbf{A}-\lambda\mathbf{I}]$ is a polynomial of degree n.

14. Show that for any $n\times n$ matrix $\mathbf{A}$

$$\det(\mathbf{A})=\prod_{i=1}^{n}\lambda_i$$

where $\lambda_1, \lambda_2, \ldots, \lambda_n$ are the (not necessarily distinct) eigenvalues of $\mathbf{A}$. (*Hint*: Consider the definition of the characteristic polynomial.)

15. The *trace* of a square $n \times n$ matrix $\mathbf{A}$ is the sum of its diagonal elements. That is,

$$\text{Trace } \mathbf{A} = \sum_{i=1}^{n} a_{ii}$$

Show that

$$\text{Trace } \mathbf{A} = \sum_{i=1}^{n} \lambda_i$$

where $\lambda_1, \lambda_2, \ldots, \lambda_n$ are the (not necessarily distinct) eigenvalues of $\mathbf{A}$. (*Hint*: Consider the coefficient a_{n-1} in the characteristic polynomial of $\mathbf{A}$.)

16. Show that for an upper or lower triangular matrix the eigenvalues are equal to the diagonal elements.

17. (a) Find the eigenvalues and eigenvectors of

$$\mathbf{A} = \begin{bmatrix} 5 & -1 & -3 & 3 \\ -1 & 5 & 3 & -3 \\ -3 & 3 & 5 & -1 \\ 3 & -3 & -1 & 5 \end{bmatrix}$$

(b) Find a matrix $\mathbf{H}$ such that $\mathbf{D} = \mathbf{H}^{-1}\mathbf{A}\mathbf{H}$ is diagonal.

18. For the following two matrices:

$$\mathbf{A} = \begin{bmatrix} -2 & 0 & -1 \\ 4 & 2 & 4 \\ 0 & 0 & -1 \end{bmatrix} \qquad \mathbf{B} = \frac{1}{8}\begin{bmatrix} 9 & 0 & -3 \\ 10 & -8 & 2 \\ 3 & 0 & -1 \end{bmatrix}$$

find (a) the characteristic polynomial; (b) the determinant and trace; (c) the eigenvalues; and (d) the right and left eigenvectors.

19. A real square matrix $\mathbf{A}$ is *symmetric* if $\mathbf{A}^T = \mathbf{A}$. Show that for a symmetric matrix (a) all eigenvalues are real; (b) if $\mathbf{e}_i$ and $\mathbf{e}_j$ are eigenvectors associated with λ_i and λ_j, where $\lambda_i \neq \lambda_j$, then $\mathbf{e}_i^T \mathbf{e}_j = 0$.

20. For the matrix

$$\mathbf{A} = \begin{bmatrix} 1 & 0 & -2 \\ \frac{1}{2} & 2 & 1 \\ \frac{1}{2} & 0 & 3 \end{bmatrix}$$

find (a) the characteristic polynomial; (b) all eigenvalues; (c) all eigenvectors; and (d) the Jordan form of $\mathbf{A}$.

21. If two matrices $\mathbf{A}$ and $\mathbf{B}$ are related by a change of basis (that is, if $\mathbf{B} = \mathbf{P}^{-1}\mathbf{A}\mathbf{P}$ for some $\mathbf{P}$), then the matrices are said to be *similar*. Prove that similar matrices have the same characteristic polynomial.

22. The members of the basis associated with the Jordan Canonical Form are often referred to as occurring in *chains*—this terminology arising from the following interpretation. If the geometric multiplicity of an eigenvalue λ is m, then m linearly independent eigenvectors are part of the basis. Each of these eigenvectors satisfies $[\mathbf{A}-\lambda\mathbf{I}]\mathbf{e}=\mathbf{0}$ and each is considered as the first element of one of m separate chains associated with the eigenvalue λ. The next member of the chain associated with $\mathbf{e}$ is a vector $\mathbf{f}$ such that $[\mathbf{A}-\lambda\mathbf{I}]\mathbf{f}=\mathbf{e}$. The chain continues with a $\mathbf{g}$ satisfying $[\mathbf{A}-\lambda\mathbf{I}]\mathbf{g}=\mathbf{f}$, and so forth until the chain ends. The original m eigenvectors generate m separate chains, which may have different lengths.

    Given a matrix $\mathbf{J}$ in Jordan form with m blocks associated with the eigenvalue λ, find the m eigenvectors of $\mathbf{J}$. Also find the vectors in the chain associated with each eigenvector.

*23. *Matrix Perturbation.* Show that given any $n\times n$ matrix $\mathbf{A}$ and an arbitrary $\varepsilon>0$, it is possible to perturb the entries of $\mathbf{A}$ by an amount less than ε so that the resulting matrix is diagonalizable. (This result is useful in many theoretical developments, since it is often easiest to work out a theory for diagonalizable matrices and then extend the theory to the general case by a limiting argument.)

NOTES AND REFERENCES

There are a large number of texts on linear algebra and matrix theory that can be used to supplement this chapter as background for the remaining chapters. Some suggestions are Bellman [B6], Hoffman and Kunze [H3], Strang [S6], and Gantmacher [G2], [G3]. A brief treatment together with applications is contained in Kemeny, Mirkil, Snell, and Thompson [K10].

Section 3.1. The Leontief model is used extensively for various empirical economic investigations, and large matrices representing the economy in a given year have been constructed. For a sample of an actual large matrix see Leontief [L3].

chapter 4.

Linear State Equations

At this point the concept of dynamics, as represented by ordinary difference and differential equations, is combined with the machinery of linear algebra to begin a study of the modern approach to dynamic systems. The foundation for this approach is the notion of a system of first-order equations, either in discrete or continuous time.

4.1 SYSTEMS OF FIRST-ORDER EQUATIONS

In discrete time, an nth-order system is defined in terms of n variables $x_1(k), x_2(k), \ldots, x_n(k)$ that are each functions of the index k. These n variables are related by a system of n first-order difference equations of the following general form:

$$
\begin{aligned}
x_1(k+1) &= f_1(x_1(k), x_2(k), \ldots, x_n(k), k) \\
x_2(k+1) &= f_2(x_1(k), x_2(k), \ldots, x_n(k), k) \\
&\ \ \vdots \\
x_n(k+1) &= f_n(x_1(k), x_2(k), \ldots, x_n(k), k)
\end{aligned}
\tag{4-1}
$$

The functions f_i, $i = 1, 2, \ldots, n$ define the system. They may be simple in form or quite complex, depending on the situation the system describes. The variables $x_i(k)$, $i = 1, 2, \ldots, n$ are regarded as the unknowns whose values are determined (at least in part) by the system of equations. These variables are referred to as *state variables.*

If the system is defined for $k = 0, 1, 2, \ldots,$ then the n values $x_1(0), x_2(0), \ldots, x_n(0)$ are referred to as *initial conditions.* If the initial conditions are specified, then they may be substituted into the right-hand side of (4-1) to yield the values of $x_1(1), x_2(1), \ldots, x_n(1)$. These in turn can be substituted in the right-hand side to yield $x_1(2), x_2(2), \ldots, x_n(2)$. This recursive process can be continued to yield the unique solution corresponding to the given initial conditions. At each stage k of the recursion, the corresponding set of state variables $x_1(k), x_2(k), \ldots, x_n(k)$ serve as initial conditions for the remaining stages.

The analog of (4-1) in continuous time is a system of first-order differential equations. Such a system is defined in terms of n variables $x_1(t), x_2(t), \ldots, x_n(t)$ that are each functions of the continuous variable t. These n variables are related by a system of n equations of the following general form:*

$$
\begin{aligned}
\dot{x}_1(t) &= f_1(x_1(t), x_2(t), \ldots, x_n(t), t) \\
\dot{x}_2(t) &= f_2(x_1(t), x_2(t), \ldots, x_n(t), t) \\
&\;\;\vdots \\
\dot{x}_n(t) &= f_n(x_1(t), x_2(t), \ldots, x_n(t), t)
\end{aligned}
\tag{4-2}
$$

Again the n variables $x_1(t), x_2(t), \ldots, x_n(t)$ are referred to as *state variables.*

Some examples of systems in discrete or continuous time were presented in Chapter 1. The first-order models of geometric and exponential growth are simple examples, corresponding to the elementary case $n = 1$. The cohort population model is an excellent discrete-time example for general n. The goats and wolves model is an example of a continuous-time system, corresponding to $n = 2$. Dozens of others are presented throughout the remainder of the text.

A characteristic of systems of equations, as compared with the ordinary difference and differential equations discussed in Chapter 2, is that they simultaneously relate several variables rather than just one. This multivariable aspect is often characteristic of even the simplest situation, especially if a particular phenomenon is viewed as consisting of several components. The variables in a system might, for example, represent a decomposition of a given quantity, such as population into age groups, or economic production into commodities. In physical problems, the variables might represent components of position and velocity in various spacial dimensions. The system framework retains the distinction among the variables.

* We employ the standard "dot" notation $\dot{x}(t) \equiv \dfrac{dx(t)}{dt}$.

Linear Systems

This chapter addresses linear systems. A discrete-time system is *linear* if it has the following form:

$$
\begin{aligned}
x_1(k+1) &= a_{11}(k)x_1(k) + a_{12}(k)x_2(k) + \cdots + a_{1n}(k)x_n(k) + w_1(k) \\
x_2(k+1) &= a_{21}(k)x_1(k) + a_{22}(k)x_2(k) + \cdots + a_{2n}(k)x_n(k) + w_2(k) \\
&\vdots \\
x_n(k+1) &= a_{n1}(k)x_1(k) + a_{n2}(k)x_2(k) + \cdots + a_{nn}(k)x_n(k) + w_n(k)
\end{aligned}
$$

Again, the variables $x_1(k), x_2(k), \ldots, x_n(k)$ are the *state variables* of the system, and they may take on arbitrary real values. The values $a_{ij}(k)$, $i = 1, 2, \ldots, n$, $j = 1, 2, \ldots, n$ are fixed *parameters* or *coefficients* of the system. As indicated by their argument k, it is allowable for these parameters to depend on time, but this dependency is predetermined and independent of the values assumed by the state variables. If these parameters do not depend on k, the system is said to have *constant coefficients* or to be *time-invariant*. The values $w_i(k)$, $i = 1, 2, \ldots, n$ are also *parameters* denoting the driving or forcing terms in the system. The essential defining feature of a linear system, of course, is that all terms are linear with respect to the state variables.

The general description above is somewhat tedious to write out in detail, and matrix notation can be used to great advantage. With this notation, the system can be expressed in the equivalent form

$$
\mathbf{x}(k+1) = \mathbf{A}(k)\mathbf{x}(k) + \mathbf{w}(k)
$$

Here $\mathbf{x}(k)$ is the $n \times 1$ state vector and $\mathbf{w}(k)$ is the $n \times 1$ forcing vector. That is,

$$
\mathbf{x}(k) = \begin{bmatrix} x_1(k) \\ x_2(k) \\ \cdot \\ \cdot \\ \cdot \\ x_n(k) \end{bmatrix} \qquad \mathbf{w}(k) = \begin{bmatrix} w_1(k) \\ w_2(k) \\ \cdot \\ \cdot \\ \cdot \\ w_n(k) \end{bmatrix}
$$

The matrix $\mathbf{A}(k)$ is the square $n \times n$ matrix consisting of the coefficients $a_{ij}(k)$. It is referred to as the *system matrix*.

For continuous-time systems the situation, as usual, is entirely analogous. A continuous-time linear dynamic system of order n is described by the set of n ordinary differential equations:

$$
\begin{aligned}
\dot{x}_1(t) &= a_{11}(t)x_1(t) + a_{12}(t)x_2(t) + \cdots + a_{1n}(t)x_n(t) + w_1(t) \\
\dot{x}_2(t) &= a_{21}(t)x_1(t) + a_{22}(t)x_2(t) + \cdots + a_{2n}(t)x_n(t) + w_2(t) \\
&\vdots \\
\dot{x}_n(t) &= a_{n1}(t)x_1(t) + a_{n2}(t)x_2(t) + \cdots + a_{nn}(t)x_n(t) + w_n(t)
\end{aligned}
$$

As before, the $x_i(t)$, $i = 1, 2, \ldots, n$, are *state variables*, the $a_{ij}(t)$ are *parameters* or *coefficients*, and the $w_i(t)$, $i = 1, 2, \ldots, n$, are *forcing terms*. In order to guarantee existence and uniqueness of solution, the $a_{ij}(t)$'s are usually assumed to be continuous in t.

Just as for discrete-time systems, continuous-time linear dynamic systems are conveniently expressed in matrix notation. In this notation an nth-order system takes the form

$$\dot{\mathbf{x}}(t) = \mathbf{A}(t)\mathbf{x}(t) + \mathbf{w}(t)$$

where $\mathbf{x}(t)$ is the $n \times 1$ state vector, $\mathbf{w}(t)$ is the $n \times 1$ forcing vector, and $\mathbf{A}(t)$ is the $n \times n$ matrix of coefficients referred to as the *system matrix*.

Inputs

In most applications the forcing or driving terms in a system are derived from a single, or perhaps a few, specific inputs to the system. In some cases these inputs actually may be consciously controlled in an effort to guide the behavior of the system. In other cases, they may be fixed by the environment, but still retain an interpretation as input. When, for example, the simplest first-order model is viewed as a description of a bank balance, the forcing term corresponds to deposits into the account. Likewise, in more complex systems the forcing term is typically derived from some identifiable input source.

There is often a simple structural connection between the source of input and the resulting forcing terms. For instance, a single input source may affect all of the equations in the system, or an input may enter only a few equations. It is useful to explicitly display the particular structural relation in the formulation of the system equations. The definition of a linear system is expanded slightly to account for this additional structure.

A discrete-time linear system with inputs has the following form:

$$
\begin{aligned}
x_1(k+1) &= a_{11}(k)x_1(k) + a_{12}(k)x_2(k) + \cdots + a_{1n}(k)x_n(k) \\
&\quad + b_{11}(k)u_1(k) + \cdots + b_{1m}(k)u_m(k) \\
x_2(k+1) &= a_{21}(k)x_1(k) + a_{22}(k)x_2(k) + \cdots + a_{2n}(k)x_n(k) \\
&\quad + b_{21}(k)u_1(k) + \cdots + b_{2m}(k)u_m(k) \\
&\vdots \\
x_n(k+1) &= a_{n1}(k)x_1(k) + \cdots \qquad\qquad + a_{nn}(k)x_n(k) \\
&\quad + b_{n1}(k)u_1(k) + \cdots + b_{nm}(k)u_m(k)
\end{aligned}
$$

The variables $u_1(k), u_2(k), \ldots, u_m(k)$ are the *control variables* or the *input variables* of the system.

In matrix notation the system takes the form

$$\mathbf{x}(k+1) = \mathbf{A}(k)\mathbf{x}(k) + \mathbf{B}(k)\mathbf{u}(k)$$

where $\mathbf{B}(k)$ is an $n \times m$ matrix and $\mathbf{u}(k)$ is an $m \times 1$ input vector. The matrix $\mathbf{B}(k)$ is referred to as the *distribution matrix*, since it acts to distribute the inputs into the system. A common case is where $m = 1$, corresponding to a single control variable. In this case the $\mathbf{B}(k)$ matrix reduces to a n-dimensional column vector, and accordingly, in this case $\mathbf{B}(k)$ is usually replaced by the notation $\mathbf{b}(k)$ to explicitly indicate a column vector rather than a more general matrix.

In terms of the earlier definition, it is clear that we have simply made the replacement

$$\mathbf{w}(k) = \mathbf{B}(k)\mathbf{u}(k)$$

From a mathematical viewpoint it is somewhat irrelevant how the driving term is determined. However, for practical purposes this expanded notation is useful, since it is more closely related to the structure of the situation.

Finally, let us point out the obvious extension to continuous time. A linear system with input has the form

$$\dot{\mathbf{x}}(t) = \mathbf{A}(t)\mathbf{x}(t) + \mathbf{B}(t)\mathbf{u}(t)$$

Example (Obsolescence). Let us consider the life history of a class of goods in a country—perhaps some appliance, such as washing machines. We assume that households purchase new washing machines and keep them until they suffer a fatal breakdown or become obsolete. At any one time, therefore, there is a distribution of various aged washing machines throughout the country. We shall describe the system of equations that governs this distribution.

Let us employ a discrete-time formulation based on periods of one year. The basic assumption we make in order to develop the model is that there is a certain probability β_i that any washing machine i years old will remain in service at least one more year. This probability may be relatively high for young machines and low for old machines. We assume that no machine survives to an age of $n+1$ years.

With these assumptions, we divide the washing machines into cohorts of one-year age groups. Let $x_i(k)$ be the number of surviving washing machines of age i years during period (year) k. Then we have the equations

$$x_{i+1}(k+1) = \beta_i x_i(k) \qquad i = 0, 1, 2, \ldots, n-1$$

The number of washing machines less than one year old is equal to the number of purchases $u(k)$ during the year; that is,

$$x_0(k+1) = u(k)$$

This system of equations can be put in matrix form as

$$\begin{bmatrix} x_0(k+1) \\ x_1(k+1) \\ \cdot \\ \cdot \\ \cdot \\ x_n(k+1) \end{bmatrix} = \begin{bmatrix} 0 & 0 & \cdot & \cdot & \cdot & 0 \\ \beta_0 & 0 & \cdot & \cdot & \cdot & 0 \\ 0 & \beta_1 & \cdot & \cdot & \cdot & 0 \\ \cdot & & \cdot & & & \cdot \\ \cdot & & & \cdot & & \cdot \\ 0 & & & & \beta_{n-1} & 0 \end{bmatrix} \begin{bmatrix} x_0(k) \\ x_1(k) \\ \cdot \\ \cdot \\ \cdot \\ x_n(k) \end{bmatrix} + \begin{bmatrix} 1 \\ 0 \\ 0 \\ \\ \\ 0 \end{bmatrix} u(k)$$

which is a special case of the general form

$$\mathbf{x}(k+1) = \mathbf{A}\mathbf{x}(k) + \mathbf{b}u(k)$$

The variable $u(k)$, representing purchases, is the input to the system. In order for a solution of the system to be determined, it is of course necessary to specify an input sequence. This might be done in various ways, depending on the analysis objectives. The simplest would be to specify purchases directly, as perhaps an increasing function of k. A more realistic approach might recognize that purchases partly consist of replacements for those machines that are retired, so that $u(k)$ must ultimately be tied back to $\mathbf{x}(k)$.

This simple model is discussed again later in this chapter and in Problem 2. It is referred to as the *straight-through cohort* model, since cohorts pass directly through without influencing each other.

The State Vector and State Space

The vector $\mathbf{x}(k)$ in discrete-time systems [or $\mathbf{x}(t)$ in continuous-time systems] is referred to as the *state vector* because this vector is a complete description of the system at the time k, at least in so far as determining future behavior. As observed earlier, the state vector serves as a kind of running collection of initial conditions. Knowledge of these conditions at a given time together with a specification of future inputs is all that is necessary to specify future behavior. Indeed, in discrete time, the future behavior can be calculated recursively from the system of difference equations once the current state is known. In continuous time the future is likewise determined by the current state, although it may not be quite so easily determined as in the discrete-time case.

One often refers to *state space* as the n-dimensional space in which the state vector is defined. Accordingly, one can visualize the evolution of a dynamic system in terms of the state vector moving within the state space.

4.2 CONVERSION TO STATE FORM

Ordinary difference and differential equations, as treated in Chapter 2, can be easily converted to equivalent systems of first-order equations. The theory of systems is, therefore, a proper generalization of that earlier theory.

Consider the linear difference equation

$$y(k+n)+a_{n-1}(k)y(k+n-1)+\cdots+a_0(k)y(k)=u(k), \qquad k=0,1,2,\ldots$$

To construct a suitable system representation, define n state variables as n successive values of $y(k)$. In particular, let

$$\begin{aligned} x_1(k) &= y(k) \\ x_2(k) &= y(k+1) \\ &\vdots \\ x_n(k) &= y(k+n-1) \end{aligned}$$

With these definitions, it follows immediately that

$$\begin{aligned} x_1(k+1) &= x_2(k) \\ x_2(k+1) &= x_3(k) \\ &\vdots \\ x_{n-1}(k+1) &= x_n(k) \end{aligned}$$

The value of $x_n(k+1)$ can be found from the original difference equation as

$$x_n(k+1)=-a_0(k)x_1(k)-a_1(k)x_2(k)-\cdots-a_{n-1}(k)x_n(k)+u(k)$$

Defining the state vector $\mathbf{x}(k)$ having components $x_1(k), x_2(k), \ldots, x_n(k)$, as above, produces the linear system

$$\mathbf{x}(k+1)=\begin{bmatrix} 0 & 1 & 0 & \cdots & 0 & 0 \\ 0 & 0 & 1 & \cdots & 0 & 0 \\ \cdot & \cdot & \cdot & & \cdot & \cdot \\ \cdot & \cdot & \cdot & & \cdot & \cdot \\ \cdot & \cdot & \cdot & & \cdot & \cdot \\ 0 & 0 & 0 & \cdots & 0 & 1 \\ -a_0(k) & -a_1(k) & -a_2(k) & \cdots & -a_{n-2}(k) & -a_{n-1}(k) \end{bmatrix}\mathbf{x}(k)+\begin{bmatrix} 0 \\ 0 \\ \cdot \\ \cdot \\ \cdot \\ 0 \\ 1 \end{bmatrix}u(k)$$

with

$$y(k)=x_1(k), \qquad k=0,1,2,\ldots$$

Matrices with the special structure above, with ones along an off-diagonal and zeroes everywhere else except the bottom row, occur frequently in dynamic system theory, and are referred to as *companion matrices.*

Differential equations can be converted to state form in a similar way. In this case the state variables are taken to be the original dependent variable $y(t)$ and its first $n-1$ derivatives. The resulting structure is identical to that for difference equations.

Example 1. Consider the second-order difference equation

$$y(k+2)+2y(k+1)+3y(k)=u(k)$$

Following the procedures outlined above, we define

$$x_1(k)=y(k) \qquad x_2(k)=y(k+1)$$

In terms of these variables the system can be written as

$$x_1(k+1)=x_2(k)$$
$$x_2(k+1)=-2x_2(k)-3x_1(k)+u(k)$$

or, in matrix form,

$$\begin{bmatrix} x_1(k+1) \\ x_2(k+1) \end{bmatrix} = \begin{bmatrix} 0 & 1 \\ -3 & -2 \end{bmatrix} \begin{bmatrix} x_1(k) \\ x_2(k) \end{bmatrix} + \begin{bmatrix} 0 \\ 1 \end{bmatrix} u(k)$$

Example 2 (Newton's Laws). The conversion process for differential equations is illustrated most simply by the second-order system derived from Newton's laws, discussed in Sect. 1.3. The equation takes the form (assuming unit mass)

$$\frac{d^2x}{dt^2}=u(t)$$

It can be converted to state variable form by defining the state variables $x_1=x$, $x_2=(dx/dt)$. It follows that

$$\frac{dx_1}{dt}=x_2$$

$$\frac{dx_2}{dt}=u$$

In matrix form the system can then be written

$$\frac{d\mathbf{x}}{dt}=\begin{bmatrix} 0 & 1 \\ 0 & 0 \end{bmatrix}\mathbf{x}+\begin{bmatrix} 0 \\ 1 \end{bmatrix}u(t)$$

4.3 DYNAMIC DIAGRAMS

The mathematical device of representing a dynamic situation as a system of first-order difference or differential equations has the structural interpretation that a high-order system is just a collection of interconnected first-order systems. This interpretation often can be effectively exploited visually by displaying the interconnection pattern diagrammatically. There is a simple and useful convention for constructing such diagrams.

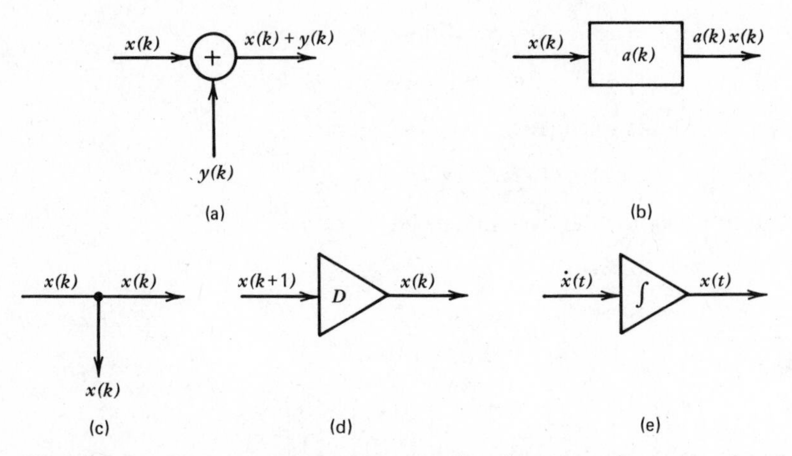

Figure 4.1. Elementary components for dynamic diagrams. (a) Summer. (b) Transmission. (c) Splitting. (d) Unit delay. (e) Integrator.

In the linear case, dynamic diagrams are built up from the five elementary components illustrated in Figs. 4.1*a*–4.1*e*. The diagrams are to be interpreted as if the scalar value runs along the lines, somewhat like voltage on a wire. The *summer* adds whatever comes into it, instantaneously producing the sum. The *transmission* multiplies the incoming scalar by the constant indicated in the box. *Splitting* refers simply to dividing a line into two lines, each of which carries the original value. The *delay* is the basic dynamic component for

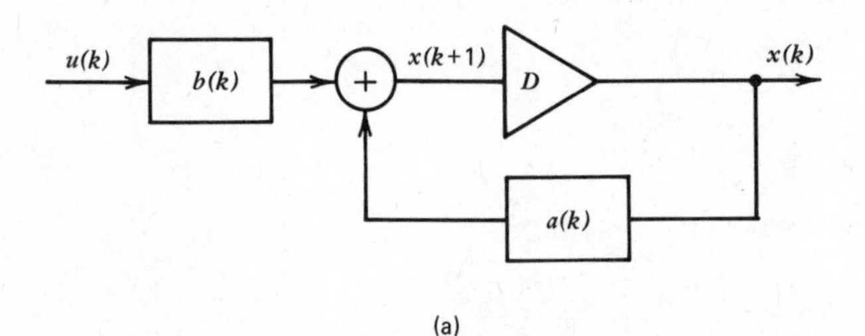

(a)

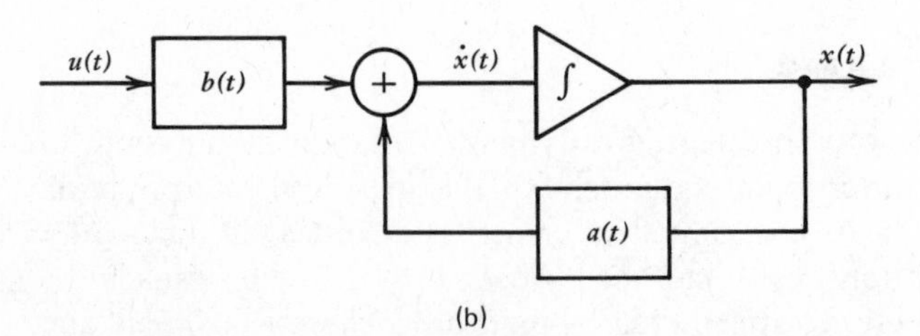

(b)

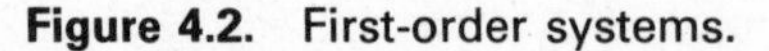

Figure 4.2. First-order systems.

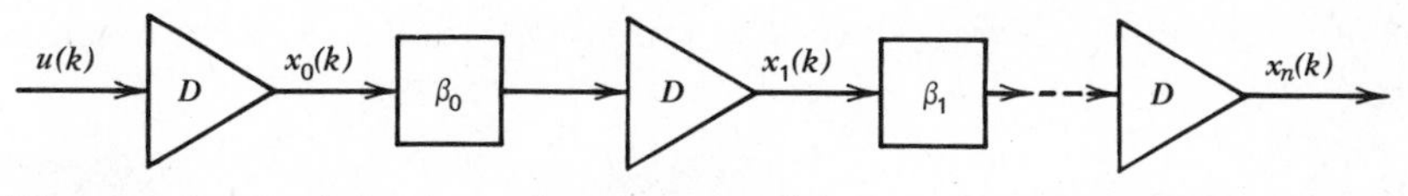

Figure 4.3. Straight-through cohort model.

discrete-time systems. Whatever comes in is delayed for one period and becomes the output for the next period. [Instead of using $x(k+1)$ as input in Fig. 4.1*d*, the reader may find it helpful to use $x(k)$ as input, in which case the output would be $x(k-1)$.] The *integrator* is the basic dynamic component for continuous-time systems. Whatever time function comes in is integrated and the resulting new function appears as output. Thus, the input is the derivative of the output. These basic components can be combined to represent any linear system.

Example 1 (First-order Systems). The first-order system

$$x(k+1)=a(k)x(k)+b(k)u(k)$$

corresponds to the diagram shown in Fig. 4.2*a*. Entering the summer are the two terms $b(k)u(k)$ and $a(k)x(k)$. They are summed, and one period later this sum appears at the output of the diagram as $x(k+1)$.

The continuous-time system

$$\dot{x}(t)=a(t)x(t)+b(t)u(t)$$

corresponds to the diagram in Fig. 4.2*b*.

Example 2 (The Straight-Through Cohort Model). The cohort model associated with washing machines is shown in diagram form in Fig. 4.3. A characteristic of the cohort model, which is obvious from just the verbal description of the system, is that without inputs the system eventually will have zero population in each of its cohorts. No formal analysis is required to deduce this. This prominent characteristic of the model is, however, somewhat masked by the matrix representation, until the structure of the matrix is examined. By contrast, this characteristic is displayed in full relief by the dynamic diagram of Fig. 4.3, where it is clear that the early state variables soon become zero if there is no input to rejuventate them.

4.4 HOMOGENEOUS DISCRETE-TIME SYSTEMS

A linear dynamic system is said to be *homogeneous* or *free* if there is no forcing term in its defining equation. In discrete-time the state vector representation for such a system is

$$\mathbf{x}(k+1)=\mathbf{A}(k)\mathbf{x}(k) \tag{4-3}$$

The term *homogeneous* derives of course from its usage in ordinary difference and differential equations. In the vernacular of dynamic system theory such systems are referred to as *free*, since they run by themselves without external control.

Solution to Free System

The free system

$$\mathbf{x}(k+1)=\mathbf{A}(k)\mathbf{x}(k)$$

can be solved recursively once an initial value of the state is specified. One simply writes, by repeated substitution,

$$\begin{aligned}\mathbf{x}(1)&=\mathbf{A}(0)\mathbf{x}(0)\\ \mathbf{x}(2)&=\mathbf{A}(1)\mathbf{x}(1)=\mathbf{A}(1)\mathbf{A}(0)\mathbf{x}(0)\\ &\cdot\\ &\cdot\\ &\cdot\end{aligned}$$

and, in general,

$$\mathbf{x}(k)=\mathbf{A}(k-1)\mathbf{A}(k-2)\cdots\mathbf{A}(0)\mathbf{x}(0)$$

In view of the expression for the solution to the free system, it is natural to define the special matrix

$$\mathbf{\Phi}(k,0)=\mathbf{A}(k-1)\mathbf{A}(k-2)\cdots\mathbf{A}(0) \tag{4-4}$$

which is called the *state-transition matrix*. Multiplication of any initial state vector by this matrix yields the state at time k.

The definition of the state-transition matrix can be generalized to account for the possibility of initiating the system at a time other than zero. The general definition is given below.

Definition. The *state-transition matrix* of the homogeneous system (4-4) is

$$\begin{aligned}\mathbf{\Phi}(k,l)&=\mathbf{A}(k-1)\mathbf{A}(k-2)\cdots\mathbf{A}(l), \qquad k>l\\ \mathbf{\Phi}(k,k)&=\mathbf{I}\end{aligned} \tag{4-5}$$

Alternatively (but equivalently), it is the matrix satisfying

$$\begin{aligned}\mathbf{\Phi}(k+1,l)&=\mathbf{A}(k)\mathbf{\Phi}(k,l), \qquad k>l\\ \mathbf{\Phi}(l,l)&=\mathbf{I}\end{aligned} \tag{4-6}$$

This general definition is consistent with an interpretation of yielding the value of $\mathbf{x}(k)$ if $\mathbf{x}(l)$ is known $(k\geq l)$. Indeed, application of the recursion process directly yields

$$\mathbf{x}(k)=\mathbf{\Phi}(k,l)\mathbf{x}(l)$$

which is referred to as the state-transition property of $\mathbf{\Phi}(k, l)$. The equivalence of the two alternative forms stated in the definition should be clear.

The state-transition matrix is defined as a product of system matrices. It is, essentially, a shorthand way of indicating the product, and therefore is of great notational value. Unfortunately, however, there is no shortcut procedure for calculating the state-transition matrix—the product form, although somewhat implicit, is the simplest general representation. However, as will be seen in the examples presented later in this section, the special structure of certain systems often enables one to calculate an explicit expression for the system transition matrix.

Fundamental Sets of Solutions

The state-transition matrix was defined above as a product of system matrices because it can be easily demonstrated that that form produces the solutions to the original difference equation. The concept of the state-transition matrix can be developed by another line of reasoning that leads to a valuable alternative interpretation. This viewpoint highlights the role of linearity, and suppresses the constructive approach to solution determination. This more indirect argument, although serving only as an alternative procedure in the discrete-time case, is essential in continuous time, where a direct constructive approach is not available. This alternative approach is based on a set of fundamental solutions, and is patterned after the classical approach to ordinary difference equations. It is also closely related to the structure represented by the alternative statement (4-6) in the original definition of the state-transition matrix.

For simplicity, it is assumed here that the system matrix $\mathbf{A}(k)$ is nonsingular for all k. This is not an essential restriction, but it makes the arguments and results cleaner.

Consider a collection of n solutions to the homogeneous equation (4-3). Let us denote these solutions by $\mathbf{x}^1(k), \mathbf{x}^2(k), \ldots, \mathbf{x}^n(k)$. Each solution $\mathbf{x}^i(k)$ is an n-dimensional vector function of k; that is, each solution is a sequence of vectors satisfying the recursive relation (4-3). Each of these n solutions might be found by starting with an initial vector and generating successive terms by recursion, or they might be found by some alternative procedure. We require, however, that these n solutions be linearly independent, consistent with the following general definition.

Definition. A set of m vector sequences $\mathbf{x}^1(k), \mathbf{x}^2(k), \ldots, \mathbf{x}^m(k)$, $k = 0, 1, 2, \ldots$ is said to be *linearly independent* if there is no nontrivial linear combination of them that is identically zero. That is, if the relation $\alpha_1\mathbf{x}^1(k) + \alpha_2\mathbf{x}^2(k) + \cdots + \alpha_m\mathbf{x}^m(k) = \mathbf{0}$, for all k, implies that all the α_i's are zero.

Note that this definition is stated with respect to all $k = 0, 1, 2, \ldots$. If a solution were defined on a sequence of finite length, the definition would be modified accordingly. Also note, however, that the definition does not require that, for each fixed k, the vectors $\mathbf{x}^1(k), \mathbf{x}^2(k), \ldots, \mathbf{x}^m(k)$ be linearly independent in the usual sense for n-dimensional vectors. It is sufficient, for example, that for one value of k (say $k=0$) they be linearly independent.

Suppose that $\mathbf{x}^1(k), \mathbf{x}^2(k), \ldots, \mathbf{x}^n(k)$ is a set of n linearly independent solutions to the original difference equation (4-3). Thus, each $\mathbf{x}^i(k)$, $i=1,2,\ldots,n$ satisfies

$$\mathbf{x}^i(k+1)=\mathbf{A}(k)\mathbf{x}^i(k) \tag{4-7}$$

Such a set is called a *fundamental set of solutions.* As we will show, every solution of (4-3) can be expressed as a linear combination of these n solutions.

In order to facilitate the required manipulations, it is convenient to stack these n solution vectors side-by-side as the n columns of a $n\times n$ matrix, denoted $\mathbf{X}(k)$; that is,

$$\mathbf{X}(k)=[\mathbf{x}^1(k)\,\mathbf{x}^2(k)\cdots\mathbf{x}^n(k)] \tag{4-8}$$

This matrix of solutions is referred to as a *fundamental matrix of solutions.*

A fundamental matrix of solutions satisfies the underlying system equation as a unit; that is,

$$\mathbf{X}(k+1)=\mathbf{A}(k)\mathbf{X}(k) \tag{4-9}$$

This is true because each column of $\mathbf{X}(k)$ is a solution, and therefore satisfies the system equation. The matrix equation is really just n separate equations—one for each of the columns.

Lemma. *A fundamental matrix of solutions $\mathbf{X}(k)$ is nonsingular for every value of k.*

Proof. This result follows from the linear independence of the solutions and the nonsingularity of $\mathbf{A}(k)$. Suppose to the contrary that $\mathbf{X}(k_0)$ were singular for some index k_0. Then, according to the fundamental lemma of linear algebra, it would follow that

$$\mathbf{X}(k_0)\boldsymbol{\alpha}=\mathbf{0}$$

for some nonzero n-vector $\boldsymbol{\alpha}$. Multiplication by $\mathbf{A}(k_0)$ would then lead to

$$\mathbf{X}(k_0+1)\boldsymbol{\alpha}=\mathbf{A}(k_0)\mathbf{X}(k_0)\boldsymbol{\alpha}=\mathbf{0}$$

while multiplication by $\mathbf{A}(k_0-1)^{-1}$ would produce

$$\mathbf{X}(k_0-1)\boldsymbol{\alpha}=\mathbf{A}(k_0-1)^{-1}\mathbf{X}(k_0)\boldsymbol{\alpha}=\mathbf{0}$$

By continuing these forward and backward processes, it would follow that $\mathbf{X}(k)\boldsymbol{\alpha}=\mathbf{0}$ for all k. This however is equivalent to

$$\alpha_1\mathbf{x}^1(k)+\alpha_2\mathbf{x}^2(k)+\cdots+\alpha_n\mathbf{x}^n(k)=\mathbf{0}$$

which contradicts the assumption that the fundamental set of solutions is linearly independent. ▮

With the above result it is easy to derive an expression for the state-transition matrix. Suppose an arbitrary solution $\mathbf{x}(k)$ to (4-3) is given. Its initial condition is $\mathbf{x}(0)$. Now consider the vector sequence $\bar{\mathbf{x}}(k)$ defined in terms of the initial condition $\mathbf{x}(0)$ by

$$\bar{\mathbf{x}}(k) = \mathbf{X}(k)\mathbf{X}(0)^{-1}\mathbf{x}(0)$$

It is clear that $\bar{\mathbf{x}}(0) = \mathbf{x}(0)$. Also, if the vector $\boldsymbol{\alpha}$ is defined by

$$\boldsymbol{\alpha} = \mathbf{X}(0)^{-1}\mathbf{x}(0)$$

we have

$$\bar{\mathbf{x}}(k) = \mathbf{X}(k)\boldsymbol{\alpha}$$

From this expression it is clear that $\bar{\mathbf{x}}(k)$ is just a linear combination of solutions, and by the linearity of the system this linear combination is itself a solution. However, since it has the initial value $\mathbf{x}(0)$, the two solutions $\mathbf{x}(k)$ and $\bar{\mathbf{x}}(k)$ must be identical (by the uniqueness of solutions); that is $\bar{\mathbf{x}}(k) = \mathbf{x}(k)$. It follows therefore that any solution $\mathbf{x}(k)$ can be expressed as

$$\mathbf{x}(k) = \mathbf{X}(k)\mathbf{X}(0)^{-1}\mathbf{x}(0)$$

The above procedure can be generalized to express $\mathbf{x}(k)$ in terms of $\mathbf{x}(l)$ rather than $\mathbf{x}(0)$. Indeed the same argument shows that

$$\mathbf{x}(k) = \mathbf{X}(k)\mathbf{X}(l)^{-1}\mathbf{x}(l)$$

In view of this relation we may state the following proposition.

Proposition. *Let $\mathbf{X}(k)$ be a fundamental matrix of solutions, corresponding to the system*

$$\mathbf{x}(k+1) = \mathbf{A}(k)\mathbf{x}(k)$$

Then the state-transition matrix is given by the expression

$$\boldsymbol{\Phi}(k, l) = \mathbf{X}(k)\mathbf{X}(l)^{-1} \tag{4-10}$$

for $k \geq l$.

The above algebraic result has a simple interpretation in terms of matching initial conditions. For simplicity let us consider the relation

$$\boldsymbol{\Phi}(k, 0) = \mathbf{X}(k)\mathbf{X}(0)^{-1}$$

Since $\mathbf{X}(k)$ is a fundamental matrix of solutions, it is also true that $\mathbf{X}(k)\mathbf{X}(0)^{-1}$ is a fundamental matrix of solutions; the columns of the latter being various linear combinations of the columns of $\mathbf{X}(k)$. Therefore, $\boldsymbol{\Phi}(k, 0)$ is a fundamental matrix of solutions. The set is normalized, however, so that $\boldsymbol{\Phi}(0, 0) = \mathbf{I}$.

Expressed in terms of its n columns, this fundamental matrix of solutions consists of n distinct solutions $\mathbf{x}^1(k)$, $\mathbf{x}^2(k), \ldots, \mathbf{x}^n(k)$, characterized by the special initial conditions

$$\mathbf{x}^i(0) = \begin{bmatrix} 0 \\ 0 \\ \cdot \\ \cdot \\ 1 \\ \cdot \\ \cdot \\ 0 \\ 0 \end{bmatrix} \qquad i = 1, 2, 3, \ldots, n$$

with the 1 in the ith coordinate position. This special fundamental set of solutions can be used to construct the solution corresponding to a given arbitrary set of initial conditions, $\mathbf{x}(0)$. The solution is expressed as a linear combination of the fundamental solutions, and the appropriate linear combination is simply $\mathbf{x}(k) = \mathbf{\Phi}(k, 0)\mathbf{x}(0)$. It is because the fundamental set has the special initial conditions, equal to the n unit basis vectors, that construction of a linear combination that agrees with $\mathbf{x}(0)$ is so easy.

Example 1 (Time-Invariant System). An extremely important special case is that of a time-invariant linear system

$$\mathbf{x}(k+1) = \mathbf{A}\mathbf{x}(k)$$

where $\mathbf{A}$ is fixed, independent of k. It is easily seen that

$$\mathbf{\Phi}(k, 0) = \mathbf{A}^k, \qquad k \geq 0$$

or, more generally,

$$\mathbf{\Phi}(k, l) = \mathbf{A}^{k-l}, \qquad k \geq l$$

If $\mathbf{A}$ is invertible, then these expressions are valid for $k < l$ as well.

In the time-invariant case, we often write $\mathbf{\Phi}(k)$ instead of $\mathbf{\Phi}(k, 0)$, since then $\mathbf{\Phi}(k, l) = \mathbf{\Phi}(k-l)$.

Example 2 (A Nonconstant System). Consider the linear homogeneous system, defined for $k \geq 0$:

$$\begin{bmatrix} x_1(k+1) \\ x_2(k+1) \end{bmatrix} = \begin{bmatrix} 1 & k+1 \\ 0 & 1 \end{bmatrix} \begin{bmatrix} x_1(k) \\ x_2(k) \end{bmatrix}$$

A fundamental set of solutions can be found by starting with two linearly independent initial conditions. It is easy to see that one solution is the sequence

$$\mathbf{x}^1(k) = \begin{bmatrix} 1 \\ 0 \end{bmatrix} \qquad \text{all } k \geq 0$$

A second solution can be constructed by repeated substitution in the system equation, yielding for $k=0, 1, 2, 3, 4, \ldots$

$$\begin{bmatrix} 0 \\ 1 \end{bmatrix}, \begin{bmatrix} 1 \\ 1 \end{bmatrix}, \begin{bmatrix} 3 \\ 1 \end{bmatrix}, \begin{bmatrix} 6 \\ 1 \end{bmatrix}, \begin{bmatrix} 10 \\ 1 \end{bmatrix}, \ldots$$

Thus, we have the second solution defined by

$$\mathbf{x}^2(k) = \begin{bmatrix} k(k+1)/2 \\ 1 \end{bmatrix}$$

These two solutions form a fundamental set of solutions, which yields the matrix

$$\mathbf{X}(k) = \begin{bmatrix} 1 & k(k+1)/2 \\ 0 & 1 \end{bmatrix}$$

If we are given an arbitrary initial condition vector

$$\mathbf{x}(0) = \begin{bmatrix} \alpha_1 \\ \alpha_2 \end{bmatrix}$$

the corresponding solution will be

$$\mathbf{x}(k) = \mathbf{X}(k)\mathbf{X}(0)^{-1}\mathbf{x}(0)$$

Since $\mathbf{X}(0)$ is the identity in this case, we can write the general solution

$$\mathbf{x}(k) = \begin{bmatrix} \alpha_1 + \alpha_2 k(k+1)/2 \\ \alpha_2 \end{bmatrix}$$

Example 3 (Natchez Indian Social Structure). Most societies are organized, either explicitly or implicitly, into class segments. The higher classes control power and the distribution of resources. Class membership is determined by inheritance, and often the higher classes practice endogamy (inner-class marriage) in order to "close" the class and prevent the dispersion of power.

The Natchez Indians in the Lower Mississippi devised an ingenious system of marriage rules apparently in an attempt to create an "open" class structure where power could to some extent be rotated rather than perpetuated within families. The society was divided into two main classes—*rulers* and *commoners*, with the commoners termed *stinkards*. The ruling class was divided into three subclasses—*Suns*, *Nobles*, and *Honoreds*. Members of the ruling class could marry only commoners. Offspring of a marriage between a female member of the ruling class and a male commoner inherited the class status of the mother, while offspring of a marriage between a male member of the ruling class dropped down a notch. Thus, the child of a Sun father became a Noble, and so forth.

The complete set of allowed marriages and offspring designations is summarized in Table 4.1. A blank in a box in the table indicates that the corresponding marriage is not allowed, and the name in a box is the class designation of the offspring of a corresponding allowed marriage.

Table 4.1. Natchez Marriage Rules

Mother \ Father	Sun	Noble	Honored	Stinkard
Sun				Sun
Noble				Noble
Honored				Honored
Stinkard	Noble	Honored	Stinkard	Stinkard

It is of interest to investigate the propagation of the class distribution of this system. For this purpose we develop a state variable description of the dynamic process implied by these rules. For purposes of developing the model, we assume that the population can be divided into distinct generations, and we make the following three simplifying assumptions:

(a) Each class has an equal number of men and women in each generation.
(b) Each individual marries once and only once, and marries someone in the same generation.
(c) Each couple has exactly one son and one daughter.

Since the number of men is equal to the number of women in each class, it is sufficient to consider only the male population. Let $x_i(k)$ denote the number of men in class i in generation k, where the classes are numbered (1) Suns, (2) Nobles, (3) Honored, and (4) Stinkards.

Since a Sun son is produced by every Sun mother and in no other way, and since the number of Sun mothers is equal to the number of Sun fathers, we may write

$$x_1(k+1) = x_1(k)$$

A Noble son is produced by every Sun father and by every Noble mother and in no other way; therefore,

$$x_2(k+1) = x_1(k) + x_2(k)$$

Similarly, an Honored son is produced by every Noble father and by every Honored mother and in no other way; therefore,

$$x_3(k+1) = x_2(k) + x_3(k)$$

Finally, the number of Stinkard sons is equal to the total number of Stinkard fathers minus the number of Stinkard fathers married to Suns or Nobles. Thus,

$$x_4(k+1) = -x_1(k) - x_2(k) + x_4(k)$$

In state variable form these equations become

$$\begin{bmatrix} x_1(k+1) \\ x_2(k+1) \\ x_3(k+1) \\ x_4(k+1) \end{bmatrix} = \begin{bmatrix} 1 & 0 & 0 & 0 \\ 1 & 1 & 0 & 0 \\ 0 & 1 & 1 & 0 \\ -1 & -1 & 0 & 1 \end{bmatrix} \begin{bmatrix} x_1(k) \\ x_2(k) \\ x_3(k) \\ x_4(k) \end{bmatrix}$$

The system matrix $\mathbf{A}$ is in this case a constant 4×4 matrix, and accordingly the system is a time-invariant, free, dynamic system of fourth order. The state-transition matrix can be found as powers of the system matrix. That is,

$$\mathbf{\Phi}(k, 0) = \mathbf{A}^k \qquad k \geq 0$$

These powers can, of course, be found numerically by brute force, but in this case it is relatively easy to find an analytic expression for them. We write the matrix $\mathbf{A}$ as $\mathbf{A} = \mathbf{I} + \mathbf{B}$, where

$$\mathbf{B} = \begin{bmatrix} 0 & 0 & 0 & 0 \\ 1 & 0 & 0 & 0 \\ 0 & 1 & 0 & 0 \\ -1 & -1 & 0 & 0 \end{bmatrix}$$

We may then use the binomial expansion to write

$$\mathbf{A}^k = (\mathbf{I} + \mathbf{B})^k = \mathbf{I}^k + \binom{k}{1}\mathbf{I}^{k-1}\mathbf{B} + \binom{k}{2}\mathbf{I}^{k-2}\mathbf{B}^2 + \cdots + \mathbf{B}^k$$

where $\binom{k}{i}$ denotes the binomial coefficient $k!/(k-i)!(i!)$. (The binomial expansion is valid in this matrix case because $\mathbf{I}$ and $\mathbf{B}$ commute—that is, because $\mathbf{IB} = \mathbf{BI}$.) The expression simplifies in this particular case because

$$\mathbf{B}^2 = \begin{bmatrix} 0 & 0 & 0 & 0 \\ 0 & 0 & 0 & 0 \\ 1 & 0 & 0 & 0 \\ -1 & 0 & 0 & 0 \end{bmatrix}$$

but $\mathbf{B}^3$ (and every higher power) is zero. Thus,

$$\mathbf{A}^k = (\mathbf{I}+\mathbf{B})^k = \mathbf{I}+k\mathbf{B}+\frac{k(k-1)}{2}\mathbf{B}^2$$

or, explicitly,

$$\mathbf{A}^k = \begin{bmatrix} 1 & 0 & 0 & 0 \\ k & 1 & 0 & 0 \\ \dfrac{k(k-1)}{2} & k & 1 & 0 \\ \dfrac{-k(k+1)}{2} & -k & 0 & 1 \end{bmatrix}$$

Once $\mathbf{A}^k$ is known, it is possible to obtain an expression for the class populations at any generation k in terms of the initial populations. Thus,

$$\mathbf{x}(k) = \begin{bmatrix} x_1(0) \\ kx_1(0)+x_2(0) \\ \frac{1}{2}k(k-1)x_1(0)+kx_2(0)+x_3(0) \\ -\frac{1}{2}k(k+1)x_1(0)-kx_2(0)+x_4(0) \end{bmatrix}$$

From this analytical solution one can determine the behavior of the social system. First, it can be directly verified that the total population of the society is constant from generation to generation. This follows from the earlier assumption (c) and can be verified by summing the components of $\mathbf{x}(k)$.

Next, it can be seen that unless $x_1(0)=x_2(0)=0$, there is no steady distribution of population among the classes. If, however, $x_1(0)=x_2(0)=0$, corresponding to no Suns or Nobles initially, there will be no Suns or Nobles in any successive generation. In this situation, the Honored and Stinkard population behave according to $x_3(k+1)=x_3(k)$, $x_4(k+1)=x_4(k)$, and therefore their populations remain fixed at their initial values.

If either the Sun or Noble class is initially populated, then the number of Stinkards will decrease with k, and ultimately there will not be enough Stinkards to marry all the members of the ruling class. At this point the social system, as defined by the given marriage rules, breaks down.

4.5 GENERAL SOLUTION TO LINEAR DISCRETE-TIME SYSTEMS

We turn now to consideration of the forced system

$$\mathbf{x}(k+1) = \mathbf{A}(k)\mathbf{x}(k)+\mathbf{B}(k)\mathbf{u}(k) \tag{4-11}$$

As before $\mathbf{x}(k)$ is an n-dimensional state vector, $\mathbf{A}(k)$ is an $n\times n$ system matrix, $\mathbf{B}(k)$ is an $n\times m$ distribution matrix, and $\mathbf{u}(k)$ is an m-dimensional input vector. The general solution to this system can be expressed quite simply in terms of the state-transition matrix defined in Sect. 4.4. The solution can be established easily by algebraic manipulation, and we shall do this first. Interpretation of the solution is, however, just as important as the algebraic verification, and a major part of this section is devoted to exposition of that interpretation.

Proposition. *The solution of the system* (4-11), *in terms of the initial state* $\mathbf{x}(0)$ *and the inputs, is*

$$\mathbf{x}(k)=\mathbf{\Phi}(k,0)\mathbf{x}(0)+\sum_{l=0}^{k-1}\mathbf{\Phi}(k,l+1)\mathbf{B}(l)\mathbf{u}(l) \tag{4-12}$$

Proof. To verify that the proposed expression (4-12) does represent the solution, it is only necessary to verify that it satisfies the basic recursion (4-11) and the initial condition. An important relation for this purpose is

$$\mathbf{\Phi}(k+1,l+1)=\mathbf{A}(k)\mathbf{\Phi}(k,l+1) \tag{4-13}$$

from the basic definition of the state-transition matrix. We note first that the proposed solution is correct for $k=0$, since it reduces to $\mathbf{x}(0)=\mathbf{x}(0)$. The verification can therefore proceed by induction from $k=0$.

The proposed solution (4-12) when written with $k+1$ replacing k is

$$\mathbf{x}(k+1)=\mathbf{\Phi}(k+1,0)\mathbf{x}(0)+\sum_{l=0}^{k}\mathbf{\Phi}(k+1,l+1)\mathbf{B}(l)\mathbf{u}(l)$$

The last term in the summation (corresponding to $l=k$) can be separated from the summation sign to produce

$$\mathbf{x}(k+1)=\mathbf{\Phi}(k+1,0)\mathbf{x}(0)+\sum_{l=0}^{k-1}\mathbf{\Phi}(k+1,l+1)\mathbf{B}(l)\mathbf{u}(l)+\mathbf{B}(k)\mathbf{u}(k)$$

Using relation (4-13) this becomes

$$\mathbf{x}(k+1)=\mathbf{A}(k)\mathbf{\Phi}(k,0)\mathbf{x}(0)+\mathbf{A}(k)\sum_{l=0}^{k-1}\mathbf{\Phi}(k,l+1)\mathbf{B}(l)\mathbf{u}(l)+\mathbf{B}(k)\mathbf{u}(k)$$

This, in turn, with the proposed form for $\mathbf{x}(k)$, becomes

$$\mathbf{x}(k+1)=\mathbf{A}(k)\mathbf{x}(k)+\mathbf{B}(k)\mathbf{u}(k)$$

showing that the proposed solution in fact satisfies the defining difference equation. ▮

Superposition

The linearity of the system (4-11) implies that the solution can be computed by the principle of *superposition*. Namely, the total response due to several inputs is the sum of their individual responses, plus an initial condition term. This leads to a useful interpretation of the solution formula (4-12).

Let us investigate each term in the general solution (4-12). The first term $\mathbf{\Phi}(k, 0)\mathbf{x}(0)$ is the contribution to $\mathbf{x}(k)$ due to the initial condition $\mathbf{x}(0)$. It is the response of the system as if it were free. When nonzero inputs are present this term is not eliminated, other terms are simply added to it.

The second term, which is the first of the terms represented by the summation sign, is that associated with the first input. The term is $\mathbf{\Phi}(k, 1)\mathbf{B}(0)\mathbf{u}(0)$. To see how this term arises, let us look again at the underlying system equation

$$\mathbf{x}(k+1) = \mathbf{A}(k)\mathbf{x}(k) + \mathbf{B}(k)\mathbf{u}(k)$$

At $k=0$ this becomes

$$\mathbf{x}(1) = \mathbf{A}(0)\mathbf{x}(0) + \mathbf{B}(0)\mathbf{u}(0)$$

If we assume for the moment that $\mathbf{x}(0) = \mathbf{0}$ [which we might as well assume, since we have already discussed the contribution due to $\mathbf{x}(0)$], then we have

$$\mathbf{x}(1) = \mathbf{B}(0)\mathbf{u}(0)$$

This means that the short-term effect of the input $\mathbf{u}(0)$ is to set the state $\mathbf{x}(1)$ equal to the vector $\mathbf{B}(0)\mathbf{u}(0)$. Even if there were no further inputs, the system would continue to respond to this value of $\mathbf{x}(1)$ in a manner similar to its response to an initial condition. Indeed, the vector $\mathbf{x}(1)$ acts exactly like an initial condition, but at $k=1$ rather than $k=0$. From our knowledge of the behavior of free systems we can therefore easily deduce that the corresponding response, for $k > 1$, is

$$\mathbf{x}(k) = \mathbf{\Phi}(k, 1)\mathbf{x}(1)$$

In terms of $\mathbf{u}(0)$, which produced this $\mathbf{x}(1)$, the response is

$$\mathbf{x}(k) = \mathbf{\Phi}(k, 1)\mathbf{B}(0)\mathbf{u}(0)$$

which is precisely the term in the expression for the general solution corresponding to $\mathbf{u}(0)$.

For an input at another time, say at time l, the analysis is virtually identical. In the absence of initial conditions or other inputs, the effect of the input $\mathbf{u}(l)$ is to transfer the state from zero at time l to $\mathbf{B}(l)\mathbf{u}(l)$ at time $l+1$. From this point the response at $k > l+1$ is determined by the free system, leading to

$$\mathbf{x}(k) = \mathbf{\Phi}(k, l+1)\mathbf{B}(l)\mathbf{u}(l)$$

as the response due to $\mathbf{u}(l)$.

The total response of the system is the superposition of the separate responses considered above; the response to each individual input being calculated as a free response to the instantaneous change it produces. We see, therefore, in terms of this interpretation, that the total solution (4-12) to the system can be regarded as a sum of free responses initiated at different times.

Time-Invariant Systems (Impulse Response)

If the system (4-11) is time-invariant, the general solution and its interpretation can be slightly simplified. This leads to the formal concept of the *impulse response* of a linear time-invariant system that is considered in greater detail in Chapter 8.

Corresponding to the linear time-invariant system

$$\mathbf{x}(k+1)=\mathbf{A}\mathbf{x}(k)+\mathbf{B}\mathbf{u}(k) \tag{4-14}$$

the state-transition matrix takes the simple form

$$\boldsymbol{\Phi}(k, l+1)=\mathbf{A}^{k-l-1} \tag{4-15}$$

Therefore, the general solution corresponding to (4-14) is

$$\mathbf{x}(k)=\mathbf{A}^{k}\mathbf{x}(0)+\sum_{l=0}^{k-1}\mathbf{A}^{k-l-1}\mathbf{B}\mathbf{u}(l) \tag{4-16}$$

Everything said about the more general time-varying solution certainly applies to this special case. To obtain further insight in this case, however, let us look more closely at the response due to a single input. For simplicity assume that the input is scalar-valued (i.e., one-dimensional). In that case we write the distribution matrix $\mathbf{B}$ as $\mathbf{b}$ to indicate that it is in fact an n-vector.

The response due to an input $u(0)$ at time $k=0$ is

$$\mathbf{x}(k)=\mathbf{A}^{k-1}\mathbf{b}u(0)$$

If $u(0)=1$, corresponding to a unit input at time $k=0$, the response takes the form

$$\mathbf{x}(k)=\mathbf{A}^{k-1}\mathbf{b}$$

This response is termed the impulse response of the system. It is defined as the response due to a unit input at time $k=0$.

The importance of the impulse response is that for linear time-invariant systems it can be used to determine the response to later inputs as well. For example, let us calculate the response to an input $u(l)$. Because the system is time-invariant, the response due to an input at time l is identical to that due to one of equal magnitude at time zero, except that it is shifted by l time units. Thus the response is

$$\mathbf{x}(k)=\mathbf{A}^{k-1-l}\mathbf{b}u(l) \quad \text{for } k\geq l+1$$

Of course, the response for $k<l$ is zero.

The response of a linear time-invariant system to an arbitrary sequence of inputs is made up from the basic response pattern of the impulse response. This basic response pattern is initiated at various times with various magnitudes by inputs at those times; the magnitude of an input directly determining the proportionate magnitude of the corresponding response pattern. The total response, which may appear highly complex, is just the sum of the individual (shifted) response patterns.

Example (First-Order System). Consider the system

$$x(k+1)=ax(k)+u(k)$$

where $0<a<1$. The general solution to the system is

$$x(k)=a^k x(0)+\sum_{l=0}^{k-1} a^{k-l-1}u(l)$$

The impulse response is a geometric sequence. The total response to any input is just a combination of delayed versions of this basic geometric sequence, each with a magnitude equal to that of the corresponding input term.

The interpretation of the solution in terms of the impulse response is illustrated in Fig. 4.4*a*–4.4*c*. Part (a) shows the impulse response, (b) shows a hypothetical input sequence, and (c) shows the composite response made up from the components.

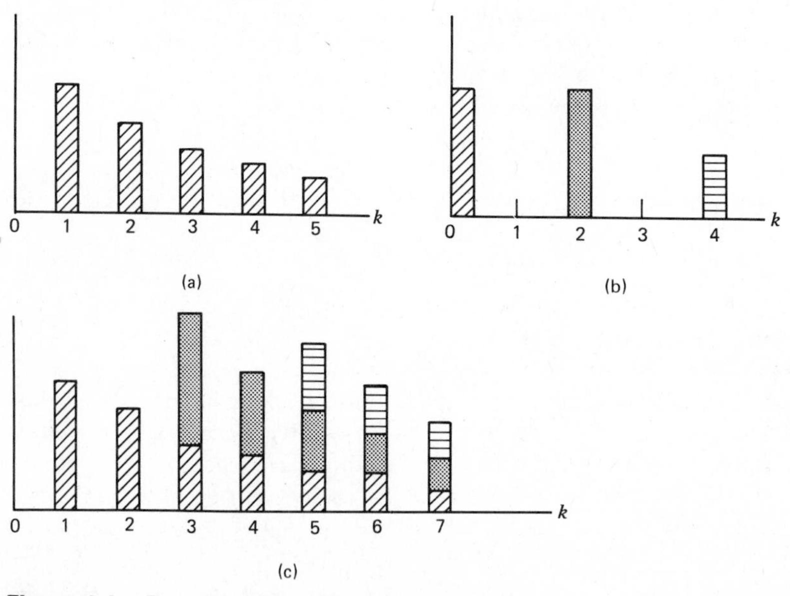

Figure 4.4. Decomposition of response.

4.6 HOMOGENEOUS CONTINUOUS-TIME SYSTEMS

A solution procedure almost parallel to that for discrete time is applicable to linear continuous-time dynamic systems. As in the discrete-time case, it is best to first consider in some detail the homogeneous or free system, which in continuous time has the form

$$\dot{\mathbf{x}}(t) = \mathbf{A}(t)\mathbf{x}(t) \tag{4-17}$$

If the elements of the matrix $\mathbf{A}(t)$ are continuous functions of t, the system (4-17) will possess a unique solution corresponding to each initial state vector. Unlike the discrete-time case, however, where it is possible to write down an explicit expression for the solution in terms of the initial state vector, no such general expression exists in the continuous-time case. Although this is perhaps disappointing, it does not seriously inhibit progress toward the goal of paralleling the development of the discrete-time case. The concepts of *state-transition matrix* and *fundamental sets of solutions* are still applicable, and form the basis of a very satisfactory theory.

The theory itself is concerned essentially with relations among different solutions rather than with the issues of whether solutions actually exist. Accordingly, it is assumed that given a time τ, if $\mathbf{x}(\tau)$ is specified, there is a unique solution to (4-17) having this value at $t=\tau$. It is then possible to define the state-transition matrix indirectly as a matrix solution to the original differential equation.

Definition. The *state-transition matrix* $\mathbf{\Phi}(t, \tau)$ corresponding to the homogeneous system (4-17) is the $n\times n$ matrix function satisfying

$$\frac{d}{dt}\mathbf{\Phi}(t, \tau) = \mathbf{A}(t)\mathbf{\Phi}(t, \tau) \tag{4-18}$$

$$\mathbf{\Phi}(\tau, \tau) = \mathbf{I} \tag{4-19}$$

Let us examine this definition to see, at least in principle, how the state-transition matrix might be found. Let us fix τ. The n columns of $\mathbf{\Phi}(t, \tau)$ are then each vector functions of t, and to satisfy (4-18) each of these columns must be solutions to the original equation (4-17). In addition, these solutions must satisfy the special condition at $t=\tau$ implied by $\mathbf{\Phi}(\tau, \tau)=\mathbf{I}$; that is, at $t=\tau$ each solution vector must be equal to one of the standard unit basis vectors. By finding these n solutions to (4-17), the matrix $\mathbf{\Phi}(t, \tau)$ (for the fixed value of τ) can be constructed. This (conceptual) procedure is then repeated for all values of τ.

The state-transition matrix as defined above has the important state-transition property. Suppose $\mathbf{x}(t)$ is any solution to (4-17). Let τ be fixed. Then it is easy to see that for any t

$$\mathbf{x}(t) = \mathbf{\Phi}(t, \tau)\mathbf{x}(\tau) \tag{4-20}$$

This is true because the right-hand side is a linear combination of the columns of $\mathbf{\Phi}(t, \tau)$, and hence is itself a solution. This solution is equal to $\mathbf{x}(t)$ at $t=\tau$, and hence it must be equal to it for all t.

Fundamental Sets of Solutions

Just as in the discrete-time case, the matrix $\mathbf{\Phi}(t, \tau)$ can be expressed in terms of an arbitrary fundamental set of solutions. This procedure helps to clarify the structure of the state-transition matrix.

Analogously to the discrete-time case, we say that a set of vector-valued time functions $\mathbf{x}^1(t)$, $\mathbf{x}^2(t), \ldots, \mathbf{x}^m(t)$ is *linearly independent* if there is no nontrivial linear combination of them that is identically zero. Let $\mathbf{x}^1(t)$, $\mathbf{x}^2(t), \ldots, \mathbf{x}^n(t)$ be a linearly independent set of solutions to the homogeneous system (4-17). We ignore the issue of how to obtain these n solutions (since as we have said it is impossible to prescribe a general method), but simply assume that they can be found. These solutions form a *fundamental set of solutions.* If arranged as the columns of an $n\times n$ matrix $\mathbf{X}(t)$, the resulting matrix is a *fundamental matrix of solutions.* The matrix satisfies the matrix differential equation

$$\dot{\mathbf{X}}(t)=\mathbf{A}(t)\mathbf{X}(t) \tag{4-21}$$

Lemma. *A fundamental matrix of solutions $\mathbf{X}(t)$ is nonsingular for all t.*

Proof. Suppose to the contrary that for a specific τ the matrix $\mathbf{X}(\tau)$ were singular. This would imply the existence of a nonzero vector $\boldsymbol{\alpha}$ such that

$$\mathbf{X}(\tau)\boldsymbol{\alpha}=\mathbf{0}$$

Defining

$$\mathbf{x}(t)=\mathbf{X}(t)\boldsymbol{\alpha}$$

it would follow that $\mathbf{x}(t)$ is a solution to (4-17), equal to zero at $t=\tau$. It then would follow by the uniqueness of solutions that $\mathbf{x}(t)=\mathbf{0}$ for all $t\geq\tau$. The same argument can be applied backward in time (by setting $t'=-t$) to conclude that $\mathbf{x}(t)=0$ for $t\leq\tau$. Hence, $\mathbf{X}(t)\boldsymbol{\alpha}=\mathbf{0}$ for all t. But this contradicts the assumption of linear independence of the original fundamental set of solutions. ∎

Proposition. *Let $\mathbf{X}(t)$ be a fundamental matrix of solutions corresponding to the system*

$$\dot{\mathbf{x}}(t)=\mathbf{A}(t)\mathbf{x}(t)$$

Then the state-transition matrix is given by the expression

$$\mathbf{\Phi}(t, \tau)=\mathbf{X}(t)\mathbf{X}(\tau)^{-1} \tag{4-22}$$

Proof. For fixed τ the right-hand side of (4-22) is itself a fundamental matrix of solutions, and hence satisfies (4-18). The matrix $\mathbf{X}(\tau)^{-1}$ simply combines the various columns in $\mathbf{X}(t)$ so that the particular solutions have the special unit basis conditions at $t=\tau$, satisfying (4-19). ▮

Example 1. To illustrate these concepts, let us consider a simple example in which analytic expressions for solutions can be easily found. Consider the two-dimensional system

$$\dot{\mathbf{x}}(t)=\begin{bmatrix} 0 & 0 \\ t & 1/t \end{bmatrix}\mathbf{x}(t)$$

(defined for $t \geq 1$, so that all terms are finite).

From the first of the two individual equations, it is clear that $x_1(t)$ is constant; $x_1(t)=c$. Then making the substitution $z(t)=x_2(t)/t$, the second equation reduces to

$$t\dot{z}(t)+z(t)=ct+z(t)$$

This collapses to $\dot{z}(t)=c$ and hence we deduce that $x_2(t)=ct^2+dt$, where c and d are arbitrary constants.

In view of the above, one possible fundamental matrix of solutions is

$$\mathbf{X}(t)=\begin{bmatrix} 1 & 0 \\ t^2 & t \end{bmatrix}$$

Accordingly,

$$\mathbf{\Phi}(t, \tau)=\mathbf{X}(t)\mathbf{X}(\tau)^{-1}=\begin{bmatrix} 1 & 0 \\ t^2 & t \end{bmatrix}\begin{bmatrix} 1 & 0 \\ -\tau & 1/\tau \end{bmatrix}$$

$$\mathbf{\Phi}(t, \tau)=\begin{bmatrix} 1 & 0 \\ t(t-\tau) & t/\tau \end{bmatrix}$$

Time-Invariant Systems

Consider the system

$$\dot{\mathbf{x}}(t)=\mathbf{A}\mathbf{x}(t) \tag{4-23}$$

We can show, for this general time-invariant, or constant coefficient, system, that the fundamental matrix of solutions satisfying

$$\dot{\mathbf{X}}(t)=\mathbf{A}\mathbf{X}(t) \tag{4-24}$$

$$\mathbf{X}(0)=\mathbf{I} \tag{4-25}$$

can be expressed as a power-series expansion in the form

$$\mathbf{X}(t)=\mathbf{I}+\mathbf{A}t+\frac{\mathbf{A}^2t^2}{2!}+\frac{\mathbf{A}^3t^3}{3!}+\cdots+\frac{\mathbf{A}^kt^k}{k!}+\cdots \tag{4-26}$$

Before verifying this formula, we point out that a series of this kind, expressed as a sum of matrices, can be regarded as a collection of separate series, one for each component of the matrix it defines. That is, the ijth element of the matrix on the left-hand side of the equation is the sum of the series defined by the ijth elements of each matrix on the right-hand side of the equation. The particular series in (4-26) defines an $n \times n$ matrix for each fixed value of t. In general, in forming series of this type, one must carefully delineate conditions under which convergence to a limit is guaranteed, or, specifically, when convergence of the individual series for each element is guaranteed. In this particular case it can be shown (see Problem 20) that the series defined by (4-26) converges for any matrix $\mathbf{A}$ and all values of t.

To verify that the matrix $\mathbf{X}(t)$ defined by the series (4-26) satisfies (4-24) and (4-25) is simple. Substituting $t=0$ into the series leads immediately to $\mathbf{X}(0)=\mathbf{I}$. To verify that the differential equation is satisfied, we differentiate each term of the series, obtaining

$$\begin{aligned}\dot{\mathbf{X}}(t) &= \mathbf{0}+\mathbf{A}+\mathbf{A}^2 t+\frac{\mathbf{A}^3 t^2}{2!}+\cdots \\ &= \mathbf{A}\left(\mathbf{I}+\mathbf{A}t+\frac{\mathbf{A}^2 t^2}{2!}+\cdots\right)=\mathbf{A}\mathbf{X}(t)\end{aligned}$$

The series used to define $\mathbf{X}(t)$ is the matrix analog of the series for e^{at} in the familiar scalar case. For this reason it is appropriate, and very convenient, to denote the series (actually its limit) as an exponential. Thus we define, for any t, the matrix exponential

$$e^{\mathbf{A}t}=\mathbf{I}+\mathbf{A}t+\frac{\mathbf{A}^2 t^2}{2!}+\cdots+\frac{\mathbf{A}^k t^k}{k!}+\cdots \tag{4-27}$$

which is itself a square matrix the same size as $\mathbf{A}$.

The state-transition matrix of the time-invariant system is

$$\mathbf{\Phi}(t, \tau)=\mathbf{X}(t)\mathbf{X}(\tau)^{-1}$$

This can be written as

$$\begin{aligned}\mathbf{\Phi}(t, \tau) &= e^{\mathbf{A}t}e^{-\mathbf{A}\tau} \\ &= e^{\mathbf{A}(t-\tau)}\end{aligned}$$

Thus $\mathbf{\Phi}(t, \tau)$ depends, in the time-invariant case, only on the difference $t-\tau$. For this reason, it is customary when working with time-invariant systems to suppress the double index and define

$$\mathbf{\Phi}(t)=e^{\mathbf{A}t}$$

Example 2 (Harmonic Motion). Let us consider the equation of harmonic motion, as defined in Chapter 2. The motion is defined by the second-order homogeneous equation

$$\frac{d^2x}{dt^2}+\omega^2 x=0 \tag{4-28}$$

where ω is a fixed positive constant.

In state-variable form this system can be written

$$\begin{bmatrix}\dot{x}_1\\ \dot{x}_2\end{bmatrix}=\begin{bmatrix}0 & 1\\ -\omega^2 & 0\end{bmatrix}\begin{bmatrix}x_1\\ x_2\end{bmatrix} \tag{4-29}$$

where $x=x_1$ and $\dot{x}=x_2$. The state-transition matrix corresponding to this time-invariant system is

$$\mathbf{\Phi}(t)=e^{\mathbf{A}t}$$

where $\mathbf{A}$ is the coefficient matrix in (4-29). We can easily calculate that

$$\mathbf{A}^2=\begin{bmatrix}-\omega^2 & 0\\ 0 & -\omega^2\end{bmatrix}=-\omega^2\mathbf{I}$$

From this we can conclude that, if k is even,

$$\mathbf{A}^k=(-1)^{k/2}\omega^k\mathbf{I}$$

$$\mathbf{A}^{k+1}=(-1)^{k/2}\omega^k\mathbf{A}$$

This leads to an explicit expression for the series representation of the state-transition matrix. For example, the element in the upper left-hand corner of the matrix is

$$1-\frac{\omega^2t^2}{2!}+\frac{\omega^4t^4}{4!}-\frac{\omega^6t^6}{6!}+\cdots=\cos\omega t$$

Similar expressions can be found for the other elements, leading finally to

$$\mathbf{\Phi}(t)=e^{\mathbf{A}t}=\begin{bmatrix}\cos\omega t & (\sin\omega t)/\omega\\ -\omega\sin\omega t & \cos\omega t\end{bmatrix} \tag{4-30}$$

Any solution to the original equation, therefore, will be a combination of sine and cosine terms.

Example 3 (The Lanchester Model of Warfare). A famous model of warfare was developed by Lanchester in 1916. In this model, members of a fighting force are characterized as having a *hitting power*, determined by their military technology. The hitting power is defined to be the number of casualities per unit time (on the average) that one member can inflict on the enemy.

Suppose N_1 units of one force, each with hitting power α, are engaged with N_2 units of a second force, each with hitting power β. Suppose further that the hitting power of the first force is directed equally against all units of the second, and vice versa. The dynamic model for the engagement, determining the reduction in forces, is

$$\dot{N}_1(t) = -\beta N_2(t)$$
$$\dot{N}_2(t) = -\alpha N_1(t)$$

When expressed as a system, these equations correspond to the system matrix

$$\mathbf{A} = \begin{bmatrix} 0 & -\beta \\ -\alpha & 0 \end{bmatrix}$$

The state-transition matrix can be found in a manner quite analogous to that used for harmonic motion. We have

$$\mathbf{A}^2 = \begin{bmatrix} \alpha\beta & 0 \\ 0 & \alpha\beta \end{bmatrix} = \alpha\beta\mathbf{I}$$

and thus, in general, if k is even

$$\mathbf{A}^k = (\alpha\beta)^{k/2}\mathbf{I}$$
$$\mathbf{A}^{k+1} = (\alpha\beta)^{k/2}\mathbf{A}$$

The expansion

$$e^{\mathbf{A}t} = \mathbf{I} + \mathbf{A}t + \frac{\mathbf{A}^2t^2}{2!} + \cdots$$

can be expressed in terms of the hyperbolic functions $\sinh(\sqrt{\alpha\beta}\, t)$ and $\cosh(\sqrt{\alpha\beta}\, t)$, but we leave the details to the reader. The Lanchester model is discussed further in Problem 15.

4.7 GENERAL SOLUTION TO LINEAR CONTINUOUS-TIME SYSTEMS

We turn now to the solution of the general linear continuous-time system

$$\dot{\mathbf{x}}(t) = \mathbf{A}(t)\mathbf{x}(t) + \mathbf{B}(t)\mathbf{u}(t) \tag{4-31}$$

where as usual $\mathbf{x}(t)$ is an n-dimensional state vector, $\mathbf{A}(t)$ is an $n \times n$ matrix, $\mathbf{u}(t)$ is an m-dimensional vector of inputs, and $\mathbf{B}(t)$ is an $n \times m$ distribution matrix. Just as in the discrete-time case, the solution to the general nonhomogeneous system can be relatively easily expressed in terms of the state-transition matrix associated with the homogeneous (or free) system.

Again, it is simplest to first propose and verify the solution directly.

Proposition. *The solution of the system* (4-31) *in terms of the initial state* $\mathbf{x}(0)$ *and the inputs is*

$$\mathbf{x}(t)=\mathbf{\Phi}(t, 0)\mathbf{x}(0)+\int_0^t \mathbf{\Phi}(t, \tau)\mathbf{B}(\tau)\mathbf{u}(\tau)\,d\tau \tag{4-32}$$

Proof. Before beginning the actual proof, the integral sign in this formula deserves a bit of explanation. Let us fix some $t>0$. Then $\mathbf{x}(t)$ is an n-dimensional vector determined by the right-hand side of the equation. The first term on the right is just a matrix times a vector, so it is a vector. The integrand, the expression inside the integral sign, is likewise an n vector for each value of τ. The integral is a (continuous) summation of these vectors, and is therefore itself a vector. The integration can be performed componentwise, each component being an ordinary function of the variable τ.

To verify that (4-32) is in fact the solution to the system (4-31), we differentiate with respect to t. Differentiation of the integral produces two terms—one corresponding to differentiation with respect to t inside the integral sign, and the other corresponding to differentiation with respect to the upper limit of the integral. Thus,

$$\begin{aligned}\frac{d}{dt}\mathbf{x}(t)=&\frac{d}{dt}\mathbf{\Phi}(t, 0)\mathbf{x}(0)\\ &+\int_0^t \frac{d}{dt}\mathbf{\Phi}(t, \tau)\mathbf{B}(\tau)\mathbf{u}(\tau)\,d\tau+\mathbf{\Phi}(t, t)\mathbf{B}(t)\mathbf{u}(t)\end{aligned}$$

Using the basic properties of the state-transition matrix

$$\frac{d}{dt}\mathbf{\Phi}(t, \tau)=\mathbf{A}(t)\mathbf{\Phi}(t, \tau)$$

$$\mathbf{\Phi}(t, t)=\mathbf{I}$$

the above reduces to

$$\begin{aligned}\frac{d}{dt}\mathbf{x}(t)&=\mathbf{A}(t)\mathbf{\Phi}(t, 0)\mathbf{x}(0)\\ &\quad+\int_0^t \mathbf{A}(t)\mathbf{\Phi}(t, \tau)\mathbf{B}(\tau)\mathbf{u}(\tau)\,d\tau+\mathbf{B}(t)\mathbf{u}(t)\\ &=\mathbf{A}(t)\mathbf{x}(t)+\mathbf{B}(t)\mathbf{u}(t)\end{aligned}$$

which shows that the proposed solution satisfies the system equation. ▮

Superposition

The principle of superposition applies to linear continuous-time systems the same as it does to discrete-time systems. The overall effect due to several

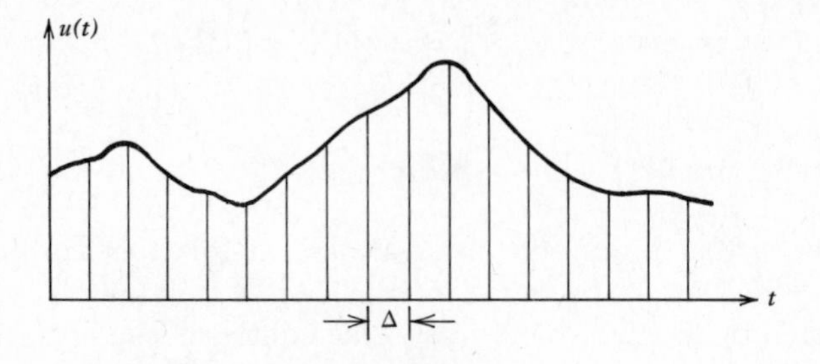

Figure 4.5. Decomposition of input.

different inputs is the sum of the effects that would be produced by the individual inputs (if the initial condition were zero). This idea can be used to interpret the formula for the general solution.

The first term on the right-hand side of the solution (4-32) represents the response due to the initial condition $\mathbf{x}(0)$. This response is determined directly by the state-transition matrix developed for the free system, and it is a component of every solution.

To interpret the second term, imagine the input function $\mathbf{u}(t)$ as being broken up finely into a sequence of individual pulses of width Δ, as illustrated in Fig. 4.5. At time τ the pulse will have an (approximate) height of $\mathbf{u}(\tau)$. If the pulse at τ were the only input, and if the initial state were zero, then the immediate effect of this pulse would be to transfer the state from zero, just prior to the pulse, to $\Delta\mathbf{B}(\tau)\mathbf{u}(\tau)$ just after it. This is because the resulting value of the state is the integral of the pulse.

After the state has been transferred from zero to $\Delta\mathbf{B}(\tau)\mathbf{u}(\tau)$, the longer-term response, in the absence of further inputs, is determined by the free system. Therefore, for $t>\tau$ the response due to the pulse at τ would be $\Delta\mathbf{\Phi}(t, \tau)\mathbf{B}(\tau)\mathbf{u}(\tau)$. The total effect due to the whole sequence of pulses is the sum of the individual responses, as represented in the limit by the integral term on the right-hand side of the solution formula (4-32).

Example (First-Order Decay). Consider a first-order system governed by the equation

$$\dot{x}(t) = -rx(t) + u(t) \tag{4-33}$$

where $r>0$. This is referred to as a decay system, since in the absence of inputs the solution is

$$x(t) = e^{-rt}x(0) \tag{4-34}$$

which decays to zero exponentially.

Suppose the system is initially at rest, at time $t=0$, and an input u of unit magnitude is applied starting at time $t=0$. Let us calculate the resulting time

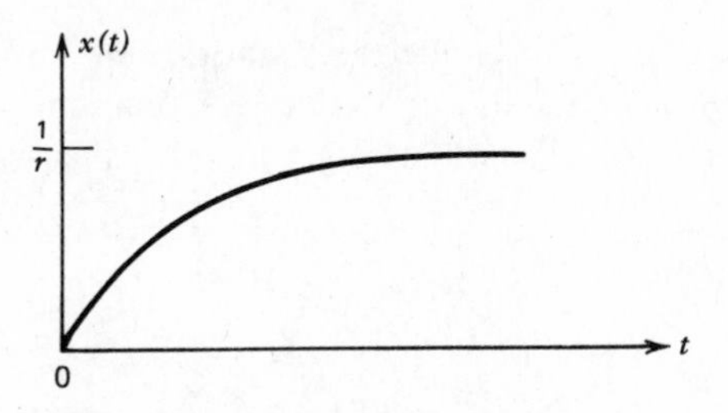

Figure 4.6. Response of decay system.

response. The state-transition matrix (which is 1×1 in this case) is

$$\mathbf{\Phi}(t)=e^{-rt}$$

The solution with zero initial condition and unity input is, therefore,

$$x(t)=\int_0^t e^{-r(t-\tau)}\,d\tau=\frac{e^{-r(t-\tau)}}{r}\bigg|_0^t$$

$$=\frac{1}{r}[1-e^{-rt}]$$

This response is illustrated in Fig. 4.6.

*4.8 EMBEDDED STATICS

In formulating equations to describe a dynamic situation, the equations one writes may not initially be in the standard state variable form. It is, however, often most convenient to transform the equations to the standard form. This procedure is usually not difficult; indeed, in many instances it is so straightforward that one performs the necessary operations without hesitation. Nevertheless, it is worthwhile to recognize that this transformation is in fact a necessary step.

A general form that is likely to arise (arbitrarily expressed in discrete-time just for specificity) is

$$\mathbf{Ex}(k+1)=\mathbf{Ax}(k)+\mathbf{Bu}(k) \tag{4-35}$$

where $\mathbf{E}$ and $\mathbf{A}$ are $n\times n$ matrices and $\mathbf{B}$ is an $n\times m$ matrix. These matrices may in general depend on k without affecting the essence of our discussion. If $\mathbf{E}$ is nonsingular, it is simple to transform the equations by multiplying by the inverse of $\mathbf{E}$. This yields the standard state vector form

$$\mathbf{x}(k+1)=\mathbf{E}^{-1}\mathbf{Ax}(k)+\mathbf{E}^{-1}\mathbf{Bu}(k) \tag{4-36}$$

If $\mathbf{E}$ is *not* invertible the situation is more interesting. The system then consists of a mixture of static and dynamic equations; the static equations being in some sense embedded within the dynamic framework. Under rather general conditions (see Problem 21) such a system with embedded statics can be transformed to a state vector dynamic system having an order less than the dimension of the original system of equations. The following examples illustrate this point.

Example 1. Consider the system defined by

$$x_1(k+1)+x_2(k+1)=x_1(k)+2x_2(k)+u(k) \tag{4-37}$$

$$0=2x_1(k)+x_2(k)+u(k) \tag{4-38}$$

This has the form of (4-35) with

$$\mathbf{E}=\begin{bmatrix}1 & 1\\ 0 & 0\end{bmatrix}$$

which is singular. To obtain the reduced form for this particular system, we add the two equations to produce

$$x_1(k+1)+x_2(k+1)=3[x_1(k)+x_2(k)]+2u(k) \tag{4-39}$$

This shows that the variable

$$z(k)=x_1(k)+x_2(k) \tag{4-40}$$

can serve as a state variable for the system. The dynamic portion of the system takes the form

$$z(k+1)=3z(k)+2u(k) \tag{4-41}$$

The original variables x_1 and x_2 can be expressed in terms of z and u by solving (4-38) and (4-40) simultaneously. This leads to

$$x_1(k)=-z(k)-u(k) \tag{4-42}$$

$$x_2(k)=2z(k)+u(k) \tag{4-43}$$

Example 2 (National Economics—The Harrod–Type Model). A dynamic model of the national economy was proposed in Sect. 1.3. In terms of variables that have a specific economic meaning, the basis for the model is the following three equations:

$$Y(k)=C(k)+I(k)+G(k)$$

$$C(k)=mY(k)$$

$$Y(k+1)-Y(k)=rI(k)$$

In these equations only the variable $G(k)$ is an input variable. The others are derived variables that, at least in some measure, describe the condition of the system. In a vector-matrix format the defining equations take the form

$$\begin{bmatrix} 0 & 0 & 0 \\ 0 & 0 & 0 \\ 0 & 0 & 1 \end{bmatrix} \begin{bmatrix} I(k+1) \\ C(k+1) \\ Y(k+1) \end{bmatrix} = \begin{bmatrix} 1 & 1 & -1 \\ 0 & -1 & m \\ r & 0 & 1 \end{bmatrix} \begin{bmatrix} I(k) \\ C(k) \\ Y(k) \end{bmatrix} + \begin{bmatrix} 1 \\ 0 \\ 0 \end{bmatrix} G(k)$$

In this form it is clear that the original equations can be regarded as a dynamic system with embedded statics. This particular system is easy to reduce to a first-order system by a series of substitutions, as carried out in Sect. 1.3. This leads to the first-order dynamic system

$$Y(k+1) = [1 + r(1-m)]Y(k) - rG(k)$$

The other variables can be recovered by expressing them in terms of $Y(k)$ and $G(k)$. In particular,

$$C(k) = mY(k)$$
$$I(k) = (1-m)Y(k) - G(k)$$

Example 3 (National Economics—Another Version). The dynamic model of the national economy presented above can be regarded as being but one of a whole family of possible (and plausible) models. Other forms that are based on slightly different hypotheses can result in distinct dynamic structures. The relationships between these different models is most clearly perceived in the nonreduced form; that is,in the form that contains embedded statics.

Samuelson proposed a model of the national economy based on the following assumptions. National income $Y(k)$ is equal to the sum of consumption $C(k)$, investment $I(k)$, and government expenditure $G(k)$. Consumption is proportional to the national income of the preceding year; and investment is proportional to the increase in consumer spending of that year over the preceding year.

In equation form, the Samuelson model is

$$Y(k) = C(k) + I(k) + G(k)$$
$$C(k+1) = mY(k)$$
$$I(k+1) = \mu[C(k+1) - C(k)]$$

In our generalized matrix form, the system becomes

$$\begin{bmatrix} 0 & 0 & 0 \\ 0 & 1 & 0 \\ 1 & -\mu & 0 \end{bmatrix} \begin{bmatrix} I(k+1) \\ C(k+1) \\ Y(k+1) \end{bmatrix} = \begin{bmatrix} 1 & 1 & -1 \\ 0 & 0 & m \\ 0 & -\mu & 0 \end{bmatrix} \begin{bmatrix} I(k) \\ C(k) \\ Y(k) \end{bmatrix} + \begin{bmatrix} 1 \\ 0 \\ 0 \end{bmatrix} G(k)$$

This system can be reduced to a second-order system in standard form.

4.9 PROBLEMS

1. *Moving Average.* There are many situations where raw data is subjected to an averaging process before it is displayed or used for decision making. This smoothes the data sequence, and often highlights the trends while suppressing individual deviations.

   Suppose a sequence of raw data is denoted $u(k)$. A simple four-point averager produces a corresponding sequence $y(k)$ such that each $y(k)$ is the average of the data points $u(k)$, $u(k-1)$, $u(k-2)$, $u(k-3)$. Find a representation for the averager of the form

$$\mathbf{x}(k+1)=\mathbf{A}\mathbf{x}(k)+\mathbf{b}u(k)$$
$$y(k)=[\mathbf{c}^T\mathbf{x}(k)+du(k)]$$

   where $\mathbf{x}(k)$ is three-dimensional, $\mathbf{A}$ is a 3×3 matrix, $\mathbf{b}$ is a 3×1 (column) vector, $\mathbf{c}^T$ is a 1×3 (row) vector, and d is a scalar.

2. *Cohort Model.* Suppose that the input $u(k)$ of new machines in the example in Sect. 4.1 is chosen to exactly equal the number of machines going out of service that year. Write the corresponding state space model and show that it is a special case of the general cohort population model described in Chapter 1. Repeat under the assumption that in addition to replacements there are new purchases amounting to y percent of the total number of machines in service.

3. Consider the linear difference equation

$$y(k+n)+a_{n-1}y(k+n-1)+\cdots+a_0y(k)$$
$$=b_{n-1}u(k+n-1)+b_{n-2}u(k+n-2)+\cdots+b_0u(k)$$

   Show that this equation can be put in state space form

$$\mathbf{x}(k+1)=\mathbf{A}x(k)+\mathbf{b}u(k)$$

   by defining

$$\begin{aligned}
x_1(k)&=-a_0y(k-1)+b_0u(k-1)\\
x_2(k)&=-a_0y(k-2)+b_0u(k-2)\\
&\quad-a_1y(k-1)+b_1u(k-1)\\
&\vdots\\
x_{n-1}(k)&=-a_0y(k-n+1)+b_0u(k-n+1)\\
&\quad-a_1y(k-n+2)+b_1u(k-n+2)\\
&\quad\vdots\\
&\quad-a_{n-2}y(k-1)+b_{n-2}u(k-1)\\
x_n(k)&=y(k)
\end{aligned}$$

4. *Nonlinear Systems.* Consider the nonlinear difference equation of the form

$$y(k+n) = F[y(k+n-1), \ldots, y(k), u(k+n-1), \ldots, u(k), k]$$

(a) Find a state space representation of the difference equation. (*Hint:* The representation will be more than n-dimensional.)
(b) Find an n-dimensional representation in the case where F has the special form

$$F = \sum_{i=1}^{n} f_i(y(k+n-i), u(k+n-i))$$

5. *Labor–Management Negotiations.* Consider a wage dispute between labor and management. At each stage of the negotiations, labor representatives submit a wage demand to management that, in turn, presents a counter offer. Since the wage offer will be usually less than the wage demand, further negotiations are required. One can formulate this situation as a dynamic system, where at each period management "updates" its previous offer by the addition of some fraction α of the difference between last period's demand and offer. Labor also "updates" its previous demand by the subtraction of some fraction β of the difference between the demand and offer of the last period. Let x_1 equal the management offer and x_2 equal the labor demand. Write the dynamic state equations (in matrix form) for the situation described above.

6. Consider the two social systems whose marriage rules are summarized in Fig. 4.7. In each system there are four social classes, and every child born to a certain class combination becomes a member of the class designated in the table. The assumptions (a), (b), and (c) of the Natchez Indian example hold as well. For each system: (a) write the state equation for the social system; and (b) compute the solution to the state equations.

7. *A Simple Puzzle.* We have four timepieces whose performance is described as follows: The wall clock loses two minutes in an hour. The table clock gets two

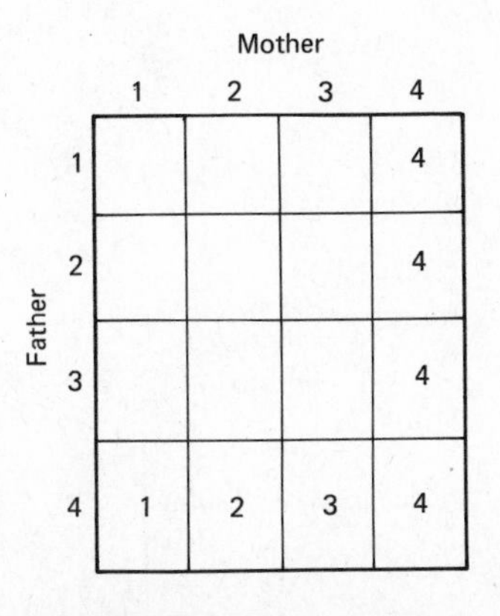

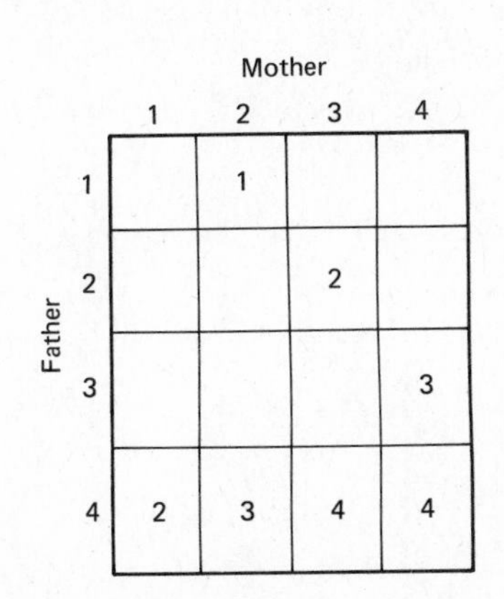

Figure 4.7. Social systems.

minutes ahead of the wall clock in an hour of real time. The alarm clock falls two minutes behind the table clock in an hour. The last piece, the wristwatch, gets two minutes ahead of the alarm clock in an hour. At noon all four timepieces were set correctly. Let x_1 equal the wall clock reading, x_2 be the table clock reading, x_3 the alarm clock reading, and x_4 the wristwatch reading, and consider noon as the starting time (i.e., $k = 0$).

(a) Write the dynamic equations corresponding to the four given statements about performance. Directly translate into the form

$$\mathbf{E}\mathbf{x}(k+1) = \mathbf{C}\mathbf{x}(k) + \mathbf{c}$$

(b) Convert the system to the standard form

$$\mathbf{x}(k+1) = \mathbf{A}\mathbf{x}(k) + \mathbf{b}$$

[*Hint:* $(\mathbf{I}-\mathbf{B})^{-1} = \mathbf{I} + \mathbf{B} + \mathbf{B}^2 + \cdots + \mathbf{B}^k + \cdots$ whenever $\mathbf{B}$ is such that the series converges.]

(c) Find the state-transition matrix $\mathbf{\Phi}(k)$.

(d) Find a simple general formula for $\mathbf{x}(k)$. What time will the wristwatch show at 7:00 p.m. (i.e., at $k = 7$)?

8. *A Classic Puzzle.* Repeat Problem 7 above for the alternate description: The wall clock loses two minutes in an hour. The table clock gets two minutes ahead of the wall clock *for every hour registered* on the wall clock. The alarm clock falls two minutes behind the table clock for every hour registered on the table clock. The wristwatch gets two minutes ahead of the alarm clock for every hour registered on the alarm clock.

9. *Properties of State-Transition Matrix.* Let $\mathbf{\Phi}(t, \tau)$ be the state-transition matrix corresponding to the linear system

$$\dot{\mathbf{x}}(t) = \mathbf{A}(t)\mathbf{x}(t)$$

Show:

(a) $\mathbf{\Phi}(t_2, t_0) = \mathbf{\Phi}(t_2, t_1)\mathbf{\Phi}(t_1, t_0)$

(b) $\mathbf{\Phi}(t, \tau)^{-1} = \mathbf{\Phi}(\tau, t)$

(c) $\dfrac{d}{d\tau}\mathbf{\Phi}(t, \tau) = -\mathbf{\Phi}(t, \tau)\mathbf{A}(\tau)$

10. A model of a satellite has the form

$$\dot{\mathbf{x}}(t) = \mathbf{A}\mathbf{x}(t)$$

where $\mathbf{A}$ is a 4×4 matrix. Given that the state-transition matrix for this system is

$$\mathbf{\Phi}(t, 0) = \begin{bmatrix} 4-3\cos\omega t & \sin\omega t/\omega & 0 & 2(1-\cos\omega t)/\omega \\ 3\omega\sin\omega t & \cos\omega t & 0 & 2\sin\omega t \\ 6(-\omega t+\sin\omega t) & -2(1-\cos\omega t)/\omega & 1 & (4\sin\omega t-3\omega t)/\omega \\ 6\omega(-1+\cos\omega t) & -2\sin\omega t & 0 & 4\cos\omega t-3 \end{bmatrix}$$

find the matrix $\mathbf{A}$.

11. *Linearization.* Suppose one has a system of n *nonlinear* first-order differential equations:

$$\dot{y}_1 = f_1(y_1, y_2, \ldots, y_n, u, t)$$
$$\dot{y}_2 = f_2(y_1, y_2, \ldots, y_n, u, t)$$
$$\vdots$$
$$\dot{y}_n = f_n(y_1, y_2, \ldots, y_n, u, t)$$

where $u(t)$ is an input or forcing term. Further, suppose that for a *particular set* of initial conditions and input $\{\bar{y}_1(t_0), \bar{y}_2(t_0), \ldots, \bar{y}_n(t_0), \bar{u}(t)\}$ there is a known solution

$$\bar{\mathbf{y}}(t) = \begin{bmatrix} \bar{y}_1(t) \\ \bar{y}_2(t) \\ \cdot \\ \cdot \\ \cdot \\ \bar{y}_n(t) \end{bmatrix}$$

We wish to investigate the behavior of the system when the input function and initial conditions are changed slightly. That is, we require a solution to the system when $y_i(t_0) = \bar{y}_i(t_0) + z_i(t_0)$, $i = 1, 2, \ldots, n$ and $u(t) = \bar{u}(t) + v(t)$. To satisfy the original system of equations, we seek a solution of the form $\mathbf{y}(t) = \bar{\mathbf{y}}(t) + \mathbf{z}(t)$. So the system becomes:

$$(\dot{\bar{y}}_1 + \dot{z}_1) = f_1(\bar{y}_1 + z_1, \bar{y}_2 + z_2, \ldots, \bar{y}_n + z_n, \bar{u} + v, t)$$
$$(\dot{\bar{y}}_2 + \dot{z}_2) = f_2(\bar{y}_1 + z_1, \bar{y}_2 + z_2, \ldots, \bar{y}_n + z_n, \bar{u} + v, t)$$
$$\vdots$$
$$(\dot{\bar{y}}_n + \dot{z}_n) = f_n(\bar{y}_1 + z_1, \bar{y}_2 + z_2, \ldots, \bar{y}_n + z_n, \bar{u} + v, t)$$

(a) Assume that v and the $z_i(t_0)$'s are small (so that the original conditions are slightly different from the conditions for which a solution exists) and expand the new system of equations in a Taylor series about the original conditions and input. Neglect terms of order 2 or higher.

(b) From part (a), write down a system of time-varying linear differential equations that express the effect of the perturbation on the original (known) solution $\bar{\mathbf{y}}$. Thus, the new solution will consist of $\bar{\mathbf{y}}$ plus the contribution found in this section. What is the time-varying system matrix?

12. *Application of Linearization.* Consider the nonlinear system

$$\dot{x}_1 = x_2$$
$$\dot{x}_2 = 2x_1^3 - u(t)x_2$$

(a) For the particular initial conditions $x_1(1) = 1$, $x_2(1) = -1$ and the input function $u(t) = 0$, find the solution $\bar{\mathbf{x}}(t)$. (*Hint:* Try powers of t.)

(b) Using the results of Problem 11 above, find a time-varying linear system that describes the behavior of small perturbations.
(c) Find a linear equation for an approximate solution to the original system, corresponding to $x_1(1) = 1.5, x_2(1) = 0.5, u(t) = 0.5$.

13. Consider the first-order dynamic system

$$\dot{x}(t) = ax(t) + b$$

with initial condition $x(0) = 1$.
(a) Find the solution $x(t)$ using the general formula for solutions to nonhomogeneous systems.
(b) Note that the system can be expressed as a second-order homogeneous system in the form

$$\dot{x}(t) = ax(t) + by(t)$$

$$\dot{y}(t) = 0$$

with $x(0) = 1$, $y(0) = 1$. Denoting the corresponding 2×2 system matrix by $\mathbf{\hat{A}}$, calculate $e^{\mathbf{\hat{A}}t}$. Use this to find $x(t)$ and $y(t)$.

14. Consider the time-varying differential equation

$$\ddot{y} + \frac{4}{t}\dot{y} + \frac{2}{t^2}y = u(t)$$

for $t > 0$.
(a) Define state variables $x_1 = y$ and $x_2 = \dot{y}$ and find a representation of the form

$$\dot{\mathbf{x}}(t) = \mathbf{A}(t)\mathbf{x}(t) + \mathbf{b}u(t)$$

(b) Find two linearly independent solutions to the homogeneous scalar equation. (*Hint:* Try $y = t^k$.)
(c) Construct the matrix $\mathbf{X}(t)$, a fundamental matrix of solutions, based on the results of part (b).
(d) Find the state-transition matrix $\mathbf{\Phi}(t, \tau)$.

15. *Lanchester Model.* A fair amount of information can be deduced directly from the form of the Lanchester equations, without actually solving them.
(a) What condition on the sizes of the two forces must be met in order that the fractional loss rate of both sides be equal? [Fractional loss rate $= \dot{N}/N$.]
(b) Find a relation of the form $F(N_1, N_2) = c$, where c is a constant, which is valid throughout the engagement. [The result should be a simple quadratic expression similar to the answer in (a).]
(c) If $N_1(0)$ and $N_2(0)$ are known, who will win? (The battle continues until one side is totally depleted.)
(d) As a function of time, the size of either side will have the form

$$N_i(t) = A_i \sinh \omega t + B_i \cosh \omega t$$

What is the value of ω?

16. *Floquet Theory.* Consider a system of differential equations of the form

$$\dot{\mathbf{x}}(t)=\mathbf{A}(t)\mathbf{x}(t)$$

where the $n\times n$ matrix $\mathbf{A}(t)$ is periodic with a period T. That is,

$$\mathbf{A}(t+T)=\mathbf{A}(t)$$

for every t. Let $\mathbf{X}(t)$ be a fundamental matrix of solutions, with the normalizing property that $\mathbf{X}(0)=\mathbf{I}$.

(a) Observe that $\mathbf{X}(t+T)$ is also a fundamental matrix of solutions and write an equation relating $\mathbf{X}(t)$ and $\mathbf{X}(t+T)$.

(b) Let the matrix $\mathbf{C}$ be defined such that

$$\mathbf{X}(T)=e^{\mathbf{C}T}$$

and define $\mathbf{B}(t)$ such that

$$\mathbf{X}(t)=\mathbf{B}(t)e^{\mathbf{C}t}$$

Show that in this representation for $\mathbf{X}(t)$, the matrix $\mathbf{B}(t)$ is periodic with period T.

17. *Variation of Parameters.* The classic derivation of the general solution formula for a homogeneous system of differential equations is based on a method called *variation of parameters.* Consider the system

$$\dot{\mathbf{x}}(t)=\mathbf{A}(t)\mathbf{x}(t)+\mathbf{b}(t)$$

Let $\mathbf{X}(t)$ be a fundamental matrix of solutions for the corresponding homogeneous equation. It is known that any solution to the homogeneous equation can be written as $\mathbf{X}(t)\mathbf{y}$, where $\mathbf{y}$ is an n-vector of (fixed) parameters. It is conceivable then that it might be helpful to express the general solution to the nonhomogeneous equation in the form $\mathbf{x}(t)=\mathbf{X}(t)\mathbf{y}(t)$, where now $\mathbf{y}(t)$ is an n-vector of varying parameters.

(a) Using the suggested form as a trial solution, find a system of differential equations that $\mathbf{y}(t)$ must satisfy.

(b) Solve the system in (a) by integration and thereby obtain the general solution to the original system.

(c) Convert the result of (b) to one using the state-transition matrix, and verify that it is the same as the result given in the text.

18. *Time-Varying System.* A closed-form expression for the transition matrix of a time-varying system

$$\dot{\mathbf{x}}(t)=\mathbf{A}(t)\mathbf{x}(t)$$

can be found only in special cases. Show that if $\mathbf{A}(t)\mathbf{A}(\tau)=\mathbf{A}(\tau)\mathbf{A}(t)$ for all t, τ, then the transition matrix may be written as

$$\mathbf{\Phi}(t,\tau)=\exp\int_\tau^t \mathbf{A}(\xi)\,d\xi$$

(*Hint:* $\exp\int_\tau^t \mathbf{A}(\xi)\,d\xi=\mathbf{I}+\int_\tau^t \mathbf{A}(\xi)\,d\xi+\frac{1}{2}\int_\tau^t \mathbf{A}(\xi)\,d\xi\int_\tau^t \mathbf{A}(\xi)\,d\xi+\cdots$)

19. Find the impulse response of the straight-through cohort model.

20. Given a real $n \times n$ matrix $\mathbf{A}$, let M be a bound for the magnitude of its elements, that is, $|a_{ij}| \leq M$ for all i, j.
    (a) Find a bound for the elements of $\mathbf{A}^k$.
    (b) Let b_k and c_k, $k = 0, 1, 2, \ldots$ be two sequences of real numbers with $b_k > 0$. If the series $\sum_{k=0}^{\infty} b_k$ converges and if $|c_k| \leq b_k$ for all k, then the series $\sum_{k=0}^{\infty} c_k$ also converges. Given that the series $\sum_{k=0}^{\infty} a^k/k!$ converges (to e^a) for any value of a, show that the matrix series $\sum_{k=0}^{\infty} \mathbf{A}^k t^k/k!$ converges.

21. *Embedded Statics.* Suppose a system is described by a set of equations of the form

$$\begin{bmatrix} \mathbf{T} \\ \mathbf{0} \end{bmatrix} \mathbf{x}(k+1) = \begin{bmatrix} \mathbf{C} \\ \mathbf{D} \end{bmatrix} \mathbf{x}(k) + \begin{bmatrix} \mathbf{u}(k) \\ \mathbf{v}(k) \end{bmatrix}$$

where $\mathbf{x}(k)$ is an n-dimensional vector, $\mathbf{T}$ and $\mathbf{C}$ are $m \times n$ matrices, $\mathbf{D}$ is an $(n-m) \times n$ matrix, and $\mathbf{u}(k)$ and $\mathbf{v}(k)$ are m and $(n-m)$-dimensional vectors, respectively. Assume that the $n \times n$ matrix

$$\begin{bmatrix} \mathbf{T} \\ \mathbf{D} \end{bmatrix}$$

is nonsingular. Following the steps below, it is possible to convert this system to state vector form.
    (a) Define $\mathbf{y}(k) = \mathbf{T}\mathbf{x}(k)$ and show that with this definition, and the lower part of the system equation, one may express $\mathbf{x}(k)$ in the form

$$\mathbf{x}(k) = \mathbf{H}\mathbf{y}(k) - \mathbf{G}\mathbf{v}(k)$$

    Give an explicit definition of $\mathbf{G}$ and $\mathbf{H}$.
    (b) Show that the top part of the original system can be written in the state vector form

$$\mathbf{y}(k+1) = \mathbf{R}\mathbf{y}(k) + \mathbf{B}\mathbf{v}(k) + \mathbf{u}(k)$$

    and give expressions for $\mathbf{R}$ and $\mathbf{B}$. Note that $\mathbf{x}(k)$ can be recovered from $\mathbf{y}(k)$ using part (a).
    (c) Apply this procedure to Example 3, Sect. 4.8.

NOTES AND REFERENCES

General. As with the material in Chapter 2, the theoretical content of this chapter is quite standard and much of it is contained in the references mentioned at the end of Chapter 2. This chapter, however, begins to incorporate more explicitly the viewpoint of dynamics, as opposed to simply difference and differential equations. This is manifested most importantly by the concept of the state vector, by the explicit recognition of inputs, by the introduction of a state-transition matrix relating the states at two time instances, and by the view of the general solution to a linear equation as being composed of a series of free responses. This viewpoint is represented by books such as DeRusso, Roy, and Close [D1], Kwakernaak and Sivan [K16], and Rugh [R7].

Section 4.4. Codification of the exact class inheritance rules of the Natchez Indian social structure is not available. Rather, the rules presented here have been inferred from observations and writings of early French explorers. There is a possibility that the actual scheme differed somewhat. See White, Murdock, and Scaglion [W4]. Presentation in this form (originally due to Robert Busch) is contained in Goldberg [G8].

Section 4.6. For the Lanchester model see Lanchester [L1] or Saaty [S1].

Section 4.8. See Luenberger [L12] for a general theory of this type. For general background on dynamic economic models see Allen [A1], Baumol [B4], and Gandolfo [G1]. A concise statement of several possible versions, including the ones referred to here, is contained in Papandreou [P1]. For the specific model of the example see Samuelson [S2].

Section 4.9. The classic puzzle, Problem 8, is contained in Kordemsky [K15].

chapter 5.

Linear Systems with Constant Coefficients

The subject of linear systems with constant coefficients is in some sense the core of dynamic systems theory. These systems have a rich full theory, and they provide a natural format for penetrating analyses of many important dynamic phenomena.

A linear system with constant coefficients is described in discrete time as

$$\mathbf{x}(k+1)=\mathbf{A}\mathbf{x}(k)+\mathbf{B}\mathbf{u}(k)$$

and in continuous time as

$$\dot{\mathbf{x}}(t)=\mathbf{A}\mathbf{x}(t)+\mathbf{B}\mathbf{u}(t)$$

In either case, it is known from the general results of Chapter 4 that a major role is played by the corresponding homogeneous equation. Since the homogeneous equation is defined entirely by the associated system matrix $\mathbf{A}$, it can be expected that much of the theory of linear systems with constant coefficients is derived directly from matrix theory. Indeed, this is the case, and a good portion of the theory in this chapter is based on the results from linear algebra presented in Chapter 3.

The most important concept discussed in this chapter is that of system eigenvalues and eigenvectors, defined by the matrix $\mathbf{A}$. These eigenvectors define special first-order dynamic systems, embedded within the overall system, that behave independently of the rest of the system. The original complex system can be decomposed into a collection of simpler systems associated with various eigenvalues. This decomposition greatly facilitates analysis.

In addition to theory, this chapter also contains several extended examples and applications. They are included to illustrate the theory, broaden our scope of classic models, and in some cases, to illustrate the kind of analysis that can be achieved with the theory. In each of the examples, one objective is to illustrate how mathematical analysis can supplement and guide our intuitive reasoning, without displacing it.

5.1 GEOMETRIC SEQUENCES AND EXPONENTIALS

A special role is played in the study of linear time-invariant systems by geometric sequences in discrete time, and by exponential functions in continuous time. They have a unique reproducing property when applied as input functions, and this property helps explain why these functions occur as natural solutions to homogeneous systems.

In discrete time the basic dynamic element is the unit delay. If a sequence $u(k)$ is applied as input to a unit delay, then the output $x(k)$ is governed by

$$x(k+1)=u(k)$$

Now if $u(k)$ is a geometric sequence, say $u(k)=a^k$, $k=0, 1, 2, \ldots$, the output is $x(k)=a^{k-1}=a^k/a$, $k=1, 2, \ldots$. Thus, for any $k \geq 1$, the output is just a multiple of the input. In other words, the effect of a delay is simply to multiply a geometric sequence by a constant.

Any nth-order linear constant-coefficient system in discrete time consists of a combination of n unit delay elements and a number of scalar multiplications, as depicted in Fig. 5.1. If a geometric sequence is applied as an input at any point, it will pass through the various delays and constant multiples to which it is connected without changing its form. Thus, this particular geometric sequence will be a component of the overall response.

Similarly, it can be seen that geometric sequences occur in the homogeneous system. In the homogeneous system the response at any point serves as the input to other parts of the system. If this response is a geometric sequence, it will travel around the system and eventually return to the original point with the same form. For consistency, however, it must return with the same magnitude, as well as general form, as it started. Only certain geometric sequences (that is, only certain values of the parameter a in a^k) give this result. Such a geometric sequence is part of the homogeneous solution.

As a specific example, consider the first-order system

$$x(k+1)=ax(k)+bu(k) \tag{5-1}$$

which is depicted in Fig. 5.2. The homogeneous response is

$$x(k)=a^k x(0)$$

The geometric sequence a^k can pass from $x(k)$ through the multiplication by a, then through the unit delay, returning to the original point with the same form and magnitude.

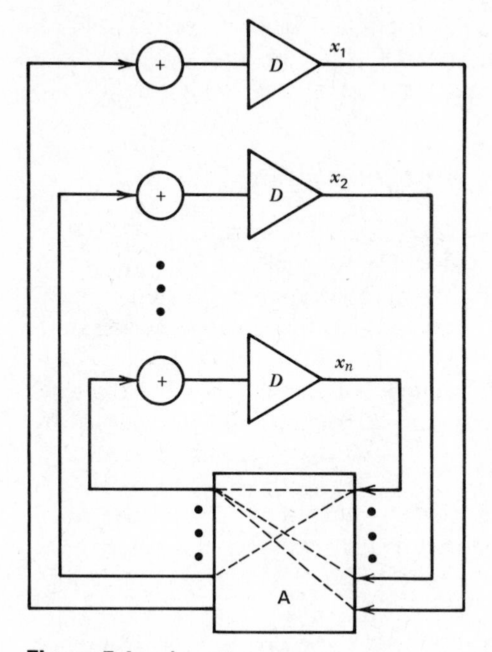

Figure 5.1. An nth-order system.

Suppose now that an input function is applied, which is itself a geometric sequence but with a ratio $\mu \neq a$. That is, $u(k) = \mu^k$. The solution $x(k)$ in this case consists of the sum of two geometric sequences, one of ratio a and one of ratio μ. Specifically, the solution is

$$x(k) = \left(x(0) - \frac{b}{\mu - a}\right) a^k + \left(\frac{b}{\mu - a}\right) \mu^k \tag{5-2}$$

The geometric sequence input passes through the system with unaltered form but with changed magnitude, and in addition a component of the homogeneous response is superimposed.

This discussion, of course, has an exact analog in the continuous-time case. There, exponential functions of the form e^{-at} have the reproducing property.

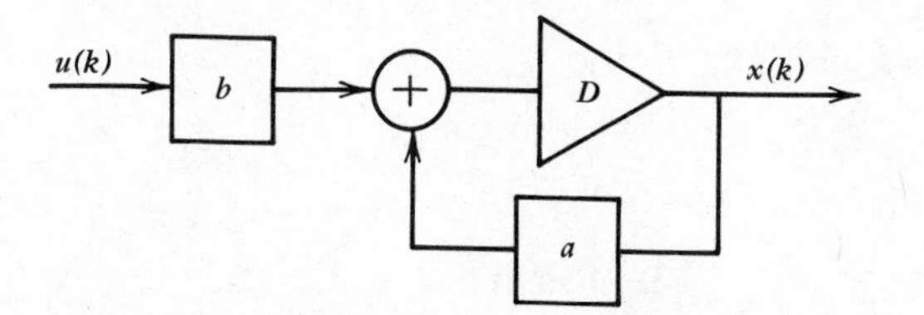

Figure 5.2. Discrete-time first-order system.

An nth-order linear constant coefficient system is composed of n differentiators and a number of scalar multipliers. An exponential retains its form when acted on by any of these.

The discussion in this section shows that even without detailed analysis it can be inferred that geometric sequences and exponential functions are fundamental components of the solution to time-invariant systems. Carrying this observation further, and developing general techniques for determining which geometric sequences or exponential functions occur in a given system, leads to a study of system eigenvalues and eigenvectors.

5.2 SYSTEM EIGENVECTORS

In this section it is shown that an eigenvector of the system matrix defines an independent first-order subsystem of the total system. It follows that if a complete set of n linearly independent eigenvectors can be found, the overall system can be decomposed into n separate first-order systems.

Consider the homogeneous discrete-time system

$$\mathbf{x}(k+1) = \mathbf{A}\mathbf{x}(k) \tag{5-3}$$

where $\mathbf{x}(k)$ is an n-dimensional state vector and $\mathbf{A}$ is an $n \times n$ matrix. Suppose the vector $\mathbf{e}_i$ is an eigenvector of $\mathbf{A}$ with associated eigenvalue λ_i. That is,

$$\mathbf{A}\mathbf{e}_i = \lambda_i \mathbf{e}_i \tag{5-4}$$

Suppose also that the initial state vector is set equal to a scalar multiple of this eigenvector; that is, $\mathbf{x}(0) = \alpha \mathbf{e}_i$. Then from (5–3) and (5-4) it follows that

$$\mathbf{x}(1) = \mathbf{A}\mathbf{x}(0) = \alpha \mathbf{A}\mathbf{e}_i = \lambda_i \alpha \mathbf{e}_i = \lambda_i \mathbf{x}(0)$$

Therefore, in this special case, the next state is just λ_i times the initial state. Furthermore, it is easy to see that all successive states are also various scalar multiples of the initial state.

The above observation shows that once the state is equal to a multiple of an eigenvector it remains a multiple of the eigenvector. Therefore, once this situation is achieved, the state can be characterized at any particular time by specifying the corresponding multiplying factor. That is, one writes

$$\mathbf{x}(k) = z_i(k)\mathbf{e}_i$$

where $z_i(k)$ is the appropriate scalar multiplying coefficient.

The multiplying coefficient itself satisfies a first-order difference equation. To see this, note that the system equation yields

$$\mathbf{x}(k+1) = \mathbf{A}\mathbf{x}(k) = \lambda_i z_i(k)\mathbf{e}_i$$

which, by the definition of $z_i(k+1)$, shows that $z_i(k+1)=\lambda_i z_i(k)$. These results are summarized by the following theorem.

Theorem. *If the state vector of a linear homogeneous constant coefficient system is initially aligned with an eigenvector, it continues to be aligned in subsequent time periods. The coefficient, $z_i(k)$, determining the multiple of the eigenvector at a time k, satisfies the first-order equation*

$$z_i(k+1)=\lambda_i z_i(k) \tag{5-5}$$

Example. Consider the second-order difference equation

$$2y(k+2)=3y(k+1)-y(k)$$

By defining

$$x_1(k)=y(k)$$
$$x_2(k)=y(k+1)$$

the original equation can be converted to the system

$$\begin{bmatrix} x_1(k+1) \\ x_2(k+1) \end{bmatrix} = \begin{bmatrix} 0 & 1 \\ -\frac{1}{2} & \frac{3}{2} \end{bmatrix} \begin{bmatrix} x_1(k) \\ x_2(k) \end{bmatrix}$$

One eigenvalue and associated eigenvector for this system is

$$\lambda_1 = \tfrac{1}{2}, \quad \mathbf{e}_1 = \begin{bmatrix} 1 \\ \frac{1}{2} \end{bmatrix}$$

If the state vector is initiated with x_1 and x_2 in the proportions 1 to $\frac{1}{2}$, the state vector will continue to have these same proportions for all k. The state vector will be reduced at each step by a multiplicative factor equal to $\lambda_1=\frac{1}{2}$.

In terms of the original difference equation, the system state is equal to this eigenvector if two successive $y(k)$'s are in the proportions 1 to $\frac{1}{2}$. If that condition holds for any k it will hold for all future k's. That means that each successive pair of $y(k)$'s will be in the proportions 1 to $\frac{1}{2}$, and hence that for every k, $y(k+1)=\frac{1}{2}y(k)$. In this case, therefore, the eigenvector translates immediately into a simple relation among successive values of the variable that defines the difference equation. (See Problem 3 for a generalization.)

5.3 DIAGONALIZATION OF A SYSTEM

Again consider the homogeneous system

$$\mathbf{x}(k+1)=\mathbf{A}\mathbf{x}(k) \tag{5-6}$$

Suppose that the system matrix $\mathbf{A}$ has a complete set of n linearly independent eigenvectors $\mathbf{e}_1, \mathbf{e}_2, \ldots, \mathbf{e}_n$ with corresponding eigenvalues $\lambda_1, \lambda_2, \ldots, \lambda_n$. The eigenvalues may or may not be distinct. We shall show how these n eigenvectors can be used to define n separate first-order systems. This procedure is sometimes of direct computational benefit, but it is perhaps most important as a conceptual aid.

Let an arbitrary value of the state $\mathbf{x}(k)$ be specified. Since there are n eigenvectors, this state can be expressed as a linear combination of the eigenvectors in the form

$$\mathbf{x}(k)=z_1(k)\mathbf{e}_1+z_2(k)\mathbf{e}_2+\cdots+z_n(k)\mathbf{e}_n \tag{5-7}$$

where $z_i(k)$, $i=1,2,\ldots,n$ are scalars. Using the fact that $\mathbf{A}\mathbf{e}_i=\lambda_i\mathbf{e}_i$, multiplication of (5-7) by the matrix $\mathbf{A}$ yields

$$\begin{aligned}\mathbf{x}(k+1)&=\mathbf{A}\mathbf{x}(k)\\&=\lambda_1 z_1(k)\mathbf{e}_1+\lambda_2 z_2(k)\mathbf{e}_2+\cdots+\lambda_n z_n(k)\mathbf{e}_n\end{aligned}$$

Therefore, expressing $\mathbf{x}(k+1)$ as a linear combination of eigenvectors in the form

$$\mathbf{x}(k+1)=z_1(k+1)\mathbf{e}_1+z_2(k+1)\mathbf{e}_2+\cdots+z_n(k+1)\mathbf{e}_n \tag{5-8}$$

we see that the scalar coefficients z_i satisfy the first-order equations

$$\begin{aligned}z_1(k+1)&=\lambda_1 z_1(k)\\z_2(k+1)&=\lambda_2 z_2(k)\\&\ \ \vdots\\z_n(k+1)&=\lambda_n z_n(k)\end{aligned} \tag{5-9}$$

The state vector, therefore, can be considered at each time instant to consist of a linear combination of the n eigenvectors. As time progresses, the weighting coefficients change (each independently of the others) so that the relative weights may change. Consequently, the system can be viewed as n separate first-order systems, each governing the coefficient of one eigenvector.

Change of Variable

The above analysis can be transformed directly into a convenient manipulative technique through the formal introduction of a change of variable. Let $\mathbf{M}$ be the modal matrix of $\mathbf{A}$. That is, $\mathbf{M}$ is the $n\times n$ matrix whose n columns are the eigenvectors of $\mathbf{A}$. For a given $\mathbf{x}(k)$, we define the vector $\mathbf{z}(k)$ by

$$\mathbf{x}(k)=\mathbf{M}\mathbf{z}(k) \tag{5-10}$$

This is, of course, just the vector representation of the earlier equation (5–7) with the components of the vector $\mathbf{z}(k)$ equal to the earlier $z_i(k)$'s. Substitution of this change of variable in the system equation yields

$$\mathbf{Mz}(k+1)=\mathbf{AMz}(k)$$

or, equivalently,

$$\mathbf{z}(k+1)=\mathbf{M}^{-1}\mathbf{AMz}(k) \tag{5-11}$$

This defines a new system that is related to the original system by a change of variable.

The new system matrix $\mathbf{M}^{-1}\mathbf{AM}$ is the system matrix corresponding to the system governing the $z_i(k)$'s as expressed earlier by (5-9). Accordingly, we may write $\mathbf{M}^{-1}\mathbf{AM}=\mathbf{\Lambda}$, where $\mathbf{\Lambda}$ is the diagonal matrix with the eigenvalues of $\mathbf{A}$ on the diagonal. The modal matrix $\mathbf{M}$ defines a new coordinate system in which $\mathbf{A}$ is represented by the diagonal matrix $\mathbf{\Lambda}$. (See Sect. 3.7.) When written out in detail (5-11) becomes

$$\begin{bmatrix} z_1(k+1) \\ z_2(k+1) \\ \\ \cdot \\ \cdot \\ \cdot \\ z_n(k+1) \end{bmatrix} = \begin{bmatrix} \lambda_1 & 0 & 0 & \dots & 0 \\ 0 & \lambda_2 & 0 & \dots & 0 \\ 0 & & \lambda_3 & & \\ \cdot & & \cdot & & \\ \cdot & & \cdot & & \\ \cdot & & \cdot & & \\ 0 & & & & \lambda_n \end{bmatrix} \begin{bmatrix} z_1(k) \\ z_2(k) \\ \\ \cdot \\ \cdot \\ \cdot \\ z_n(k) \end{bmatrix} \tag{5-12}$$

which explicitly displays the diagonal form obtained by the change of variable.

Calculation of $\mathbf{A}^k$

The state-transition matrix of a constant coefficient discrete-time system is $\mathbf{A}^k$. This matrix can be calculated easily by first converting $\mathbf{A}$ to diagonal form. The basic identity

$$\mathbf{M}^{-1}\mathbf{AM}=\mathbf{\Lambda} \tag{5-13}$$

can be rewritten as

$$\mathbf{A}=\mathbf{M\Lambda M}^{-1} \tag{5-14}$$

which provides a representation for $\mathbf{A}$ in terms of its eigenvalues and eigenvectors. It follows that

$$\begin{aligned}\mathbf{A}^2&=(\mathbf{M\Lambda M}^{-1})(\mathbf{M\Lambda M}^{-1})=\mathbf{M\Lambda}(\mathbf{M}^{-1}\mathbf{M})\mathbf{\Lambda M}^{-1}\\&=\mathbf{M\Lambda}^2\mathbf{M}^{-1}\end{aligned}$$

because $\mathbf{M}^{-1}\mathbf{M}=\mathbf{I}$. In a similar way it follows that for any $k\geq 0$

$$\mathbf{A}^k=\mathbf{M\Lambda}^k\mathbf{M}^{-1} \tag{5-15}$$

Therefore, calculation of $\mathbf{A}^k$ is transferred to the calculation of $\mathbf{\Lambda}^k$. However, since $\mathbf{\Lambda}$ is diagonal, one finds immediately that

$$\mathbf{\Lambda}^k = \begin{bmatrix} \lambda_1^k & 0 & 0 & \cdots & 0 \\ 0 & \lambda_2^k & & & \cdot \\ & & & & \cdot \\ 0 & & \lambda_3^k & & \cdot \\ \cdot & & & \cdot & \\ \cdot & & & & \\ \cdot & & & & \\ 0 & & & & \lambda_n^k \end{bmatrix} \tag{5-16}$$

Calculation of $\mathbf{A}^k$ for all k is thus accomplished easily once the transformation to diagonal form is determined. From the viewpoint of dynamic systems, the operation represented by (5-15) can be interpreted as one of first transforming to diagonal form, solving the system in those terms, and then transforming back.

As a result of this calculation, it is clear that when $\mathbf{A}^k$ is expressed as a function of k, each element is a linear combination of the geometric sequences λ_i^k, $i = 1, 2, \ldots, n$. This in turn is reflected into the form of solution to the original homogeneous system. It is made up of these same geometric sequences.

Continuous-Time Systems

Exactly the same sort of analysis can be applied to continuous-time systems. Suppose the system is governed by

$$\dot{\mathbf{x}}(t) = \mathbf{A}\mathbf{x}(t) \tag{5-17}$$

where $\mathbf{A}$ is an $n \times n$ matrix with n linearly independent eigenvectors. With $\mathbf{M}$ the modal matrix as before, the change of variable

$$\mathbf{x}(t) = \mathbf{M}\mathbf{z}(t) \tag{5-18}$$

transforms the system to

$$\dot{\mathbf{z}}(t) = \mathbf{M}^{-1}\mathbf{A}\mathbf{M}\mathbf{z}(t) \tag{5-19}$$

When written out in detail this is

$$\begin{bmatrix} \dot{z}_1(t) \\ \dot{z}_2(t) \\ \cdot \\ \cdot \\ \cdot \\ \dot{z}_n(t) \end{bmatrix} = \begin{bmatrix} \lambda_1 & 0 & \cdots & 0 \\ 0 & \lambda_2 & & 0 \\ & \cdot & & \cdot \\ & & \cdot & \cdot \\ & & & \cdot \\ 0 & 0 & \cdots & \lambda_n \end{bmatrix} \begin{bmatrix} z_1(t) \\ z_2(t) \\ \cdot \\ \cdot \\ \cdot \\ z_n(t) \end{bmatrix} \tag{5-20}$$

The state vector at any time is a linear combination of the n eigenvectors. In the continuous-time case, the coefficients $z_i(t)$ of the eigenvectors each satisfy a simple first-order differential equation. Hence, again the system can be considered to be n separate first-order systems.

Calculation of $e^{\mathbf{A}t}$

Calculation of the state-transition matrix $e^{\mathbf{A}t}$ is also greatly facilitated by diagonalization. It has already been observed that

$$\mathbf{A}^k = \mathbf{M}\boldsymbol{\Lambda}^k\mathbf{M}^{-1} \tag{5-21}$$

Therefore, it follows that the series

$$e^{\mathbf{A}t} = \mathbf{I} + \mathbf{A}t + \frac{\mathbf{A}^2t^2}{2!} + \cdots$$

can be written as

$$e^{\mathbf{A}t} = \mathbf{I} + \mathbf{M}\boldsymbol{\Lambda}\mathbf{M}^{-1}t + \mathbf{M}\boldsymbol{\Lambda}^2\mathbf{M}^{-1}\frac{t^2}{2!} + \cdots \tag{5-22}$$

Factoring out $\mathbf{M}$ and $\mathbf{M}^{-1}$ produces

$$\begin{aligned} e^{\mathbf{A}t} &= \mathbf{M}\left(\mathbf{I} + \boldsymbol{\Lambda}t + \frac{\boldsymbol{\Lambda}^2t^2}{2!} + \cdots\right)\mathbf{M}^{-1} \\ &= \mathbf{M}e^{\boldsymbol{\Lambda}t}\mathbf{M}^{-1} \end{aligned} \tag{5-23}$$

Therefore, calculation of $e^{\mathbf{A}t}$ is transferred to the calculation of $e^{\boldsymbol{\Lambda}t}$. However, since $\boldsymbol{\Lambda}$ is a diagonal matrix, each matrix $\boldsymbol{\Lambda}^k$ is also diagonal, with ith diagonal element λ_i^k. Therefore, $e^{\boldsymbol{\Lambda}t}$ is also diagonal, with ith diagonal element $1 + \lambda_i t + \cdots$. Thus,

$$e^{\boldsymbol{\Lambda}t} = \begin{bmatrix} e^{\lambda_1 t} & 0 & & 0 \\ 0 & e^{\lambda_2 t} & & \cdot \\ \cdot & \cdot & & \\ \cdot & & \cdot & \cdot \\ \cdot & & & \cdot \\ 0 & & & e^{\lambda_n t} \end{bmatrix}$$

This can be substituted into (5-23) to obtain an explicit expression for $e^{\mathbf{A}t}$.

Example. Consider the system

$$\dot{\mathbf{x}}(t) = \begin{bmatrix} 2 & 1 \\ 2 & 3 \end{bmatrix}\mathbf{x}(t)$$

The system matrix has eigenvalues $\lambda_1 = 1$, $\lambda_2 = 4$ and corresponding modal matrix

$$\mathbf{M} = \begin{bmatrix} 1 & 1 \\ -1 & 2 \end{bmatrix}, \qquad \mathbf{M}^{-1} = \frac{1}{3}\begin{bmatrix} 2 & -1 \\ 1 & 1 \end{bmatrix}$$

as shown in Chapter 3 (Example 1, Sect. 3.6).

Introduction of the change of variable

$$z_1 = \tfrac{1}{3}(2x_1 - x_2)$$
$$z_2 = \tfrac{1}{3}(x_1 + x_2)$$

leads directly to the differential equations

$$\dot{z}_1(t) = z_1(t)$$
$$\dot{z}_2(t) = 4z_2(t)$$

which is the diagonal form.

The state-transition matrix of the original system can be computed as

$$e^{\mathbf{A}t} = \mathbf{M}\begin{bmatrix} e^t & 0 \\ 0 & e^{4t} \end{bmatrix}\mathbf{M}^{-1} = \frac{1}{3}\begin{bmatrix} 2e^t + e^{4t} & e^{4t} - e^t \\ 2e^{4t} - 2e^t & e^t + 2e^{4t} \end{bmatrix}$$

Diagram Interpretation

The diagram interpretation of the diagonalization process is straightforward and useful. When expressed in the new coordinates (with components z_i) the diagram of the system breaks apart into separate systems. The result is illustrated in Fig. 5.3 for discrete-time systems, but exactly the same diagram applies in continuous time with delays replaced by integrators. The z_i's are the coefficients of the various eigenvectors as they combine to produce the state vector. The eigenvectors themselves do not show up explicitly in this diagram, although they must be used to obtain it.

Finally, it should be emphasized that the role of the diagonalization process is at least as much conceptual as it is computational. Although calculation of the state-transition matrix can be facilitated if the eigenvectors are known, the problem of computing the eigenvalues and eigenvectors for a large system is itself a formidable task. Often this form of detailed analysis is not justified by the scope of the motivating study. Indeed, when restricted to numerical methods it is usually simplest to evaluate a few particular solutions directly by recursion. A full collection of eigenvectors in numerical form is not always very illuminating.

On the other hand, from a conceptual viewpoint, the diagonalization process is invaluable, for it reveals an underlying simplicity of linear systems. Armed with this concept, we know, when faced with what appears to be a

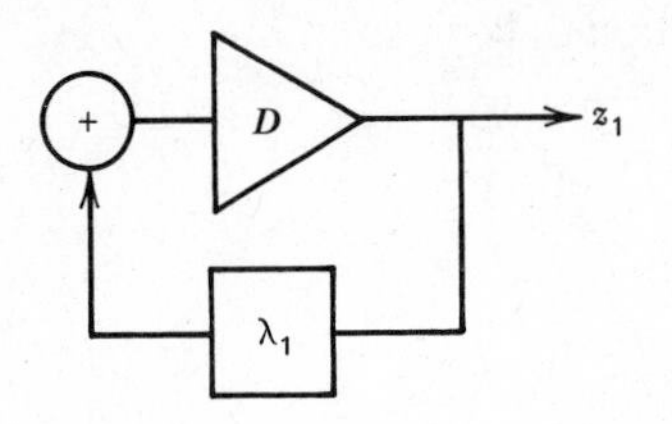

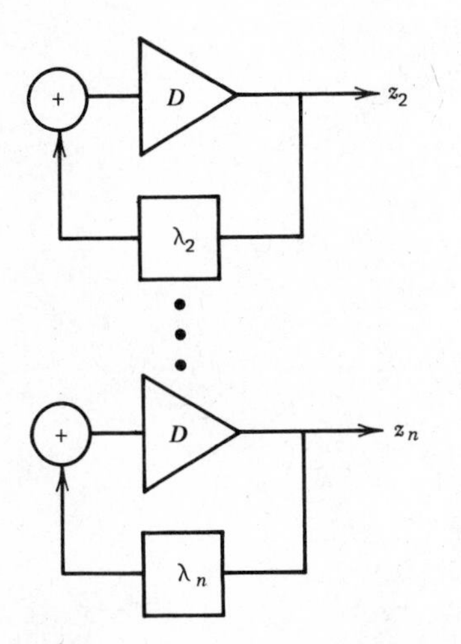

Figure 5.3. Diagonal diagram.

complex interconnected system, that there is a way to look at it, through a kind of distorted lenses which changes variables, so that it appears simply as a collection of first-order systems. Even if we never find the diagonalizing transformation, the knowledge that one exists profoundly influences our perception of a system and enriches our analysis methodology.

5.4 DYNAMICS OF RIGHT AND LEFT EIGENVECTORS

The diagonalization of a system as discussed in Sect. 5.3 can be clarified further through the relation of right and left eigenvectors. When referring simply to "eigenvector" we have meant a *right eigenvector* defined by

$$\mathbf{A}\mathbf{e}_i = \lambda_i \mathbf{e}_i \tag{5-24}$$

However, there are also *left eigenvectors* defined by

$$\mathbf{f}_i^T \mathbf{A} = \lambda_i \mathbf{f}_i^T \tag{5-25}$$

The two are dual concepts that play together in the diagonalization process.

In terms of a dynamic system of the form

$$\mathbf{x}(k+1)=\mathbf{A}\mathbf{x}(k) \tag{5-26}$$

the right and left eigenvectors have distinct interpretations. The *right eigenvector* is most naturally regarded as a vector in the state space. If the state is set equal to a right eigenvector then successive states will be scalar multiples of this eigenvector. Thus, it is proper to regard a right eigenvector as a special value of the state. And, in the diagonalizable case, any value of the state vector is a linear combination of the various eigenvectors.

A *left eigenvector* is more naturally regarded as a scalar-valued function of the state, rather than as a vector in the state space. A left eigenvector $\mathbf{f}^T$ defines the scalar function

$$z(k)=\mathbf{f}^T\mathbf{x}(k) \tag{5-27}$$

It associates a scalar with each value of the state vector. If, for example, $\mathbf{f}^T$ were the vector $[1\,0\,0\ldots0]$, then the corresponding function would be $z(k)=x_1(k)$; that is, it would be the value of the first component of $\mathbf{x}(k)$. If, as another example, $\mathbf{f}^T$ were the vector $[1\,1\ldots1]$, the corresponding function would be $z(k)=x_1(k)+x_2(k)+\cdots+x_n(k)$; that is, it would be the sum of the components of $\mathbf{x}(k)$. In general, a left eigenvector defines a certain linear combination of the components of the state vector. As the state vector evolves in time, the associated value of the linear combination also evolves.

Suppose $\mathbf{f}_i^T$ is a left eigenvector with corresponding eigenvalue λ_i. From the system equation

$$\mathbf{x}(k+1)=\mathbf{A}\mathbf{x}(k) \tag{5-28}$$

it follows that for the corresponding z_i

$$z_i(k+1)=\mathbf{f}_i^T\mathbf{x}(k+1)=\mathbf{f}_i^T\mathbf{A}\mathbf{x}(k)=\lambda_i\mathbf{f}_i^T\mathbf{x}(k)=\lambda_i z_i(k) \tag{5-29}$$

Thus, the associated scalar function of the state satisfies a first-order difference equation.

If $\mathbf{A}$ has distinct eigenvalues then (as shown in Sect. 3.8) the right and left eigenvalues satisfy the biorthogonality relation

$$\mathbf{f}_i^T\mathbf{e}_j=0 \tag{5-30}$$

for all $i\neq j$. In this case it is natural to normalize the left eigenvectors with respect to the right eigenvectors so that

$$\mathbf{f}_i^T\mathbf{e}_i=1 \tag{5-31}$$

for each $i=1,2,\ldots,n$. With this normalization the $z_i(k)$'s defined above are exactly the same as the $z_i(k)$'s that serve as the coefficients in the eigenvector expansion of the state. Specifically, for each k,

$$\mathbf{x}(k) = z_1(k)\mathbf{e}_1 + z_2(k)\mathbf{e}_2 + \cdots + z_n(k)\mathbf{e}_n \tag{5-32}$$

To verify this, we multiply both sides by $\mathbf{f}_i^T$ to obtain

$$\begin{aligned}\mathbf{f}_i^T\mathbf{x}(k) &= z_1(k)\mathbf{f}_i^T\mathbf{e}_1 + z_2(k)\mathbf{f}_i^T\mathbf{e}_2 + \cdots + z_n(k)\mathbf{f}_i^T\mathbf{e}_n \\ &= z_i(k)\mathbf{f}_i^T\mathbf{e}_i\end{aligned} \tag{5-33}$$

where the second equality follows from the biorthogonality relation (5-30). In view of the normalization (5-31), the above reduces to

$$\mathbf{f}_i^T\mathbf{x}(k) = z_i(k) \tag{5-34}$$

which coincides with the original definition of $z_i(k)$.

The interpretation of the two types of eigenvectors should be visualized simultaneously. The right eigenvectors define special directions in the state space. Once the state vector points in one of these directions, it continues to point in the same direction, although its magnitude may change. In the case of distinct eigenvalues, the state vector can always be expressed as a linear combination of the various right eigenvectors—the various weighting coefficients each changing with time. Each left eigenvector, on the other hand, defines a system variable that behaves according to a first-order equation. The two concepts are intimately related. If the state points in the direction of a right eigenvector, all the variables defined by the left eigenvectors of different eigenvalues are zero. More generally, the weighting coefficients of the various eigenvectors that make up the state are the variables defined by the corresponding left eigenvectors. These coefficients are each governed by a first-order equation, and as their values change the state vector moves correspondingly.

The migration example of the next section illustrates these concepts. It shows that within the context of a given application the right and left eigenvectors can have a strong intuitive meaning. This example should help clarify the abstract relations presented in this section.

5.5 EXAMPLE: A SIMPLE MIGRATION MODEL

Assume that the population of a country is divided into two distinct segments: rural and urban. The natural yearly growth factors, due to procreation, in both segments are assumed to be identical and equal to α (that is, the population at year $k+1$ would be α times the population at year k). The population distribution, however, is modified by migration between the rural and urban segments. The rate of this migration is influenced by the need for a base of rural activity that is adequate to support the total population of the country—the optimal rural base being a given fraction γ of the total population. The yearly level of migration itself, from rural to urban areas, is proportional to the excess of rural population over the optimal rural base.

If the rural and urban populations at year k are denoted by $r(k)$ and $u(k)$, respectively, then the total population is $r(k)+u(k)$, the optimal rural base is $\gamma[r(k)+u(k)]$, and thus the excess rural population is $\{r(k)-\gamma[r(k)+u(k)]\}$. A simple dynamic model of the migration process, based on the above assumptions, is then

$$r(k+1)=\alpha r(k)-\beta\{r(k)-\gamma[r(k)+u(k)]\}$$
$$u(k+1)=\alpha u(k)+\beta\{r(k)-\gamma[r(k)+u(k)]\}$$

In this model, the growth factor α is positive (and usually greater than unity). The migration factor β is positive, and is assumed to be less than α. The parameter γ is the ideal fraction of the total population that would be rural in order to support the total population. This parameter is a measure of rural productivity. Each of these parameters might normally change with time, but they are assumed constant for purposes of this example.

The model can be easily put in the state vector form

$$\mathbf{x}(k+1)=\mathbf{A}\mathbf{x}(k)$$

where

$$\mathbf{A}=\begin{bmatrix}\alpha-\beta(1-\gamma) & \beta\gamma \\ \beta(1-\gamma) & \alpha-\beta\gamma\end{bmatrix}$$

and

$$\mathbf{x}(k)=\begin{bmatrix}r(k)\\ u(k)\end{bmatrix}$$

At this point one might normally proceed by writing the characteristic polynomial as a first step toward finding the eigenvalues and eigenvectors of the matrix $\mathbf{A}$, but in this case at least some of that information can be deduced by simple reasoning. Because the natural growth rates of both regions are identical, it is clear that the total population grows at the common rate. Migration simply redistributes the population; it does not influence the overall growth. Therefore, we expect that the growth factor α is one eigenvalue of $\mathbf{A}$ and that the row vector $\mathbf{f}_1^T=[1\ 1]$ is a *left* eigenvector of $\mathbf{A}$, because $\mathbf{f}_1^T\mathbf{x}(k)=r(k)+u(k)$ is the total population. Indeed, checking this mathematically, we find

$$[1\ \ 1]\begin{bmatrix}\alpha-\beta(1-\gamma) & \beta\gamma \\ \beta(1-\gamma) & \alpha-\beta\gamma\end{bmatrix}=\alpha[1\ \ 1]$$

which verifies our conjecture. This left eigenvector tells us what variable within the system (total population in this case) always grows by the factor α.

The corresponding *right* eigenvector of $\mathbf{A}$ defines the distribution of population that is necessary in order for *all* variables to grow by the factor α. Again, this vector can be deduced by inspection. In order for *both* rural and urban population to grow by the factor α it is necessary that there be no net migration. Therefore, it is necessary that $r(k)=\gamma[r(k)+u(k)]$; or, equivalently, that rural and urban population be in the balanced proportions $\gamma:(1-\gamma)$. Therefore, we expect that

$$\mathbf{e}_1=\begin{bmatrix}\gamma\\1-\gamma\end{bmatrix}$$

is a *right* eigenvector of $\mathbf{A}$ corresponding to the eigenvalue α. Again, a simple check verifies this conjecture.

By exploiting the simple observation about total population growth, we have deduced one eigenvalue together with its corresponding left and right eigenvectors. We can find the remaining eigenvalue and its corresponding left and right eigenvectors by use of the biorthogonality relations that hold among right and left eigenvectors.

The second *right* eigenvector must be orthogonal to the first *left* eigenvector. Thus,

$$\mathbf{e}_2=\begin{bmatrix}1\\-1\end{bmatrix}$$

Multiplying this by $\mathbf{A}$ verifies that it is an eigenvector and shows that the corresponding eigenvalue is $\alpha-\beta$. It represents the population distribution corresponding to zero total population. This eigenvector has both positive and negative components and hence is not a physically realizable state since population must be nonnegative. However, this eigenvector (or a multiple of it) will in general be a component of the actual population vector.

To determine what variable in the system grows by the factor $\alpha-\beta$ it is necessary to find the second *left* eigenvector. It will be orthogonal to the first *right* eigenvector and hence

$$\mathbf{f}_2^T=[1-\gamma \quad -\gamma]$$

This left eigenvector corresponds to the net rural population imbalance, since $\mathbf{f}_2^T\mathbf{x}(k)=r(k)-\gamma[r(k)+u(k)]$.

The various eigenvectors associated with this system together with their interpretations and interrelations are summarized in Table 5.1. A simple diagram in diagonal form summarizing the analysis is shown in Fig. 5.4.

The above analysis can be translated into a fairly complete verbal description of the general behavior of the migration model. Overall population grows by the factor α each year. If there is an initial imbalance of rural versus urban population, with say more than a fraction γ of the population in the rural

Table 5.1. Eigenvectors for Migration Example

Left Eigenvector	Right Eigenvector
$\mathbf{f}_1^T = [1\ 1]$	$\mathbf{e}_1 = \begin{bmatrix} \gamma \\ 1-\gamma \end{bmatrix}$
corresponds to total population	corresponds to the condition of zero rural imbalance
$\mathbf{f}_2^T = [1-\gamma\ \ -\gamma]$	$\mathbf{e}_2 = \begin{bmatrix} 1 \\ -1 \end{bmatrix}$
corresponds to net rural imbalance	corresponds to the condition of zero population

sector, then there is migration, and it grows by the factor $\alpha - \beta$. The rural imbalance also grows by this factor. Since $\alpha > \beta > 0$, the growth factor of rural imbalance is always less than that of population, so eventually the relative imbalance (that is, the ratio of imbalance to total population) tends to disappear. If $\beta > \alpha$, then $\alpha - \beta < 0$ and migration oscillates, being from rural to urban one year and from urban to rural the next. If $\beta < \alpha$, migration is in the same direction each year.

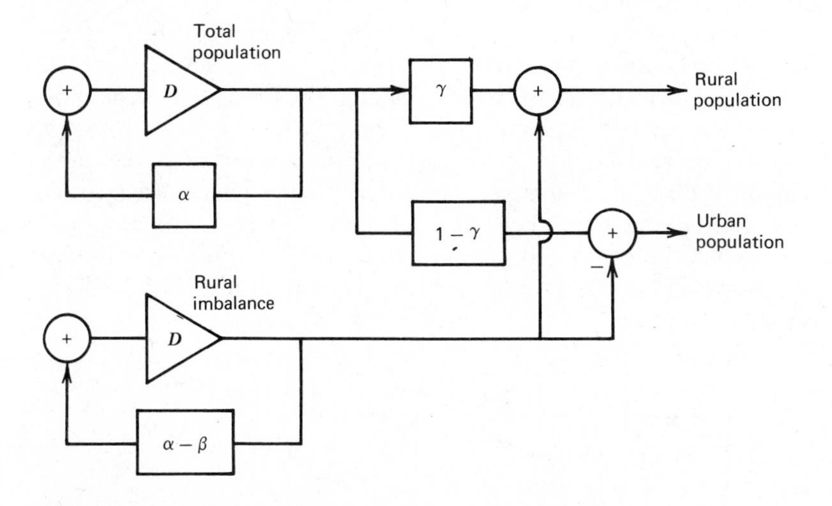

Figure 5.4. Diagonal form for migration example.

5.6 MULTIPLE EIGENVALUES

If the system matrix **A** has repeated roots of its characteristic polynomial corresponding to the presence of multiple eigenvalues, it may not be possible to find a complete set of eigenvectors. If this is the case, the matrix **A** cannot be diagonalized, and the system cannot be reduced to a set of uncoupled first-order systems. The characteristic response of such systems is somewhat different than that of systems which can be diagonalized.

The situation with multiple eigenvalues is typified by the system

$$\begin{bmatrix} x_1(k+1) \\ x_2(k+1) \end{bmatrix} = \begin{bmatrix} a & 1 \\ 0 & a \end{bmatrix} \begin{bmatrix} x_1(k) \\ x_2(k) \end{bmatrix} \tag{5-35}$$

As determined by the characteristic polynomial, this system has both eigenvalues equal to a, but there is only one (linearly independent) eigenvector. The state-transition matrix can be computed directly in this case, and is found to be

$$\mathbf{\Phi}(k) = \mathbf{A}^k = \begin{bmatrix} a^k & ka^{k-1} \\ 0 & a^k \end{bmatrix} \tag{5-36}$$

The response due to an arbitrary initial condition will therefore generally contain two kinds of terms: those of the form a^k and those of the form ka^{k-1}.

This is a general conclusion for the multiple root case. If there is not a full set of eigenvectors, the system response will contain terms involving powers of k times the normal geometric terms associated with the eigenvalues. First there will be terms of the ka^{k-1}, and then there may be terms of the form k^2a^{k-2}, and so forth. The exact number of such terms required is related to the Jordan form of the matrix **A** (see Chapter 3, Sect. 9).

One way to visualize the repeated root situation is in terms of the diagram of Fig. 5.5 that corresponds to the system (5-35). The system can be regarded as two first-order systems with one serving as input to the other. The system on the left generates a geometric sequence of the form a^k (times a constant that is determined by the initial condition). This sequence in turn serves as the input to the system on the right. If this second system had a characteristic root different from that of the first, then, as shown in Sect. 5.1, the response generated at x_1 would be composed of two geometric series corresponding to the two roots. However, in this case the output will contain terms of the form ka^{k-1} as well as a^k.

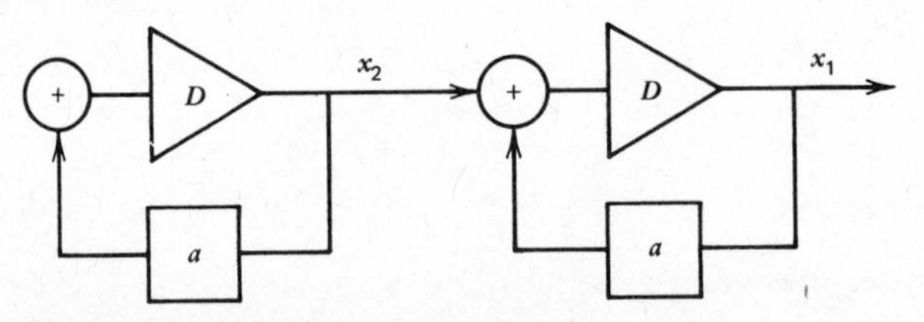

Figure 5.5. Multiple-root example.

The general situation can be described in this same way. The system may contain (either directly or after a change of variable) some chains of first-order systems with a common root, the output of one serving as the input to another. If the maximum length chain is only one, the response associated with this root will consist only of terms of the form a^k. If the maximum length is two, terms of the form a^k and ka^{k-1} will appear, and so forth for longer chains. The highest-order term of this kind that could ever occur in an nth-order system is $k^{n-1}a^{k-n+1}$

Perturbation Analysis

A system with multiple roots can be perturbed, by introducing a slight change in some coefficients, to produce a system with distinct roots. Indeed, the original system can be regarded as the limit of a collection of systems with distinct roots. For example, consider the collection of systems

$$\begin{bmatrix} x_1(k+1) \\ x_2(k+1) \end{bmatrix} = \begin{bmatrix} a & 1 \\ 0 & b \end{bmatrix} \begin{bmatrix} x_1(k) \\ x_2(k) \end{bmatrix} \tag{5-37}$$

with $b \neq a$. The original system (5-35) is obtained as the limiting case where $b = a$. For $b \neq a$ the system has distinct eigenvalues a, b and therefore can be diagonalized. The modal matrix $\mathbf{M}$ and its inverse are easily found to be

$$\mathbf{M} = \begin{bmatrix} 1 & 1 \\ 0 & b-a \end{bmatrix} \qquad \mathbf{M}^{-1} = \begin{bmatrix} 1 & 1/(a-b) \\ 0 & 1/(b-a) \end{bmatrix}$$

and $\mathbf{M}^{-1}\mathbf{A}\mathbf{M} = \mathbf{\Lambda}$, where

$$\mathbf{\Lambda} = \begin{bmatrix} a & 0 \\ 0 & b \end{bmatrix}$$

It follows that $\mathbf{A}^k = \mathbf{M}\mathbf{\Lambda}^k\mathbf{M}^{-1}$, which when written out is

$$\mathbf{A}^k = \begin{bmatrix} 1 & 1 \\ 0 & b-a \end{bmatrix} \begin{bmatrix} a^k & 0 \\ 0 & b^k \end{bmatrix} \begin{bmatrix} 1 & 1/(a-b) \\ 0 & 1/(b-a) \end{bmatrix}$$

or, finally,

$$\mathbf{A}^k = \begin{bmatrix} a^k & (b^k - a^k)/(b-a) \\ 0 & b^k \end{bmatrix} \tag{5-38}$$

When $a \neq b$, the right-hand side of (5-38) is well defined and it shows explicitly that all terms are combinations of a^k and b^k. The value of the right-hand side of (5-38) in the limiting case of $b = a$ can be found from the identity

$$b^k - a^k = (b-a)(b^{k-1} + b^{k-2}a + \cdots + ba^{k-2} + a^{k-1})$$

In the limit as $b \to a$ this yields

$$\frac{b^k - a^k}{b-a} \to ka^{k-1}$$

Therefore, when $b=a$

$$\mathbf{A}^k=\begin{bmatrix} a^k & ka^{k-1} \\ 0 & a^k \end{bmatrix}$$

which agrees with our earlier calculation.

Continuous-Time Systems

The analysis is virtually identical for a continuous-time system defined by

$$\dot{\mathbf{x}}(t)=\mathbf{A}\mathbf{x}(t)$$

For example, if as before,

$$\mathbf{A}=\begin{bmatrix} a & 1 \\ 0 & a \end{bmatrix}$$

we find that

$$e^{\mathbf{A}t}=\begin{bmatrix} e^{at} & te^{at} \\ 0 & e^{at} \end{bmatrix}$$

Therefore, in addition to the expected exponential term e^{at}, the response contains terms of the form te^{at}. Longer chains of a common eigenvalue produce additional terms of the form t^2e^{at}, t^3e^{at}, and so forth.

5.7 EQUILIBRIUM POINTS

In many situations the natural rest points of a dynamic system are as much of interest as the mechanisms of change. Accordingly, we introduce the following definition.

Definition. A vector $\bar{\mathbf{x}}$ is an *equilibrium point* of a dynamic system if it has the property that once the system state vector is equal to $\bar{\mathbf{x}}$ it remains equal to $\bar{\mathbf{x}}$ for all future time.

Thus, an equilibrium point is just what the term implies. It is a point where the state vector is in equilibrium and does not move. It is a general definition, applying to discrete- and continuous-time systems, and to nonlinear as well as linear systems. Our present interest, of course, is in linear constant-coefficient systems.

Homogeneous Discrete-Time Systems

The homogeneous system

$$\mathbf{x}(k+1)=\mathbf{A}\mathbf{x}(k) \tag{5-39}$$

always has the origin (the point $\bar{\mathbf{x}}=\mathbf{0}$) as an equilibrium point. Once the state is equal to $\mathbf{0}$ it will not change.

In some cases there are other equilibrium points. An equilibrium point must satisfy the condition

$$\bar{\mathbf{x}}=\mathbf{A}\bar{\mathbf{x}} \tag{5-40}$$

and this condition is identical to the statement that $\bar{\mathbf{x}}$ is an eigenvector of $\mathbf{A}$ with corresponding eigenvalue equal to unity. Therefore, if unity is an eigenvalue of $\mathbf{A}$, any corresponding eigenvector is an equilibrium point. If unity is not an eigenvalue of $\mathbf{A}$, the origin is the only equilibrium point of system (5-39).

Nonhomogeneous Discrete-Time Systems

We now consider linear systems that have constant coefficients and a constant input term: specifically, systems of the form

$$\mathbf{x}(k+1)=\mathbf{A}\mathbf{x}(k)+\mathbf{b} \tag{5-41}$$

An equilibrium point of such a system must satisfy the equation

$$\bar{\mathbf{x}}=\mathbf{A}\bar{\mathbf{x}}+\mathbf{b} \tag{5-42}$$

If unity is *not* an eigenvalue of $\mathbf{A}$, then the matrix $\mathbf{I}-\mathbf{A}$ is nonsingular, and there is a unique solution

$$\bar{\mathbf{x}}=[\mathbf{I}-\mathbf{A}]^{-1}\mathbf{b} \tag{5-43}$$

If unity is an eigenvalue there may be no equilibrium point or an infinity of such points, depending on whether or not (5-42) represents a consistent set of equations. In most cases of interest, there is a unique equilibrium point given by (5-43).

Continuous-Time Systems

For continuous-time systems the situation is similar. The homogeneous system

$$\dot{\mathbf{x}}(t)=\mathbf{A}\mathbf{x}(t) \tag{5-44}$$

always has the origin as an equilibrium point. If the matrix $\mathbf{A}$ is singular (corresponding to $\mathbf{A}$ having zero as an eigenvalue), then there are other equilibrium points. If, on the other hand, $\mathbf{A}$ is nonsingular, then the origin is the unique equilibrium point of (5-44).

For the system with constant input

$$\dot{\mathbf{x}}(t)=\mathbf{A}\mathbf{x}(t)+\mathbf{b} \tag{5-45}$$

an equilibrium point $\bar{\mathbf{x}}$ must satisfy the equation

$$\mathbf{0} = \mathbf{A}\bar{\mathbf{x}} + \mathbf{b} \tag{5-46}$$

Thus, if $\mathbf{A}$ is nonsingular, there is a unique equilibrium point

$$\bar{\mathbf{x}} = -\mathbf{A}^{-1}\mathbf{b}$$

If $\mathbf{A}$ is singular there may or may not be equilibrium points.

We see then that, for both discrete- and continuous-time systems, zero is always an equilibrium point of a homogeneous system. An eigenvalue of unity is critical in the general determination of equilibrium points for discrete-time systems. And an eigenvalue of zero is critical in the general determination of equilibrium points for continuous-time systems. In either case, significant information can often be obtained by determining how an equilibrium point changes if various parameters in the system are changed. The example of the next section illustrates this point.

5.8 EXAMPLE: SURVIVAL THEORY IN CULTURE

McPhee has argued that "survival of the fittest" applies to cultures, creative ideas, and esthetics as well as to biological species. Through audience testing with an associated box office criterion, or through passage from generation to generation through apprenticeship and teaching, every cultural form is subject to screening that favors the best material.

There are seemingly obvious disparities in the quality mix of different cultures associated with different cultural media. Classical music and classical literature seem to be of higher caliber than, say, television programming. Established scientific theories seem to be of higher caliber than motion pictures. These differences might initially be thought to be influenced primarily by the differences in creative talent applied to the media, or by differences in audience discrimination. But, in fact, it is argued here that the observed quality is most profoundly influenced by the inherent structure of a media and its screening mechanism. Mass cultures distributed by mass media have a severe disadvantage when viewed from this perspective, as compared with cultures that are distributed on a more individualized basis.

We present McPhee's elementary theory that captures the essence of the survival argument. The reader should be able to see how the primitive version can be modified to incorporate additional factors.

The basic characteristic of any cultural media is that of repetitive screening. The total offering at any one time, or during any one season, consists of a mix of new material and older material that has survived previous screenings. In turn, not all of the current offering will survive to the following season. The chances for survival are, at least to some extent, related to quality. Typically, however, the screening process is not perfect. Some poor quality material survives, when perhaps it should not, and some high quality material is lost. But on an average basis, the chances for survival are usually better for high quality material than for low. The proportion of good quality material in any season is determined by the dynamics of the imperfect screening process.

To be specific, suppose that within a given culture it is possible to distinguish three distinct quality levels—1, 2, and 3—with 1 being the best, 2 the second best, and 3 the worst. Assume that the proportion of the ith quality material offered in a season that survives to the next season is α_i. Clearly, by definition, $0 \le \alpha_i \le 1$ for each i.

Now suppose an amount b_i of new material of the ith level is introduced at the beginning of each season, $i = 1, 2, 3$. The total amount of ith level material offered in a season is, therefore, the sum of that which survived from the past season and the new material. In other words, denoting the amount of the ith level material presented in season k by $x_i(k)$ the media can be described by

$$x_i(k+1) = \alpha_i x_i(k) + b_i \tag{5-47}$$

The overall mechanism can be described in matrix form as

$$\begin{bmatrix} x_1(k+1) \\ x_2(k+1) \\ x_3(k+1) \end{bmatrix} = \begin{bmatrix} \alpha_1 & 0 & 0 \\ 0 & \alpha_2 & 0 \\ 0 & 0 & \alpha_3 \end{bmatrix} \begin{bmatrix} x_1(k) \\ x_2(k) \\ x_3(k) \end{bmatrix} + \begin{bmatrix} b_1 \\ b_2 \\ b_3 \end{bmatrix} \tag{5-48}$$

Because the system matrix is diagonal, the system itself can be regarded as three separate first-order systems operating simultaneously.

Let us determine the equilibrium quality make-up of this system. Because of the diagonal structure, the three equilibrium values can be found separately. They are

$$\begin{aligned} x_1 &= \frac{b_1}{1-\alpha_1} \\ x_2 &= \frac{b_2}{1-\alpha_2} \\ x_3 &= \frac{b_3}{1-\alpha_3} \end{aligned} \tag{5-49}$$

Some interesting conclusions can be drawn from this simple result. First, as expected, the proportion of highest quality material is enhanced in the limiting mixture as compared to the input composition. The screening process has a positive effect. Second, and most important, as the α_i's increase, the effectiveness of the screening process is enhanced. Suppose, for example, that $\alpha_1 = 2\alpha_2$, meaning that the chance of first quality material surviving is twice that of the second quality grade. At $\alpha_1 = .2$, $\alpha_2 = .1$, long-term screening factors are

$$\frac{1}{1-\alpha_1} = 1.25 \qquad \frac{1}{1-\alpha_2} = 1.11$$

While for $\alpha_1 = .8$, $\alpha_2 = .4$ they are 5.00 and 1.67, respectively. We conclude, therefore, that the benefit of screening is greatest when the chances for survival of all material is relatively high.

This is in accord with our general observations of cultural media. Television programming is of relatively poor quality because the overall survival rate is low; only a small percentage of programs survive even a single season. Therefore, the composition almost identically corresponds to the input composition. On the other hand, the overall survival rate in symphonic music is high, the offerings changing little from season to season. The composition, therefore, is more a reflection of the screening discrimination than of current creative talent.

Unfortunately, the answer for improving quality does not spring directly from this analysis. Although the low survival rate in mass media is in a sense responsible for low overall quality, we cannot advocate increasing the survival rate. The nature of the media itself makes that inappropriate. The particular medium used to distribute a culture is, to a large extent, dictated by the characteristics of the culture itself. Nevertheless, this analysis does yield insight as to the underlying processes that inevitably lead to quality characteristics.

5.9 STABILITY

The term *stability* is of such common usage that its meaning, at least in general terms, is well known. In the setting of dynamic systems, stability is defined with respect to a given equilibrium point. An equilibrium point is stable if when the state vector is moved slightly away from that point, it tends to return to it, or at least does not keep moving further away. The classic example is that of a stick perfectly aligned with the vertical. If the stick is balanced on its bottom end, it is in unstable equilibrium. If on the other hand it is hanging from a support at the top end, it is in stable equilibrium.

Here we consider stability of equilibrium points corresponding to linear time-invariant systems of the form

$$\mathbf{x}(k+1) = \mathbf{A}\mathbf{x}(k) + \mathbf{b} \tag{5-50}$$

or

$$\dot{\mathbf{x}}(t) = \mathbf{A}\mathbf{x}(t) + \mathbf{b} \tag{5-51}$$

In order to make the discussion precise, the following definition is introduced.

Definition. An equilibrium point $\bar{\mathbf{x}}$ of a linear time-invariant system (5-50) or

(5-51) is *asymptotically stable* if for any initial condition the state vector tends to $\bar{\mathbf{x}}$ as time increases. The point is *unstable* if for some initial condition the corresponding state vector tends toward infinity.

It is important to observe that stability issues for (5-50) and (5-51) are tied directly to the corresponding homogeneous equations. Suppose, for example, that $\bar{\mathbf{x}}$ is an equilibrium point of (5-50). Then we have

$$\mathbf{x}(k+1)-\bar{\mathbf{x}}=\mathbf{A}\mathbf{x}(k)-\mathbf{A}\bar{\mathbf{x}}+\mathbf{b}-\mathbf{b}$$

and thus

$$\mathbf{x}(k+1)-\bar{\mathbf{x}}=\mathbf{A}(\mathbf{x}(k)-\bar{\mathbf{x}})$$

It is clear that the condition for $\mathbf{x}(k)$ to tend to $\bar{\mathbf{x}}$ in (5-50) is identical to that for $\mathbf{z}(k)$ to tend to $\mathbf{0}$ in the homogeneous system

$$\mathbf{z}(k+1)=\mathbf{A}\mathbf{z}(k)$$

(A similar argument holds in continuous time.) Therefore, in the case of the linear systems (5-50) and (5-51), asymptotic stability or instability does not depend explicitly on the equilibrium point, but instead is determined by the properties of the homogeneous equation.

Another way to deduce the above conclusion is this. The complete solution to, say, (5-50) consists of a constant $\bar{\mathbf{x}}$ (a particular solution) and a solution to the homogeneous equation. Asymptotic stability holds if every homogeneous solution tends to zero.

The character of the solutions to the homogeneous equation is determined by the eigenvalues of the matrix $\mathbf{A}$. The discrete- and continuous-time cases are considered separately below.

Discrete-Time Systems

Consider first the system

$$\mathbf{x}(k+1)=\mathbf{A}\mathbf{x}(k) \tag{5-52}$$

To obtain conditions for asymptotic stability assume initially that $\mathbf{A}$ can be diagonalized. Then there is a matrix $\mathbf{M}$ such that

$$\mathbf{A}=\mathbf{M}\mathbf{\Lambda}\mathbf{M}^{-1}$$

where

$$\mathbf{\Lambda}=\begin{bmatrix} \lambda_1 & & & & \\ & \lambda_2 & & & \\ & & \cdot & & \\ & & & \cdot & \\ & & & & \cdot \\ & & & & & \lambda_n \end{bmatrix}$$

and the λ_i's are the eigenvalues of $\mathbf{A}$. Furthermore,

$$\mathbf{A}^k = \mathbf{M}\mathbf{\Lambda}^k\mathbf{M}^{-1} = \mathbf{M}\begin{bmatrix} \lambda_1^k & & & & \\ & \cdot & & & \\ & & \lambda_2^k & & \\ & & & \cdot & \\ & & & & \cdot \\ & & & & & \lambda_n^k \end{bmatrix}\mathbf{M}^{-1}$$

The requirement of asymptotic stability is equivalent to the requirement that the matrix $\mathbf{A}^k$ tend toward the zero matrix as k increases, since otherwise some initial condition could be found that had a corresponding solution not tending toward zero. This requirement, in turn, is equivalent to requiring that all the terms λ_j^k tend toward zero.

All terms will tend toward zero if and only if $|\lambda_j|<1$ for every j. Thus, a necessary and sufficient condition for asymptotic stability is that all eigenvalues of $\mathbf{A}$ lie inside the unit circle of the complex plane.

There is a less direct, but perhaps more intuitive, way to see this result in terms of the decomposition of the system into uncoupled first-order systems. A first-order system of the form

$$z(k+1) = \lambda z(k)$$

has solution $z(k) = \lambda^k z(0)$. It is asymptotically stable, by the reasoning above, if and only if the magnitude of λ is less than one. When $\mathbf{A}$ is diagonalizable, the system can be transformed to a set of uncoupled first-order systems, as shown in Sect. 5.3, with the eigenvalues of $\mathbf{A}$ determining the behavior of the individual systems. For the whole system to be asymptotically stable, each of the subsystems must be asymptotically stable.

If the matrix $\mathbf{A}$ has multiple eigenvalues, the conclusion is unchanged. A multiple eigenvalue λ introduces terms of the form λ^k, $k\lambda^{k-1}$, $k^2\lambda^{k-2}$, and so forth into the response. The highest-order term of this form possible for an nth-order system is $k^{n-1}\lambda^{k-n+1}$. As long as λ has magnitude less than one, however, the decreasing geometric term outweighs the increase in k^i for any i, and the overall term tends toward zero for large k. Therefore, the existence of multiple roots does not change the qualitative behavior for large k.

It is easy to deduce a partial converse result concerning instability. If the magnitude of any eigenvalue is greater than one, then there will be a vector $\mathbf{x}(0)$, the corresponding eigenvector, which leads to a solution that increases geometrically toward infinity. Thus, the existence of an eigenvalue with a magnitude greater than one is sufficient to indicate instability. Summarizing all

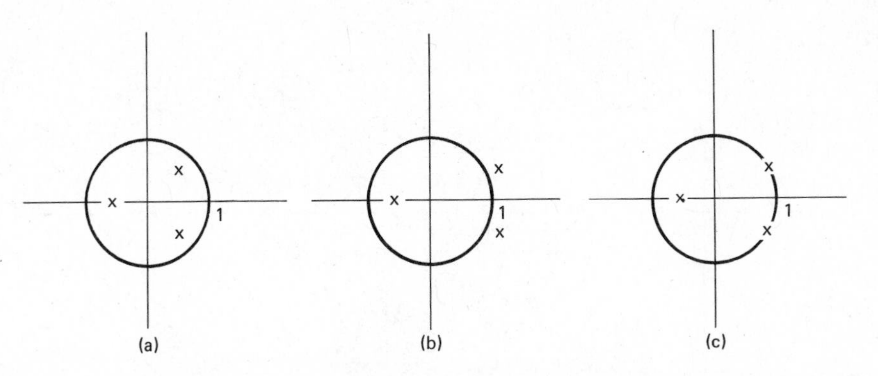

Figure 5.6. Eigenvalues and stability in discrete-time systems. (a) Asymptotically stable. (b) Unstable. (c) Marginally stable.

of the above, and accounting for the fact that the homogeneous system determines stability, we may state the following theorem. (Also see Fig. 5.6.)

Theorem 1. *A necessary and sufficient condition for an equilibrium point of the system (5-50) to be asymptotically stable is that the eigenvalues of* **A** *all have magnitude less than one (that is, the eigenvalues must all lie inside the unit circle in the complex plane). If at least one eigenvalue has magnitude greater than one, the equilibrium point is unstable.*

Because the stability of an equilibrium point of a linear discrete-time system depends only on the structure of the system matrix **A**, it is common to refer to stability of the system itself, or stability of the matrix **A**. Thus, we say "the system is asymptotically stable" or "the matrix **A** it asymptotically stable" when, in the context of a discrete-time linear system, the matrix **A** has all of its eigenvalues inside the unit circle.

Continuous-Time Systems

The condition for asymptotic stability for the continuous-time system (5-51) is similar in nature to that for a discrete-time system. Assuming first that the system matrix **A** is diagonalizable, the system is reduced to a set of first-order equations, each of the form

$$\dot{z}(t) = \lambda z(t)$$

where λ is an eigenvalue of **A**. The solutions to this diagonal system are

$$z(t) = e^{\lambda t} z(0)$$

Each λ can be written as a sum of a real and imaginary part as $\lambda = \mu + i\omega$, where $i = \sqrt{-1}$ and μ and ω are real. Accordingly, $e^{\lambda t} = e^{\mu t} e^{i\omega t}$, and this exponential tends to zero if and only if $\mu < 0$. In other words, the real part of λ

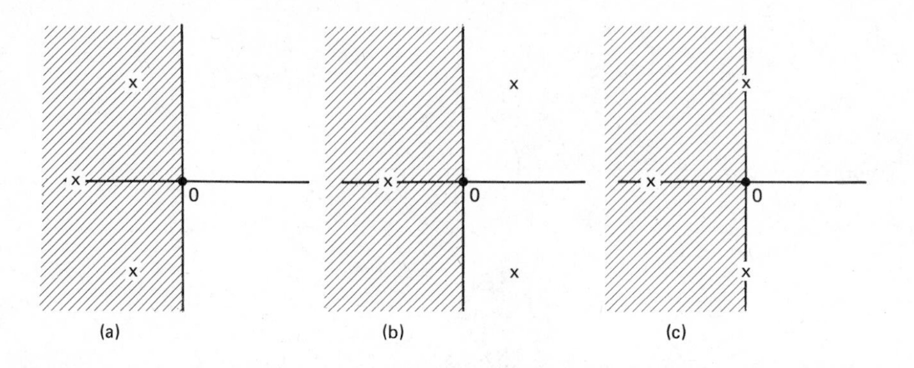

Figure 5.7. Eigenvalues and stability in continuous-time systems. (a) Asymptotically stable. (b) Unstable. (c) Marginally stable.

must be negative. Therefore, for asymptotic stability it is necessary and sufficient that the real part of each λ be negative. Or, in perhaps more visual terminology, each λ must lie in the left half of the plane of complex numbers.

Again, the presence of multiple roots does not change the conclusion. Such roots introduce terms of the form $e^{\lambda t}$, $te^{\lambda t}$, $t^2e^{\lambda t}, \ldots, t^{n-1}e^{\lambda t}$, which each go to zero as $t \to \infty$ provided that the real part of λ is negative. In conclusion, therefore, the companion to Theorem 1 is stated below. (Also see Fig. 5.7.)

Theorem 2. *A necessary and sufficient condition for an equilibrium point of the continuous-time system* (5-51) *to be asymptotically stable is that the eigenvalues of* $\mathbf{A}$ *all have negative real part (that is, the eigenvalues must lie in the left half of the complex plane). If at least one eigenvalue has positive real part, the point is unstable.*

As in the discrete-time case, stability of an equilibrium point of a linear continuous-time system depends only on the structure of the matrix $\mathbf{A}$ and not explicitly on the equilibrium point itself. Thus, we say "the system is asymptotically stable" or "the matrix $\mathbf{A}$ is asymptotically stable" when, in the context of a continuous-time linear system, the eigenvalues of $\mathbf{A}$ are all inside the left-half plane.

Marginal Stability

For discrete- and continuous-time systems, there is an intermediate case that is not covered by the above stability theorems. For discrete time this case is where no eigenvalues are outside the unit circle but one or more is exactly on the boundary of the circle. For continuous time, it is where no eigenvalues are in the right half of the complex plane, but one or more have real part exactly equal to zero. (See Figs. 5.6 and 5.7.)

In these intermediate situations further analysis is required to characterize the long-term behavior of solutions. If the boundary eigenvalues all have a complete set of associated eigenvectors, then the only terms they produce are of the form λ^k in discrete time, or $e^{i\omega t}$ in continuous time. The terms have constant absolute value. Therefore, the state vector neither tends to infinity nor to zero. The system is neither unstable, nor asymptotically stable. This special intermediate situation is referred to as *marginal stability*. It is an important special case, arising in several applications.

Example. Some common two-dimensional systems have the form

$$\dot{\mathbf{x}}(t) = \mathbf{A}\mathbf{x}(t)$$

where the $\mathbf{A}$ matrix has the off-diagonal form

$$\mathbf{A} = \begin{bmatrix} 0 & c_1 \\ c_2 & 0 \end{bmatrix}$$

For instance, the mass and spring model has this form with $c_1 = -\omega^2$, $c_2 = 1$. The Lanchester model of warfare has this form with c_1 and c_2 both negative.

The characteristic equation associated with this $\mathbf{A}$ matrix is

$$\lambda^2 = c_1 c_2$$

The relationship of the parameters c_1 and c_2 to stability can be deduced directly from this equation. There are really only three cases.

CASE 1. The parameters c_1 and c_2 are nonzero and have the same sign.

In this case $c_1 c_2$ is positive, and therefore the two eigenvalues are $\lambda_1 = \sqrt{c_1 c_2}$ and $\lambda_2 = -\sqrt{c_1 c_2}$. The first of these is itself always positive and therefore the system is unstable.

CASE 2. The parameters c_1 and c_2 are nonzero and have opposite signs.

In this case (of which the mass and spring is an example) $c_1 c_2$ is negative and therefore the two eigenvalues are imaginary (lying on the axis between the left and right halves of the complex plane). This implies oscillatory behavior and a system which is marginally stable.

CASE 3. Either c_1 or c_2 is zero.

If both c_1 and c_2 are zero the system is marginally stable. If only one of the two parameters c_1 and c_2 is zero, the system has zero as a repeated root in a chain of length two. Some components will grow linearly with k and, therefore, the system is unstable.

In view of these three cases, it is clear that an off-diagonal system can never be stable (except marginally).

5.10 OSCILLATIONS

A good deal of qualitative information about solutions can be inferred directly from the eigenvalues of a system, even without calculation of the corresponding eigenvectors. For instance, as discussed in Sect. 5.9, stability properties are determined entirely by the eigenvalues. Other general characteristics of the solution can be deduced by considering the placement of the eigenvalues in the complex plane. Each eigenvalue defines both a characteristic growth rate and a characteristic frequency of oscillation. These relations are examined in this section.

It is sufficient to consider only homogeneous systems, since their solutions underlie the general response of linear systems. Even in this case, however, the complete pattern of solution can be quite complex, for it depends on the initial state vector and on the time patterns of each of the eigenvectors that in conjunction comprise the solution. To decompose the solution into components, it is natural to consider the behavior associated with a single eigenvalue, or a complex conjugate pair of eigenvalues. Each of these acts separately and has a definite characteristic pattern.

Continuous Time

Let λ be an eigenvalue of a continuous-time system. It is convenient to express λ in the form $\lambda = \mu + i\omega$. The characteristic response associated with this eigenvalue is $e^{\lambda t} = e^{\mu t}e^{i\omega t}$. The coefficient that multiplies the associated eigenvector varies according to this characteristic pattern.

If λ is real, then $\lambda = \mu$ and $\omega = 0$. The coefficient $e^{\lambda t}$ is then always of the same sign. No oscillations are derived from an eigenvalue of this type.

If λ is complex, then its complex conjugate $\bar{\lambda} = \mu - i\omega$ must also be an eigenvalue. The solution itself is always real, so, overall, the imaginary numbers cancel out. The contribution due to λ and $\bar{\lambda}$ in any component therefore contains terms of the form $(A \sin \omega t + B \cos \omega t)e^{\mu t}$. Such terms oscillate with a frequency ω and have a magnitude that either grows or decays exponentially according to $e^{\mu t}$. In summary, for continuous-time systems the frequency of oscillation (in radians per unit time) due to an eigenvalue is equal to its imaginary part. The rate of exponential growth (or decay) is equal to its real part. (See Fig. 5.8*a*.)

Discrete Time

Let λ be an eigenvalue of a discrete-time system. In this case it is convenient to express λ in the form $\lambda = re^{i\theta} = r(\cos\theta + i \sin\theta)$. The characteristic response due to this eigenvalue is $\lambda^k = r^k e^{ik\theta} = r^k \cos k\theta + ir^k \sin k\theta$. The coefficient that multiplies the associated eigenvector varies according to this characteristic pattern.

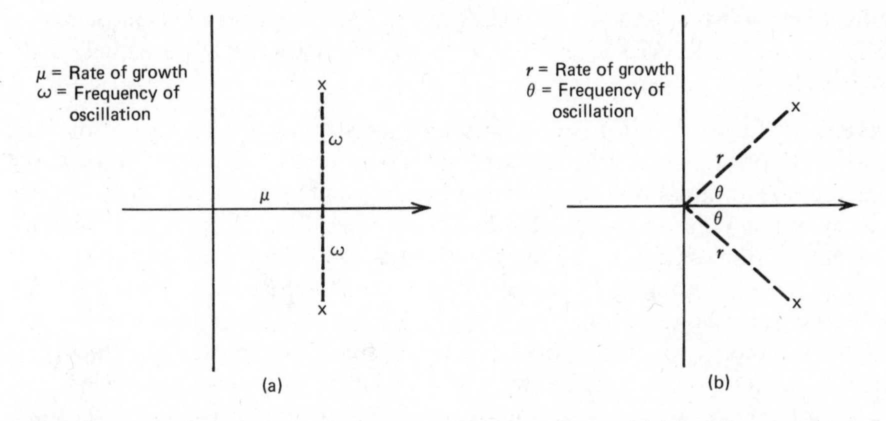

Figure 5.8. Relation of eigenvalue location to oscillation. (a) Continuous time. (b) Discrete time.

If λ is real and *positive*, the response pattern is the geometric sequence r^k, which increases if $r>1$ and decreases if $r<1$. No oscillations are derived from a positive eigenvalue. If λ is real and *negative*, the response is an alternating geometric sequence.

If λ is complex, it will appear with its complex conjugate. The real response due to both eigenvalues is of the form $r^k(A \sin k\theta + B \cos k\theta)$. If $\theta \neq 0$, the expression within the parentheses will change sign as k varies. However, the exact pattern of variation may not be perfectly regular. In fact, if θ is an irrational multiple of 2π, the expression does not have a finite cycle length since each value of k produces a different value in the expression. This is illustrated in Fig. 5.9*a*. Because of this phenomenon, it is useful to superimpose a quasi-continuous solution on the discrete solution by allowing k to be continuous. The period of oscillation then can be measured in the standard way, and often will be some nonintegral multiple of k. This is illustrated in Fig. 5.9*b*. In summary, for discrete-time systems the frequency of oscillation (in

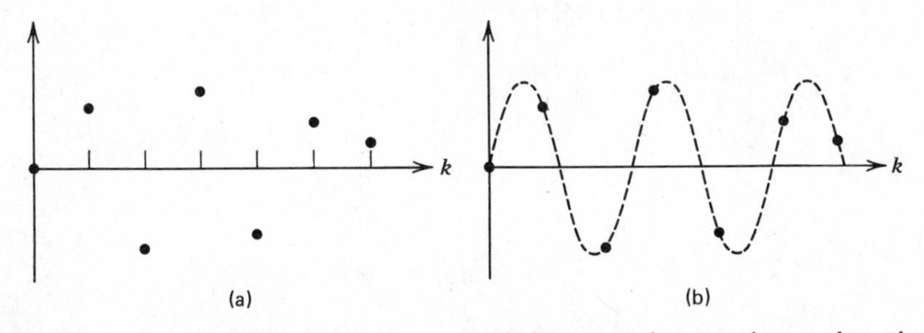

Figure 5.9. A discrete-time pattern and its superimposed quasicontinuous pattern.

radians per unit time) due to an eigenvalue is equal to its angle as measured in the complex plane. The rate of geometric growth (or decay) is equal to the magnitude of the eigenvalue. (See Fig. 5.8*b*.)

Example (The Hog Cycle). For nearly a century it has been observed that the production of hogs is characterized by a strong, nearly regular cycle of about four years' duration. Production alternately rises and falls, forcing prices to cycle correspondingly in the opposite direction. (See Fig. 5.10). The economic hardship to farmers caused by these cycles has motivated a number of government policies attempting to smooth them, and has focused attention on this rather curious phenomenon.

One common explanation of the cycle is based on the cobweb theory of supply and demand interaction, the basic argument being that the cycles are a result of the farmers' use of current prices in production decisions. By responding quickly to current price conditions, hog producers introduce the characteristic cobweb oscillations. However, as shown below, when the pure cobweb theory is applied to hog production, it predicts that the hog cycle would have a period of only two years. Thus, this simple view of the behavior of hog producers is not consistent with the observed cycle length. It seems appropriate therefore to revise the assumptions and look for an alternative explanation. One possible approach is discussed in the second part of this example.

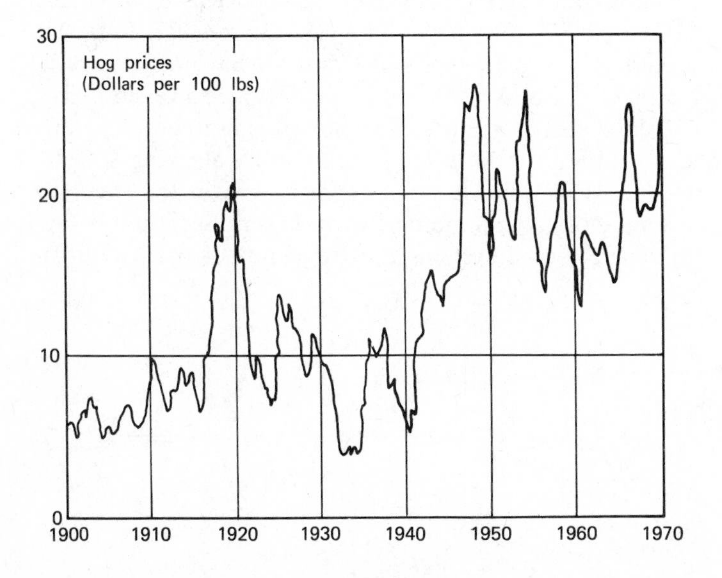

Figure 5.10. The U.S. hog cycle.

The Cobweb Model

A somewhat simplified description of the hog production process is as follows: There are two hog production seasons a year corresponding to spring and fall breeding. It is about six months from breeding to weaning, and five or six months of growth from weaning to slaughter. Very few sows are required for the breeding stock, so farmers generally slaughter and market essentially their whole stock of mature hogs. Translated into analytical terms, it is reasonable to develop a discrete-time model with a basic period length of six months. Production can be initiated at the beginning of each period, but takes two periods to complete. In this discrete-time framework we denote the number of mature hogs produced at the beginning of period k by $h(k)$, and the corresponding unit price by $p(k)$.

As is standard, we assume that demand for hogs in any period is determined by price according to a linear demand curve. In particular, demand at period k is given by

$$d(k) = d_0 - ap(k) \tag{5-53}$$

In a similar way, we assume that there is a supply curve of the form

$$\hat{h} = s_0 + b\hat{p} \tag{5-54}$$

The quantity $\hat{h}$ represents the level of seasonal production that a farmer would maintain if his estimate of future hog prices were $\hat{p}$.

In the cobweb model, it is assumed that at the beginning of period k a farmer decides how many hogs to breed on the basis of current price. This leads to the equations

$$h(k+2) = s_0 + bp(k) \tag{5-55}$$

$$d(k) = d_0 - ap(k) \tag{5-56}$$

where $h(k+2)$ corresponds to the breeding decision at period k, which specifies the ultimate number of mature hogs at period $k+2$. Equating supply and demand [i.e., setting $d(k) = h(k)$] and eliminating $p(k)$ leads to

$$h(k+2) = -\rho h(k) + \rho d_0 + s_0 \tag{5-57}$$

where $\rho = b/a$. The characteristic polynomial of this difference equation is

$$\lambda^2 + \rho = 0 \tag{5-58}$$

which has the two imaginary roots $\lambda = \pm i\sqrt{\rho}$.

This result is slightly different in form than that obtained in the usual cobweb model since there are two imaginary roots rather than a single negative root. This is because the production interval is now divided into two periods.

However, the squares of these imaginary roots are equal to the root corresponding to a full production duration.

The interpretation of the results of this model is that hog cycles would have a period of four time intervals (since the roots are imaginary) or, equivalently, of *two* years. Furthermore, hog production is a side-line operation on many farms that is easily expanded or contracted, depending on economic incentives. This suggests that $\rho>1$ and hence the cobweb model predicts an unstable situation.

Smoothed Price Prediction Model

As an alternative model, we assume that farmers are aware of the inherent instability and accompanying oscillatory nature of the hog market. Accordingly, they act more conservatively than implied by the pure cobweb model by basing their estimate of future price on an average of past prices. For example, since the oscillation is roughly four years in duration, it is reasonable to average over a time span of at least half this long. A uniform average over two years would have the form

$$\hat{p}(k)=\tfrac{1}{5}[p(k)+p(k-1)+p(k-2)+p(k-3)+p(k-4)] \tag{5-59}$$

Using this estimated price in the supply equation (5-54) and equating supply and demand leads to

$$\begin{aligned} h(k+2)=&-\frac{\rho}{5}[h(k)+h(k-1)+h(k-2)+h(k-3)+h(k-4)] \\ &+\rho d_0+s_0 \end{aligned} \tag{5-60}$$

This sixth-order difference equation has the characteristic equation

$$\lambda^6+\frac{\rho}{5}[\lambda^4+\lambda^3+\lambda^2+\lambda+1]=0 \tag{5-61}$$

It is of course difficult to compute the roots of this sixth-order polynomial. However, assuming $\rho=2.07$, it can be verified that the roots of largest magnitude are $\lambda=\sqrt{2}/2\pm i\sqrt{2}/2$. These roots have a magnitude of one and correspond to a cycle period of *exactly* four years. The response due to these eigenvalues is thus an oscillation of a period of four years that persists indefinitely, neither increasing or decreasing in magnitude.

One can argue of course that different predictors would lead to different roots, and hence to different cycle lengths. However, we argue that in the aggregate, producers average in such a way so as to just maintain (marginal) stability. A short-term average would lead to shorter cycles but instability. An excessively long-term average would lead to long cycle lengths, but a sluggish system in which farmers would not respond to the strong economic incentives to adjust to price trends. As a group they tend to walk the fine line between stability and instability. If $\rho \simeq 2$, then to reach this balance point, the price estimate used must be close in form to the one presented by (5-59), and consequently the cycle length must be close to four years.

5.11 DOMINANT MODES

The long-term behavior of a linear time-invariant system often is determined largely by only one or two of its eigenvalues and corresponding eigenvectors. These *dominant* eigenvalues and eigenvectors are therefore of special interest to the analyst.

Discrete-Time System

Consider the discrete-time system

$$\mathbf{x}(k+1) = \mathbf{A}\mathbf{x}(k) \tag{5-62}$$

Suppose the matrix $\mathbf{A}$ has eigenvalues $\lambda_1, \lambda_2, \ldots, \lambda_n$ with $|\lambda_1| > |\lambda_2| \geq |\lambda_3| \cdots \geq |\lambda_n|$. The eigenvalue λ_1 of greatest magnitude is the *dominant eigenvalue.* For simplicity, let us assume that there is a complete set of eigenvectors, and that there is only one eigenvalue of greatest magnitude. If there were two greatest of equal magnitude (as there would be if the eigenvalues were complex), then both would be considered dominant.

Any initial condition vector can be expressed as a linear combination of all eigenvectors in the form

$$\mathbf{x}(0) = \alpha_1 \mathbf{e}_1 + \alpha_2 \mathbf{e}_2 + \cdots + \alpha_n \mathbf{e}_n \tag{5-63}$$

Correspondingly, the solution to (5-62) at an arbitrary time $k > 0$ is

$$\mathbf{x}(k) = \alpha_1 \lambda_1^k \mathbf{e}_1 + \alpha_2 \lambda_2^k \mathbf{e}_2 + \cdots + \alpha_n \lambda_n^k \mathbf{e}_n \tag{5-64}$$

Since λ_1^k grows faster than λ_i^k for $i = 2, 3, \ldots, n$, it follows that for large k

$$|\alpha_1 \lambda_1^k| \gg |\alpha_i \lambda_i^k| \qquad \text{for } i = 2, 3, \ldots, n$$

as long as $\alpha_1 \neq 0$. In other words, for large values of k, the coefficient of the first eigenvector in the expansion (5-64) is large relative to the other coefficients. Hence, for large values of k, the state vector $\mathbf{x}(k)$ is essentially aligned with the eigenvector $\mathbf{e}_1$.

If $\alpha_1 = 0$, the first coefficient in (5-64) is zero for all values of k, and the state vector will not line up with $\mathbf{e}_1$. Theoretically, in this case the eigenvalue of next greatest magnitude would determine the behavior of the system for large values of k. In practice, however, the dominant eigenvalue almost always takes hold—at least ultimately—for a slight perturbation at any step introduces a small $\alpha_1 \neq 0$ that grows faster than any other term.

From the above discussion, it can be deduced that if there is a single dominant eigenvalue, the state vector tends to align itself with the corresponding eigenvector. If there is not a single dominant eigenvalue, but two that are complex conjugates, it can be similarly inferred that the state vector tends toward the two-dimensional space defined by the two corresponding eigenvectors. Typically, in this case, oscillations are generated, characteristic of the complex dominate eigenvalues.

A similar analysis applies to nonhomogeneous systems such as

$$\mathbf{x}(k+1)=\mathbf{A}\mathbf{x}(k)+\mathbf{b}$$

If there is an asymptotically stable equilibrium point $\bar{\mathbf{x}}$, then the state converges to $\bar{\mathbf{x}}$. The rate at which it converges is governed essentially by the eigenvalue of greatest magnitude (which is, however, less than one). The error vector $\mathbf{x}(k)-\bar{\mathbf{x}}$ will be closely aligned with the corresponding eigenvector, as it tends to zero.

Continuous-Time Systems

Consider the system

$$\dot{\mathbf{x}}(t)=\mathbf{A}\mathbf{x}(t) \tag{5-65}$$

Suppose the matrix $\mathbf{A}$ has eigenvalues $\lambda_1, \lambda_2, \ldots, \lambda_n$ ordered now according to $\text{Re}(\lambda_1)>\text{Re}(\lambda_2)\geq\text{Re}(\lambda_3)\cdots\geq\text{Re}(\lambda_n)$. The eigenvalue λ_1 with greatest real part is the *dominant eigenvalue* in this case.

As before let us suppose there is a single dominant eigenvalue and that there is a complete set of eigenvectors. Paralleling the previous analysis, any initial state vector $\mathbf{x}(0)$ can be expressed in terms of the eigenvectors in the form

$$\mathbf{x}(0)=\alpha_1\mathbf{e}_1+\alpha_2\mathbf{e}_2+\cdots+\alpha_n\mathbf{e}_n \tag{5-66}$$

The corresponding solution to (5-65) is

$$\mathbf{x}(t)=\alpha_1 e^{\lambda_1 t}\mathbf{e}_1+\alpha_2 e^{\lambda_2 t}\mathbf{e}_2+\cdots+\alpha_n e^{\lambda_n t}\mathbf{e}_n \tag{5-67}$$

Writing each eigenvalue in terms of its real and imaginary parts as

$$\lambda_k=\mu_k+i\omega_k$$

it is easy to see that $|\alpha_k e^{\lambda_k t}|=|\alpha_k|e^{\mu_k t}$. Thus, since the real part determines the rate of growth of the exponential, it is clear that as $t\to\infty$

$$|\alpha_1 e^{\lambda_1 t}|\gg|\alpha_i e^{\lambda_i t}|, \qquad i=2,3,\ldots,n$$

provided that $\alpha_1\neq0$. Therefore, the first term in (5-67) dominates all the others for large values of t, and hence for large values of t the vector $\mathbf{x}(t)$ is closely aligned with the dominant eigenvector $\mathbf{e}_1$.

Subdominant Modes

The long-term behavior of a system is determined most directly by the dominant eigenvalue and associated eigenvector. However, the other eigenvalues and eigenvectors also play important roles. These *subdominant* eigenvalues can be ordered in a natural way (by their magnitudes for discrete-time systems, or by their real parts for continuous-time systems) corresponding to their order of importance for long-term behavior.

The subdominant eigenvalues essentially determine how quickly the system state moves toward the eigenvector corresponding to the dominant mode. A cohort population system, for example, may have a dominant eigenvalue that is real and greater than one indicating that there is long-term growth. The corresponding eigenvector defines the population distribution to which the system tends as it grows. The rate at which the actual population distribution approaches this distribution is determined largely by the second greatest eigenvalue.

A special situation in which the first subdominant eigenvalue has a particularly important role is when the dominant eigenvalue is equal to one (or zero in the continuous-time case). The system is then marginally stable. The state vector tends toward the corresponding dominant eigenvector, and in the long term it neither grows nor contracts. Thus, the state converges to a fixed vector, and it is the eigenvalue of second largest magnitude that determines how fast it approaches the limiting vector. The following example illustrates this phenomenon.

Example (Fixed Capacity of a Cultural System). Suppose that in the survival theory of culture discussed in Sect. 5.8 there is a limit on the capacity of the system. Such limits might be representative of television programming or theater where only a fixed number of offerings are feasible in any season. Let us consider whether the imposition of a capacity constraint changes the equilibrium quality distribution, and whether it speeds up or slows down the rate at which equilibrium is achieved.

For simplicity let us distinguish only two levels of quality in the culture: good and bad. In this case the original system, without a capacity constraint, is represented by the second-order system

$$\begin{bmatrix} x_1(k+1) \\ x_2(k+1) \end{bmatrix} = \begin{bmatrix} \alpha_1 & 0 \\ 0 & \alpha_2 \end{bmatrix} \begin{bmatrix} x_1(k) \\ x_2(k) \end{bmatrix} + \begin{bmatrix} b_1 \\ b_2 \end{bmatrix}$$

It is assumed that x_1 represents the amount of "good" material in the system, while x_2 represents the amount of "bad" material. According to this specification, it should hold that $\alpha_1 > \alpha_2$, indicating that good material has a better chance to survive than bad material. The parameters must also satisfy $0<\alpha_1<1$, $0<\alpha_2<1$.

The equilibrium point of the system is

$$x_1 = b_1/(1-\alpha_1)$$
$$x_2 = b_2/(1-\alpha_2)$$

The eigenvalues of the system are α_1 and α_2 and therefore the equilibrium is stable. Since the dominant eigenvalue is α_1, the rate at which the equilibrium is approached is essentially $\alpha_1^{\,k}$.

Now let us modify the system to incorporate the capacity constraint that $x_1(k)+x_2(k)$ must be a fixed constant, independent of k. To satisfy this requirement, the amount of new material admitted to the system each season must be varied, depending on the space available for new entries. We assume, however, that the basic screening parameters and the input quality distribution remain unchanged. Only the level of input is varied.

The new system is then governed by

$$\begin{bmatrix} x_1(k+1) \\ x_2(k+1) \end{bmatrix} = \begin{bmatrix} \alpha_1 & 0 \\ 0 & \alpha_2 \end{bmatrix} \begin{bmatrix} x_1(k) \\ x_2(k) \end{bmatrix} + \begin{bmatrix} b_1 \\ b_2 \end{bmatrix} u(k)$$

where $u(k)$ varies in order to satisfy the capacity requirement. For notational simplicity, we assume that $b_1+b_2=1$. This entails no loss of generality since a constant multiple can be incorporated into the definition of $u(k)$. To find an explicit expression for $u(k)$ we write

$$\begin{aligned} x_1(k+1)+x_2(k+1) &= \alpha_1 x_1(k)+\alpha_2 x_2(k)+(b_1+b_2)u(k) \\ &= \alpha_1 x_1(k)+\alpha_2 x_2(k)+u(k) \end{aligned}$$

The requirement $x_1(k+1)+x_2(k+1)=x_1(k)+x_2(k)$ then yields

$$u(k)=(1-\alpha_1)x_1(k)+(1-\alpha_2)x_2(k)$$

This shows how $u(k)$ must be adjusted each period on the basis of what is already in the system. Substituting this value, the new system takes the explicit form

$$\begin{bmatrix} x_1(k+1) \\ x_2(k+1) \end{bmatrix} = \begin{bmatrix} \alpha_1+b_1(1-\alpha_1) & b_1(1-\alpha_2) \\ b_2(1-\alpha_1) & \alpha_2+b_2(1-\alpha_2) \end{bmatrix} \begin{bmatrix} x_1(k) \\ x_2(k) \end{bmatrix}$$

It should be noted that the new system is homogeneous while the original system is not. Total volume in the system is $x_1(k)+x_2(k)=[1\ 1]\mathbf{x}(k)$. Since the volume is constant, it follows that $[1\ 1]$ is a left eigenvector of the system with the corresponding eigenvalue equal to one. As we shall verify shortly, this is also the dominant eigenvalue.

Because one is an eigenvalue, the corresponding *right* eigenvector (or any multiple of it) is an equilibrium point for the system. The right eigenvector corresponding to the eigenvalue of one must satisfy

$$[\alpha_1+b_1(1-\alpha_1)]x_1+b_1(1-\alpha_2)x_2=x_1$$
$$b_2(1-\alpha_1)x_1+[\alpha_2+b_2(1-\alpha_2)]x_2=x_2$$

A solution is

$$x_1=b_1/(1-\alpha_1)$$
$$x_2=b_2/(1-\alpha_2)$$

Thus, the equilibrium points of the new system are just scalar multiples of that of the old system. This can be explained by the fact that $u(k)$ approaches a constant as $k\to\infty$. Thus, for large k the new system looks very much like the original system.

The speed at which the system converges to the equilibrium is now governed by the eigenvalue of second largest absolute value. In our second-order example this is, of course, the only other eigenvalue. To find this eigenvalue, we write the corresponding characteristic polynomial that (after a bit of algebra) reduces to

$$\lambda^2-(1+b_2\alpha_1+b_1\alpha_2)\lambda+b_2\alpha_1+b_1\alpha_2$$

Recalling that $\lambda_1=1$ is an eigenvalue of the system, we can easily factor this polynomial obtaining the characteristic equation

$$(\lambda-1)(\lambda-b_2\alpha_1-b_1\alpha_2)=0$$

The second eigenvalue is therefore

$$\lambda_2=b_2\alpha_1+b_1\alpha_2$$

Since this second eigenvalue is a weighted average of α_1 and α_2, it follows that $\alpha_2<\lambda_2<\alpha_1$.

Two conclusions can be drawn from this result. First, since $0<\lambda_2<1$ the system is marginally stable, with $\lambda_1=1$ being the dominant eigenvalue. Second, the speed at which the system converges toward equilibrium is governed by this second eigenvalue. Since this value is less than α_1, this new system converges toward equilibrium faster than the original system.

Finally, as should be routine when completing an analysis, let us interpret this result by checking its consistency with less formal reasoning. Suppose that the system is not in equilibrium, with a higher proportion of bad material than there would be in equilibrium. Since bad material is screened out relatively rapidly, this means that more than a normal level of input is required in order to satisfy the capacity requirement. This high level of input tends to quickly infuse more good material. Later, as the proportions move toward equilibrium, the level of input decreases since good material remains in the system a long time. The initial high level of input, however, acts to move the system more rapidly than would the constant equilibrium input level. Thus, we do expect the capacity constraint to speed convergence.

5.12 THE COHORT POPULATION MODEL

The cohort population model, introduced briefly in Chapter 1, is an excellent example for detailed study. It has a wealth of structure that can be translated into interesting and significant statements about its dominant eigenvalue and eigenvector. The cohort model deserves attention, however, not only as an application of theory, but also as a structure important in many application areas where items are categorized by age.

The basic, single sex, cohort population model presented in Chapter 1 is defined by the difference equations

$$x_0(k+1)=\alpha_0 x_0(k)+\alpha_1 x_1(k)+\alpha_2 x_2(k)+\cdots+\alpha_n x_n(k)$$
$$x_{i+1}(k+1)=\beta_i x_i(k), \qquad i=0,1,2,\ldots,n-1$$

where α_i is the birthrate of the ith age group and β_i is the survival rate of the ith age group. In matrix form the general cohort model becomes

$$\begin{bmatrix} x_0(k+1) \\ x_1(k+1) \\ \\ \cdot \\ \cdot \\ \cdot \\ x_n(k+1) \end{bmatrix} = \begin{bmatrix} \alpha_0 & \alpha_1 & \alpha_2 & \cdots & \alpha_n \\ \beta_0 & 0 & 0 & \cdots & 0 \\ 0 & \beta_1 & 0 & \cdots & 0 \\ \cdot & \cdot & & & \cdot \\ \cdot & & \cdot & & \cdot \\ \cdot & & & & \cdot \\ 0 & 0 & \cdots & \beta_{n-1} & 0 \end{bmatrix} \begin{bmatrix} x_0(k) \\ \\ x_1(k) \\ \cdot \\ \cdot \\ \cdot \\ x_n(k) \end{bmatrix} \tag{5-68}$$

We write this in abbreviated form as

$$\mathbf{x}(k+1)=\mathbf{P}\mathbf{x}(k) \tag{5-69}$$

Change of Variable

It is convenient to introduce a simple change of variable that preserves the general structure of the cohort model but simplifies it by making the coefficients along the lower diagonal (the β_i's) all equal to one. For this purpose define

$$\begin{aligned} l_0 &= 1 \\ l_k &= \beta_0\beta_1\cdots\beta_{k-1}, \qquad k=1,2,\ldots,n \end{aligned} \tag{5-70}$$

The number l_k can be interpreted as the probability of a newborn surviving to reach the kth age group.

In terms of these numbers, define the diagonal matrix

$$\mathbf{D} = \begin{bmatrix} l_0 & & & & & & \\ & l_1 & & & & & \\ & & l_2 & & & & \\ & & & \cdot & & & \\ & & & & \cdot & & \\ & & & & & \cdot & \\ & & & & & & l_n \end{bmatrix} \tag{5-71}$$

and define the change of variable from $\mathbf{x}(k)$ to $\mathbf{y}(k)$ by

$$\mathbf{x}(k) = \mathbf{D}\mathbf{y}(k)$$

With respect to the vector $\mathbf{y}(k)$, the system takes the form

$$\mathbf{y}(k+1) = \bar{\mathbf{P}}\mathbf{y}(k) \tag{5-72}$$

where $\bar{\mathbf{P}} = \mathbf{D}^{-1}\mathbf{P}\mathbf{D}$.

The reader can verify the explicit form

$$\mathbf{D}^{-1}\mathbf{P}\mathbf{D} = \begin{bmatrix} \alpha_0 & l_1\alpha_1 & l_2\alpha_2 & \cdots & & l_n\alpha_n \\ 1 & & & & & 0 \\ & 1 & & & & 0 \\ & & \cdot & & & \cdot \\ & & & \cdot & & \cdot \\ & & & & \cdot & \cdot \\ & & & & 1 & 0 \end{bmatrix}$$

Therefore, defining $a_i = l_i\alpha_i$, one has

$$\bar{\mathbf{P}} = \begin{bmatrix} a_0 & a_1 & \cdots & & & a_n \\ 1 & & & & & 0 \\ & 1 & & & & 0 \\ & & \cdot & & & \cdot \\ & & & \cdot & & \cdot \\ & & & & \cdot & \cdot \\ & & & & 1 & 0 \end{bmatrix} \tag{5-73}$$

This new matrix has the same structure as the original, but with unity coefficients below the main diagonal.

The new variables $y_i(k)$ and the new matrix $\bar{\mathbf{P}}$ have simple interpretations in terms of the population system. The variable $y_i(k)$ is equal to number of original members in the cohort that now occupies the ith age span. This number includes the deceased as well as the presently living members of this cohort. With the variables defined in this way, there is no possibility of leaving a cohort, and therefore the appropriate new survival factors are all unity. The new birthrates account for the fact that only the living members of a cohort can possibly have children. Since the number of living members of group i is $l_i y_i(k)$, the new effective birthrate is $l_i \alpha_i = a_i$, as indicated in $\bar{\mathbf{P}}$. Therefore, the new system is a representation where all members of a cohort, dead or alive, are accounted for through all $n+1$ age span periods.

Characteristic Polynomial

The special structure of the population matrix enables one to deduce quite explicit results concerning the dominant eigenvalue and corresponding eigenvector. It is simplest to first work with the matrix $\bar{\mathbf{P}}$, derived above, and then transfer the results back to $\mathbf{P}$ through the similarity transformation. (The characteristic polynomial of $\bar{\mathbf{P}}$ is the same as that of $\mathbf{P}$ since $\bar{\mathbf{P}}$ and $\mathbf{P}$ are similar.)

The characteristic polynomial of $\bar{\mathbf{P}}$ is easily found to be

$$(-1)^{n+1}(\lambda^{n+1} - a_0\lambda^n - a_1\lambda^{n-1} - \cdots - a_n)$$

Accordingly, the characteristic equation is

$$\lambda^{n+1} = a_0\lambda^n + a_1\lambda^{n-1} + \cdots + a_n \tag{5-74}$$

Under a rather mild assumption, it is possible to show that the dominant root of the characteristic equation (the root of largest absolute value) is a real positive number and the corresponding eigenvector has positive components. The only assumption required is that at least two consecutive a_i's are strictly positive.

Theorem. *Suppose that $a_i \geq 0$, $i = 0, 1, 2, \ldots, n$, and that for some m, $0 < m \leq n$, there holds $a_{m-1} > 0$, $a_m > 0$. Then the characteristic equation* (5-74) *has a unique eigenvalue λ_0 of greatest absolute value. This eigenvalue is positive and simple.*

Proof. For $\lambda \neq 0$, (5-74) can be rewritten equivalently as

$$1 = a_0\lambda^{-1} + a_1\lambda^{-2} + \cdots + a_n\lambda^{-(n+1)} \tag{5-75}$$

which is of the form

$$1 = f(\lambda)$$

Since each of the a_i's is nonnegative, it is clear that $f(\lambda)$ increases strictly monotonically from zero to infinity as λ varies from infinity to zero. Therefore, there is exactly one positive real root of the equation, and that root has algebraic multiplicity one. Let us denote this positive eigenvalue by λ_0.

To show that λ_0 is the dominant eigenvalue, suppose λ is any other nonzero solution to the characteristic equation. This solution can be written in the form $\lambda^{-1}=re^{i\theta}$, where $r>0$. Substitution of this expression into (5-75) gives

$$1=a_0re^{i\theta}+a_1r^2e^{i2\theta}+\cdots+a_nr^{(n+1)}e^{i(n+1)\theta}$$

Taking the real value of this equation gives

$$1=a_0r\cos\theta+a_1r^2\cos 2\theta+\cdots+a_nr^{(n+1)}\cos(n+1)\theta$$

Since $\cos\theta\leq 1$, it is clear that the right-hand side of this equation is no greater than $a_0r+a_1r^2+\cdots+a_nr^{(n+1)}$, and, therefore, in view of the monotonic behavior of this expression, $r\geq r_0$, where $r_0=\lambda_0^{-1}$. In fact, the only way that equality can be achieved is for all the cosine terms corresponding to nonzero a_i's to be unity. However, since it is assumed that for some m that $a_{m-1}>0$ and $a_m>0$, it would follow that $\cos m\theta=1$ and $\cos(m+1)\theta=1$. But this would imply that $\cos\theta=1$ and accordingly that λ were positive. Therefore, for any $\lambda\neq\lambda_0$ it must follow that $r>r_0$, which means $|\lambda|<\lambda_0$. Therefore, the dominant eigenvalue λ_0 is positive and of algebraic multiplicity one. ∎

Dominant Eigenvector

The dominant eigenvector, corresponding to the positive dominant eigenvalue λ_0, is easy to calculate. The eigenvectors $\bar{\mathbf{e}}_0$ and $\mathbf{e}_0$ for the matrices $\bar{\mathbf{P}}$ and $\mathbf{P}$, respectively, are

$$\bar{\mathbf{e}}_0=\begin{bmatrix}\lambda_0^n\\ \lambda_0^{n-1}\\ \cdot\\ \cdot\\ \cdot\\ \lambda_0\\ 1\end{bmatrix},\qquad \mathbf{e}_0=\begin{bmatrix}l_0\lambda_0^n\\ l_1\lambda_0^{n-1}\\ \cdot\\ \cdot\\ \cdot\\ l_{n-1}\lambda_0\\ l_n\end{bmatrix}$$

Validity of the above form for $\bar{\mathbf{e}}_0$ follows because the first component of $\bar{\mathbf{P}}\bar{\mathbf{e}}_0$ is $a_0\lambda_0^n+\cdots+a_n$, which since λ_0 satisfies the characteristic equation is equal to λ_0^{n+1}. It is then easy to check that $\bar{\mathbf{P}}\bar{\mathbf{e}}_0=\lambda_0\bar{\mathbf{e}}_0$. It should be noted that since $\lambda_0>0$, the components of $\bar{\mathbf{e}}_0$ are all positive. The corresponding dominant eigenvector of $\mathbf{P}$ is found from the relation $\mathbf{e}_0=\mathbf{D}\bar{\mathbf{e}}_0$, where $\mathbf{D}$ is defined by (5-71). Thus, the components of $\mathbf{e}_0$ are also all positive if the β_i's are all positive.

The dominant eigenvalue and eigenvector have natural interpretations within the context of the population system. The eigenvalue λ_0 is the *natural growth rate* of the system. The population ultimately tends to grow at this rate as the system state vector becomes closely aligned with the dominant eigenvector. This rate is always positive, although it may be less than one. The eigenvector $\mathbf{e}_0$ represents the stable distribution of population. Once this distribution is achieved, all age-group populations simultaneously grow by the factor λ_0 each period—the population may grow as a whole, but the relative distribution remains fixed.

*5.13 THE SURPRISING SOLUTION TO THE NATCHEZ PROBLEM

We recall the rather unhappy prediction concerning the fate of a social culture organized along the lines of the Natchez Indians. The society was divided into four classes, with the requirement that everyone in the ruling class must select a marriage partner from the lowest class. The class designation of the off-spring of a marriage was the higher of (a) the mother's class, or (b) the class one level below the father's class. By assuming a uniform birthrate, an equal number of men and women in each class, and no inter-generation marriages, it was discovered that eventually the society would deplete its lowest class, and therefore could not continue.

In terms of historical record the Natchez structure did not have the opportunity to play itself out long enough to test the validity of this analysis. However, the culture had apparently survived for several generations in essentially this form before they were discovered and largely destroyed by French explorers. It may be that the system actually did work well and that, accordingly, some aspect of our earlier analysis is crucially defective, or it may be that the society had not existed long enough for the inevitable consequences of the dynamic relations to bring on the predicted disaster.

But let us set aside, as probably unresolvable, the question of what might have happened to the Natchez. Let us instead ask whether the hypotheses of our earlier analysis can be altered in order to produce a system that avoids collapse.

In particular, let us investigate a potential resolution based on an assumption that there are different birth rates in the different classes. This is a plausible hypothesis since such differences are known to occur in many class systems. Indeed, to set the stage for this pursuit, let us formally pose the *Natchez problem* as follows: Is there a collection of birthrates for each of the allowable marriage combinations so that the resulting dynamic class structure can sustain a stable distribution?

One plausible answer suggests itself immediately. Since the original system failed by depleting the lowest class, a balance could probably be attained by increasing the birthrate in the lowest class. This seems only logical. We increase the regeneration rate of the critical resource that otherwise tends to be depleted. In fact, however, this is *not* a solution to the dilemma. Such an increase cannot produce a balanced situation. The problem, nevertheless, does have a solution and that solution is achieved, in part, by *decreasing* the

Table 5.2 Marriage Rules and Birth-rates

		Father A	Father B	Father C
Mother	A			A α_1
	B			B α_3
	C	B α_2	C α_4	C α_5

birthrate in the lowest class relative to the other classes.

Many readers will at first find this result surprising, but the analysis that follows should be convincing. When the total solution is derived, and when it is discussed in the light of the analysis, it should seem not at all surprising. Indeed it probably will seem to be the obvious solution after all.

In order to simplify the analysis a three-class system sufficient to capture the general characteristics of the Natchez structure, rather than the full four-class system, is investigated. The system is defined by Table 5.2. The three classes are now designated by the relatively bland labels A, B, and C. If a label appears in a box in the table it indicates that that box corresponds to an allowable marriage. The label in the box indicates the class designation of the offspring. The parameter α_i in the box indicates the (average) birthrate associated with that type of marriage, expressed in number of male children per marriage.

Making the usual assumptions that (1) there are an equal number of men and women in each class every generation, and (2) there are no intergeneration marriages, it is straightforward to write the dynamic equations implied by this table. Denote by $x_1(k)$, $x_2(k)$, and $x_3(k)$ the number of males in classes A, B, and C, respectively, in generation k. Then we can write immediately,

$$x_1(k+1) = \alpha_1 x_1(k)$$

Likewise, since class B children result from every class A father and every class B mother, we may write

$$x_2(k+1) = \alpha_2 x_1(k) + \alpha_3 x_2(k)$$

Finally, class C children are produced by all class B fathers and by all class C fathers except those who marry class A or class B women. Therefore,

$$x_3(k+1) = \alpha_4 x_2(k) + \alpha_5[x_3(k) - x_1(k) - x_2(k)]$$

In matrix form the system is

$$\begin{bmatrix} x_1(k+1) \\ x_2(k+1) \\ x_3(k+1) \end{bmatrix} = \begin{bmatrix} \alpha_1 & 0 & 0 \\ \alpha_2 & \alpha_3 & 0 \\ -\alpha_5 & (\alpha_4 - \alpha_5) & \alpha_5 \end{bmatrix} \begin{bmatrix} x_1(k) \\ x_2(k) \\ x_3(k) \end{bmatrix} \tag{5-76}$$

which we abbreviate by $\mathbf{x}(k+1) = \mathbf{A}\mathbf{x}(k)$.

Before progressing with the general analysis, let us briefly study the special case where $\alpha_i = 1$ for all i, in order to verify that this reduced class system is subject to the same disaster as the four-class Natchez system. In this case

$$\mathbf{A} = \begin{bmatrix} 1 & 0 & 0 \\ 1 & 1 & 0 \\ -1 & 0 & 1 \end{bmatrix}$$

Paralleling our earlier analysis of the full Natchez system, we write $\mathbf{A} = \mathbf{I} + \mathbf{B}$ and note that $\mathbf{B}^2 = \mathbf{0}$. Thus, $\mathbf{A}^k = \mathbf{I} + k\mathbf{B}$, which in explicit form is

$$\mathbf{A}^k = \begin{bmatrix} 1 & 0 & 0 \\ k & 1 & 0 \\ -k & 0 & 1 \end{bmatrix}$$

Therefore, if $x_1(0) > 0$, the solution eventually reaches a stage where $x_3(k) < 0$, corresponding to breakdown. Thus, this reduced system captures the essential features of the Natchez system.

Now let us return to the general model with variable birthrates. It is helpful to spell out what is required of an acceptable solution to the *Natchez problem.* First, there must be an eigenvector of the system whose components are all positive and whose associated eigenvalue is positive. Such an eigenvector would represent a population distribution that, once achieved, would not change with time. The total population would grow at a rate equal to the corresponding eigenvalue, but the relative value of the components would not change.

A second requirement on a solution is that the eigenvalue of the eigenvector described above must be the dominant eigenvalue of the system. Only in this way can it be assured that the population distribution, if perturbed from this distritution, tends to return to it. Finally, as a third requirement, the stable distribution, defined in terms of class members in each generation, must have enough members in the lowest class to supply marriage partners for the higher classes. In view of these requirements, the problem is reduced to determining birthrates such that the $\mathbf{A}$ matrix has a suitable dominant eigenvector.

Since the $\mathbf{A}$ matrix is triangular, its eigenvalues are the same as the diagonal elements α_1, α_3, α_5. (See Problem 16, Chapter 3.) Let us suppose first that $\alpha_5 > \alpha_3$ and $\alpha_5 > \alpha_1$, so that α_5 is the dominant eigenvalue. This corresponds to the proposal of increasing the birthrate of the lowest class. The eigenvector associated with this eigenvalue must satisfy the matrix equation

$$\begin{bmatrix} \alpha_1 - \alpha_5 & 0 & 0 \\ \alpha_2 & (\alpha_3 - \alpha_5) & 0 \\ \alpha_5 & (\alpha_4 - \alpha_5) & 0 \end{bmatrix} \begin{bmatrix} x_1 \\ x_2 \\ x_3 \end{bmatrix} = \begin{bmatrix} 0 \\ 0 \\ 0 \end{bmatrix}$$

The first line implies immediately, however, that $x_1 = 0$, and hence any eigenvector associated with α_5 cannot have all positive components. Therefore, α_5 cannot be the dominant eigenvalue and simultaneously satisfy the other conditions imposed by the problem formulation. The same argument can be easily seen to apply to the eigenvalue α_3.

The only remaining possibility, therefore, is to arrange things so that α_1 is the dominant eigenvalue. Let us assume, accordingly, that $\alpha_1 > \alpha_3$ and $\alpha_1 > \alpha_5$. The eigenvector associated with the eigenvalue α_1 must satisfy the equations

$$\begin{bmatrix} 0 & 0 & 0 \\ \alpha_2 & (\alpha_3 - \alpha_1) & 0 \\ -\alpha_5 & (\alpha_4 - \alpha_5) & (\alpha_5 - \alpha_1) \end{bmatrix} \begin{bmatrix} x_1 \\ x_2 \\ x_3 \end{bmatrix} = \begin{bmatrix} 0 \\ 0 \\ 0 \end{bmatrix}$$

The second equation (which is the first nontrivial equality) yields

$$x_2 = \frac{\alpha_2}{(\alpha_1 - \alpha_3)} x_1 \tag{5-77}$$

Since $\alpha_1 > \alpha_3$, it follows that $x_2 > 0$ if $x_1 > 0$. The last equation then yields

$$x_3 = \frac{1}{\alpha_1 - \alpha_5} \left(-\alpha_5 + \alpha_2 \frac{(\alpha_4 - \alpha_5)}{\alpha_1 - \alpha_3} \right) x_1 \tag{5-78}$$

Since $\alpha_1 - \alpha_5 > 0$, it follows that $x_3 > 0$ if and only if

$$-\alpha_5(\alpha_1 - \alpha_3) + \alpha_2(\alpha_4 - \alpha_5) > 0$$

Equivalently, for the eigenvector to have all positive components we require

$$\begin{gathered} \alpha_1 > \alpha_3 \\ \alpha_1 > \alpha_5 \\ \alpha_2(\alpha_4 - \alpha_5) > \alpha_5(\alpha_1 - \alpha_3) \end{gathered} \tag{5-79}$$

If these conditions are satisfied, α_1 will be the dominant eigenvalue and its corresponding eigenvector will have positive components.

It is clear that if $\alpha_1 > \alpha_3$, and $\alpha_2 > 0$, $\alpha_4 > 0$, the other two required inequalities will be satisfied for *small* α_5. This corresponds to having a *low* birthrate in the lowest class. Thus, the Natchez problem is solved by having a relatively low lower-class birthrate.

The above relations, however, are not complete. By themselves they do not constitute an acceptable solution to our problem—and it is here that our original intuition, that there should be an increase in production of the lower class, is in a sense validated. It is necessary to check that in every generation there are enough class C members to supply marriage partners for the upper classes. In particular, it must hold that

$$x_3 \geq x_1 + x_2 \tag{5-80}$$

This inequality is an additional requirement to be imposed on the eigenvector representing the stable population distribution. Substituting (5-77) and (5-78) into (5-80) produces, after a bit of algebra, the condition

$$\frac{(\alpha_4-\alpha_5)}{(\alpha_1-\alpha_5)}-\frac{\alpha_5\,(\alpha_1-\alpha_3)}{\alpha_2\,(\alpha_1-\alpha_5)}-\frac{(\alpha_1-\alpha_3)}{\alpha_2}-1\geq 0 \tag{5-81}$$

All terms on the left-hand side of this inequality, except the first, are negative. Therefore, a necessary condition is easily seen to be $\alpha_4>\alpha_1$. Thus, the value of α_4, which represents the indirect production of class C males from the next higher class, must be increased in order to compensate for the reduction in the direct production.

A solution to our problem is given by any set of birthrates satisfying the set of inequalities above. For example, $\alpha_1=1.1$, $\alpha_2=1.0$, $\alpha_3=1.0$, $\alpha_4=1.3$, and $\alpha_5=.9$ works, and has a stable population distribution with proportions 1, 10, and 15.5. A set of numbers of this form does represent a solution to the formal problem posed, although it is, at best, perhaps questionable that a society would be able to arrange a set of societal norms producing average birthrates that vary among marriage types in a fashion even approximating this peculiar pattern.

Now that the analysis is completed and the answer to our problem has been derived, let us review the essence of the analysis to discover that the answer is really not so surprising after all. The key to the situation is the simple equation $x_1(k+1)=\alpha_1 x_1(k)$. The highest class can grow only at the rate α_1, no matter how the other classes grow or what values are prescribed for other parameters. Therefore, growth in the upper class constrains (in fact determines) the ultimate growth rate of any population distribution. Or, put another way, if one is seeking a steady growth situation (with nonzero x_1), the growth rate must be α_1.

The fallacy in selecting a large value for α_5, the birthrate of the lowest class, is that if the birthrate is increased enough to avoid depletion, the lowest class will ultimately grow at that large birthrate and leave the populations of the other classes behind. In relative terms, the other classes will be depleted, for they are unable to keep up. There will not be a balanced distribution in which all classes grow at the same rate.

The direct growth rates of all classes must not exceed the growth rate α_1 of the constraining class A. On the other hand, to insure that there are sufficient members of the lowest class, their indirect production, as offspring of the next higher class, must be high enough to compensate for the low birthrate in the class itself. This brief analysis suggests, therefore, that a solution obtained by reducing α_3 and α_5 with respect to α_1, and increasing α_4, is not terribly surprising after all!

5.14 PROBLEMS

1. Verify Eq. (5-2) of Sect. 5.1.

2. Given the system

$$\mathbf{x}(k+1)=\mathbf{A}\mathbf{x}(k)$$

consider a solution of the form $\mathbf{x}(k)=\lambda^k\mathbf{x}(0)$ for some fixed vector $\mathbf{x}(0)$ and some constant λ. By substituting this form of solution into the system equation, find conditions that must be satisfied by λ and $\mathbf{x}(0)$.

3. Consider the difference equation

$$y(k+n)+a_{n-1}y(k+n-1)+\cdots+a_0y(k)=0$$

As defined in Chapter 2, this equation has the characteristic equation

$$\lambda^n+a_{n-1}\lambda^{n-1}+\cdots+a_0=0$$

and if λ_i is a root of this equation, then $y(k)=\lambda_i^k$ is a solution to the difference equation. Write the difference equation in state vector form. Show that the characteristic polynomial of the resulting system matrix is identical to that given above. Find an eigenvector corresponding to λ_i and show that it is equivalent to the earlier solution.

4. Find the eigenvalues and eigenvectors of the harmonic motion system and the Lanchester model (Examples 2 and 3, Sect. 4.6). Use this information to (re)calculate the state-transition matrix in each case.

*5. *Coordinate Symmetry.* Suppose a system has the property that its system matrix is unchanged if the state variables are permuted in some way. Suppose in particular that $\mathbf{P}^{-1}\mathbf{A}\mathbf{P}=\mathbf{A}$, where $\mathbf{P}$ represents a change of variable. Show that if $\mathbf{e}$ is a system eigenvector, then so is $\mathbf{Pe}$. As an example consider the mass and spring system shown in Fig. 5.11. Formulate the system in state variable form. Identify a symmetry and find $\mathbf{P}$. Find the eigenvectors.

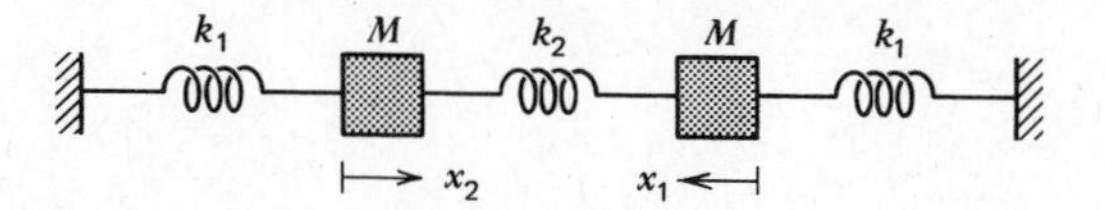

Figure 5.11. Symmetric system.

6. Convert the migration model of Sect. 5.5 to diagonal form and find the state-transition matrix.

7. For the migration model of Sect. 5.5, show that the condition

$$0\le\beta\le\min\left(\frac{\alpha}{1-\gamma},\frac{\alpha}{\gamma}\right)$$

is necessary and sufficient in order to guarantee that both urban and rural populations remain nonnegative given any nonnegative initial conditions. Show that $0 \le \beta \le 2\alpha$ is an equivalent condition.

8. Consider the following modification of the migration model discussed in Sect. 5.5. A third sector, suburban population, is added, and we assume that it is governed by $s(k+1) = \alpha s(k) + \delta u(k)$. That is, each year a fraction δ of the urban population moves to suburbia; there is no possibility of a return flow. We modify the other two equations to

$$r(k+1) = \alpha r(k) - \beta\{r(k) - \gamma[r(k) + u(k) + s(k)]\}$$
$$u(k+1) = \alpha u(k) + \beta\{r(k) - \gamma[r(k) + u(k) + s(k)]\} - \delta u(k)$$

Find the eigenvalues and right and left eigenvectors of this three-sector model and interpret the results.

9. Assume that the $n \times n$ matrix $\mathbf{A}$ has n distinct eigenvalues, $\lambda_1, \lambda_2, \ldots, \lambda_n$. Let $\mathbf{M} = [\mathbf{e}_1, \mathbf{e}_2, \ldots, \mathbf{e}_n]$ be the modal matrix of $\mathbf{A}$, where $\mathbf{e}_1, \mathbf{e}_2, \ldots, \mathbf{e}_n$ are the *right* eigenvectors.

(a) Show that

$$\mathbf{M}^{-1} = \begin{bmatrix} \mathbf{f}_1^T \\ \mathbf{f}_2^T \\ \cdot \\ \cdot \\ \cdot \\ \mathbf{f}_n^T \end{bmatrix}$$

where $\mathbf{f}_1^T, \mathbf{f}_2^T, \ldots, \mathbf{f}_n^T$ are the corresponding normalized left eigenvectors.

(b) Show that $\mathbf{A}$ can be expressed as

$$\mathbf{A} = \sum_{i=1}^{n} \lambda_i \mathbf{e}_i \mathbf{f}_i^T$$

(c) Show that

$$\mathbf{A}^k = \sum_{i=1}^{n} \lambda_i^k \mathbf{e}_i \mathbf{f}_i^T$$

10. Find the eigenvalues and eigenvectors for the Natchez Indian system (Example 3, Sect. 4.4). (There are only two eigenvectors.) From the explicit formula for $\mathbf{A}^k$ infer what the lengths of the two Jordan chains must be.

11. *Cournot Theory of Duopoly.* Duopoly refers to a market structure in which two firms compete to serve the industry demand. Since price varies with total production, it is clear that each firm must account for the actions of the other when determining its own best production level. Various dynamic processes result, depending on the particular strategies employed by the two firms.

Suppose the total industry demand in any period is

$$d(k)=200-2p(k)$$

where $p(k)$ is current price. Let $q_1(k)$, $q_2(k)$ denote the output levels of the two firms. Assuming that the price adjusts so as to sell all current output, it follows that

$$p(k)=100-\tfrac{1}{2}[q_1(k)+q_2(k)]$$

Suppose further that the total cost of production for the two firms are

$$C_1(k)=5q_1(k)$$
$$C_2(k)=\tfrac{1}{2}q_2(k)^2$$

Both firms know the demand curve, but each knows only its own cost curve. The current profits for the two firms are revenue minus cost; that is, $p(k)q_1(k)-C_1(k)$ and $p(k)q_2(k)-C_2(k)$, respectively.

The Cournot theory is based on the assumption that each firm selects its output level to maximize its own profit, using some estimate of its competitor's output. The corresponding function, for each firm, expressing the best output level as a function of the estimate of the competitor's output, is called a *reaction function.*

(a) Assuming each firm estimates its competitor's output by using its competitor's actual output of the previous period, derive two reaction functions in the form of two first-order linear difference equations.
(b) Find the equilibrium outputs of each firm.
(c) Derive a general solution for *even* periods and verify your answer to part (b).
(d) Suppose that both firms estimate each others' output as a simple average of the previous two periods in an effort to smooth out the oscillatory response. Show that for arbitrary initial conditions, the convergence to equilibrium need not be more rapid. (*Hint*: You need not factor, but you must analyze, the new characteristic polynomial.)

12. *Stackelberg Theory of Duopoly* (see above). Stackelberg suggested that one firm might increase its profits if it were able to observe its competitor's reaction function. Such a firm would then use its competitor's reaction function as its estimate of its competitor's output.

(a) Show that if the first firm substitutes the second firm's reaction function of part (a) of the previous problem into its profit equation and maximizes with respect to $q_1(k+1)$, then the equilibrium is unchanged.
(b) In a Stakelberg strategy the two firms are designated leader and follower. The leader, knowing the follower's reaction function, reveals its actual planned output to the follower. The follower then uses the value supplied by the leader in its reaction function to determine its output level. The leader's original plan anticipates the follower's response to it. Show that if the first firm is the leader, its equilibrium profit is higher than in the Cournot equilibrium.

13. *The Routh Test.* There is a simple but powerful test to determine whether a given polynomial has any roots in the right half of the complex plane. Consider the nth-order polynomial $p(\lambda)=a_n\lambda^n+a_{n-1}\lambda^{n-1}+\cdots+a_1\lambda+a_0$. The Routh array is

then constructed as shown below:

$$\begin{array}{cccc} a_0 & a_2 & a_4 & \cdots \\ a_1 & a_3 & a_5 & \cdots \\ b_1 & b_2 & b_3 & \cdots \\ c_1 & c_2 & c_3 & \cdots \\ \cdot & & & \\ \cdot & & & \\ \cdot & & & \end{array}$$

In this array the even-numbered coefficients are placed in sequence in the top row and the odd-numbered ones in the second row. The elements of the third row are found by the cross multiplication formulas

$$b_1 = \frac{a_1 a_2 - a_3 a_0}{a_1}$$

$$b_2 = \frac{a_1 a_4 - a_5 a_0}{a_1}$$

$$b_3 = \frac{a_1 a_6 - a_7 a_0}{a_1}$$

and so forth. Successive rows are computed from the two proceeding rows using the same formula. Finally, when no more rows can be defined, the number of sign changes in the first column of the array is equal to the number of roots of $p(\lambda)$ in the right half-plane.

As an example, the array for $P(\lambda) = -4\lambda^4 + \lambda^3 + 2\lambda^2 + \lambda + 4$ is

$$\begin{array}{ccc} 4 & 2 & -4 \\ 1 & 1 & \\ -2 & -4 & \\ -1 & & \\ -4 & & \end{array}$$

This array has one change of sign in the first column, indicating that there is exactly one root in the right half-plane. For each of the following polynomials determine how many roots are in the right half-plane:

(a) $\lambda^2 - 2\lambda + 1$
(b) $\lambda^3 + 4\lambda^2 + 5\lambda + 2$
(c) $-2\lambda^5 - 4\lambda^4 + \lambda^3 + 2\lambda^2 + \lambda + 4$

14. (a) Show that reversing the order of coefficients (replacing a_i by a_{n-i}) in the Routh test must give the same result.
(b) Show that this can be helpful for testing $\lambda^6 + \lambda^5 + 3\lambda^4 + 2\lambda^3 + 4\lambda^2 + a\lambda + 8$ where a is a parameter.

15. For stability analysis of discrete-time systems it is necessary to determine whether the characteristic equation has all its roots inside the unit circle in the complex plane. Consider the transformation from z to λ

$$\lambda = \frac{z+1}{z-1}$$

(a) How does this transform the unit circle?
(b) Use this transformation together with the Routh test (Problem 13) to determine how many roots of the polynomial $z^4 + 2z^2 + z + 1$ lie outside the unit circle.

16. *Multidimensional Cobweb Theory.* The demands for various commodities are rarely independent. If the price of one rises then it is likely that its demand will decrease, and the demand for similar commodities, which are partial substitutes, will rise. As an example, demand for two commodities might be described by the relation

$$\mathbf{d} = \mathbf{A}\mathbf{p} + \mathbf{d}_0$$

where $\mathbf{d}$ is a two-dimensional vector of demands, $\mathbf{p}$ is a two-dimensional vector of corresponding prices, $\mathbf{d}_0$ is a constant two-dimensional vector, and

$$\mathbf{A} = \begin{bmatrix} -\alpha & \beta \\ \gamma & -\delta \end{bmatrix}$$

where α, β, γ, δ are positive, and $\alpha\delta - \beta\gamma \neq 0$.

As in the standard cobweb model, assume that supply is determined from price with a one period delay

$$\mathbf{s}(k+1) = \mathbf{s}_0 + \mathbf{E}\mathbf{p}(k)$$

Assume also that the supplies of the two commodities are independent, and thus that $\mathbf{E}$ is a diagonal matrix with positive diagonal elements. For simplicity let $\mathbf{E} = \mathbf{I}$.

(a) Equate supply and demand and obtain a dynamic equation for $\mathbf{p}(k)$.
(b) Find the characteristic polynomial of the matrix $\mathbf{A}$, and determine whether the roots are real or complex.
(c) What condition must be satisfied by the eigenvalues of $\mathbf{A}$ in order for the system in (a) to be asymptotically stable?
(d) A bundle of goods (x_1, x_2) consists of x_1 units of the first commodity and x_2 units of the second. The price of the bundle is $q(k) = p_1(k)x_1 + p_2(k)x_2$. For $\alpha = 4$, $\beta = 1$, $\gamma = 2$, and $\delta = 3$, find the bundles whose prices are governed by first-order dynamic equations, and display the corresponding equations. *Hint*: A bundle can be represented as a row vector.

*17. Suppose that hog farmers replace (5-59) in Sect. 5.10 with a price estimator of the form

$$\hat{p}(k) = \tfrac{1}{7}[p(k) + p(k-1) + p(k-2) + p(k-3) + p(k-4) + p(k-5) + p(k-6)]$$

(a) Show that this leads to the characteristic equation

$$\lambda^8 + \frac{\rho}{7}[\lambda^6 + \lambda^5 + \lambda^4 + \lambda^3 + \lambda^2 + \lambda + 1] = 0$$

(b) Given that ρ is such that the largest root of this equation has magnitude equal to one, what is the cycle length corresponding to the root? *Hint*: Analyze the characteristic equation geometrically, on the unit circle.

(c) What is ρ?

18. *Analysis of Structures.* Mechanical structures (bridges, aircraft frames, buildings, etc.) are built from materials having some elasticity, and are therefore dynamic systems. To determine their response to varying forces due to traffic, winds, earthquakes, and so forth, it is important to calculate the natural frequencies of oscillation. As an example, consider a four story building as illustrated in Fig. 5.12. As an idealization, assume that the elements of the frame are inextensible and that the mass is concentrated in the floors. The floors can be displaced laterally with respect to each other, but the bending elasticity of the building frame then generates restoring forces. Assuming that all frame members have the same elasticity, the force vector $\mathbf{f}$ is related to the displacement vector $\mathbf{x}$ by $\mathbf{f} = \mathbf{K}\mathbf{x}$, where $\mathbf{K}$ is the *stiffness matrix*. By Newton's laws, the force also satisfies $-\mathbf{f} = \mathbf{M}\ddot{\mathbf{x}}(t)$, where $\mathbf{M}$ is the *mass matrix*. In this case

$$\mathbf{K} = k\begin{bmatrix} 1 & -1 & 0 \\ -1 & 2 & -1 \\ 0 & -1 & 2 \end{bmatrix} \qquad \mathbf{M} = m\begin{bmatrix} 1 & 0 & 0 \\ 0 & 1 & 0 \\ 0 & 0 & 2 \end{bmatrix}$$

(a) For this structure, show that the natural frequencies of the system are of the form $\omega = \sqrt{\lambda}$ where λ is an eigenvalue of $\mathbf{M}^{-1}\mathbf{K}$.

(b) In order to be immune to earthquakes, the building should have all its natural frequencies well above 10 cycles/sec. If $m = 10^4$ Kg and $k = 10^7$ Newtons/meter, is this an acceptable design? [Note: Using these units, ω of part (a) is in radians/sec.]

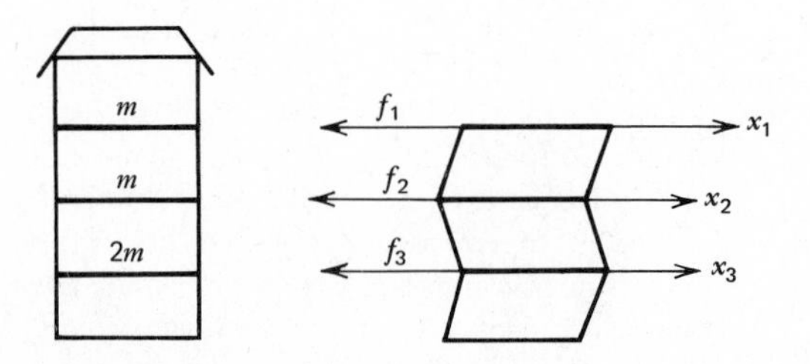

Figure 5.12. A four-story building.

19. Make the following simplifications in the population model:
—women have children between the ages of 15 and 44 inclusive;
—each woman has exactly one female child;
—each woman lives to be at least 45 years old.

Using a time interval of 15 years with these assumptions, and considering only those females less than 45 years old, gives the following model:

$$\begin{bmatrix} x_1(k+1) \\ x_2(k+1) \\ x_3(k+1) \end{bmatrix} = \begin{bmatrix} 0 & \alpha & 1-\alpha \\ 1 & 0 & 0 \\ 0 & 1 & 0 \end{bmatrix} \begin{bmatrix} x_1(k) \\ x_2(k) \\ x_3(k) \end{bmatrix}$$

where $0 \leq \alpha \leq 1$.

(a) By finding the eigenvalues, determine the growth rate and how quickly the population moves to equilibrium.

(b) What happens as α varies from 0 to 1?

20. Suppose that a cohort population matrix $\mathbf{P}$ has been transformed to $\bar{\mathbf{P}}$, as described in Sect. 5.12. If $a_0 + a_1 + \cdots + a_n \simeq 1$ (i.e., each woman has approximately one female child in her lifetime), it can be expected that the eigenvalue λ_0 is close to one. Setting $\lambda_0 = 1 + \varepsilon$, and using the fact that $(1+\varepsilon)^m \simeq 1 + m\varepsilon$ when ε is small, substitute in the characteristic equation to find an approximate value for ε.
Estimate ε if

$$\bar{\mathbf{P}} = \begin{bmatrix} .01 & .78 & .25 & .01 \\ 1 & 0 & 0 & 0 \\ 0 & 1 & 0 & 0 \\ 0 & 0 & 1 & 0 \end{bmatrix}$$

*21. *Reproductive Value.* Associated with the dominant eigenvalue λ_0 of a cohort population matrix $\mathbf{P}$, there is a left eigenvector $\mathbf{f}_0^T = [f_0, f_1, f_2, \ldots, f_n]$. Show that

$$f_k = \alpha_k + \sum_{i=k-1}^{n} (\beta_k \beta_{k+1} \cdots \beta_{i-1}) \alpha_i \lambda_0^{k-i}$$

Show that f_k can be interpreted as a measure of the total number of (female) children that a woman in age group k will have, on the average, during the remainder of her lifetime. Children born in the future are discounted at a rate λ_0, the natural growth rate of the population. Therefore, a child who will be born i time periods later is counted now, not as 1, but as the fraction λ_0^{-i}.
The function value

$$V = \mathbf{f}_0^T \mathbf{x}(k) = \sum_{i=0}^{n} f_i x_i(k)$$

is the total reproductive value of the current population. It is the total discounted number of potential children of the present population. Since $\mathbf{f}_0^T$ is a left eigenvector, it follows that the reproductive value increases by the factor λ_0 each time period, *no matter what the population distribution.* Verify this in terms of the interpretation of the reproductive value.

22. *Lattice Vibrations.* Many of the observed properties of a solid, such as specific heat, dielectric properties, optical and sound transmission properties, and electrical

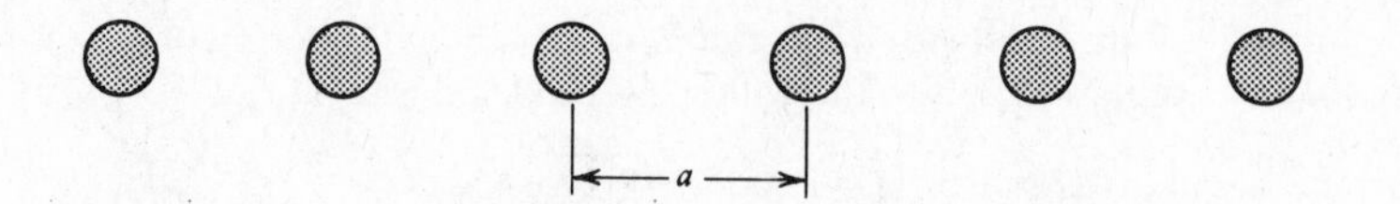

Figure 5.13 A monatomic lattice.

resistance, are determined in large part by the characteristic oscillations of the atoms that comprise the solid. In a crystal the atoms are arranged in a regular pattern, or lattice, that determines the characteristics of the lattice vibrations. As a simple example of the kind of relations that can be deduced from the lattice structure, consider a one-dimensional monatomic lattice consisting of an infinite linear chain of equally spaced identical atoms. (See Fig. 5.13.) Each atom has a point mass of m, and the nominal spacing between atoms is a. If two adjacent atoms have a spacing different than a, there is an electric force proportional to the error in separation that tends to restore the nominal spacing. The proportionality constant is denoted by β.

Let $x_k(t)$ denote the position of the kth atom relative to its nominal position. Then using Newton's laws

$$M\ddot{x}_k(t)=\beta[x_{k+1}(t)-x_k(t)]+\beta[x_{k-1}(t)-x_k(t)]$$

for $k=\ldots,-1,0,1,2,\ldots$.

As a trial solution to this infinite set of equations, it is reasonable to suppose that each atom vibrates with a common frequency but with different phases. Thus, we seek solutions of the form

$$x_k(t)=e^{i\omega t}e^{ik\theta a}$$

(a) Find the relation between ω and θ (a dispersion relation).
(b) What is the maximum frequency that can be propagated?
(c) The velocity of wave propagation is ω/θ. What is the velocity for small frequencies?

NOTES AND REFERENCES

Section 5.5. The migration model was developed in collaboration with Stephen Haas.

Section 5.8. This example is based on McPhee [Mc1].

Section 5.10. The hog cycle example was developed in collaboration with Thomas Keelin. For another approach see Meadows [M3]. For background on expectations and cycles see Arrow and Nerlove [A3].

Section 5.12. The matrix approach to population analysis is credited to Leslie [L4] and the matrix in the model is itself often called a *Leslie matrix*. For an excellent treatment of population dynamics see Keyfitz [K13].

Section 5.13. The possibility of changing the birthrates in the Natchez model to avoid collapse has been suggested by Fischer [F3]. The analysis in this section, however, is new.

Section 5.14. For an introductory discussion of (nondynamic) aspects of Cournot and Stakelberg equilibria (problems 11 and 12) see Henderson and Quandt [H2]. The Routh test (Problems 13–15) is sometimes referred to as the Routh–Hurwitz test. Routh and Hurwitz obtained different (but essentially equivalent) solutions to the problem of determining stability in about 1880.

The analysis of structures as briefly introduced in Problem 18 is a large field of application of dynamic systems. See Fertis [F2]. For more on lattice vibrations (Problem 22) see Donovan and Angress [D2].

chapter 6.

Positive Linear Systems

6.1 INTRODUCTION

A positive linear system is a linear system in which the state variables are always positive (or at least nonnegative) in value. Such systems arise frequently since the state variables of many real systems represent quantities that may not have meaning unless they are nonnegative. In a cohort population model, for example, each variable remains nonnegative and corresponds to the population in a cohort class. In many economic systems the variables corresponding to quantities of goods remain nonnegative. And in an arms race the level of defense is nonnegative. A positive linear system automatically preserves the nonnegativity of the state variables.

The theory of positive systems is deep and elegant—and yet pleasantly consistent with intuition. Practically everything that one might hope to extrapolate from an understanding of simple first-order systems does, in fact, carry over to positive linear systems of arbitrary order. Indeed, just the knowledge that the system is positive allows one to make some fairly strong statements about its behavior; these statements being true no matter what values the parameters may happen to take. It is for positive systems, therefore, that dynamic systems theory assumes one of its most potent forms.

To explain the concept of positive linear systems more fully and more precisely, consider the homogeneous discrete-time dynamic system

$$\mathbf{x}(k+1)=\mathbf{A}\mathbf{x}(k) \tag{6-1}$$

If the state vector $\mathbf{x}(k)$ is nonnegative but otherwise arbitrary [that is, if every component of $\mathbf{x}(k)$ is nonnegative], under what circumstances can one be certain that the new state vector $\mathbf{x}(k+1)$ will also be nonnegative? Only if the elements in the matrix $\mathbf{A}$ are each nonnegative. To see this, suppose an element a_{ij} were negative. Then for the nonnegative vector $\mathbf{x}(k)=(0, 0, \ldots 0, 1, 0, 0, \ldots 0)$ with the one in the ith component, the new vector would have its jth component equal to a_{ij}, which is negative. Thus, in order to guarantee that a nonnegative state leads in turn to a new nonnegative state, the $\mathbf{A}$ matrix itself must have nonnegative entries. The converse proposition is even easier to see. If the elements of $\mathbf{A}$ are all nonnegative, then $\mathbf{x}(k+1)=\mathbf{A}\mathbf{x}(k)$ is certainly nonnegative for every nonnegative $\mathbf{x}(k)$.

A discrete-time linear homogeneous system is therefore defined to be positive (or nonnegative) if the elements of its system matrix $\mathbf{A}$ are all nonnegative. It logically follows that the theory of such systems is built upon the theory of positive matrices—a theory that is remarkably rich. The cornerstone of this theory is the famous Frobenius–Perron theorem, which is presented in the next section. This theorem plays a fundamental role in mathematical economics, dynamics, probability theory, and any linear theory involving positivity.

Preview of Frobenius–Perron Theorem

The main result of the Frobenius–Perron theorem is that for a matrix $\mathbf{A}$, all of whose elements are strictly positive, there is an eigenvalue of largest absolute value and this eigenvalue is in fact positive and simple. Furthermore, there is a positive eigenvector corresponding to this positive eigenvalue.

A major portion of this result can be quickly deduced from a knowledge of the general theory of linear time-invariant systems. In terms of dynamic system theory, the eigenvalue of largest absolute value corresponds to the dominant eigenvalue. Assume that there is a simple, dominant eigenvalue of the matrix $\mathbf{A}$. For large k and almost any initial condition, the solution $\mathbf{x}(k)$ to the system (6-1) tends to be aligned with the corresponding eigenvector. Since for any positive initial state vector the subsequent state vectors will all be positive, it follows immediately that the dominant eigenvector, to which the system converges, must have positive components and the corresponding eigenvalue must be positive. Thus, if there is a simple dominant eigenvalue, it is easy to see that it must be positive and must have an associated positive eigenvector.

The Frobenius–Perron theorem is a refinement of this basic result. It guarantees that there is in fact a simple dominant eigenvalue. In view of the importance of dominant eigenvalues, the Frobenius–Perron result is clearly of great value to dynamic system analysis.

Some Results on Positive Dynamic Systems

As stated earlier in this section, the theory of positive linear systems is both deep and elegant. It also has strong intuitive content, especially within the context of a given application. It is perhaps helpful, before confronting the somewhat lengthy proof of the Frobenius–Perron theorem, to preview, in broad terms at least, some of the major results that hold for positive systems. Perhaps these results and the variety of classical examples that fall in this category will serve as adequate motivation to master the foundation theory.

The first important property of positive linear systems is that concerning the dominant eigenvalue and its associated eigenvector, as described by the Frobenius–Perron theorem. The existence and positivity of the dominant eigenvalue and eigenvector essentially reduces the computation of long-term performance to an eigenvector calculation.

The second property of positive systems is a remarkable connection between stability and positivity. An inhomogeneous positive system, in general, may or may not possess an equilibrium point that itself is nonnegative. From the viewpoint of applications only a nonnegative equilibrium is of any real interest. There remains, however, the issue as to whether the equilibrium point is stable. For positive systems there is a direct correspondence between existence of a positive equilibrium point and stability. Thus, if a positive equilibrium point is found, it is stable. Conversely, if the system is stable, the corresponding equilibrium point is nonnegative.

A third major result comes under the heading of comparative statics, which is applicable to stable systems. Consider a stable system at rest at its equilibrium point. If some parameter of the system is slightly modified, the system moves to a new equilibrium point. Comparative statics refers to the question of how the change in equilibrium point is related to the parameter change that produced it. For general linear systems, of course, not much can be said to describe this relationship. For positive systems, however, it can be shown that positive changes (such as increasing a term in the **A** matrix) produce corresponding positive changes (increases) in the components of the equilibrium points. This result, and others of a similar nature, mean that significant qualitative conclusions can be inferred about the behavior of a positive system even though the values of the parameters may not be known precisely.

6.2 POSITIVE MATRICES

In this section the fact that a positive matrix has a positive dominant eigenvalue is established. Before proceeding directly into that development, however, it is convenient to introduce some notation for distinguishing positive (and non-negative) vectors and matrices.

If $\mathbf{A} = [a_{ij}]$ is a matrix, we write:

(i) $\mathbf{A} > \mathbf{0}$ if $a_{ij} > 0$ for all i,j

(ii) $\mathbf{A} \geq \mathbf{0}$ if $a_{ij} \geq 0$ for all i,j and $a_{ij} > 0$ for at least one element

(iii) $\mathbf{A} \geqq \mathbf{0}$ if $a_{ij} \geq 0$ for all i,j.

These cases are distinguished verbally by stating that (i) $\mathbf{A}$ is *strictly positive* if all its elements are strictly greater than zero, (ii) $\mathbf{A}$ is *positive* or *strictly nonnegative* if all elements of $\mathbf{A}$ are nonnegative but at least one element is nonzero, and (iii) $\mathbf{A}$ is *nonnegative* if all elements are nonnegative. This same terminology can be applied to matrices of any dimension (whether square or not), but attention is focused in this section on square matrices and on row and column vectors.

These definitions are also used to imply meaning for inequalities of the form $\mathbf{A} \geq \mathbf{B}$. We say $\mathbf{A} \geq \mathbf{B}$ for two matrices $\mathbf{A}$ and $\mathbf{B}$ of the same dimension if $\mathbf{A}-\mathbf{B} \geq \mathbf{0}$. Similar definitions apply to the other forms of positivity.

The proof of the existence of a positive dominant eigenvalue of a positive (square) matrix $\mathbf{A}$ is somewhat indirect. First a positive number is associated with the matrix $\mathbf{A}$ through examination of a set of inequalities, and it is subsequently shown that this number is in fact an eigenvalue of $\mathbf{A}$.

Let $\mathbf{A}$ be a nonnegative $n \times n$ matrix; that is, $\mathbf{A} \geqq \mathbf{0}$. Consider the set of real numbers λ such that

$$\mathbf{A}\mathbf{x} \geqq \lambda \mathbf{x}$$

for some $\mathbf{x} \geq \mathbf{0}$ (that is, for some strictly nonnegative vector $\mathbf{x}$). First, it can be noted that one number that always works is $\lambda = 0$, because any $\mathbf{x} \geq \mathbf{0}$ when multiplied by $\mathbf{A} \geqq \mathbf{0}$ must yield a nonnegative vector. Second, it can be seen that λ cannot be arbitrarily large. This is true because each component of the vector $\mathbf{A}\mathbf{x}$ is always less than Mx_i, where M is the sum of the elements of $\mathbf{A}$ and x_i is the maximum component of $\mathbf{x}$. Thus, for $\lambda > M$ there can be no $\mathbf{x} \geq \mathbf{0}$ with $\mathbf{A}\mathbf{x} \geqq \lambda \mathbf{x}$. Define λ_0 as the maximum of the real numbers λ for which $\mathbf{A}\mathbf{x} \geqq \lambda \mathbf{x}$ is satisfied for some $\mathbf{x} \geq \mathbf{0}$. In explicit terms*

$$\lambda_0 = \max\{\lambda: \quad \mathbf{A}\mathbf{x} \geqq \lambda \mathbf{x} \quad \text{some} \quad \mathbf{x} \geq 0\} \tag{6-2}$$

In view of the earlier observation, it follows that $0 \leq \lambda_0 < \infty$.

The next theorem below shows that, in the case where $\mathbf{A}$ is strictly positive, the value of λ_0 defined by (6-2) is a dominant eigenvalue for $\mathbf{A}$. This is equivalent to the statement that the inequality in the defining relation for λ_0 is satisfied by equality.

The proof of this theorem is substantially more difficult than most others in the book, and the mechanics of the proof are not essential for later developments. It may be appropriate, at first reading, to carefully study the theorem statement itself and then proceed directly to the next subsection where the statements of Theorems 2 and 3 can be read for comparison.

Theorem 1 (Frobenius-Perron). *If* $\mathbf{A} > \mathbf{0}$, *then there exists* $\lambda_0 > 0$ *and* $\mathbf{x}_0 > \mathbf{0}$ *such that* (a) $\mathbf{A}\mathbf{x}_0 = \lambda_0 \mathbf{x}_0$; (b) if $\lambda \neq \lambda_0$ *is any other eigenvalue of* $\mathbf{A}$, *then* $|\lambda| < \lambda_0$; (c) λ_0 *is an eigenvalue of geometric and algebraic multiplicity* 1.

* A continuity argument can be used to show that the maximum always exists.

Proof. (a) Let λ_0 be defined by (6-2), and let $\mathbf{x}_0 \geq \mathbf{0}$ be a vector corresponding to λ_0 in definition (6-2). That is, let $\mathbf{x}_0$ satisfy $\mathbf{A}\mathbf{x}_0 \geqq \lambda_0\mathbf{x}_0$. Clearly $\lambda_0 > 0$. Since $\mathbf{A} > \mathbf{0}$, it follows that $\mathbf{A}\mathbf{x} > \mathbf{0}$ for any $\mathbf{x} \geq \mathbf{0}$. Therefore, $\mathbf{A}[\mathbf{A}\mathbf{x}_0 - \lambda_0\mathbf{x}_0] > \mathbf{0}$ unless $\mathbf{A}\mathbf{x}_0 = \lambda_0\mathbf{x}_0$. Suppose $\mathbf{A}\mathbf{x}_0 \geq \lambda_0\mathbf{x}_0$; then for $\mathbf{y}_0 = \mathbf{A}\mathbf{x}_0 > \mathbf{0}$ it would follow that $\mathbf{A}\mathbf{y}_0 - \lambda_0\mathbf{y}_0 > \mathbf{0}$, or equivalently $\mathbf{A}\mathbf{y}_0 > \lambda_0\mathbf{y}_0$. But if this were true λ_0 could be increased slightly without violating the inequality, which contradicts the definition of λ_0. Therefore, it follows that $\mathbf{A}\mathbf{x}_0 = \lambda_0\mathbf{x}_0$. Also since $\mathbf{x}_0 \geq \mathbf{0}$ implies $\mathbf{A}\mathbf{x}_0 > \mathbf{0}$, the equation $\mathbf{A}\mathbf{x}_0 = \lambda_0\mathbf{x}_0$ implies that $\mathbf{x}_0 > \mathbf{0}$.

(b) Let $\lambda \neq \lambda_0$ be an eigenvalue of $\mathbf{A}$, and let a corresponding nonzero eigenvector be $\mathbf{y}$, $\mathbf{A}\mathbf{y} = \lambda\mathbf{y}$. Let $|\mathbf{y}|$ denote the vector whose components are the absolute values of the components of $\mathbf{y}$ and consider the vector $\mathbf{A}|\mathbf{y}|$. The first component of this vector is $a_{11}|y_1| + a_{12}|y_2| + \cdots + a_{1n}|y_n|$. Since the a_{ij}'s are positive, this sum is greater than or equal to $|a_{11}y_1 + a_{12}y_2 + \cdots + a_{1n}y_n|$. Since a similar result holds for the other components, it follows that $\mathbf{A}|\mathbf{y}| \geqq |\mathbf{A}\mathbf{y}|$. Thus, $\mathbf{A}|\mathbf{y}| \geqq |\lambda\mathbf{y}| = |\lambda|\,|\mathbf{y}|$. From the definition of λ_0 it follows immediately that $|\lambda| \leq \lambda_0$.

In order to prove that strict inequality holds, consider the matrix $\mathbf{A}_\delta = \mathbf{A} - \delta\mathbf{I}$, where $\delta > 0$ is chosen small enough so that $\mathbf{A}_\delta$ is still strictly positive. From the equation $(\lambda - \delta)\mathbf{I} - \mathbf{A}_\delta = \lambda\mathbf{I} - \mathbf{A}$, it follows that $\lambda_0 - \delta$ and $\lambda - \delta$ are eigenvalues of $\mathbf{A}_\delta$. Furthermore, because $\mathbf{A}_\delta$ is strictly positive, it follows that $|\lambda - \delta| \leq \lambda_0 - \delta$, since $\lambda_0 - \delta$ is the largest eigenvalue of $\mathbf{A}_\delta$. However, if $|\lambda| = \lambda_0$, $\lambda \neq \lambda_0$ (so that only λ_0 is positive), it follows by direct computation of the absolute value (see Fig. 6.1) that $|\lambda - \delta| > |\lambda_0 - \delta|$. (The subtraction of δ affects the absolute value of a real number more than a complex one.) This is a contradiction.

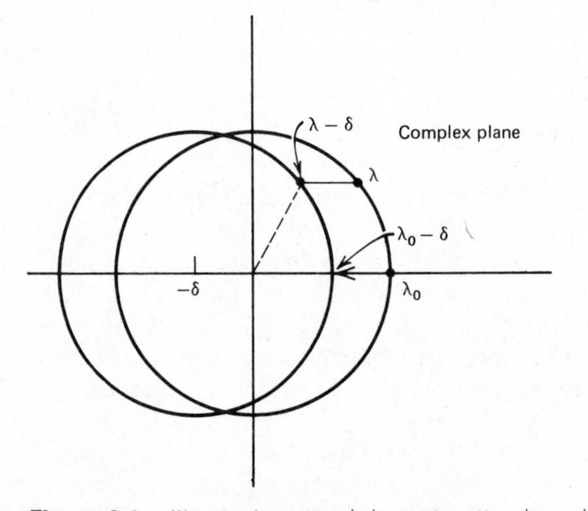

Figure 6.1. Illustration that $|\lambda| = \lambda_0$ implies $|\lambda - \delta| > |\lambda_0 - \delta|$.

(c) To prove that the geometric multiplicity is 1 it must be shown that (to within a scalar multiple) $\mathbf{x}_0$ is the only eigenvector associated with λ_0. Suppose there were another. Then, since $\mathbf{A}$ is real, there will be a real eigenvector $\mathbf{y}_0$ that is linearly independent of $\mathbf{x}_0$. Since $\mathbf{x}_0 > \mathbf{0}$, it is possible to find a linear combination $\mathbf{w} = \alpha\mathbf{x}_0 + \mathbf{y}_0$ such that $\mathbf{w} \geq \mathbf{0}$, but not $\mathbf{w} > \mathbf{0}$. However, since $\mathbf{A}\mathbf{w} = \lambda_0\mathbf{w}$ is strictly positive (because $\mathbf{A}$ is), we have a contradiction. Therefore, the geometric multiplicity of λ_0 is 1.

Finally, suppose the algebraic multiplicity is greater than 1. Then, since the geometric multiplicity is 1, there must be a Jordan chain of length at least two associated with λ_0. (See Problem 22, Chapter 3.) Thus, there is a vector $\mathbf{z}$ such that $(\mathbf{A} - \lambda_0\mathbf{I})\mathbf{z} = \mathbf{y}$ and $(\mathbf{A} - \lambda_0\mathbf{I})\mathbf{y} = \mathbf{0}$. In view of what was shown above, $\mathbf{y}$ must be a multiple of $\mathbf{x}_0$, and thus without loss of generality it can be assumed that $(\mathbf{A} - \lambda_0\mathbf{I})\mathbf{z} = \mathbf{x}_0$. Now let $\mathbf{f}_0$ be the strictly positive eigenvector of $\mathbf{A}^T$ corresponding to λ_0. Then $\mathbf{f}_0^T$ is a *left* eigenvector of $\mathbf{A}$ and we have

$$0 = \mathbf{f}_0^T(\mathbf{A} - \lambda_0\mathbf{I})\mathbf{z} = \mathbf{f}_0^T\mathbf{x}_0$$

But $\mathbf{f}_0^T\mathbf{x}_0$ is positive because both $\mathbf{f}_0$ and $\mathbf{x}_0$ are strictly positive. We have a contradiction, and can therefore conclude that the algebraic multiplicity is 1. ▮

Extensions to Nonnegative Matrices

Many of the results presented in the above theorem for strictly positive matrices can be extended to nonnegative matrices. We state two alternative formulations without proof.

The first is a direct extension showing that strict positivity of $\mathbf{A}$ can be replaced by strict positivity of a power of $\mathbf{A}$.

Theorem 2. *Let* $\mathbf{A} \geq \mathbf{0}$ *and suppose* $\mathbf{A}^m > \mathbf{0}$ *for some positive integer* m. *Then conclusions* (a), (b), *and* (c) *of Theorem* 1 *apply to* $\mathbf{A}$.

The second alternative is the corresponding theorem for an arbitrary nonnegative matrix. In this case the conclusions are weaker.

Theorem 3. *Let* $\mathbf{A} \geqq \mathbf{0}$. *Then there exists* $\lambda_0 \geq 0$ *and* $\mathbf{x}_0 \geq \mathbf{0}$ *such that* (a) $\mathbf{A}\mathbf{x}_0 = \lambda_0\mathbf{x}_0$; (b) *if* $\lambda \neq \lambda_0$ *is any other eigenvalue of* $\mathbf{A}$, *then* $|\lambda| \leq \lambda_0$.

There are other versions of this type of theorem that impose requirements lying somewhere between simple nonnegativity and strict positivity of a power of $\mathbf{A}$. (See, for example, Problem 10, Chapter 7.)

Example 1. Consider the following five nonnegative matrices:

$$\mathbf{A}_1 = \begin{bmatrix} 1 & 1 \\ 1 & 1 \end{bmatrix} \quad \mathbf{A}_2 = \begin{bmatrix} 0 & 1 \\ 1 & 1 \end{bmatrix}$$

$$\mathbf{A}_3 = \begin{bmatrix} 1 & 0 \\ 0 & 1 \end{bmatrix} \quad \mathbf{A}_4 = \begin{bmatrix} 0 & 1 \\ 1 & 0 \end{bmatrix} \quad \mathbf{A}_5 = \begin{bmatrix} 0 & 1 \\ 0 & 0 \end{bmatrix}$$

The matrix $\mathbf{A}_1$ is strictly positive so Theorem 1 applies. Indeed $\lambda_0=2$ is the simple dominant eigenvalue. The matrix $\mathbf{A}_2$ is not strictly positive, but $\mathbf{A}_2^2>\mathbf{0}$. Thus, Theorem 2 applies and indeed $\lambda_0=(1+\sqrt{5})/2$ is the simple dominant eigenvalue. In the last three cases, no power of $\mathbf{A}$ is strictly positive. For $\mathbf{A}_3$ the dominant eigenvalue is $\lambda_0=1$, but there is not a unique eigenvector. For $\mathbf{A}_4$ the two eigenvalues are $\lambda=-1,1$, so the positive eigenvalue is not strictly greater in absolute value than the other eigenvalue. For $\mathbf{A}_5$ the dominant eigenvalue is $\lambda_0=0$, and it has geometric multiplicity 1 but algebraic multiplicity 2.

Bounds on λ_0

The eigenvalue $\lambda_0 \geq 0$ associated with a nonnegative matrix $\mathbf{A}$ in Theorems 1, 2, and 3 above is referred to as the *Frobenius–Perron eigenvalue* of $\mathbf{A}$. A useful set of bounds can be derived for the value of λ_0, expressed in terms of either the row or column sums of the matrix $\mathbf{A}$. These bounds can be used to obtain a quick numerical estimate of λ_0 for a given nonnegative matrix, and they are useful in some theoretical investigations.

Let $\mathbf{A}$ be a nonnegative matrix with largest eigenvalue λ_0. Let $\mathbf{x}_0=(x_1, x_2, \ldots, x_n)$ be the corresponding positive eigenvector, and for convenience assume that this eigenvector is normalized such that $\sum_{i=1}^n x_i=1$. We have $\mathbf{A}\mathbf{x}_0=\lambda_0\mathbf{x}_0$, or in detail

$$
\begin{aligned}
a_{11}x_1+a_{12}x_2+\cdots+a_{1n}x_n &= \lambda_0 x_1\\
a_{21}x_1+a_{22}x_2+\cdots+a_{2n}x_n &= \lambda_0 x_2\\
&\;\;\vdots\\
a_{n1}x_1+a_{n2}x_2+\cdots+a_{nn}x_n &= \lambda_0 x_n
\end{aligned}
$$

Summing these n equations we obtain

$$\Delta_1 x_1+\Delta_2 x_2+\cdots+\Delta_n x_n=\lambda_0(x_1+x_2+\cdots+x_n)$$

where Δ_i denotes the sum of the elements in the ith column of $\mathbf{A}$. Recalling the normalization of $\mathbf{x}_0$, there results

$$\lambda_0=\Delta_1 x_1+\Delta_2 x_2+\cdots+\Delta_n x_n$$

Therefore, λ_0 is a weighted average of the column sums. Since the average must lie between the two extremes, we obtain the useful bounds

$$\operatorname*{Min}_i \Delta_i \leq \lambda_0 \leq \operatorname*{Max}_i \Delta_i$$

The same argument can be applied to the matrix $\mathbf{A}^T$, which has the same λ_0 as $\mathbf{A}$. This yields the bounds

$$\underset{i}{\text{Min}}\ \delta_i \leq \lambda_0 \leq \underset{i}{\text{Max}}\ \delta_i$$

where now δ_i is the sum of elements in the ith *row* of $\mathbf{A}$.

Example 2. Consider the strictly positive matrix

$$\mathbf{A} = \begin{bmatrix} 1 & 2 \\ 4 & 3 \end{bmatrix}$$

Since the column sums are both 5 it follows that $\lambda_0 = 5$. The row sums are 3 and 7. The eigenvector associated with λ_0 can be found from (either of) the equations

$$-4x_1 + 2x_2 = 0$$
$$4x_1 - 2x_2 = 0$$

Thus, $\mathbf{x}_0 = (1, 2)$.

Example 3. Consider the positive matrix

$$\mathbf{A} = \begin{bmatrix} 0 & 0 & 6 \\ 1 & 0 & 4 \\ 0 & 1 & 1 \end{bmatrix}$$

which is in companion form. The column sums show that $1 \leq \lambda_0 \leq 11$. The row sums, however, yield the tighter bounds $2 \leq \lambda_0 \leq 6$. Actually, in this case $\lambda_0 = 3$. The corresponding eigenvector is $\mathbf{x}_0 = (2,2,1)$.

6.3 POSITIVE DISCRETE-TIME SYSTEMS

The theory of positive matrices can be applied directly to positive linear dynamic systems, and yields some surprisingly strong conclusions.

Dominant Eigenvector Analysis

An obvious role of the Frobenius–Perron theorem in dynamic system analysis is its guarantee that a positive system has a nonnegative dominant eigenvalue. The general theory of linear time-invariant systems, discussed in Chapter 5, reveals the importance of this eigenvalue (and its eigenvector) as determining the long-term behavior of the system.

Consider, in particular, a homogeneous positive discrete-time system of the form

$$\mathbf{x}(k+1) = \mathbf{A}\mathbf{x}(k) \tag{6-3}$$

with $\mathbf{A}$ strictly positive, or $\mathbf{A}^m$ strictly positive for some $m \geq 1$. As $k \to \infty$ the solution to (6-3) is approximately equal to

$$\mathbf{x}(k) = \alpha \lambda_0^k \mathbf{x}_0$$

where λ_0 is the Frobenius–Perron eigenvalue of $\mathbf{A}$, $\mathbf{x}_0$ is the corresponding positive eigenvector, and α depends on the initial condition. As an example, the system matrix for the cohort population model is nonnegative. (Indeed under suitable assumptions some power of the matrix is strictly positive.) We can conclude immediately (as in Sect. 5.12) that it must have a positive dominant growth rate and a corresponding positive population eigenvector.

Nonhomogeneous Systems and Stability

Consider now a nonhomogeneous system of the form

$$\mathbf{x}(k+1) = \mathbf{A}\mathbf{x}(k) + \mathbf{b} \tag{6-4}$$

Such a system is said to be *nonnegative* if $\mathbf{A} \geqq \mathbf{0}$ and $\mathbf{b} \geqq \mathbf{0}$. It is easy to see that these conditions on the matrix $\mathbf{A}$ and the vector $\mathbf{b}$ exactly correspond to the condition that the solution to (6-4) be nonnegative for any nonnegative initial vector $\mathbf{x}(0)$. It is pretty clear that these conditions are sufficient. And, to see that $\mathbf{b} \geqq \mathbf{0}$ is necessary, just assume $\mathbf{x}(0) = \mathbf{0}$. To see that $\mathbf{A} \geqq \mathbf{0}$ is necessary, consider large positive $\mathbf{x}(0)$'s. Therefore, the requirement that $\mathbf{A} \geqq \mathbf{0}$ and $\mathbf{b} \geqq \mathbf{0}$ in (6-4) is consistent with our earlier motivation.

The nonhomogeneous system (6–4) may have an equilibrium point $\bar{\mathbf{x}}$ satisfying

$$\bar{\mathbf{x}} = \mathbf{A}\bar{\mathbf{x}} + \mathbf{b} \tag{6-5}$$

Indeed if $\mathbf{I} - \mathbf{A}$ is nonsingular, there is always a unique equilibrium point. However, an equilibrium point satisfying (6-5) may not itself be a nonnegative vector. Clearly, within the context of positive systems, interest focuses mainly on nonnegative equilibrium points, and accordingly an important issue is to characterize those positive systems having nonnegative equilibrium points.

A related, but apparently separate issue, is whether a given equilibrium point is stable. In fact, however, for positive systems these issues are essentially identical: if the system (6-4) is asymptotically stable, its equilibrium point is nonnegative; and conversely if there is a nonnegative equilibrium point for some $\mathbf{b} > \mathbf{0}$, the system is asymptotically stable. This remarkable result is stated formally below.

Theorem 1. *Given* $\mathbf{A} \geqq \mathbf{0}$ *and* $\mathbf{b} > \mathbf{0}$, *the matrix* $\mathbf{A}$ *has all of its eigenvalues strictly within the unit circle of the complex plane if and only if there is an* $\bar{\mathbf{x}} \geqq \mathbf{0}$ *satisfying*

$$\bar{\mathbf{x}} = \mathbf{A}\bar{\mathbf{x}} + \mathbf{b} \tag{6-5}$$

Proof. Suppose first that $\mathbf{A}$ has all eigenvalues inside the unit circle. Then $\mathbf{I}-\mathbf{A}$ is nonsingular and a unique $\bar{\mathbf{x}}$ satisfies (6-5). Since the system (6-4) is asymptotically stable, any solution sequence converges to $\bar{\mathbf{x}}$. However, given any $\mathbf{x}(0) \geqq \mathbf{0}$ the solution sequence satisfies $\mathbf{x}(k) \geqq \mathbf{0}$ for all $k \geq 0$. Thus $\bar{\mathbf{x}} \geqq 0$. (Note that this part of the theorem is true with $\mathbf{b} \geqq \mathbf{0}$.)

Conversely, suppose there is $\bar{\mathbf{x}} \geqq \mathbf{0}$ satisfying (6-5). Since $\mathbf{b} > \mathbf{0}$, it must follow that actually $\bar{\mathbf{x}} > \mathbf{0}$. Let λ_0 be the Frobenius–Perron eigenvalue of $\mathbf{A}$, and let $\mathbf{f}_0^T$ be a corresponding *left* eigenvector, with $\mathbf{f}_0^T \geq \mathbf{0}$. Multiplication of (6-5) by $\mathbf{f}_0^T$ yields

$$(1-\lambda_0)\mathbf{f}_0^T\bar{\mathbf{x}} = \mathbf{f}_0^T\mathbf{b}$$

Since both $\mathbf{f}_0^T\bar{\mathbf{x}}$ and $\mathbf{f}_0^T\mathbf{b}$ are positive, it follows that $\lambda_0 < 1$. Thus, all eigenvalues of $\mathbf{A}$ are strictly within the unit circle of the complex plane. ▮

Although the above theorem is quite strong in that no similar statement holds for general linear time-invariant systems, the result is consistent with elementary reasoning. If a positive system is asymptotically stable, its equilibrium point must be nonnegative since every solution converges to it, and yet every solution that is initially nonnegative remains nonnegative.

Example 1 (Population Model with Immigration). Suppose

$$\mathbf{x}(k+1) = \mathbf{A}\mathbf{x}(k) + \mathbf{b}$$

represents a cohort population model. The $\mathbf{A}$ matrix corresponds to the usual cohort matrix and is nonnegative. The vector $\mathbf{b}$ represents immigration. The components of $\mathbf{b}$ are positive and represent the one-period inflow of the various cohort groups.

If $\mathbf{A}$ is asymptotically stable (corresponding to a population system that without immigration would eventually be reduced to zero), then there will be a positive equilibrium population distribution when there is immigration. If $\mathbf{A}$ is unstable (corresponding to an inherently growing population system), there will be no nonnegative equilibrium population distribution when there is immigration. (The system equations may have an equilibrium point—but it will not be nonnegative.) The results of this simple example are in accord with our intuitive understanding of the behavior of a population system. Theorem 1 translates this intuition into a general result.

Inverses

A property of positive matrices, which is closely related to the stability result above, is that, for real values λ, the matrix $\lambda\mathbf{I}-\mathbf{A}$ has an inverse that is itself a positive matrix provided that $\lambda > \lambda_0$. This result, which is an important and useful part of the theory of positive matrices, can be regarded as an instance where positive matrices are the analog of positive numbers. In this case the numerical statement being: if $a > 0$, then $[\lambda - a]^{-1} > 0$ if and only if $\lambda > a$.

Let us first establish a useful lemma.

Lemma on Series Expansion of Inverse. *If* $\mathbf{A}$ *is a matrix with all eigenvalues strictly inside the unit circle, then*

$$[\mathbf{I}-\mathbf{A}]^{-1} = \mathbf{I}+\mathbf{A}+\mathbf{A}^2+\mathbf{A}^3+\cdots \tag{6-6}$$

Proof. Assume first that the series on the right converges to a matrix $\mathbf{B}$. Then we find that

$$\begin{aligned}[\mathbf{I}-\mathbf{A}]\mathbf{B} &= \mathbf{I}+\mathbf{A}+\mathbf{A}^2+\cdots\\ &\quad -\mathbf{A}-\mathbf{A}^2-\mathbf{A}^3-\cdots\\ &= \mathbf{I}\end{aligned}$$

Thus, $\mathbf{B} = [\mathbf{I}-\mathbf{A}]^{-1}$.

To see that the series does converge, consider first the case where $\mathbf{A}$ can be put in diagonal form, say

$$\mathbf{M}^{-1}\mathbf{A}\mathbf{M} = \mathbf{\Lambda}$$

where $\mathbf{\Lambda}$ is diagonal. Then $\mathbf{A}^k = \mathbf{M}\mathbf{\Lambda}^k\mathbf{M}^{-1}$ and the series (6-6) is made up of various geometric series of the form λ_i^k, where the λ_i's are the eigenvalues of $\mathbf{A}$. These series converge, since it is assumed that $|\lambda_i| < 1$. If $\mathbf{A}$ is not diagonalizable, there may be series of the form λ_i^k, $k\lambda_i^k$, $k^2\lambda_i^k, \ldots, k^{n-1}\lambda_i^k$. Again these converge. ▮

Theorem 2. *Let* $\mathbf{A}$ *be a nonnegative matrix with associated Frobenius–Perron eigenvalue* λ_0. *Then the matrix* $[\lambda\mathbf{I}-\mathbf{A}]^{-1}$ *exists and is positive if and only if* $\lambda > \lambda_0$.

Proof. Suppose first that $\lambda > \lambda_0$. Clearly $\lambda > 0$, since $\lambda_0 \geq 0$. The matrix $\bar{\mathbf{A}} = \mathbf{A}/\lambda$ has its eigenvalues all less than 1 in absolute value. By the above lemma

$$[\lambda\mathbf{I}-\mathbf{A}]^{-1} = \frac{1}{\lambda}[\mathbf{I}-\bar{\mathbf{A}}]^{-1} = \frac{1}{\lambda}\left(\mathbf{I}+\frac{\mathbf{A}}{\lambda}+\frac{\mathbf{A}^2}{\lambda^2}+\cdots\right)$$

Thus, $[\lambda\mathbf{I}-\mathbf{A}]^{-1}$ exists, and it is positive since every term of the series expansion is nonnegative.

To prove the converse statement, suppose $\lambda < \lambda_0$. Let $\mathbf{x}_0 \geq \mathbf{0}$ be an eigenvector corresponding to λ_0. Then $\mathbf{A}\mathbf{x}_0 \geq \lambda\mathbf{x}_0$. Or equivalently $[\lambda\mathbf{I}-\mathbf{A}]\mathbf{x}_0 + \mathbf{p} = \mathbf{0}$ for some $\mathbf{p} \geq \mathbf{0}$. If $[\lambda\mathbf{I}-\mathbf{A}]^{-1}$ exists, then $[\lambda\mathbf{I}-\mathbf{A}]^{-1}\mathbf{p} = -\mathbf{x}_0$. Thus, since $\mathbf{p} \geq \mathbf{0}$, $[\lambda\mathbf{I}-\mathbf{A}]^{-1}$ cannot be positive. ▮

Example 2. For the matrix

$$\mathbf{A} = \begin{bmatrix} 1 & 2\\ 4 & 3\end{bmatrix}$$

considered in Example 2 of Sect. 2, we know $\lambda_0 = 5$. We can compute

$$[6\mathbf{I}-\mathbf{A}]^{-1} = \begin{bmatrix} 5 & -2 \\ -4 & 3 \end{bmatrix}^{-1} = \frac{1}{7}\begin{bmatrix} 3 & 2 \\ 4 & 5 \end{bmatrix}$$

which is positive as expected, since $6 > \lambda_0$. However,

$$[4\mathbf{I}-\mathbf{A}]^{-1} = \begin{bmatrix} 3 & -2 \\ -4 & 1 \end{bmatrix}^{-1} = -\frac{1}{5}\begin{bmatrix} 1 & 2 \\ 4 & 3 \end{bmatrix}$$

which is not positive, since $4 < \lambda_0$.

6.4 QUALITY IN A HIERARCHY—THE PETER PRINCIPLE

A simple, although somewhat satirical, observation concerning organizational hierarchies is the famous Peter Principle: "A man rises until he reaches his level of incompetence." This principle has been used to explain our frequently frustrating perception that most important jobs seem to be held by incompetent individuals. According to the Peter Principle, the hierarchical promotion structure effectively guides people toward incompetent positions, where they remain. Thus, by the nature of the hierarchical structure, a perception of incompetent performance is inevitable.

A hierarchy can be considered to be a positive dynamic system. Once this notion is formalized, it is possible to introduce detailed assumptions on the hierarchical promotion patterns and logically deduce the corresponding quality pattern.

From a dynamic systems viewpoint the study of these structures is somewhat novel. Rather than focusing on the time evolution of the dynamic system, consideration is directed at movement up the hierarchy. In other words, the steps of the hierarchy serve the role that steps in time usually serve. Quality is considered as the state, which changes as it moves up the hierarchical structure.

Consider a hierarchy consisting of n levels. Level 1 is the lowest, while level n is the highest. Within each level there are m types of individuals, characterized by m degrees of performance, or degrees of quality, at that level. Here the indexing is in the opposite direction, with 1 being the best type and m the poorest. (See Fig. 6.2.) A given individual in any given year k is therefore characterized by his level i within the overall hierarchical ladder and by his quality index j, which rates him with his colleagues at the same hierarchical level.

During the course of one year, each individual may either remain at a given level, be promoted to the next higher level, or leave the system. During that year he may also have changed his quality type. For example, if he remains in the same hierarchical level, he might rise in performance due to longer experience; or, if he is promoted, his performance level might fall, since he is (presumably) in a more difficult job.

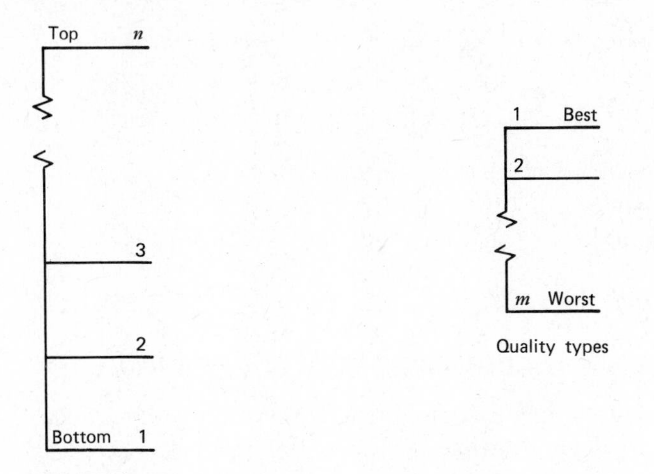

Figure 6.2. Structure of hierarchy

Corresponding to year k we let $\mathbf{x}(k)$ be the vector with $n \times m$ components that describes the population (or the expected population) in each level and quality type. The vector $\mathbf{x}(k)$ can be written in partitioned form as

$$\mathbf{x}(k) = \begin{bmatrix} \mathbf{x}_1(k) \\ \mathbf{x}_2(k) \\ \cdot \\ \cdot \\ \cdot \\ \mathbf{x}_n(k) \end{bmatrix}$$

where the ith component vector

$$\mathbf{x}_i(k) = \begin{bmatrix} x_{i1}(k) \\ x_{i2}(k) \\ \cdot \\ \cdot \\ \cdot \\ x_{im}(k) \end{bmatrix}$$

has components specifying the number of individuals at level i of various performance type. Thus $x_{ij}(k)$ denotes the number of individuals in year k at level i and of quality type j. During the course of a year, the population at level i is modified according to two transition matrices. First, there is a *recycling matrix* $\mathbf{R}_i$ that defines the new quality distribution of those who are not promoted. The matrix $\mathbf{R}_i$ is a nonnegative matrix whose elements are the fractions of those individuals of various quality type who remain in the ith level and end up with new quality designations. The two extreme cases are $\mathbf{R}_i = \mathbf{I}$, meaning everyone remains at level i with exactly the same quality designation, and $\mathbf{R}_i = 0$, meaning everyone either leaves or is promoted. In general, $\mathbf{R}_i\mathbf{x}_i(k)$ represents the contribution to $\mathbf{x}_i(k+1)$ from those who were not promoted.

Second, there is a *promotion matrix* $\mathbf{P}_{i+1}$ that, in a similar way, defines the proportions of individuals of various quality types who are promoted from level i and end up with new quality designations at level $i+1$. This matrix is also nonnegative.

The population vector corresponding to level i is, according to these definitions, governed by $\mathbf{x}_i(k+1) = \mathbf{P}_i\mathbf{x}_{i-1}(k) + \mathbf{R}_i\mathbf{x}_i(k)$. Therefore, the entire process is governed by the transition matrix

$$\mathbf{A} = \begin{bmatrix} \mathbf{R}_1 & & & & \\ \mathbf{P}_2 & \mathbf{R}_2 & & & \\ & \mathbf{P}_3 & \mathbf{R}_3 & & \\ & & \ddots & \ddots & \\ & & & \mathbf{P}_n & \mathbf{R}_n \end{bmatrix} \tag{6-7}$$

If we assume that new individuals enter the system only at the lowest level with quality distribution described by the m-dimensional vector $\mathbf{v}_0$, then the entire process is governed by the dynamic equation

$$\mathbf{x}(k+1) = \mathbf{A}\mathbf{x}(k) + \mathbf{b} \tag{6-8}$$

where

$$\mathbf{b} = \begin{bmatrix} \mathbf{v}_0 \\ \mathbf{0} \\ \mathbf{0} \\ \cdot \\ \cdot \\ \cdot \\ \mathbf{0} \end{bmatrix}$$

This is the general model, and it is a positive linear system.

The sum of the elements in the jth column of the component matrix

$$\begin{bmatrix} \mathbf{R}_i \\ \mathbf{P}_{i+1} \end{bmatrix}$$

represents the total fraction of those individuals in level i and of quality type j

who either remain in level i or are promoted. We assume that there is always some attrition, so the column sums are less than unity. It follows from the bound of λ_0 in terms of column sums (in Sect. 2) that $\lambda_0<1$. Therefore, this system is asymptotically stable. It then follows (by either Theorem 1 or 2 of Sect. 3) that the solution tends toward a unique equilibrium distribution

$$\bar{\mathbf{x}}=(\mathbf{I}-\mathbf{A})^{-1}\mathbf{b} \tag{6-9}$$

Quality Distribution

In order to address the issues raised at the beginning of this section, we are less concerned with the time pattern on the way to equilibrium than with the quality distribution of the equilibrium itself. With this in mind, we set out to define a recursive relation between quality at one level and quality at the next higher level.

Assume now that the recycling and promotion matrices do not change with level. That is, $\mathbf{R}_i=\mathbf{R}$, $\mathbf{P}_i=\mathbf{P}$ for all i. The equilibrium distribution $\bar{\mathbf{x}}$ is made up of component m-vectors; that is, $\bar{\mathbf{x}}=(\bar{\mathbf{x}}_1, \bar{\mathbf{x}}_2, \ldots, \bar{\mathbf{x}}_n)$, where each $\bar{\mathbf{x}}_i$ is the equilibrium population vector of quality types at the ith level. Writing out (6-8) one finds that the component equilibrium vectors satisfy

$$\begin{aligned}
(\mathbf{I}-\mathbf{R})\bar{\mathbf{x}}_1 &= \mathbf{v}_0\\
-\mathbf{P}\bar{\mathbf{x}}_1+(\mathbf{I}-\mathbf{R})\bar{\mathbf{x}}_2 &= \mathbf{0}\\
-\mathbf{P}\bar{\mathbf{x}}_2+(\mathbf{I}-\mathbf{R})\bar{\mathbf{x}}_3 &= \mathbf{0}\\
&\vdots\\
-\mathbf{P}\bar{\mathbf{x}}_{n-1}+(\mathbf{I}-\mathbf{R})\bar{\mathbf{x}}_n &= \mathbf{0}
\end{aligned}$$

Thus, the $\bar{\mathbf{x}}_i$'s are defined recursively by

$$\begin{aligned}
\bar{\mathbf{x}}_1 &= (\mathbf{I}-\mathbf{R})^{-1}\mathbf{v}_0\\
\bar{\mathbf{x}}_{i+1} &= (\mathbf{I}-\mathbf{R})^{-1}\mathbf{P}\bar{\mathbf{x}}_i
\end{aligned} \tag{6-10}$$

This is a linear dynamic system (indexed by hierarchy level) with system matrix $(\mathbf{I}-\mathbf{R})^{-1}\mathbf{P}$.

Since the column sums in $\mathbf{R}$ are each less than one, it follows (from Theorem 2, Sect. 3) that $(\mathbf{I}-\mathbf{R})^{-1}$ is positive. Then since $\mathbf{P}$ is nonnegative, the product $(\mathbf{I}-\mathbf{R})^{-1}\mathbf{P}$ is nonnegative. In most cases of interest, this matrix, or some positive power of it, will in fact be strictly positive. Then, the Frobenius–Perron eigenvalue of this matrix is strictly dominant, and thus for large values of i, the vector $\mathbf{x}_i$ will be approximately aligned with the corresponding eigenvector $\mathbf{e}_0$ of $(\mathbf{I}-\mathbf{R})^{-1}\mathbf{P}$. The magnitude of $\mathbf{e}_0$ is somewhat arbitrary and really not too important. We are mainly interested in the relative distribution of quality types (the relative values of the components of $\mathbf{e}_0$) rather than the absolute magnitudes.

Example. Suppose that at each hierarchical level there are two types of performance quality, competent and incompetent. Suppose the recycling and promotion matrices are

$$\mathbf{R} = \begin{bmatrix} .6 & .1 \\ 0 & .7 \end{bmatrix}, \qquad \mathbf{P} = \begin{bmatrix} .2 & 0 \\ .1 & .1 \end{bmatrix}$$

This would seem to be a fairly effective system. As is seen by summing the elements in the columns of **P**, three times as many competent individuals are promoted as incompetents. And once a competent individual is promoted, there is a two to one chance that he immediately will be competent at the higher level. An incompetent person can become competent with an extra year's experience, but there is a seven to one chance that he will not. Since the sum of the first column elements in **R** and **P** is .9 and the sum of the second column elements is also .9, it follows that 10% of both the competent and the incompetent individuals leave the system each year (except from the highest level in the organization where more leave). The system seems to do a reasonable job of holding back incompetence and rewarding competence. Let us carry out a detailed analysis.

By direct calculation we find

$$\mathbf{I} - \mathbf{R} = \tfrac{1}{10} \begin{bmatrix} 4 & -1 \\ 0 & 3 \end{bmatrix}$$

$$[\mathbf{I} - \mathbf{R}]^{-1} = \tfrac{10}{12} \begin{bmatrix} 3 & 1 \\ 0 & 4 \end{bmatrix}$$

$$[\mathbf{I} - \mathbf{R}]^{-1}\mathbf{P} = \tfrac{1}{12} \begin{bmatrix} 7 & 1 \\ 4 & 4 \end{bmatrix}$$

This last matrix, which is the interlevel transition matrix, has characteristic polynomial

$$\begin{aligned}(7-12\lambda)(4-12\lambda)-4 &= (12\lambda)^2 - 11(12\lambda) + 24 \\ &= (12\lambda - 3)(12\lambda - 8)\end{aligned}$$

Thus, the largest eigenvalue is

$$\lambda_0 = 8/12$$

(This can also be inferred directly from the fact that in this example both row sums of $[\mathbf{I} - \mathbf{R}]^{-1}\mathbf{P}$ are equal to 8/12.) The corresponding eigenvector can be found to be

$$\mathbf{e}_0 = \begin{bmatrix} 1 \\ 1 \end{bmatrix}$$

Therefore, at the highest levels of the hierarchy one can expect that the quality distribution is about equally divided as between competent and incompetent individuals. At the lower levels, on the other hand, the distribution is determined by the character of the input of new recruits $\mathbf{v}_0$. In particular, the distribution in the first level is

$$\mathbf{x}_1 = (\mathbf{I} - \mathbf{R})^{-1}\mathbf{v}_0$$

Because the ratio of the second largest to the largest eigenvalue is $\frac{3}{8}$, the equilibrium quality distribution will be achieved quite rapidly—that is, at fairly low levels. Assuming

$$\mathbf{v}_0 = \begin{bmatrix} 10 \\ 1 \end{bmatrix}$$

the resulting equilibrium population is shown in Table 6.1. We see that despite the fact that three times as many competents are promoted as incompetents, the average number of competents is only 50% at the highest levels.

Table 6.1 Quality Distribution in a Hierarchy

	Level					
	0	1	2	3	4	5
Competents	10	25.83	15.35	9.76	6.39	4.23
Incompetents	1	3.33	9.72	8.36	6.04	4.14
Total	11	29.16	25.07	18.12	12.43	8.37
% Competents	90.9	88.6	61.2	53.9	51.4	50.5

6.5 CONTINUOUS-TIME POSITIVE SYSTEMS

Practically everything derived for discrete-time positive systems has a direct analogy in continuous time. The structure of continuous-time positive systems is slightly different because the system matrix relates the state to the derivative of the state (and the derivative need not be positive) but the character of the results is virtually identical in the two cases.

In continuous time, attention focuses on Metzler matrices. A real $n \times n$ matrix $\mathbf{A}$ is a *Metzler matrix* if $a_{ij} \geq 0$ for all $i \neq j$. In other words, $\mathbf{A}$ is a Metzler matrix if all nondiagonal elements are nonnegative.

We say that a homogeneous continuous-time system

$$\dot{\mathbf{x}}(t) = \mathbf{A}\mathbf{x}(t) \tag{6-11}$$

is positive if $\mathbf{A}$ is a Metzler matrix. This condition on (6-11) is equivalent to the requirement that the system preserve nonnegativity of the state vector. To verify this we note that to insure that $\mathbf{x}(t)$ remains nonnegative it is necessary that $\dot{x}_i(t) \geq 0$ whenever $x_i(t)=0$ for $i=1,2,\ldots,n$. That is, if $\mathbf{x}(t)$ is on the boundary of the positive region its direction of movement cannot be such as to take it outside that region. This imposes the requirement that $a_{ij} \geq 0$, $i \neq j$. Thus, $a_{ij} \geq 0$, $i \neq j$ is a *necessary* condition for nonnegativity of solutions. To show that this condition is also sufficient, we note that the stronger condition $a_{ij} > 0$, $i \neq j$ is certainly sufficient, for in that case $x_i(t)=0$ implies $\dot{x}_i(t) > 0$ [unless $\mathbf{x}(t)=0$]. Therefore, the solution, starting from any nonnegative initial condition, will remain nonnegative. Since the solution depends continuously on the parameters a_{ij}, it follows that the weaker condition $a_{ij} \geq 0$, $i \neq j$ is also sufficient. Thus, the requirement that $\mathbf{A}$ be a Metzler matrix represents the natural extension of positivity to continuous-time systems.

Metzler matrices are obviously closely related to nonnegative matrices. Suppose $\mathbf{A}$ is a Metzler matrix. Then for some suitable constant $c > 0$, the matrix $\mathbf{P} = c\mathbf{I} + \mathbf{A}$ is a nonnegative matrix. The matrix $\mathbf{P}$ has a Frobenius–Perron eigenvalue $\lambda_0 \geq 0$ and a corresponding positive eigenvector $\mathbf{x}_0$. It follows immediately that $\mu_0 = \lambda_0 - c$ is an eigenvalue of the Metzler matrix $\mathbf{A}$ and that $\mathbf{x}_0$ is a corresponding eigenvector. The eigenvalue μ_0 is real, and in fact it readily follows from the nature of λ_0 that μ_0 is the eigenvalue of $\mathbf{A}$ with largest real part. (See Fig. 6.3.) By adding $c\mathbf{I}$ to $\mathbf{A}$ as above, it is possible to translate virtually all results for nonnegative matrices to equivalent results for Metzler matrices. In particular, one obtains:

Theorem 1. *Let* $\mathbf{A}$ *be a Metzler matrix. Then there exists a real* μ_0 *and an* $\mathbf{x}_0 \geq \mathbf{0}$ *such that* (a) $\mathbf{A}\mathbf{x}_0 = \mu_0 \mathbf{x}_0$; (b) *if* $\mu \neq \mu_0$ *is any other eigenvalue of* $\mathbf{A}$, *then* $\mathrm{Re}(\mu) < \mu_0$.

Proof. As outlined above, this follows from Theorem 3, Sect. 2. The strict inequality in (b) holds even if $|\lambda| = \lambda_0$ in Fig. 6.3. ∎

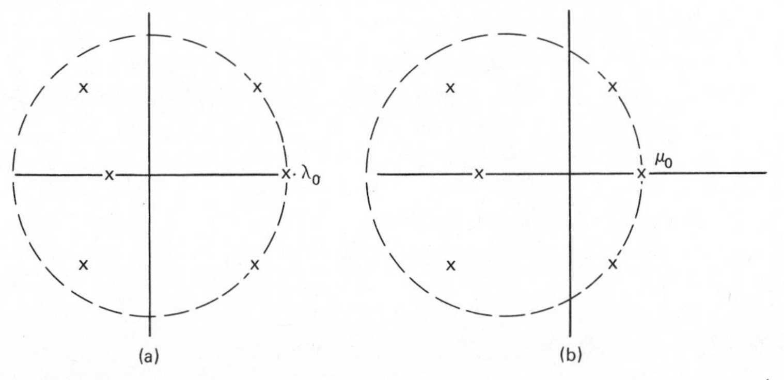

Figure 6.3. Illustration that μ_0 is eigenvalue of largest real part. (a) Eigenvalues of $\mathbf{P}$. (b) Eigenvalues of $\mathbf{A}$.

As with positive matrices, stronger versions apply if strict positivity assumptions are introduced.

The real eigenvalue μ_0 of Theorem 1 above, being the eigenvalue of greatest real part, is the dominant eigenvalue of $\mathbf{A}$ in the sense of continuous-time systems. Therefore, this eigenvalue and its corresponding positive eigenvector play roles analogous to the Frobenius–Perron eigenvalue and eigenvector in discrete-time systems.

Equilibrium Points and Stability

The results relating equilibrium points and stability for nonhomogeneous continuous-time positive systems are again analogous to the discrete-time case. A system of the form

$$\dot{\mathbf{x}}(t) = \mathbf{A}\mathbf{x}(t) + \mathbf{b} \tag{6-12}$$

is positive if and only if $\mathbf{A}$ is a Metzler matrix and the vector $\mathbf{b}$ is nonnegative. Again these requirements correspond directly to the condition that the system preserves nonnegativity of the state vector.

The system (6-12) may have an equilibrium point $\bar{\mathbf{x}}$ satisfying

$$\mathbf{0} = \mathbf{A}\bar{\mathbf{x}} + \mathbf{b} \tag{6-13}$$

As in the discrete-time case, however, interest focuses mainly on nonnegative equilibrium points, and, just as before, it turns out that existence of a nonnegative equilibrium point satisfying (6-13) essentially corresponds to the asymptotic stability of the continuous-time system (6-12). This connection between stability and existence is a powerful and useful result. Its validity for both discrete- and continuous-time systems makes it a fundamental result of positive systems.

We simply state the continuous-time version of the stability theorems. They can be easily proved by translating $\mathbf{A}$ to a nonnegative matrix $\mathbf{P} = c\mathbf{I} + \mathbf{A}$ and then using the results in Sect. 3.

Theorem 2. *Given a Metzler matrix $\mathbf{A}$ and a $\mathbf{b} > \mathbf{0}$, the matrix $\mathbf{A}$ has all of its eigenvalues strictly within the left half of the complex plane if and only if there is an $\bar{\mathbf{x}} \geq \mathbf{0}$ satisfying*

$$\mathbf{0} = \mathbf{A}\bar{\mathbf{x}} + \mathbf{b}$$

Theorem 3. *Let $\mathbf{A}$ be a Metzler matrix. Then $-\mathbf{A}^{-1}$ exists and is a positive matrix if and only if $\mathbf{A}$ has all of its eigenvalues strictly within the left half of the complex plane.*

6.6 RICHARDSON'S THEORY OF ARMS RACES

It has often been argued vehemently in national forums that armaments are insurance against war. A defenseless nation, so the argument goes, invites war by its weakness. Argued by both sides of two potentially opposing nations, this type of perhaps good-intentioned defensive logic can lead to a spiral of defense build-ups, one nation responding for its own protection to the armament build-ups of another nation. This is the arms race phenomenon.

Lewis F. Richardson, in a thirty-year study of war, proposed and developed a series of linear dynamic models of arms races. His study has become a classic example of the potential of dynamic system analysis in the study of social phenomena. It is, in fact, one of the earliest comprehensive applications of linear dynamic systems to the study of a social system.

The theory, and indeed the entire approach, is not without what are perhaps severe limitations. The proposed dynamic model is merely a crude summary description of what is by nature a complex pattern of many individual actions, and the description may not be valid in every case. However, the approach does have a certain element of validity and it provides an avenue for the formal development and exploration of some important concepts related to arms races.

The Two-Nation Model

In this model two competing nations (or perhaps two competing coalitions of nations) are denoted X and Y. The variables $x(t)$ and $y(t)$ represent, respectively, the armament levels of the nations X and Y at time t. The general form of the model is

$$\dot{x}(t) = ky(t) - \alpha x(t) + g$$
$$\dot{y}(t) = lx(t) - \beta y(t) + h$$

In this model, the terms g and h are called "grievances." They encompass the wide assortment of psychological and strategic motivations for changing armament levels, which are independent of existing levels of either nation. Roughly speaking, they are motives of revenge or dissatisfaction, and they may be due to dissatisfaction with treaties or other past political negotiations. The terms k and l are called "defense" coefficients. They are nonnegative constants that reflect the intensity of reaction by one nation to the current armament level of the other nation. It is these coefficients, associated with the motives of fear and rivalry, that can cause the exponential growth of armaments commonly associated with arms races. Finally, α and β are called "fatigue" coefficients. They are nonnegative constants that represent the fatigue and expense of expanding defenses. The corresponding terms in the dynamic equations have the effect of causing a nation to tend to retard the growth of its own armament level; the retardation effect increasing as the level increases.

The system matrix is

$$\mathbf{A} = \begin{bmatrix} -\alpha & k \\ l & -\beta \end{bmatrix}$$

which is a Metzler matrix. We therefore can apply the theory of positive systems to this case.

Let us first ask whether the system possesses an equilibrium point; that is, whether there are levels of defenses for the two nations that, if established, will be maintained indefinitely. For the two-nation model we can carry out the required calculations explicitly. We set the time derivatives to zero, arriving at the equations

$$0 = ky_0 - \alpha x_0 + g$$
$$0 = lx_0 - \beta y_0 + h$$

for the equilibrium values x_0 and y_0. These equations can be solved to yield

$$x_0 = \frac{\beta g + kh}{\alpha\beta - kl}$$

$$y_0 = \frac{\alpha h + lg}{\alpha\beta - kl}$$

This shows that unless $\alpha\beta = kl$ there is an equilibrium point. We note, however, that if all constants are nonnegative (most particularly the "grievances" g and h, since the others are always assumed to be nonnegative), the equilibrium point will correspond to nonnegative defense levels if and only if $\alpha\beta - kl > 0$.

According to the general theory of positive systems, we expect this condition to be identical to the condition of stability. Let us verify directly that this is so. The eigenvalues of the system are the roots of the characteristic equation

$$\begin{vmatrix} \lambda + \alpha & -k \\ -l & \lambda + \beta \end{vmatrix} = 0$$

or

$$(\lambda + \alpha)(\lambda + \beta) - kl = 0$$

or

$$\lambda^2 + (\alpha + \beta)\lambda + \alpha\beta - kl = 0$$

As the reader can verify (using for example the Routh test, Problem 13, Chapter 5), both roots are in the left half-plane if and only if all coefficients are positive. In this case this condition reduces to $\alpha\beta - kl > 0$. Thus, the model is stable if and only if $\alpha\beta - kl > 0$.

This stability condition has an intuitive meaning within the context of the arms race situation. The quantity kl is the product of the two "defense" coefficients, and represents the tendency to perpetrate the race. The term $\alpha\beta$ is the product of the two "fatigue" coefficients, and represents the tendency to limit defense build-ups. Instability results when the overall "defense" term outweighs the "fatigue" term.

If the grievances g and h are nonnegative, then the condition for stability is equivalent to the condition that the process have a nonnegative equilibrium point. Thus, according to this model of an arms race between two nations, if one seeks to avoid instability, it is sufficient to try to promote changes which guarantee a nonnegative equilibrium, for that equilibrium will automatically be stable.

Multi-Nation Theory

The multi-nation theory is a straightforward generalization of the two-nation model. However, because explicit computation in terms of the parameters is infeasible in the general case, the general theory of positive systems is invaluable. Also, in the multi-nation model there is the possibility of coalitions, either explicit or implicit, that enriches the scope of analysis.

Suppose there are n nations. Let x_i denote the defense level of the ith nation. The model for defense dynamics is

$$\dot{x}_i(t) = g_i + \sum_{j=1}^{n} k_{ij} x_j(t) \qquad i = 1, 2, \ldots, n$$

As before the numbers g_i represent "grievances." The k_{ij} for $i \neq j$ are "defense" coefficients, and the k_{ii} represent "fatigue." For convenience we often denote $k_{ii} = -\alpha_i$. It is assumed that $k_{ij} \geq 0$ for $i \neq j$ and that $\alpha_i > 0$. In vector notation the model takes the form

$$\dot{\mathbf{x}}(t) = \mathbf{g} + \mathbf{K}\mathbf{x}(t)$$

where the notation should be obvious.

It is easy to solve for the equilibrium point by equating the time derivatives of the model to zero. This yields the equilibrium $\mathbf{x}_0$,

$$\mathbf{x}_0 = -\mathbf{K}^{-1}\mathbf{g}$$

which exists provided that $\mathbf{K}$ is nonsingular. This equilibrium point may or may not, however, correspond to nonnegative defenses.

The rate of growth of the vector $\mathbf{x}$ in the model is governed primarily by the eigenvalue of $\mathbf{K}$ with the largest real part. Since the system is positive, the eigenvalue of greatest real part is real and has a corresponding nonnegative eigenvector. If this eigenvalue is positive the system is unstable. Furthermore, if $\mathbf{g} > \mathbf{0}$, it follows from the general theory that the condition for asymptotic stability exactly corresponds to that for the existence of a nonnegative equilibrium.

Coalitions

Let us consider the eigenvalue having the smallest real part. This real part must be negative since the trace of the matrix **K** is negative. (See Problem 15, Chapter 3.) For the moment we assume that this root is real (though it need not be) and that it is a simple nonrepeated root. Let us denote this eigenvalue by λ_n. Let $\mathbf{f}_n^T$ be the corresponding *left* eigenvector. This eigenvector defines a composite normal variable $z_n(t)=\mathbf{f}_n^T\mathbf{x}(t)$ that behaves according to a first-order dynamic system. Indeed, multiplication of the system equation by $\mathbf{f}_n^T$ yields $\dot{z}_n(t)=\lambda_n z_n(t)+\mathbf{f}_n^T\mathbf{g}$. The exponential associated with this variable is $e^{\lambda_n t}$, which is the fastest term in the behavior of the total system. Thus, $z_n(t)$ approaches its equilibrium value $-\mathbf{f}_n^T\mathbf{g}/\lambda_n$ quickly, relative to the speed at which the system as a whole approaches equilibrium. Therefore, as an approximation, the variable z_n can be considered to be constant throughout the duration of the arms race.

Since the eigenvector $\mathbf{f}_n^T$ must be orthogonal to the right eigenvector corresponding to the dominant eigenvalue, and that eigenvector is nonnegative, we can deduce that the eigenvector $\mathbf{f}_n^T$ cannot have all positive components. Generally, then, the components divide into two groups—corresponding to positive and negative components (zero valued components are ignored). These groups can be thought of as each comprising an alliance, with each country weighted by the coefficient in the left eigenvector. As the system progresses, the arms levels of these two groups increase together, at the same rate, maintaining a constant *difference* in total defense capability. Each group keeps up with the other group. Thus, the arms race can be considered as primarily being a race between the two alliances, with additional adjustments within an alliance.

A Ten-Nation Example

Richardson has suggested several ten-nation models, each assumed to be a reasonable representation of the world in 1935. The **K** matrix for one model is shown in Table 6.2. In this example, the eigenvalue of greatest real part is $\lambda_1=0.2687$. The corresponding (normalized) left eigenvector is [0.17011, 0.18234, 0.14102, 0.51527, 0.23095, 0.42807, 0.30557, 0.09696, 0.12624, 0.54510]. This is clearly an unstable arms race.

The eigenvalue with smallest real part is $\lambda_{10}=-2.25$. Its corresponding left eigenvector has weights as follows: France, 0.588; U.S.S.R., 0.449; Czechoslovakia, 0.179; Great Britain and Northern Ireland, 0.140; U.S.A., 0.046; Poland, −0.006; China, −0.015; Japan, −0.178; Italy, −0.238; Germany, −0.557. These weights seem to give a fairly accurate picture of the alignment of nations in 1935 with France at one pole, and Germany at the other.

Table 6.2. Richardson's Ten-Nation Model

		1	2	3	4	5	6	7	8	9	10
1	Czecho-slovakia	−10	0	0	2	0	0	0	1	0	0
2	China	0	−10	0	0	0	0	4	0	0	2
3	France	0	0	−18	4	0	4	0	0	0	0
4	Germany	4	0	4	−10	2	0	0	1	0	8
5	G. Britain and N.I.	0	0	0	4	−15	6	2	0	0	0
6	Italy	0	0	2	0	4	−5	0	0	0	2
7	Japan	0	4	0	0	0	0	−10	0	4	4
8	Poland	1	0	0	1	0	0	0	−10	0	1
9	U.S.A.	0	0	0	2	2	2	4	0	−7	2
10	U.S.S.R.	0	2	0	8	2	2	4	1	0	−10

6.7 COMPARATIVE STATICS FOR POSITIVE SYSTEMS

The term *comparative statics* refers to an analysis procedure that focuses on the equilibrium point of a dynamic system, and how that equilibrium point changes when various system parameters are changed. This form of analysis ignores the actual path by which the state of the dynamic system moves to its new equilibrium; it is concerned only with how the new point is related to the old point, not on the means by which it is attained. Essentially, in this approach, the dynamic aspect of the system is suppressed, and the analysis reduces to a study of the static equations that determine the equilibrium point.

One aspect of the actual dynamic structure of the system, however, must be considered if a comparative statics analysis is to be meaningful. Namely, the system must be stable. If the new equilibrium point is not stable, it is patently inconsistent to ignore the path to the new equilibrium point—that is, to regard the path as something of a detail—when without stability the state may not even tend toward the new equilibrium point. Therefore, implicit in any comparative statics investigation is an assumption of stability. In this connection the intimate relationship between stability and existence of positive equilibrium points for positive systems discussed earlier, in Sects. 6.3 and 6.5, forms a kind of backdrop for the results on comparative statics of this section.

Positive Change

Consider the linear time-invariant system

$$\mathbf{x}(k+1)=\mathbf{A}\mathbf{x}(k)+\mathbf{b} \tag{6-14}$$

where $\mathbf{A} \geqq \mathbf{0}$, $\mathbf{b} \geqq \mathbf{0}$. Assume that the system is asymptotically stable and has the unique nonnegative equilibrium point $\bar{\mathbf{x}}$. We investigate how $\bar{\mathbf{x}}$ changes if some elements of either the matrix $\mathbf{A}$ or the vector $\mathbf{b}$ (or both) are increased.

As a concrete example, let us consider an arms race, as discussed in Sect. 6.6. Suppose all nations are in equilibrium, and then one nation increases its grievance coefficient or one of its defense coefficients, but not by so much as to destroy stability. It seems pretty clear that after all nations readjust, and a new equilibrium is achieved, the arms levels of all nations will be no less than before. That is, the change in equilibrium is nonnegative. This is indeed the case, and is a rather simple consequence of positivity. The general result is stated below in Theorems 1 and 1′ for discrete-time and continuous-time systems.

Theorem 1. *Suppose $\bar{\mathbf{x}}$ and $\bar{\mathbf{y}}$ are, respectively, the equilibrium points of the two positive systems*

$$\mathbf{x}(k+1) = \mathbf{A}\mathbf{x}(k) + \mathbf{b} \tag{6-15}$$

$$\mathbf{y}(k+1) = \hat{\mathbf{A}}\mathbf{y}(k) + \hat{\mathbf{b}} \tag{6-16}$$

Suppose also that both systems are asymptotically stable and that $\hat{\mathbf{A}} \geqq \mathbf{A}$, $\hat{\mathbf{b}} \geqq \mathbf{b}$. Then $\bar{\mathbf{y}} \geqq \bar{\mathbf{x}}$.

Proof. Actually it is simplest to prove this result by considering the full dynamic process, rather than just the equilibrium points. Suppose for some k there holds $\mathbf{y}(k) \geqq \mathbf{x}(k) \geqq \mathbf{0}$. Then

$$\begin{aligned}\mathbf{y}(k+1) &= \hat{\mathbf{A}}\mathbf{y}(k) + \hat{\mathbf{b}} \\ &\geqq \hat{\mathbf{A}}\mathbf{y}(k) + \mathbf{b} \\ &\geqq \mathbf{A}\mathbf{y}(k) + \mathbf{b} \\ &\geqq \mathbf{A}\mathbf{x}(k) + \mathbf{b} = \mathbf{x}(k+1)\end{aligned}$$

Thus, $\mathbf{y}(k) \geqq \mathbf{x}(k)$ implies $\mathbf{y}(k+1) \geqq \mathbf{x}(k+1)$. Suppose, therefore, that the two systems (6-15) and (6-16) are initiated at a common point, say $\mathbf{0}$. Then $\mathbf{y}(k) \geqq \mathbf{x}(k)$ for all k. Since both systems are asymptotically stable, these sequences converge to the respective unique equilibrium points $\bar{\mathbf{y}}$ and $\bar{\mathbf{x}}$. Clearly $\bar{\mathbf{y}} \geqq \bar{\mathbf{x}}$. ∎

Theorem 1′. *Suppose $\bar{\mathbf{x}}$ and $\bar{\mathbf{y}}$ are, respectively, the equilibrium points of the two positive systems*

$$\dot{\mathbf{x}}(t) = \mathbf{A}\mathbf{x}(t) + \mathbf{b}$$

$$\dot{\mathbf{y}}(t) = \hat{\mathbf{A}}\mathbf{y}(t) + \hat{\mathbf{b}}$$

Suppose also that both systems are asymptotically stable and that $\hat{\mathbf{A}} \geqq \mathbf{A}$, $\hat{\mathbf{b}} \geqq \mathbf{b}$. Then $\bar{\mathbf{y}} \geqq \bar{\mathbf{x}}$.

Proof. Let $c>0$ be determined such that $\mathbf{A}+c\mathbf{I} \geqq \mathbf{0}$ and $\hat{\mathbf{A}}+c\mathbf{I} \geqq \mathbf{0}$. Define $\mathbf{P}=\mathbf{I}+\mathbf{A}/c$, $\hat{\mathbf{P}}=\mathbf{I}+\hat{\mathbf{A}}/c$. If μ_0 is the dominant eigenvalue of $\mathbf{A}$, then $\lambda_0 = 1+\mu_0/c$ is the Frobenius–Perron eigenvalue of $\mathbf{P}$. Since $\mu_0<0$, it follows that $\lambda_0<1$ and thus the positive matrix $\mathbf{P}$ has all its eigenvalues strictly within the unit circle. Likewise, so does $\hat{\mathbf{P}}$.

The equation $\mathbf{0}=\mathbf{A}\bar{\mathbf{x}}+\mathbf{b}$ implies $\mathbf{0}=(\mathbf{A}\bar{\mathbf{x}}+\mathbf{b})/c$, which can be written $\bar{\mathbf{x}}=\mathbf{P}\bar{\mathbf{x}}+\mathbf{b}/c$. Likewise $\bar{\mathbf{y}}=\hat{\mathbf{P}}\bar{\mathbf{y}}+\hat{\mathbf{b}}/c$. The conclusion now follows from Theorem 1. ∎

These two companion results, although somewhat simple in terms of mathematical content, embody a good portion of the theory of positive systems, and put it in a form that is easy to apply and that has strong intuitive content. The fact that the results have a qualitative character, rather than a computational character, means that the statements can be applied to general classes of applications, without the need for numerical parameter values.

Component of Greatest Change

It is possible to develop a much stronger version of the above result. The stronger result gives qualitative information on the relative changes of the various components of the state vector for certain kinds of parameter changes.

An nth-order positive dynamic system is defined in terms of n difference or differential equations—one for each state variable. Suppose now that some parameters in the ith equation are increased, while the parameters in all other equations remain unchanged. From Theorem 1, above, if the system remains stable, all variables will be at least as large at the new equilibrium as at the old equilibrium. In addition, we might expect that the ith variable, corresponding to the equation in which the parameters were increased, might in some sense be more greatly affected than other variables.

As an example, let us again consider the arms race situation. If a grievance or defense coefficient of one nation is increased, this will induce nonnegative changes in the arms levels of all nations. It can be expected, however, that the nation which experienced the direct change in its coefficient will be more greatly affected than other nations. That is, the change in arms level of that nation should in some sense be larger than for other nations. Actually, the *percentage* increase of that nation is at least as great as the percentage change of other nations, and it is this conclusion that generalizes to arbitrary asymptotically stable positive systems. In the theorem below, this is expressed by stating that the ratio of new component values to old component values is greatest for the component corresponding to the modified equation.

Theorem 2. *Let $\bar{\mathbf{x}}>\mathbf{0}$ and $\bar{\mathbf{y}}>\mathbf{0}$ be, respectively, the equilibrium points of the two* *positive and asymptotically stable systems*

$$\mathbf{x}(k+1)=\mathbf{A}\mathbf{x}(k)+\mathbf{b}$$
$$\mathbf{y}(k+1)=\hat{\mathbf{A}}\mathbf{y}(k)+\hat{\mathbf{b}}$$

in discrete time, or of

$$\dot{\mathbf{x}}(t)=\mathbf{A}\mathbf{x}(t)+\mathbf{b}$$
$$\dot{\mathbf{y}}(t)=\hat{\mathbf{A}}\mathbf{y}(t)+\hat{\mathbf{b}}$$

in continuous time. Assume that in either case $\hat{\mathbf{A}}\geqq\mathbf{A}$, $\hat{\mathbf{b}}\geqq\mathbf{b}$ *but that* $\hat{a}_{ij}=a_{ij}$ *and* $\hat{b}_i=b_i$ *for all j and all* $i\neq r$. *Then*

$$\frac{\bar{y}_r}{\bar{x}_r}\geq\frac{\bar{y}_i}{\bar{x}_i}\quad \textit{for all } i$$

Proof. We prove the theorem only for the discrete-time case, leaving the continuous-time version to the reader.

Assume first that $\mathbf{b}>\mathbf{0}$. For each $j=1,2,\ldots,n$ define $\lambda_j=\bar{y}_j/\bar{x}_j$. From Theorem 1, $\lambda_j\geq 1$ for all j. Let $\lambda=\max\{\lambda_j,\ j=1,2,\ldots,n\}$. We must show that $\lambda_r=\lambda$. Suppose to the contrary that for some i, $\lambda_i=\lambda>\lambda_r\geq 1$. Then by the definition of $\bar{\mathbf{y}}$,

$$\bar{y}_i=\sum_{j=1}^{n}\hat{a}_{ij}\bar{y}_j+\hat{b}_i$$
$$=\sum_{j=1}^{n}a_{ij}\bar{y}_j+b_i$$

or

$$\lambda_i\bar{x}_i=\sum_{j=1}^{n}a_{ij}\lambda_j\bar{x}_j+b_i$$

Thus,

$$\bar{x}_i=\sum_{j=1}^{n}a_{ij}\frac{\lambda_j}{\lambda_i}\bar{x}_j+\frac{b_i}{\lambda_i}$$

However, since $\lambda_j/\lambda_i\leq 1$ for all j and since $\lambda_i>1$, it follows that

$$\bar{x}_i<\sum_{j=1}^{n}a_{ij}\bar{x}_j+b_i$$

which contradicts the definition of $\bar{x}_i$.

If $\mathbf{b}$ is not strictly positive, consider the equations

$$\bar{\mathbf{x}}(\varepsilon)=\mathbf{A}\bar{\mathbf{x}}(\varepsilon)+\mathbf{b}+\varepsilon\mathbf{p}$$
$$\bar{\mathbf{y}}(\varepsilon)=\hat{\mathbf{A}}\bar{\mathbf{y}}(\varepsilon)+\hat{\mathbf{b}}+\varepsilon\mathbf{p}$$

where $\mathbf{p}>\mathbf{0}$. The solutions $\bar{\mathbf{x}}(\varepsilon)$ and $\bar{\mathbf{y}}(\varepsilon)$ depend continuously on ε. For any $\varepsilon>0$ the earlier result gives

$$\frac{\bar{y}_r(\varepsilon)}{\bar{x}_r(\varepsilon)} \geq \frac{\bar{y}_i(\varepsilon)}{\bar{x}_i(\varepsilon)} \quad \text{for all } i$$

Then by continuity this is true also for $\varepsilon=0$. ∎

6.8 HOMANS–SIMON MODEL OF GROUP INTERACTION

An important contribution to sociology is the theory of group interaction originally developed in qualitative and verbal form by George C. Homans, and later interpreted in the sense of a dynamic system model by Herbert A. Simon. It is a good example of how mathematical formulation can play an effective role as a medium for theory development even though the associated model probably cannot be used in explicit form in a given situation.

Consider a social group whose behavior at any time t can be characterized by the following variables:

$I(t)$—the intensity of *interaction* among the members
$F(t)$—the level of *friendliness* among the members
$A(t)$—the amount of *activity* carried on by members within the group
$E(t)$—the amount of activity imposed on the group by the *external environment.*

These variables are, of course, at best aggregate characterizations of certain aspects of group behavior. Interaction, for example, is composed of the various interactions among members in subgroups of two or more, and it like all other variables is somewhat difficult to quantify in terms of observed phenomena. Nevertheless, the verbal definitions associated with these variables are sufficiently descriptive to elicit a general notion of what is meant. Thus, in viewing two groups of the same size it is possible to tell whether one displays significantly more friendliness than another.

Homan's original verbal postulates relating these variables are:

(a) group interaction is produced by group friendliness and by group activity;
(b) friendliness tends to increase if group interaction is greater than that which would normally be associated with the existing level of friendliness; and
(c) the level of activity carried on by the group tends to increase if either the actual friendliness or the imposed activity is greater than that which is "appropriate" to the existing level of activity.

A dynamic model translating these postulates is defined by the following three equations:

$$I(t)=a_1F(t)+a_2A(t) \tag{6-17a}$$

$$\dot{F}(t)=b[I(t)-\beta F(t)] \tag{6-17b}$$

$$\dot{A}(t)=c_1[F(t)-\gamma A(t)]+c_2[E(t)-A(t)] \tag{6-17c}$$

All constants in these equations are assumed to be positive. This particular model is linear, and although more complex versions can be constructed, this version is sufficient to illustrate the type of conclusions that can be inferred.

The Equilibrium Point

As posed, the system is not quite in standard form. It consists of two dynamic and one static equation. It is a simple matter, however, to eliminate the variable $I(t)$ to obtain an equivalent description in terms of two dynamic equations:

$$\dot{F}(t)=b[(a_1-\beta)F(t)+a_2A(t)] \tag{6-18a}$$

$$\dot{A}(t)=c_1F(t)-(c_1\gamma+c_2)A(t)+c_2E(t) \tag{6-18b}$$

This is a positive linear system, since all nondiagonal coefficients are positive.

For this system, interest focuses most particularly on the equilibrium point associated with a constant E. Accordingly, assume that the parameters of the system are related in such a way that a positive equilibrium point exists. The equations for the equilibrium point (F, A) are

$$0=-(\beta-a_1)F+a_2A \tag{6-19a}$$

$$0=c_1F-(c_1\gamma+c_2)A+c_2E \tag{6-19b}$$

The first can be solved to yield

$$A=\frac{(\beta-a_1)F}{a_2} \tag{6-20}$$

This shows that for a positive equilibrium point one must first of all impose the condition

$$\beta>a_1 \tag{6-21}$$

Equation (6-20) can be substituted into (6-19b) to yield

$$\left((c_1\gamma+c_2)\frac{(\beta-a_1)}{a_2}-c_1\right)F=c_2E$$

Therefore, the additional condition for a positive equilibrium point is

$$(c_1\gamma+c_2)(\beta-a_1)-a_2c_1>0 \tag{6-22}$$

From the general theory of positive systems, it follows that together (6-21) and (6-22) also guarantee that the equilibrium point is asymptotically stable.

Comparative Statics

As an explicit model of a particular group this model is obviously of limited value. The model is intended, rather, to be a general, but modest, theoretical statement on the dynamics of group behavior. If one agrees that some model of this general character is representative of group behavior, even though specific parameter values cannot be assigned, it is possible to apply the theory of positive systems to obtain some interesting general qualitative conclusions.

Let us suppose that a certain group is in equilibrium, with positive values of friendliness F and activity A. The equilibrium must be asymptotically stable. Suppose now that the externally imposed activity E is increased. How will the new equilibrium point compare to the original one? First, by Theorem 1′ of Sect. 6.7, both group friendliness and activity are expected to increase. Second, by Theorem 2, activity will most likely increase more (on a percentage basis) than friendliness.

6.9 PROBLEMS

1. Let $\mathbf{A} \geqq \mathbf{B} \geqq \mathbf{0}$, and let $\lambda_0(\mathbf{A})$ and $\lambda_0(\mathbf{B})$ denote the corresponding Frobenius–Perron eigenvalues. Show that $\lambda_0(\mathbf{A}) \geq \lambda_0(\mathbf{B})$.

2. Find the largest eigenvalues and corresponding eigenvector of the matrix

$$\begin{bmatrix} 1 & 1 & 0 \\ 1 & 2 & 1 \\ 0 & 1 & 1 \end{bmatrix}$$

3. A certain psycho-physio-economist has developed a new universal theory. The theory hinges on the properties of a "universal matrix" whose entries have the sign structure illustrated below:

$$\mathbf{V} = \begin{bmatrix} + & + & + & - \\ + & + & + & - \\ + & + & + & - \\ - & - & - & + \end{bmatrix}$$

He has examined scores of specific matrices of this structure (using real data!) and has found in each case that there was a positive eigenvalue. Will this be true in general? [*Hint*: Look for a change of variable.]

4. Use the lemma on series expansion of the inverse to evaluate

$$\begin{bmatrix} 1 & -2 & 1 \\ 0 & 1 & 3 \\ 0 & 0 & 1 \end{bmatrix}^{-1}$$

*5. Let $\mathbf{A}$ be a strictly positive $n\times n$ matrix. Define μ_0 as the minimum of real numbers for which

$$\mu\mathbf{x}\geqq\mathbf{Ax}$$

for some $\mathbf{x}\geq\mathbf{0}$. That is,

$$\mu_0=\min\{\mu:\mu\mathbf{x}\geqq\mathbf{Ax},\text{some}\,\mathbf{x}\geq\mathbf{0}\}$$

Show that $\mu_0=\lambda_0$, the Frobenius–Perron eigenvalue of $\mathbf{A}$.

6. It can be shown that a nonnegative matrix $\mathbf{A}$ has $\lambda_0<1$ if and only if all principal minors of $\mathbf{B}=\mathbf{I}-\mathbf{A}$ are positive; that is,

$$b_{11}>0,\qquad \begin{vmatrix} b_{11} & b_{12}\\ b_{21} & b_{22}\end{vmatrix}>0,\qquad \begin{vmatrix} b_{11} & b_{12} & b_{13}\\ b_{21} & b_{22} & b_{23}\\ b_{31} & b_{32} & b_{33}\end{vmatrix}>0,\ldots|\mathbf{B}|>0$$

Use this criterion to determine whether the matrix

$$\mathbf{A}=\begin{bmatrix} \frac{1}{2} & 0 & \frac{1}{2}\\ 1 & \frac{1}{2} & 1\\ \frac{1}{2} & 0 & \frac{1}{4}\end{bmatrix}$$

has all eigenvalues inside the unit circle.

7. The Leontief input–output model is described by a set of equations of the form

$$\mathbf{x}=\mathbf{Ax}+\mathbf{c}$$

(see Sect. 3.1). Show that there is a feasible production vector $\mathbf{x}\geq\mathbf{0}$ corresponding to any demand vector $\mathbf{c}\geq\mathbf{0}$ if and only if the Frobenius–Perron eigenvalue of $\mathbf{A}$ is less than unity.

8. Let $\mathbf{A}$ be a positive $n\times n$ matrix and $\mathbf{b}$ be a positive n-vector. Suppose the system

$$\mathbf{x}(k+1)=\mathbf{Ax}(k)+\mathbf{b}$$

has an equilibrium point $\bar{\mathbf{x}}$. Suppose also that the system is initiated at a point $\mathbf{x}(0)$ such that $\mathbf{x}(0)\leqq\bar{\mathbf{x}}$. Show that $\mathbf{x}(k)\leqq\bar{\mathbf{x}}$ for all $k\geq0$.

9. Show that in Theorem 1, Sect. 6.3, the hypotheses on $\mathbf{A}$ and $\mathbf{b}$ can be changed to $\mathbf{A}>\mathbf{0}$, $\mathbf{b}\geq\mathbf{0}$.

10. *Moving Average.* Suppose two initial numbers are given and successive numbers are computed as the average of last two. For example, with 1 and 3, we generate the sequence

$$1, 3, 2, 2\tfrac{1}{2}, 2\tfrac{1}{4}, 2\tfrac{3}{8}, \ldots$$

Note that the sequence seems to be converging, but not to the average of the original two numbers (which is 2).

(a) Formulate this process in state-space form.

(b) The system matrix **A** you obtained should be positive but not strictly positive. What is the Frobenius–Perron eigenvalue λ_0?
(c) Is the system asymptotically stable?
(d) What are the right and left eigenvectors corresponding to λ_0?
(e) What is the general formula for the limiting value of a sequence of this kind, expressed in terms of its first two terms? For the example given, what is the limiting value?
(f) Now generalize to an nth-order average. That is, n initial numbers are given and successive numbers are the averages of the last n numbers. What is the formula for the limit?

11. In purest form the Peter Principle seems to imply two special features of a promotional system: (a) once reaching a level of incompetence an individual never again will be competent, and (b) an incompetent person is never promoted. Show that, in the case where the promotion and recycling matrices are constant, these two assumptions imply that the distribution of competents and incompetents is the same at every level except the first.

12. The Board of Directors of the company, having the hierarchy structure of the example in Sect. 6.4, was disheartened at the revelation that 50% of its employees at upper levels of management were incompetent. The Board therefore engaged the services of two management consulting firms to seek advice on how to improve the situation. One firm suggested that the promotion policies be tightened up to avoid promoting incompetents. They outlined a program of regular interviews, testing, and peer evaluation that they claimed would screen out essentially all incompetents from promotion. (In our terms this proposal would change **P** by replacing the element in the lower right corner by zero.)

    The second consulting firm suggested that the screening was already adequate, but that what was required was the initiation of employee training. They outlined a program of motivation enhancement, internal courses, and so forth. They estimated that, with a modest effort, they could increase the number of incompetents who become competent at their job over a one-year period from the present rate of one out of eight to a rate of two out of eight. They argued that such a program would significantly affect the quality of the upper levels. If both proposals were about equally costly to implement, which should the Board select? Will either plan drastically affect the ratio of the number of people at successive levels (near the top)?

13. *Solution of Partial Differential Equations.* Partial differential equations arise frequently in the study of fluids, electromagnetic fields, temperature distributions, and other continuously distributed quantities. As an example, the electric potential V within an enclosure is governed by Laplace's equation, which in two dimensions is

$$\frac{\partial^2 V}{\partial x^2}+\frac{\partial^2 V}{\partial y^2}=0 \qquad (*)$$

Most often V is specified on the boundary of the enclosure and $(*)$ must be solved for V inside. (See Fig. 6.4*a*.)

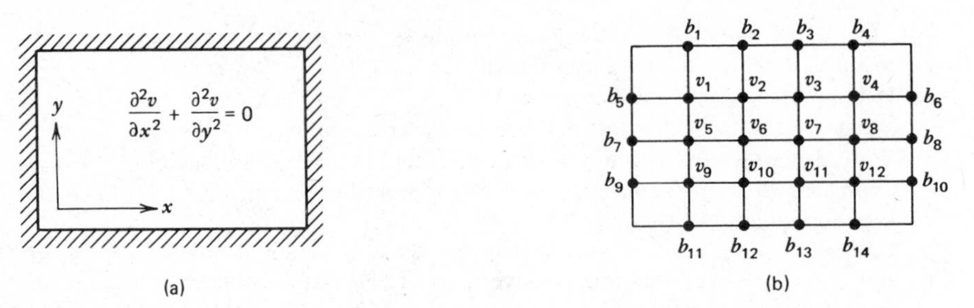

Figure 6.4. Laplace's equation. (a) Continuous. (b) Discrete grid.

To solve this problem numerically, an approximating problem is defined by establishing a finite grid of points within the enclosure, as illustrated in Fig. 6.4*b*. The second derivatives in Laplace's equation are then approximated by finite differences. Thus, at point 2 the approximation is

$$\frac{\partial^2 V}{\partial x^2} \simeq (V_3 - V_2) - (V_2 - V_1) = V_1 + V_3 - 2V_2$$

Correspondingly, at point 2,

$$\frac{\partial^2 V}{\partial x^2} + \frac{\partial^2 V}{\partial y^2} = V_1 + V_3 + V_6 + b_2 - 4V_2$$

Thus, (*) translates to the requirement that each grid value should equal the average of its four neighbors. Written at each interval grid point, the system takes the form

$$\mathbf{AV} + \mathbf{b} = \mathbf{0}$$

where $\mathbf{V}$ is the vector of unknown potentials, and $\mathbf{b}$ consists of various (known) boundary values. Note that $\mathbf{A} = \mathbf{B} - 4\mathbf{I}$, where $\mathbf{B} \geq \mathbf{0}$. Also $(\mathbf{I} + \mathbf{B})^n > \mathbf{0}$, for large n.

(a) Show that $\mathbf{b} \geq \mathbf{0}$ implies $\mathbf{V} \geq \mathbf{0}$.

(b) For large numbers of points (e.g., several hundred) it is usually best to solve the system iteratively. For example, the procedure

$$4\mathbf{V}(k+1) = \mathbf{BV}(k) + \mathbf{b}$$

is implemented by setting each new grid value equal to the average of the old grid values around it. Show that this procedure converges to the solution.

* 14. *Balanced Growth in an Expanding Economy.* Consider an economy consisting of n sectors. Its output $\mathbf{y}(k)$ in any period is used, at least in part, as input for the next period. A simple description of such an economy is that successive $\mathbf{y}(k)$'s must satisfy

$$\mathbf{y}(k) \geqq \mathbf{Ay}(k+1)$$

where $\mathbf{A}$ is a strictly positive input–output matrix, with an interpretation similar to that of the example in Sect. 3.1. Suppose that a goal is to expand the economy as rapidly as possible; that is, to increase all components of $\mathbf{y}$ as rapidly as possible.

Since there is no need to produce more output than can be used as input in the next period, the economy will satisfy

$$\mathbf{y}(k) = \mathbf{A}\mathbf{y}(k+1)$$

The expansion rate during period k is

$$r = \min_i y_i(k+1)/y_i(k)$$

and thus $\mathbf{y}(k+1) \geqq r\mathbf{y}(k)$.

(a) For a one period problem, maximal expansion is attained by selecting $\mathbf{y}(0)$ and r with $r\mathbf{A}\mathbf{y}(0) \leqq \mathbf{y}(0)$ such that r is maximal. Find r and $\mathbf{y}(0)$. (See Problem 5.)
(b) Show that the maximal one period expansion rate can be maintained for all periods, and all sectors increase at the same rate. (This growth pattern is called *balanced growth*.)

15. Let $\mathbf{B}$ be an $n \times n$ matrix with $b_{ij} \geq 0$ for $i \neq j$. Suppose there is a $\mathbf{z} \geq \mathbf{0}$ such that $\mathbf{B}\mathbf{z} < \mathbf{0}$. Show that $\mathbf{B}$ has all its eigenvalues in the left half-plane.

16. Let $\mathbf{A}$ and $\mathbf{B}$ be $m \times r$ and $r \times m$ matrices, respectively. Assume both $\mathbf{A}$ and $\mathbf{B}$ are nonnegative. Then $\mathbf{AB}$ and $\mathbf{BA}$ are square nonnegative matrices of dimension $m \times m$ and $r \times r$, respectively. Show that $\mathbf{AB}$ and $\mathbf{BA}$ have the same Frobenius–Perron eigenvalue.

*17. Suppose that in an n-nation arms race the nations are explicitly divided into two alliances, of m and $n-m$ nations. The matrix $\mathbf{K}$ of defense and fatigue coefficients has the structural form

$$\mathbf{K} = \begin{bmatrix} -\mathbf{I} & \mathbf{A} \\ \mathbf{B} & -\mathbf{I} \end{bmatrix}$$

representing the fact that defense coefficients k_{ij} are zero if i and j belong to the same alliance. The fatigue coefficients are all unity, just for simplicity. The matrices $\mathbf{A}$ and $\mathbf{B}$ are $m \times (n-m)$ and $(n-m) \times m$, respectively, and both are strictly positive.

The object of this problem is to relate this explicit structural definition of an alliance to the somewhat implicit characterization used in conjunction with the interpretation of the eigenvector corresponding to the minimal eigenvalue of $\mathbf{K}$. In particular, show, in this case, that the eigenvalue of $\mathbf{K}$ having minimal real part is in fact negative, and that the left eigenvector corresponding to this eigenvalue has the form

$$\mathbf{x}^T = [\mathbf{x}_1^T, -\mathbf{x}_2^T]$$

where $\mathbf{x}_1 > \mathbf{0}$, $\mathbf{x}_2 > \mathbf{0}$ and $\mathbf{x}_1$, $\mathbf{x}_2$ are of dimension m and $n-m$, respectively.

18. Suppose the positive system

$$\mathbf{x}(k+1) = \mathbf{A}\mathbf{x}(k) + \mathbf{b}$$

is asymptotically stable and has equilibrium point $\bar{\mathbf{x}} > \mathbf{0}$. Suppose $\mathbf{b}$ is changed by increasing a single positive component b_i to $b_i + \Delta b_i$, $\Delta b_i > 0$, and let $\bar{\mathbf{x}} + \Delta\bar{\mathbf{x}}$ denote the corresponding new equilibrium point. Show that

$$\frac{\Delta \bar{x}_i}{\bar{x}_i} \leq \frac{\Delta b_i}{b_i}$$

*19. *Embedded Statics and Positive Systems.* Consider the system

$$\dot{\mathbf{x}} = \mathbf{A}_1\mathbf{x} + \mathbf{A}_2\mathbf{y}$$
$$\mathbf{0} = \mathbf{A}_3\mathbf{x} + \mathbf{A}_4\mathbf{y}$$

where $\mathbf{x}$ is n dimensional, y is m dimensional, and $\mathbf{A}_1$, $\mathbf{A}_2$, $\mathbf{A}_3$, and $\mathbf{A}_4$ are $n \times n$, $n \times m$, $m \times n$, and $m \times m$, respectively. Assume $\mathbf{A}_4^{-1}$ exists. This system is equivalent to the reduced system

$$\dot{\mathbf{x}} = (\mathbf{A}_1 - \mathbf{A}_2\mathbf{A}_4^{-1}\mathbf{A}_3)\mathbf{x}$$

Let

$$\mathbf{A} = \begin{bmatrix} \mathbf{A}_1 & \mathbf{A}_2 \\ \mathbf{A}_3 & \mathbf{A}_4 \end{bmatrix}$$

(a) Show that $\mathbf{A}$ being Metzler *does not* imply that $\mathbf{A}_1 - \mathbf{A}_2\mathbf{A}_4^{-1}\mathbf{A}_3$ is Metzler.
(b) Show that if $\mathbf{A}$ is Metzler and stable and $\mathbf{A}_4$ is asymptotically stable, then $\mathbf{A}_1 - \mathbf{A}_2\mathbf{A}_4^{-1}\mathbf{A}_3$ is both Metzler and stable.
(c) Explain how this result relates to the Homans–Simon model of group interaction.

NOTES AND REFERENCES

Section 6.2. The theory of positive matrices goes back to Perron [P2] and Frobenius [F4]. The theory has been extended in several directions and applied in many areas, most particularly in economics. The proof presented here is due to Karlin [K7]. For another approach see Nikaido [N1]. For a good general discussion of nonnegative matrices see Gantmacher [G3].

Section 6.3. The results on the inverse of $\mathbf{I} - \mathbf{A}$ have long been an integral component of the theory of positive matrices. The form and interpretation given here, as relating stability and positivity of equilibrium points, closely ties together the theories of positive and Metzler matrices.

Section 6.4. The Peter Principle itself is described in nonmathematical terms in Peter and Hull [P3]. The quantitative model presented in this section follows Kane [K6], although the subsequent analysis is somewhat different.

Section 6.5. Matrices with positive off-diagonal elements have long been known to be implicitly included within the theory of positive matrices. It is now standard, however, following the practice in economics, to refer to them as Metzler matrices [M4].

Section 6.6. Lewis F. Richardson developed his approach to the analysis of arms races during the period of about 1919 to 1953. His book *Arms and Insecurity* [R3] was published posthumously in 1960, edited by Nicolas Rashevsky and Ernesto Trucco. In the preface to the book Kenneth E. Boulding is quoted as follows:

> Lewis Richardson's work in 'politicometrics' (if I may coin a word for it) has all the marks of what might be called, without any sense of disparagement, 'amateur genius.'

The work is unusual not only in its creative approach to the subject matter, but also its remarkable insight into principles of dynamics. The only major technical insight we might add is that embodied by the theory of positive matrices—a subject that was apparently unfamiliar to Richardson. For a summary and critique of Richardson's work see Rapoport [R2] and Saaty [S1].

Section 6.7. The main result on comparative statics, in the form of Theorem 2, is due to Morishima [M7]. Also see Sandberg [S3].

Section 6.8. See Homans [H4] and Simon [S5].

Section 6.9. The conditions in Problem 6 are generally referred to as the Hawkins–Simon conditions [H1]. Problem 14 is a simplified version of the von Neumann model of an expanding economy [V2]. The result of Problem 18 is contained in Morishima [M7].

chapter 7.

Markov Chains

Markov chains represent a special class of dynamic systems that evolve probabilistically. This class of models, which can be regarded in part as a special subclass of positive linear systems, has a wide variety of applications and a deep but intuitive body of theory. It is an important branch of dynamic systems.

A finite Markov chain can be visualized in terms of a marker that jumps around among a finite set of locations, or conditions. The transition from one location to another, however, is probabilistic rather than deterministic. A classic example is weather, which can be classified in terms of a finite number of conditions, and which changes daily from one condition to another. The possible positions for the process are termed "states." Since there are only a finite number of states, the structure of finite Markov chains appears at first to differ substantially from the standard dynamic system framework in which the state is defined over an n-dimensional continuum. However, the probabilistic evolution of a Markov chain implies that future states cannot be inferred from the present, except in terms of probability assessments. Thus, tomorrow's weather cannot be predicted with certainty, but probabilities can be assigned to the various possible conditions. Therefore, in general, future evolution of a Markov process is described by a vector of probabilities (for occurrence of the various states). This vector, and its evolution, is really the essential description of the Markov chain, and it is governed by a linear dynamic system in the sense of earlier chapters. The first part of this chapter develops this framework.

The vector of probabilities is a positive vector, and, accordingly, the dynamic system describing a Markov chain is a positive linear system. The results on positive linear systems, particularly the Frobenius–Perron theorem, thus imply important limiting properties for Markov chains. If the Frobenius–Perron eigenvector is strictly positive, all the states are visited infinitely often and with probabilities defined by the components of this eigenvector.

In many Markov chains, the Frobenius–Perron eigenvector is not strictly positive, and perhaps not unique. This has important implications in terms of the probabilistic context, and raises a series of important issues. To analyze such chains, it is necessary to systematically characterize the various possible chain structures. This leads to new insights in this useful class of models.

7.1 FINITE MARKOV CHAINS

A finite Markov chain is a discrete-step process that at any step can be in one of a finite number of conditions, or states. If the chain has n possible states, it is said to be an nth-order chain. At each step the chain may change from its state to another, with the particular change being determined probabilistically according to a given set of *transition probabilities*. Thus, the process moves stepwise but randomly among the finite number of states. Throughout this chapter only *stationary* Markov chains are considered, where the transition probabilities do not depend on the number of steps that have occurred.

Definition. An nth-order Markov chain process is determined by a set of n states $\{S_1, S_2, \ldots, S_n\}$ and a set of transition probabilities p_{ij}, $i = 1, 2, \ldots, n$, $j = 1, 2, \ldots, n$. The process can be in only one state at any time instant. If at time k the process is in state S_i, then at time $k+1$ it will be in state S_j with probability p_{ij}. An initial starting state is specified.

Example 1 (A Weather Model). The weather in a certain city can be characterized as being either sunny, cloudy, or rainy. If it is sunny one day, then sun or clouds are equally likely the next day. If it is cloudy, then there is a fifty percent chance the next day will be sunny, a twenty-five percent chance of continued clouds, and a twenty-five percent chance of rain. If it is raining, it will not be sunny the next day, but continued rain or clouds are equally likely.

Denoting the three types of weather by S, C, and R, respectively, this model can be represented by an array of transition probabilities:

	S	C	R
S	$\frac{1}{2}$	$\frac{1}{2}$	0
C	$\frac{1}{2}$	$\frac{1}{4}$	$\frac{1}{4}$
R	0	$\frac{1}{2}$	$\frac{1}{2}$

This array is read by going down the left column to the current weather condition. The corresponding row of numbers gives the probabilities associated

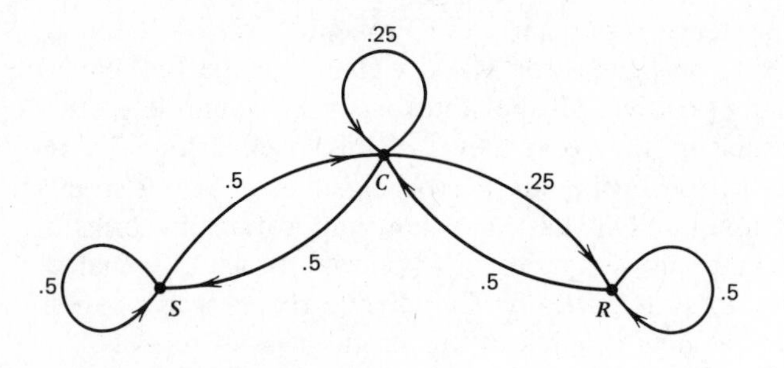

Figure 7.1. The weather chain.

with the next weather condition. The process starts with some weather condition and moves, each day, to a new condition. There is no way, however, to predict exactly which transition will occur. Only probabilistic statements, presumably based on past experience, can be made.

The weather model can be alternatively described in terms of a diagram as shown in Fig. 7.1. In general in such diagrams, nodes correspond to states, and directed paths between nodes indicate possible transitions, with the probability of a given transition labeled on the path.

Example 2 (Estes Learning Model). As a simple model of learning of an elementary task or of a small bit of information, it can be assumed that an individual is always in either of two possible states: he is in state L if he has learned the task or material, and in state N if he has not. Once the individual has learned this one thing, he will not forget it. However, if he has not yet learned it, there is a probability α, $0<\alpha<1$ that he will learn it during the next time period. This chain is illustrated in Fig. 7.2.

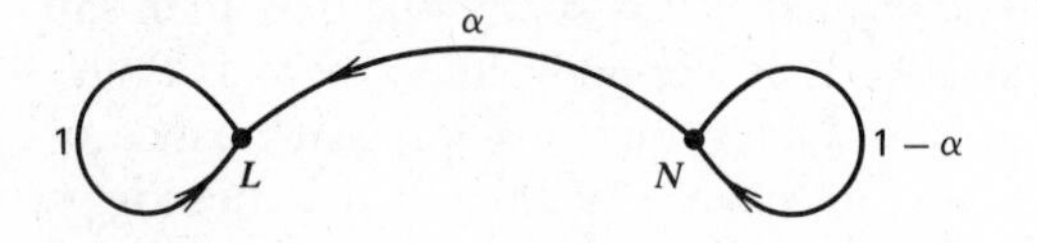

Figure 7.2. Learning model.

This idealized learning process is a two-state Markov chain having transition probabilities

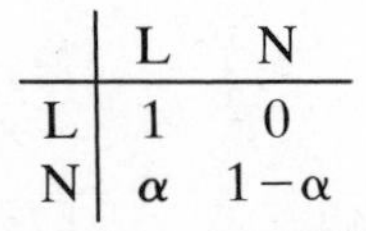

	L	N
L	1	0
N	α	$1-\alpha$

Example 3 (Gambler's Ruin). The gambler's ruin problem of Chapter 2 can be regarded as a Markov chain with states corresponding to the number of coins

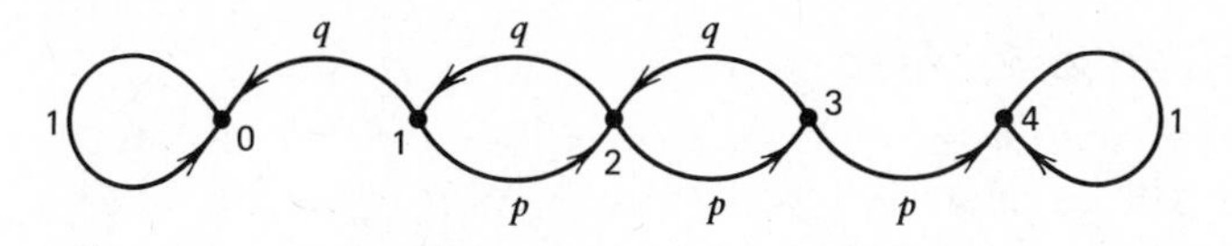

Figure 7.3. Gambler's ruin.

or chips held by player A. As a specific example, suppose both players begin with just two chips, and suppose the probability that A wins in any turn is p, while the probability that B wins is $q = 1 - p$. There are five possible states, corresponding as to whether player A has 0, 1, 2, 3, or 4 chips. The transition probabilities are

	0	1	2	3	4
0	1	0	0	0	0
1	q	0	p	0	0
2	0	q	0	p	0
3	0	0	q	0	p
4	0	0	0	0	1

and this structure is shown in Fig. 7.3. The process is initiated at the state corresponding to two chips.

Stochastic Matrices and Probability Vectors

The transition probabilities associated with a Markov chain are most conveniently regarded as the elements of an $n \times n$ matrix

$$\mathbf{P} = \begin{bmatrix} p_{11} & p_{12} & \cdots & p_{1n} \\ p_{21} & p_{22} & \cdots & p_{2n} \\ \cdot & & & \\ \cdot & & & \\ \cdot & & & \\ p_{n1} & p_{n2} & \cdots & p_{nn} \end{bmatrix}$$

It is clear that all elements of a $\mathbf{P}$ matrix associated with a Markov chain are nonnegative. Furthermore, it should be observed that the sum of elements along any row is equal to 1. This is because if the process is in state i at a given step, the probability that it goes *somewhere* during the next step must be 1. A square matrix $\mathbf{P}$ with these properties is often referred to as a *stochastic matrix.*

A vector is a *probability vector* if all its components are nonnegative and sum to 1. A fundamental relation between stochastic matrices and probability vectors is that if $\mathbf{x}^T$ is a row probability vector and $\mathbf{P}$ is a stochastic matrix, then the row vector $\mathbf{x}^T\mathbf{P}$ is also a probability vector. (The reader is asked to verify this in Problem 1.) Thus, stochastic matrices can be thought of as natural transformations in the realm of probability vectors.

The Multistep Transition Process

If an nth-order Markov chain with transition matrix $\mathbf{P}$ is initiated in state S_i, then after one step it will be in state S_j with probability p_{ij}. This fact can be characterized by stating that the probabilities of the various states after one step are the components of the row vector

$$[p_{i1} \quad p_{i2} \quad \cdots \quad p_{in}]$$

which is a probability vector. This vector is itself obtained by multiplication of $\mathbf{P}$ on the left by the special degenerate row probability vector

$$[0 \quad \cdots \quad 1 \quad \cdots \quad 0]$$

where the 1 is in the ith place.

Suppose now that we look at the Markov process after two steps. Beginning in a given initial state S_i the process will after two steps end up at some state S_j. The overall transition from S_i to S_j is governed by two applications of the underlying transition matrix. To work out the details, let $p_{ij}^{(2)}$ be the probability that starting at state S_i the process will move to state S_j after two steps. If it were known that the process would go to state S_k after the first step, we would have

$$p_{ij}^{(2)} = p_{kj}$$

However, the probability that the state is S_k after one step is p_{ik}. Summing over all possible first steps we obtain

$$p_{ij}^{(2)} = \sum_{k=1}^{n} p_{ik}p_{kj} = [\mathbf{P}^2]_{ij}$$

This calculation shows that the probability $p_{ij}^{(2)}$ is equal to the ijth element of the matrix $\mathbf{P}^2$. Thus, the two-step transition matrix is $\mathbf{P}^2$.

In a similar way, the transition probabilities for m steps are defined by the elements of the matrix $\mathbf{P}^m$. We write, for notational convenience, $p_{ij}^{(m)}$ for the ijth component of $\mathbf{P}^m$, and recognize that it is also the probability of going from S_i to S_j in m steps.

Much of the above discussion can be expressed more directly in terms of a natural association between a Markov chain and a standard dynamic system. Let $\mathbf{x}(k)^T$ be an n-dimensional *row* vector with component x_j, $j = 1, 2, \ldots, n$ corresponding to the probability that the state at step k will be equal to S_j. If the process is initiated in state S_i, then $\mathbf{x}(0)^T$ is the unit vector with a one in the ith coordinate position. Successive probability vectors are generated by the recursion

$$\mathbf{x}(k+1)^T = \mathbf{x}(k)^T\mathbf{P} \tag{7-1}$$

We recognize (7-1) as a standard, linear time-invariant system, except that it is expressed in terms of row rather than column vectors. (By using $\mathbf{P}^T$ instead of $\mathbf{P}$ the system obviously could be expressed in terms of columns. However, it is standard convention to work with the row formulation in the context of Markov chains.)

It must be emphasized that the $\mathbf{x}(k)^T$ vector is not really the state of the Markov process. At each step the state is one of the n distinct states $S_1, S_2, \ldots, S_n$. The vector $\mathbf{x}(k)^T$ gives the probabilities that the Markov process takes on specific values. Thus, the sequence of $\mathbf{x}(k)^T$ does not record a sequence of actual states, rather it is a projection of our probabilistic knowledge.

It is possible to give somewhat more substance to the interpretation of the vector $\mathbf{x}(k)^T$ by imagining a large number N of independent copies of a given Markov chain. For example, in connection with the Estes learning model we might imagine a class of N students, each governed by identical transition probabilities. Although the various chains each are assumed to have the same set of transition probabilities, the actual transitions in one chain are not influenced by those in the others. Therefore, even if all chains are initiated in the same corresponding state, they most likely will differ at later times. Indeed, if the chains all begin at state S_i, it can be expected that after one step about Np_{i1} of them will be in state S_1, Np_{i2} in state S_2, and so forth. In other words, they will be distributed among the states roughly in proportion to the transition probabilities. In the classroom example, for instance, the $\mathbf{x}(k)^T$ vector, although not a description of the evolution of any single student, is a fairly accurate description of the whole class in terms of the percentage of students in each state as a function of k. From this viewpoint Markov chains are closely related to some of our earlier models, such as population dynamics, promotions in a hierarchy, and so forth, where groups of individuals or objects move into different categories.

Analytical Issues

When viewed in terms of its successive probability vectors, a Markov chain is a linear dynamic system with a positive system matrix. Thus, it is expected that the strong limit properties of positive systems play a central role in the theory of Markov processes. Indeed this is true, and in this case these properties describe the long-term distribution of states.

In addition to characterization of the long-term distribution, there are some important and unique analytical issues associated with the study of Markov chains. One example is the computation of the average length of time for a Markov process to reach a specified state, or one of a group of states. Another is the computation of the probability that a specified state will ultimately be reached. Such problems do not have direct analogies in standard deterministic system analysis. Nevertheless, most of the analysis for Markov chains is based on the principles developed in earlier chapters.

7.2 REGULAR MARKOV CHAINS AND LIMITING DISTRIBUTIONS

As pointed out earlier, the Frobenius-Perron theorem and the associated theory of positive systems is applicable to Markov chains. For certain types of Markov chains, these results imply the existence of a unique limiting probability vector.

To begin the application of the Frobenius–Perron theorem we first observe that the Frobenius–Perron eigenvalue, the eigenvalue of largest absolute value, is always 1.

Proposition. *Corresponding to a stochastic matrix* $\mathbf{P}$ *the value* $\lambda_0=1$ *is an eigenvalue. No other eigenvalue of* $\mathbf{P}$ *has absolute value greater than* 1.

Proof. This is a special case of the argument given in Sect. 6.2, since for a stochastic matrix all row sums are equal to 1. ▮

Definition. A Markov chain is said to be *regular* if $\mathbf{P}^m > \mathbf{0}$ for some positive integer m.

This straightforward definition of regularity is perhaps not quite so innocent as it might first appear. Although many Markov chains of interest do satisfy this condition, many others do not. The weather example of the previous section is regular, for although $\mathbf{P}$ itself is not strictly positive, $\mathbf{P}^2$ is. The Estes learning model is not regular since in this case $\mathbf{P}^m$ has a zero in the upper right-hand corner for each m. Similarly, the Gambler's Ruin example is not regular. In general, recalling that $\mathbf{P}^m$ is the m-step probability transition matrix, regularity means that over a sufficiently large number of steps the Markov chain must be strictly positive. There must be a positive probability associated with every transition.

The main theorem for regular chains is stated below and consists of three parts. The first part is simply a restatement of the Frobenius–Perron theorem, while the second and third parts depend on the fact that the dominant eigenvalue associated with a Markov matrix is 1.

Theorem (Basic Limit Theorem for Markov Chains). *Let* $\mathbf{P}$ *be the transition matrix of a regular Markov chain. Then:*

(a) *There is a unique probability vector* $\mathbf{p}^T > \mathbf{0}$ *such that*

$$\mathbf{p}^T\mathbf{P} = \mathbf{p}^T$$

(b) *For any initial state i (corresponding to an initial probability vector equal to the* i*th coordinate vector* $\mathbf{e}_i^T$*) the limit vector*

$$\mathbf{v}^T = \lim_{m\to\infty} \mathbf{e}_i^T\mathbf{P}^m$$

exists and is independent of i. Furthermore, $\mathbf{v}^T$ *is equal to the eigenvector* $\mathbf{p}^T$.

(c) $\mathrm{Lim}_{m\to\infty}\,\mathbf{P}^m = \bar{\mathbf{P}}$, *where* $\bar{\mathbf{P}}$ *is the* $n\times n$ *matrix, each of whose rows is equal to* $\mathbf{p}^T$.

Proof. Part (a) follows from the Frobenius–Perron theorem (Theorem 2, Sect. 6.2) and the fact that the dominant eigenvalue is $\lambda_0 = 1$. To prove part (b) we note that since $\lambda_0 = 1$ is a simple root, it follows that $\mathbf{e}_i^T\mathbf{P}^m$ must converge to a scalar multiple of $\mathbf{p}^T$. However, since each $\mathbf{e}_i^T\mathbf{P}^m$ is a probability vector, the multiple must be 1. Part (c) is really just a restatement of (b) because part (b) shows that each row of $\mathbf{P}^m$ converges to $\mathbf{p}^T$. ∎

This result has a direct probabilistic interpretation. Parts (a) and (b) together say that starting at any initial state, after a large number of steps the probability of the chain occupying state S_i is p_i, the ith component of $\mathbf{p}^T$. The long-term probabilities are independent of the initial condition.

There are two somewhat more picturesque ways of viewing this same result. One way is to imagine starting the process in some particular state, then turning away as the process moves through many steps. Then after turning back one records the current state. If this experiment is repeated a large number of times, the state S_i will be recorded a fraction p_i of the time, no matter where the process is started.

The second way to visualize the result is to imagine many copies of the Markov chain operating simultaneously. No matter how they are started, the distribution of states tends to converge to that defined by the limit probability vector.

Finally, part (c) of the theorem is essentially an alternative way of stating the same limit property. It says that the m-step transition matrix ultimately tends toward a limit $\bar{\mathbf{P}}$. This probability matrix transforms any initial probability vector into the vector $\mathbf{p}^T$.

Example 1 (The Weather Model). The weather example of Sect. 7.1 has transition matrix

$$\mathbf{P} = \begin{bmatrix} .500 & .500 & 0 \\ .500 & .250 & .250 \\ 0 & .500 & .500 \end{bmatrix}$$

This Markov chain is certainly regular since in fact $\mathbf{P}^2 > \mathbf{0}$. Indeed, computing a few powers of the transition matrix, we find

$$\mathbf{P}^2 = \begin{bmatrix} .500 & .375 & .125 \\ .375 & .438 & .187 \\ .250 & .375 & .375 \end{bmatrix}$$

$$\mathbf{P}^4 = \begin{bmatrix} .422 & .399 & .179 \\ .398 & .403 & .199 \\ .359 & .399 & .242 \end{bmatrix}$$

$$\mathbf{P}^6 = \begin{bmatrix} .405 & .401 & .194 \\ .400 & .401 & .199 \\ .390 & .400 & .210 \end{bmatrix}$$

$$\mathbf{P}^8 = \begin{bmatrix} .401 & .401 & .198 \\ .400 & .401 & .199 \\ .397 & .401 & .202 \end{bmatrix}$$

$$\mathbf{P}^{16} = \begin{bmatrix} .400 & .400 & .200 \\ .400 & .400 & .200 \\ .400 & .400 & .200 \end{bmatrix}$$

This behavior of the powers of the probability matrix is in accordance with part (c) of Theorem 1. It follows that the equilibrium probability vector is

$$\mathbf{p}^T = [.400 \quad .400 \quad .200]$$

Indeed, as an independent check it can be verified that this is a left eigenvector of the transition matrix, corresponding to the eigenvalue of 1.

The interpretation of this vector is that in the long run the weather can be expected to be sunny 40% of the days, cloudy 40% of the days, and rainy 20% of the days.

Example 2 (Simplified Monopoly). A simple game of chance and strategy for two to four players is played on the board shown in Fig. 7.4. Each player has a marker that generally moves clockwise around the board from space to space. At each player's turn he flips a coin: if the result is "heads" he moves one space, if it is "tails" he moves two spaces. A player landing on the "Go to Jail" square, goes to "Jail" where he begins at his next turn. During the game, players may acquire ownership of various squares (except "Jail" and "Go to Jail"). If a player lands on a square owned by another player, he must pay rent in the amount shown in that square to the owner. In formulating strategy for the game it is useful to know which squares are most valuable in terms of the amount of rent that they can be expected to generate. Without some analysis, the true relative values of the various squares is not apparent.

Movement of players' markers around the board can be considered to be a

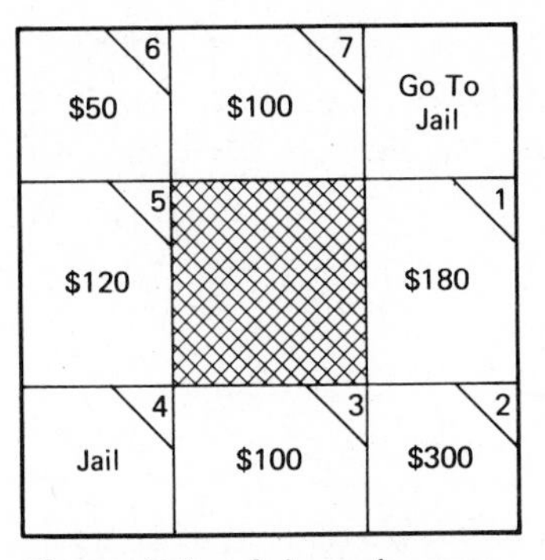

Figure 7.4. A board game.

Markov chain. The chain has seven states, corresponding to the possible landing positions on the board. The "Go to Jail" square is not counted since no piece ever really stays there, but goes instead to square number 4. The transition matrix is

$$\mathbf{P}=\begin{bmatrix} 0 & \frac{1}{2} & \frac{1}{2} & 0 & 0 & 0 & 0 \\ 0 & 0 & \frac{1}{2} & \frac{1}{2} & 0 & 0 & 0 \\ 0 & 0 & 0 & \frac{1}{2} & \frac{1}{2} & 0 & 0 \\ 0 & 0 & 0 & 0 & \frac{1}{2} & \frac{1}{2} & 0 \\ 0 & 0 & 0 & 0 & 0 & \frac{1}{2} & \frac{1}{2} \\ 0 & 0 & 0 & \frac{1}{2} & 0 & 0 & \frac{1}{2} \\ \frac{1}{2} & 0 & 0 & \frac{1}{2} & 0 & 0 & 0 \end{bmatrix}$$

After a bit of experimentation, it is seen that there is a finite probability of moving from any square to any square in seven steps; that is, $\mathbf{P}^7 > \mathbf{0}$. Therefore, it is quite clear that this Markov chain is regular, and there is an equilibrium probability vector that gives the long-term landing probabilities. To find the equilibrium probabilities we must solve the equation $\mathbf{p}^T\mathbf{P} = \mathbf{p}^T$ with $\sum_{i=1}^{n} p_i = 1$. Written out in detail the equations for the eigenvector* are

$$\begin{aligned} \tfrac{1}{2}p_7 &= p_1 \\ \tfrac{1}{2}p_1 &= p_2 \\ \tfrac{1}{2}p_1 + \tfrac{1}{2}p_2 &= p_3 \\ \tfrac{1}{2}p_2 + \tfrac{1}{2}p_3 + \tfrac{1}{2}p_6 + \tfrac{1}{2}p_7 &= p_4 \\ \tfrac{1}{2}p_3 + \tfrac{1}{2}p_4 &= p_5 \\ \tfrac{1}{2}p_4 + \tfrac{1}{2}p_5 &= p_6 \\ \tfrac{1}{2}p_5 + \tfrac{1}{2}p_6 &= p_7 \end{aligned}$$

* Remember that the coefficient matrix is $\mathbf{P}^T$ since $\mathbf{p}^T$ is a *row* vector.

These equations could be solved successively, one after the other, if p_6 and p_7 were known. If we temporarily ignore the requirement that $\sum p_i = 1$, the value of one of the p_i's can be set arbitrarily. Let us set $p_7 = 1$. Then we find

$$\begin{aligned} p_1 &= .50 \\ p_2 &= .250 \\ p_3 &= .375 \\ p_4 &= .8125 + \tfrac{1}{2}p_6 \\ p_5 &= .59375 + \tfrac{1}{4}p_6 \\ p_6 &= .703125 + \tfrac{3}{8}p_6 \\ p_7 &= 1.0 \end{aligned}$$

At this point p_6 can be found from the sixth equation and substituted everywhere else yielding

$$\begin{aligned} p_1 &= .50 \\ p_2 &= .250 \\ p_3 &= .375 \\ p_4 &= 1.375 \\ p_5 &= .875 \\ p_6 &= 1.125 \\ p_7 &= 1.0 \end{aligned}$$

The actual equilibrium probability vector is obtained by dividing each of these numbers by their sum, 5.5. Thus,

$$^{T} = [.0909 \ \ .0455 \ \ .0682 \ \ .2500 \ \ .1591 \ \ .2045 \ \ .1818]$$

Not surprisingly we find that "Jail" is visited most frequently. Accordingly, it is clear that spaces 1–3 are visited relatively infrequently. Thus, even though these spaces have high associated rents, they are not particularly attractive to own. This is verified by the relative income rates for each square, normalized with state S_7 having an income of \$100, as shown below.

State	Rent	Relative Income	Rank
S_1	\$180.	\$ 90.00	3
S_2	\$300.	\$ 75.00	4
S_3	\$100.	\$ 37.50	6
S_4	—	—	
S_5	\$120	\$105.00	1
S_6	\$ 50.	\$ 56.25	5
S_7	\$100.	\$100.00	2

7.3 CLASSIFICATION OF STATES

Many important Markov chains have a probability transition matrix containing one or more zero entries. This is the case, for instance, in all of the examples considered so far in this chapter. If the chain is regular, the zero entries are of little fundamental consequence since there is an m such that a transition between any two states is possible in m steps. In general, however, the presence of zero entries may preclude some transitions, even over an arbitrary number of steps. A first step in the development of a general theory of Markov chains is to systematically study the structure of state interconnections.

Classes of Communicating States

We say that the state S_j is *accessible* from the state S_i if by making only transitions that have nonzero probability it is possible to begin at S_i and arrive at S_j in some finite number of steps. A state S_i is always considered to be accessible from itself.

Accessibility can be determined by taking powers of the probability transition matrix. Let $p_{ij}^{(m)}$ be the ijth element of the matrix $\mathbf{P}^m$. If $p_{ij}^{(m)}>0$, then it is possible to go from S_i to S_j in m steps, since there is a positive probability that the Markov chain would make such a transition. Thus, S_j is accessible from S_i if and only if $p_{ij}^{(m)}>0$ for some integer $m \geq 0$.

The property of accessibility is not symmetric since S_j may be accessible from S_i while S_i is not accessible from S_j. The corresponding symmetric notion is termed communication and it is this property that is used to classify states.

Definition. States S_i and S_j are said to *communicate* if each is accessible from the other.

As the following proposition shows, the concept of communicating states effectively divides the states of a Markov chain into distinct classes, each with its own identity. That is, the totality of n states is partitioned into a group of classes; each state belonging to exactly one class. Later we study the classes as units, and investigate the structure of class interconnections.

Proposition. *The set of states of a Markov chain can be divided into communicating classes. Each state within a class communicates with every other state in the class, and with no other state.*

Proof. Let C_i be the set of states that communicate with S_i. If S_k and S_j belong to C_i, they also communicate, since paths of transition between them can be found in each direction by first passing through S_i. See Fig. 7.5a. Thus, all states in C_i communicate with each other.

Suppose that a state S_l outside of C_i communicated with a state S_j within C_i. Then a path from S_l to S_j could be extended to a path from S_l to S_i by

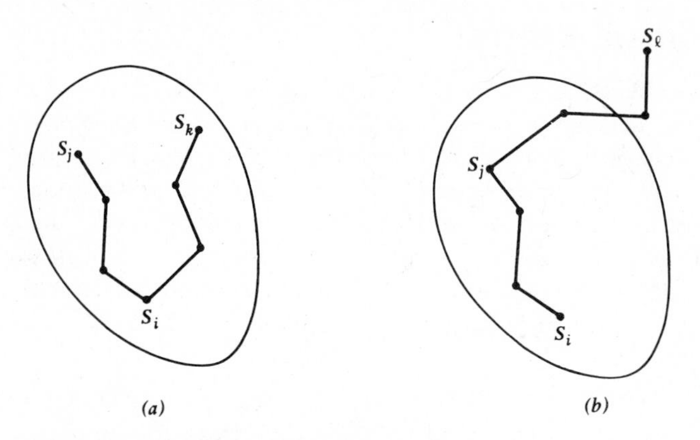

Figure 7.5. Construction for proof.

appending a path from S_j to S_i. See Fig. 7.5b. Likewise paths in the other direction could be appended. Thus, S_l would communicate with S_i as well. Hence, by contradiction, no state outside C_i can communicate with any state in C_i. Therefore different communicating classes have no common states. Every state belongs to one and only one class—the class of all states that communicate with it. ▮

An important special case is when all states communicate, in which case there is only one communicating class. If a Markov chain has only one communicating class the chain is said to be *irreducible.* Otherwise, it is *reducible.*

A regular Markov chain is irreducible, since all states communicate. However, not all irreducible Markov chains are regular. An example is the chain defined by the transition matrix

$$\mathbf{P} = \begin{bmatrix} 0 & 1 \\ 1 & 0 \end{bmatrix}$$

The chain goes from S_1 to S_2 or from S_2 to S_1 in one step. It can go from either state back to itself in two steps. However, every power of $\mathbf{P}$ contains two zero entries.

Let us apply the definitions of this section to the examples presented earlier. The weather example is irreducible, since it is possible to go from any state of weather to any other within two days. In fact, as shown earlier this chain is regular. The learning model has two states, and each is a different communicating class. Although the "*learned*" state is accessible from the "*unlearned*" state, the reverse is not true, and hence the states do not communicate. The Gambler's Ruin chain has three communicating classes. One is the state corresponding to player A having zero chips. This state corresponds to an end of the game and no other state is accessible from it.

Similarly, a second class is the state corresponding to player A having all of the chips. Finally, the third class consists of all the other states. It is possible to go from any one of these to any other (or back) in a finite number of steps.

Closed Classes

A communicating class C is said to be *closed* if there are no possible transitions from the class C to any state outside C. In other words, no state outside C is accessible from C. Thus, once the state of a Markov chain finds its way into a closed class it can never get out. Of course, the reverse is not necessarily true. It may be possible to move into the closed class from outside.

A simple example of a closed class is provided by an irreducible Markov chain. An irreducible chain has a single communicating class consisting of all states, and it is clearly closed.

Closed classes are sometimes referred to as *absorbing classes* since they tend to ultimately absorb the process. In particular, if a closed class consists of just a single state, that state is called an *absorbing state.*

In the Estes learning model, the state corresponding to "learned" is an absorbing state. In the Gambler's Ruin problem, the two end-point states are each absorbing states.

Transient Classes

A communicating class C is *transient* if some state outside of C is accessible from C. There is, therefore, a tendency for a Markov chain process to leave a transient class.

There may be allowable transitions *into* a transient class from another class as well as out. However, it is not possible for a closed path to exist that goes first outside the class and then returns, for this would imply that there were states outside of C that communicate with states in C. The connection structure between communicating classes must have an ordered flow, always terminating at some closed class. It follows of course that every Markov chain must have at least one closed class. A possible pattern of classes together with interconnections is shown in Fig. 7.6. In this figure the individual states within a class and their individual connections are not shown; the class connections illustrated are, of course, between specific states within the class. All classes in the figure are transient, except the bottom two, which are closed.

By definition, a transient class must have at least one path leading from one of its member states to some state outside the class. Thus, if the process ever reaches the state that is connected to an outside state, there is a positive probability that the process will move out of that class at the next step. Furthermore, no matter where the process begins in the transient class, there is a positive probability of reaching that exit state within a finite number of steps.

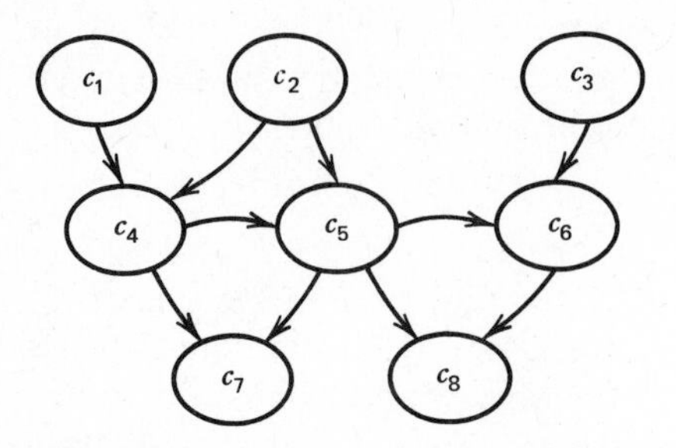

Figure 7.6. A collection of classes.

All together, over an infinite sequence of steps it seems that there is a good chance of leaving the transient class. In fact, as shown below the probability of eventually leaving is 1. (The reader can skip over the proof of this theorem without much loss of continuity.) In view of this result all states within transient classes are themselves referred to as *transient states.* The process leaves them, in favor of closed classes.

Theorem 1. *The state of a finite Markov chain is certain (with probability equal to one) to eventually enter some closed communicating class.*

Proof. From each state S_i in a transient class it is possible to reach a closed class in a finite number of steps. (See Problem 7.) Let m_i be the minimum number of steps required, and let p_i be the probability that starting at state S_i the chain will *not* reach a closed class in m_i steps. We have $p_i<1$. Now let m be the maximum of all the m_i's, and let p be the maximum of all the p_i's. Then, starting at any state, the probability of not reaching a closed class within m steps is less than or equal to p. Likewise, the probability of not reaching a closed class within km steps, where k is a positive integer, is less than or equal to p^k. Since $p<1$, p^k goes to zero as k goes to infinity. Correspondingly, the probability of reaching a closed class within km steps is at least $(1-p^k)$, which goes to 1. ∎

Relation to Matrix Structure

The classification of states as presented in this section leads to new insight in terms of the structure of the transition probability matrix $\mathbf{P}$ and the Frobenius–Perron eigenvectors. As an example, suppose that the state S_i is an absorbing state; once this state is reached, the process never leaves it. It follows immediately that the corresponding unit vector $\mathbf{e}_i^T$ (with all components zero, except the ith, which is 1) is an eigenvector. That is, $\mathbf{e}_i^T\mathbf{P}=\mathbf{e}_i^T$. It represents a (degenerate) equilibrium distribution. If there are other absorbing states, there are, correspondingly, other eigenvectors. More generally, the equilibrium eigenvectors of $\mathbf{P}$ are associated with the closed communicating classes of $\mathbf{P}$.

These relations can be expressed in terms of a *canonical form* for Markov chains. We order the states with all those associated with closed classes first, followed by those associated with transient classes. If the states are ordered this way, the transition matrix can be written in the partitioned form

$$\mathbf{P}=\begin{bmatrix}\mathbf{P}_1 & \mathbf{0}\\ \mathbf{R} & \mathbf{Q}\end{bmatrix} \tag{7-2}$$

Assuming there are r states in closed classes and $n-r$ in transient classes, the matrix $\mathbf{P}_1$ is an $r\times r$ stochastic matrix representing the transition probabilities within the closed classes; $\mathbf{Q}$ is an $(n-r)\times(n-r)$ *substochastic* matrix (at least one row sum is less than 1) representing the transition probabilities among the transients states, and $\mathbf{R}$ is an $(n-r)\times r$ matrix representing the transition probabilities from transient states to states within a closed class.

The (left) eigenvectors corresponding to the eigenvalue of 1 must have the form $\mathbf{p}^T=[\mathbf{p}_1^T, \mathbf{0}]$, where $\mathbf{p}_1^T$ is r-dimensional, representing the fact that only states in closed classes can occur with positive probability in equilibrium. (See Problem 8.) The closed classes act like separate Markov chains and have equilibrium distributions. Transient classes cannot sustain an equilibrium.

7.4 TRANSIENT STATE ANALYSIS

Many Markov chains of practical interest have transient classes and are initiated at a transient state. The Gambler's Ruin problem and the Estes learning model are two examples, which we have already discussed. When considering such chains, it is natural to raise questions related to the process of movement within the transient class before eventual absorption by a closed class. Examples of such questions are: the average length of time that the chain stays within a transient class, the average number of visits to various states, and the relative likelihood of eventually entering various closed classes.

The analysis of transient states is based on the canonical form described at the end of Sect. 7.3. We assume that the states are ordered with closed classes first, followed by transient states. The resulting canonical form is

$$\mathbf{P}=\begin{bmatrix}\mathbf{P}_1 & \mathbf{0}\\ \mathbf{R} & \mathbf{Q}\end{bmatrix} \tag{7-3}$$

We assume, as before, that there are r states in closed classes and $n-r$ transient states.

The substochastic matrix $\mathbf{Q}$ completely determines the behavior of the Markov chain within the transient classes. Thus, it is to be expected that analysis of questions concerning transient behavior is expressed in terms of $\mathbf{Q}$. Actually, a central role in transient analysis is played by the matrix $\mathbf{M}=[\mathbf{I}-\mathbf{Q}]^{-1}$—this is called the *fundamental matrix* of the Markov chain when expressed in the canonical form (7-3). As demonstrated below, it is easily established that the indicated inverse exists, so that $\mathbf{M}$ is well defined.

Proposition. *The matrix* $\mathbf{M}=[\mathbf{I}-\mathbf{Q}]^{-1}$ *exists and is positive.*

Proof. It follows from Theorem 1, Sect. 7.3, that $\mathbf{Q}^m \to \mathbf{0}$ as $m \to \infty$ since elements of $\mathbf{Q}^m$ are the m-step transition probabilities within the transient classes. Thus, the dominant eigenvalue of the nonnegative matrix $\mathbf{Q}$ is less than 1. The statement of the proposition is then a special case of Theorem 2, Sect. 6.3. ∎

The elements of the fundamental matrix have a direct interpretation in terms of the average number of visits to various transient states. Suppose that the Markov chain is initiated at the transient state S_i. Let S_j be another (or the same, if $i=j$) transient state. The probability that the process moves from S_i to S_j in one step is q_{ij}. Likewise, for any k the probability of a transition from S_i to S_j in exactly k steps is $q_{ij}^{(k)}$, the ijth element of the matrix $\mathbf{Q}^k$. If we include the zero-step transition probability $q_{ij}^{(0)}$, which is the ijth element of $\mathbf{Q}^0=\mathbf{I}$, then the sum of all these transition probabilities is

$$q_{ij}^{(0)}+q_{ij}^{(1)}+q_{ij}^{(2)}+\cdots+q_{ij}^{(k)}+\cdots$$

This sum is the average number of times that starting in state S_i the process reaches state S_j before it leaves the transient states and enters a closed class. This summation can be expressed as the ijth element of the matrix sum

$$\mathbf{I}+\mathbf{Q}+\mathbf{Q}^2+\cdots+\mathbf{Q}^k+\cdots$$

However, this in turn is equal to the fundamental matrix through the identity

$$\mathbf{M}=[\mathbf{I}-\mathbf{Q}]^{-1}=\mathbf{I}+\mathbf{Q}+\mathbf{Q}^2+\cdots+\mathbf{Q}^k+\cdots$$

(See the Lemma on Series Expansion of Inverse, Sect. 6.3.) Therefore we may state the following theorem.

Theorem 1. *The element* m_{ij} *of the fundamental matrix* $\mathbf{M}$ *of a Markov chain with transient states is equal to the mean number of times the process is in transient state* S_j *if it is initiated in transient state* S_i.

Next we observe that if we sum the terms across a row of the fundamental matrix $\mathbf{M}$, we obtain the mean number of visits to all transient states for a given starting state. This figure is the mean number of steps before being absorbed by a closed class. Formally, we conclude:

Theorem 2. *Let* $\mathbf{1}$ *denote a column vector with each component equal to* 1. *In a Markov chain with transient states, the ith component of the vector* $\mathbf{M1}$ *is equal to the mean number of steps before entering a closed class when the process is initiated in transient state* S_i.

Finally, if a chain is initiated in a transient state, it will (with probability one) eventually reach some state within a closed class. There may, however, be several possible closed class entry points. We therefore turn to the question of computing the probability that, starting at a given transient state, the chain first enters a closed class through a particular state. In the special case where the closed classes each consist of a single absorbing state, this computation gives the probabilities of terminating at the various absorbing states.

Theorem 3. *Let b_{ij} be the probability that if a Markov chain is started in transient state S_i, it will first enter a closed class by visiting state S_j. Let* $\mathbf{B}$ *be the $(n-r)\times r$ matrix with entries b_{ij}. Then*

$$\mathbf{B}=\mathbf{MR}$$

Proof. Let S_i be in a transient class and let S_j be in a closed class. The probability b_{ij} can be expressed as the probability of going from S_i to S_j directly in one step plus the probability of going first to a transient state and then ultimately to S_j. Thus,

$$b_{ij}=p_{ij}+\sum_k p_{ik}b_{kj}$$

where the summation over k is carried out over all transient states. In matrix form we have $\mathbf{B}=\mathbf{R}+\mathbf{QB}$ and hence $\mathbf{B}=[\mathbf{I}-\mathbf{Q}]^{-1}\mathbf{R}=\mathbf{MR}$. ▌

Example 1 (Learning Model). The simple Estes learning model is described by a Markov chain with transition matrix

$$\mathbf{P}=\begin{bmatrix}1 & 0\\ \alpha & 1-\alpha\end{bmatrix}$$

This matrix is already in canonical form with S_1, the "learned" state, being an absorbing state, and S_2, the "unlearned" state, being a transient state.

The $\mathbf{Q}$ matrix in this case consists of the single element $1-\alpha$. Accordingly, the fundamental matrix $\mathbf{M}$ is the single number $[1-(1-\alpha)]^{-1}=1/\alpha$.

It follows from Theorem 1 that $1/\alpha$ is the mean number of steps, starting from the unlearned state, before entering the learned state. This can vary from 1, if $\alpha=1$, to infinity, if $\alpha=0$. Theorem 2 is identical with Theorem 1 in this example, since $\mathbf{Q}$ is one-dimensional.

Theorem 3, in general, gives the probabilities of entering closed classes through various states. Since in this example the closed class consists of just a single (absorbing) state, the probability of absorption by that state should be 1. Indeed the formula of the theorem specifies this probability as $\alpha(1/\alpha)=1$.

Example 2 (A Production Line). A certain manufacturing process consists of three manufacturing stages and a completion stage. At the end of each

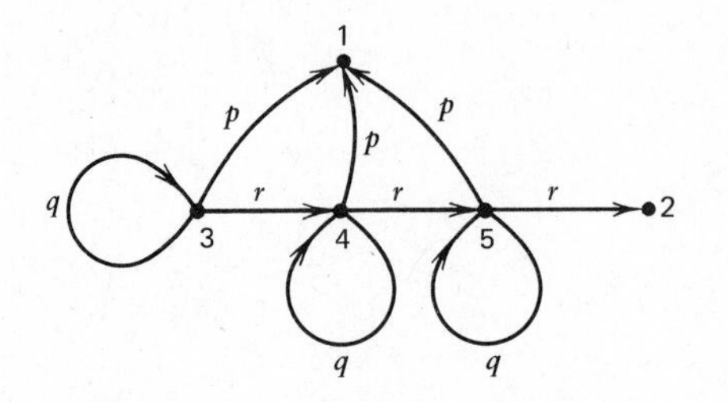

Figure 7.7. The production line chain.

manufacturing stage each item is inspected. At each inspection there is a probability p that the item will be scrapped, q that it will be sent back to that stage for reworking, and r that it will be passed on to the next stage. It is of importance to determine the probability that an item, once started, is eventually completed rather than scrapped, and to determine the number of items that must be processed through each stage. (See Fig. 7.7.)

The process can be considered to have five states:

(1) Item scrapped.
(2) Item completed.
(3) Item in first manufacturing stage.
(4) Item in second manufacturing stage.
(5) Item in third manufacturing stage.

The corresponding transition matrix is

$$\mathbf{P}=\begin{bmatrix} 1 & 0 & 0 & 0 & 0 \\ 0 & 1 & 0 & 0 & 0 \\ p & 0 & q & r & 0 \\ p & 0 & 0 & q & r \\ p & r & 0 & 0 & q \end{bmatrix}$$

The first two states are absorbing states and the other three are transient states. The transition matrix is in canonical form and the fundamental matrix is

$$\mathbf{M}=\begin{bmatrix} 1-q & -r & 0 \\ 0 & 1-q & -r \\ 0 & 0 & 1-q \end{bmatrix}^{-1}$$

It is easy to verify that

$$\mathbf{M}=\frac{1}{(1-q)^3}\begin{bmatrix} (1-q)^2 & r(1-q) & r^2 \\ 0 & (1-q)^2 & r(1-q) \\ 0 & 0 & (1-q)^2 \end{bmatrix}$$

The elements of the first row of this matrix are equal to the average number of times each item is passed through the first, second, and third manufacturing stages, respectively.

The probability of entering the two absorbing states is given by the elements of $\mathbf{B}=\mathbf{MR}$. In this case

$$\mathbf{B}=\mathbf{MR}=\frac{1}{(1-q)^3}\begin{bmatrix}(1-q)^2 & r(1-q) & r^2\\ 0 & (1-q)^2 & r(1-q)\\ 0 & 0 & (1-q)^2\end{bmatrix}\begin{bmatrix}p & 0\\ p & 0\\ p & r\end{bmatrix}$$

For example, the probability that an item is eventually completed rather than scrapped, starting from the first stage, is the first element of the second column of $\mathbf{B}$. That is, it is equal to $r^3/(1-q)^3$.

Example 3 (Gambler's Ruin). When there are a total of n coins or chips between the two players in a Gambler's Ruin game, it can be considered to be a Markov process with $n+1$ states $S_0, S_1, S_2, \ldots, S_n$, where S_i corresponds to player A having i coins. The states S_0 and S_n are each absorbing, while all others form one transient class. The kinds of questions one asks when studying this chain are those typical of transient state analysis—for example, the probability of player A winning, or the average duration of the game.

The transition matrix can be put in the canonical form

$$\mathbf{P}=\begin{bmatrix}1 & 0 & 0 & 0 & 0 & 0 & \cdots & 0\\ 0 & 1 & 0 & 0 & 0 & & & 0\\ q & 0 & 0 & p & 0 & & & \cdot\\ 0 & 0 & q & 0 & p & & & \cdot\\ \cdot & 0 & & q & 0 & p & & \cdot\\ \vdots & \vdots & & & \ddots & \ddots & \ddots & \\ 0 & 0 & & & & & & p\\ 0 & p & \cdots & \cdots & \cdots & 0 & q & 0\end{bmatrix}$$

In this representation the states are ordered $S_0, S_n, S_1, S_2, \ldots, S_{n-1}$. The matrix $\mathbf{Q}$ is the $(n-1)\times(n-1)$ matrix

$$\mathbf{Q}=\begin{bmatrix}0 & p & 0 & \cdots & 0\\ q & 0 & p & & \\ & q & 0 & \ddots & \\ & & \ddots & \ddots & p\\ & & & q & 0\end{bmatrix}$$

The fundamental matrix is $\mathbf{M}=[\mathbf{I}-\mathbf{Q}]^{-1}$, but an explicit formula for this inverse is somewhat difficult to find. It is, however, not really necessary to have an explicit representation for many purposes. Indeed, for this example, just as for many other highly structured problems, it is possible to convert the general expressions of Theorems 2 and 3 into alternative and much simpler dynamic problems.

Let us, for example, compute the probability of player A winning, starting from various states S_k. According to Theorem 3 the vector $\mathbf{x}=\mathbf{M}\mathbf{r}_2$, where $\mathbf{r}_2^T=[0\ 0\ 0\ 0\ \cdots\ p]$, has components equal to these probabilities. The vector $\mathbf{x}$ satisfies the equation

$$[\mathbf{I}-\mathbf{Q}]\mathbf{x}=\mathbf{r}_2$$

where $\mathbf{x}=(x_1, x_2, \ldots, x_{n-1})$. Written out in greater detail, this vector equation is

$$x_1-px_2=0$$
$$x_k-qx_{k-1}-px_{k+1}=0, \qquad 2\leq k\leq n-2$$
$$x_{n-1}-qx_{n-2}=p$$

Defining the additional variables $x_0=0$, $x_n=1$, the first and last equations can be expanded to have the same form as the second. In this way the above system can be expressed as the single difference equation

$$x_k-qx_{k-1}-px_{k+1}=0, \qquad k=1, 2, \ldots, n-1$$

This is the difference equation identical with that used to solve this problem in Chapter 2; and it can be solved as shown there.

In a similar fashion the average length of the game, starting from various states, can be found by application of Theorem 2. These lengths are the components of the vector $\mathbf{y}=\mathbf{M1}$. Equivalently, the vector $\mathbf{y}$ can be found as the solution to the equation $[\mathbf{I}-\mathbf{Q}]\mathbf{y}=\mathbf{1}$. Again with $\mathbf{y}=(y_1, y_2, \ldots, y_{n-1})$ and defining $y_0=0$, $y_n=0$, the vector equation for $\mathbf{y}$ can be written as the difference equation

$$y_k-qy_{k-1}-py_{k+1}=1, \qquad k=1, 2, \ldots, n-1$$

This equation can be solved by the techniques of Chapter 2. The characteristic equation is

$$\lambda-q-p\lambda^2=0$$

which has roots $\lambda=1$, $\lambda=q/p$. Assuming $p\neq q\neq\frac{1}{2}$, the general solution to the difference equation has the form

$$y_k=A+Bk+C(q/p)^k$$

The constants A and C are arbitrary, since they correspond to the general solution of the homogeneous equation. The constant B is found by temporarily setting $A=0$, $C=0$ and substituting $y_k=Bk$ into the difference equation, obtaining

$$Bk-qB(k-1)-pB(k+1)=1$$

$$-pB+qB=1$$

$$B=\frac{1}{q-p}$$

The actual values of A and C are found by setting $y_0=0$ and $y_n=0$ yielding

$$0=A+C$$

$$0=A+\frac{n}{q-p}+C\left(\frac{q}{p}\right)^n$$

Solving for A and C and substituting into the general form leads to the final result

$$y_k=\frac{n}{(q-p)[1-(q/p)^n]}\left[-1+\left(\frac{q}{p}\right)^k\right]+\frac{k}{q-p}$$

*7.5 INFINITE MARKOV CHAINS

In some applications it is natural to formulate Markov chain models having a (countably) infinite number of states $S_1, S_2, \ldots, S_k, \ldots$. An infinite Markov chain often has greater symmetry and leads to simpler formulas than a corresponding finite chain obtained by imposing an artificial termination condition. This simplicity of structure justifies the extension of concepts to infinite Markov chains.

Example (Infinite Random Walk with Reflecting Barrier). An object moves on a horizontal line in discrete unit steps. Its possible locations are given by the nonnegative integers $0, 1, 2, \ldots$. If the object is at position $i>0$, there is a probability p that the next transition will be to position $i+1$, a probability q that it will be to $i-1$, and a probability r that it will remain at i. If it is at position 0, it will move to position 1 with probability p and remain at 0 with probability $1-p$. The transition matrix for this infinite chain is

$$\mathbf{P}=\begin{bmatrix} 1-p & p & 0 & 0 & \cdots \\ q & r & p & 0 & \cdots \\ 0 & q & r & p & \cdots \\ 0 & 0 & q & r & \cdots \\ \vdots & & & & \end{bmatrix}$$

Although the technicalities associated with infinite Markov chains are somewhat more elaborate than for finite chains, much of the essence of the finite theory is extendible to the infinite chain case. Of particular importance is that there is an extended version of the basic limit theorem. This extension is presented in this section, without proof.

The concepts of accessibility, communicating classes, and irreducibility carry over directly to infinite chains. The definitions given earlier apply without change in the infinite case. For example, S_j is accessible from S_i if there is a path (of finite length) from S_i to S_j. In addition to these definitions, it is useful to introduce the concept of an aperiodic Markov chain. To illustrate this concept, consider again the two-dimensional chain with transition matrix

$$\mathbf{P}=\begin{bmatrix}0 & 1\\ 1 & 0\end{bmatrix}$$

This chain is irreducible since each of the two states is accessible from the other. However, in this chain a transition from either state back to the same state always requires an even number of steps. A somewhat more complex example is represented by the matrix

$$\mathbf{P}=\begin{bmatrix}0 & \frac{1}{2} & \frac{1}{2}\\ 1 & 0 & 0\\ 1 & 0 & 0\end{bmatrix}$$

Starting at S_1 on Step 0, S_1 will be visited on all even numbered steps, while either S_2 or S_3 are visited on odd number steps. In general, if some state in a finite or infinite Markov chain has the property that repeated visits to that state are always separated by a number of steps equal to a multiple of some integer greater than 1, that state is said to be *periodic*. A Markov chain is *aperiodic* if it has no periodic states.

The first important result for infinite chains is an extension of part (b) of the Basic Limit Theorem of Sect. 7.2.

Theorem. *For an irreducible, aperiodic Markov chain the limits*

$$v_j=\lim_{m\to\infty} p_{ij}^{(m)}$$

exist and do not depend on the initial state i.

This theorem tells us that, just as in the finite case, the process settles down with each state having a limiting probability. It is quite possible, however, in the infinite case, that the limits might all be zero.

As an example, let us refer to the infinite random walk described above. Provided that $p>0$, $q>0$, it is clear that every state communicates with every other state since there is a path of nonzero transition probabilities from any state to any other. Thus, the chain is irreducible. The chain is aperiodic since one can return to a state by resting at 0 indefinitely. Assuming $p>0$, $q>0$, the conclusion of Theorem 1 must hold. However, if $p>q$, there is a strong tendency for the moving object to drift toward the right, to higher integer points. The chance that the object returns to a specific point, say point 0, infinitely often is likely to be quite small. Indeed in this case, the limits v_i are all zero. The process drifts continually to the right so that each state has the character of a transient state.

Definition. An irreducible aperiodic Markov chain is said to be *positive recurrent* if

(a) $v_j = \lim_{m\to\infty} p_{ij}^{(m)} > 0$ for all j, and
(b) $\sum_j v_j = 1$.

According to this definition, a chain is positive recurrent if the limit probabilities form a legitimate infinite-dimensional probability vector. The next theorem establishes the relation between these limit probabilities and the existence of an infinite-dimensional eigenvector of the transition matrix.

Theorem. *Given an irreducible aperiodic Markov chain.*

(a) *It is positive recurrent if and only if there is a unique probability distribution* $\mathbf{p} = (p_1, p_2, \ldots)$ *(satisfying* $p_i > 0$ *for all* i, $\sum_i p_i = 1$*), which is a solution to*

$$p_j = \sum_i p_{ij} p_i$$

In this case,

$$p_j = v_j = \lim_{m\to\infty} p_{ij}^{(m)}$$

for all j.

(b) *If the chain is not positive recurrent, then*

$$v_j = \lim_{m\to\infty} p_{ij}^{(m)} = 0$$

for all j.

Example (continued). In the random walk, suppose $p>0$, $r>0$, and $q>0$. Let us attempt to find the v_j's by seeking an eigenvector of the transition matrix.

The v_j's should satisfy the equations

$$pv_0 - qv_1 = 0$$
$$(1-r)v_j - pv_{j-1} - qv_{j+1} = 0, \qquad j = 1, 2, 3, \ldots$$

The characteristic equation of the difference equation is $(1-r)\lambda - p - q\lambda^2 = 0$, which has roots $\lambda = 1$, p/q. If $p < q$, a solution is

$$v_j = (1-p/q)(p/q)^j$$

and this satisfies $\sum_{j=1}^{\infty} v_j = 1$. Therefore, for $p < q$ the solution is positive recurrent. If $p > q$, no solution can be found, and hence, as suspected (because of the drift toward infinity), the chain is not positive recurrent.

7.6 PROBLEMS

1. Let $\mathbf{P}$ be an $n \times n$ stochastic matrix and let $\mathbf{1}$ denote the n-dimensional column vector whose components are all 1. Show that $\mathbf{1}$ is a right eigenvector of $\mathbf{P}$ corresponding to an eigenvalue of 1. Conclude that if $\mathbf{y}^T = \mathbf{x}^T\mathbf{P}$, then $\mathbf{y}^T\mathbf{1} = \mathbf{x}^T\mathbf{1}$.

2. *Social Mobility.* The elements of the matrix below represents the probability that the son of a father in class i will be in class j. Find the equilibrium distribution of class sizes.

Upper class	.5	.4	.1
Middle class	.1	.7	.2
Lower class	.05	.55	.4

3. *Ehrenfest Diffusion Model.* Consider a container consisting of two compartments A and B separated by a membrane. There is a total of n molecules in the container. Individual molecules occasionally pass through the membrane from one compartment to the other. If at any time there are j molecules in compartment A, and $n-j$ in compartment B, then there is a probability of j/n that the next molecule to cross the membrane will be from A to B, and a probability of $(n-j)/n$ that the next crossing is in the opposite direction.
   (a) Set up a Markov chain model for this process. Is it regular?
   (b) Show that there is an equilibrium probability distribution such that the probability p_j that j molecules are in compartment A is

$$p_j = \frac{1}{2^n}\binom{n}{j}$$

4. *Languages.* The symbols of a language can be considered to be generated by a Markov process. As a simple example consider a language consisting of the symbols A, B, and S (space). The space divides the symbol sequence into words. In this language two B's or two S's never occur together. Three A's never occur together. A word never starts with AB or ends with BA. Subject to these restrictions, at any point, the next symbol is equally likely to any of the allowable possibilities.

Formulate a Markov chain for this language. What are the equilibrium symbol probabilities? (*Hint*: Let some states represent pairs of symbols.)

5. *Entropy.* The *entropy* of a probability vector $\mathbf{p}^T = (p_1, p_2, \ldots, p_n)$ is $H(\mathbf{p}) = -\sum_{i=1}^n p_i \log_2 p_i$. Entropy is a measure of the uncertainty associated with a selection of n objects, when the selection is made according to the given probabilities.

   (a) For a fixed n show that $H(p)$ is maximized by $\mathbf{p}^T = [1/n, 1/n, \ldots, 1/n]$.
   (b) For a regular finite Markov chain with transition matrix $\mathbf{P}$, the entropy is

   $$H(\mathbf{P}) = \sum_{i=1}^{n} p_i H_i$$

   where p_i is the equilibrium probability of state S_i and where H_i is the entropy of the ith row of $\mathbf{P}$. Thus, H is a weighted average of the entropy associated with the choice of the next state of the chain. Find the entropy of the weather model in Sect. 7.2.

*6. *Equivalence Classes.* Consider a set X and a relation R that holds among certain pairs of elements of X. One writes xRy if x and y satisfy the relation. The relation R is said to be an *equivalence relation* if

$$xRx \quad \text{for all } x \text{ in } X$$
$$xRy \quad \text{implies} \quad yRx$$
$$xRy \quad \text{and} \quad yRz \quad \text{implies} \quad xRz$$

   (a) Let $[x]$ denote the collection of all elements y satisfying xRy, where R is an equivalence relation. This set is called the equivalence class of x. Show that X consists of a disjoint collection of equivalence classes.
   (b) Let X be the set of states in a Markov chain and let R be the relation of communication. Show that R is an equivalence relation.

7. Show that from any state in a Markov chain it is possible to reach a closed class within a finite number of transitions having positive probability.

8. Suppose the probability transition matrix of a finite Markov chain is in the canonical form (7-2). Show that any left eigenvector corresponding to an eigenvalue of magnitude of 1 must be of the form $[\mathbf{p}_1^T, \mathbf{0}]$, where $\mathbf{p}_1^T$ is r dimensional.

*9. An $n \times n$ matrix $\mathbf{P}$ is a *permutation matrix* if for all vectors $\mathbf{x}$ the components of the vector $\mathbf{Px}$ are simply a reordering of the components of $\mathbf{x}$. Show that all elements of a permutation matrix $\mathbf{P}$ are either zero or one, and both $\mathbf{P}$ and $\mathbf{P}^T$ are stochastic matrices.

*10. *Theory of Positive Matrices.* An $n \times n$ matrix $\mathbf{A}$ is said to be *reducible* if there is a nonempty proper subset J of $\{1, 2, \ldots, n\}$ such that

$$a_{ij} = 0 \quad \text{for} \quad i \notin J, j \in J$$

(a) Show that $\mathbf{A}$ is reducible if and only if there is a permutation matrix $\mathbf{T}$ (see Problem 9) such that

$$\mathbf{T}^{-1}\mathbf{A}\mathbf{T}=\begin{bmatrix}\mathbf{A}_{11} & \mathbf{A}_{12}\\ \mathbf{0} & \mathbf{A}_{22}\end{bmatrix}$$

where $\mathbf{A}_{11}$ is square.

(b) Let $\mathbf{P}$ be a stochastic matrix associated with a finite Markov chain. Show that the chain has a single communicating class if and only if $\mathbf{P}$ is irreducible.

(c) Let $\mathbf{A}$ be a nonnegative irreducible matrix, with Frobenius–Perron eigenvalue and eigenvector λ_0, $\mathbf{x}_0$, respectively. Show that $\lambda_0>0$, $\mathbf{x}_0>\mathbf{0}$, and that λ_0 is a simple root of the characteristic polynomial of $\mathbf{A}$.

11. *Periodic Positive Matrices.* Let $\mathbf{A}$ be an irreducible positive $n\times n$ matrix (see Problem 10). Then it can be shown that there is a permutation matrix $\mathbf{T}$ such that

$$\mathbf{T}^{-1}\mathbf{A}\mathbf{T}=\begin{bmatrix}\mathbf{0} & \mathbf{0} & \cdots & \mathbf{0} & \mathbf{G}_r\\ \mathbf{G}_1 & \mathbf{0} & \cdots & \mathbf{0} & \mathbf{0}\\ \mathbf{0} & \mathbf{G}_2 & & & \\ & & \ddots & & \\ & & & \mathbf{G}_{r-1} & \mathbf{0}\end{bmatrix}$$

where the zero matrices on the main diagonal are square. Let $\lambda_0>0$ be the Frobenius–Perron eigenvalue of $\mathbf{A}$. Show that $\mathbf{A}$ has r eigenvalues of magnitude λ_0. [*Hint*: Let ω be an rth root of unity (that is, $\omega^r=1$). Show that $\lambda=\omega\lambda_0$ is an eigenvalue.]

12. *Finite Random Walk.* An object moves on a horizontal line in discrete steps. At each step it is equally likely to move one unit to right or one unit to the left. The line is a total of five units long, and there are absorbing barriers at either end. Set up the Markov chain corresponding to this random walk process. Characterize each state as transient or absorbing. Calculate the canonical form and find the fundamental matrix.

13. *First Passage Time.* Suppose $\mathbf{P}$ is the probability transition matrix of a regular Markov chain. Given an initial state $S_i\neq S_1$, show how by modifying $\mathbf{P}$ the average number of steps to reach S_1 can be computed. For the weather model, given that today is rainy, what is the expected number of days until it is sunny?

14. *Markov Chains with Reward.* You might consider your automobile and its random failures to be a Markov chain. It makes monthly transitions between the states "running well" and "not running well." When it is not running well it must be taken to a garage to be repaired, at a cost of \$50. It is possible to improve the likelihood that the automobile will continue to run well by having monthly service at a cost of \$10. Depending on your policy your automobile transitions will be governed

by either of two Markov chains, defined by

$$\mathbf{P}_1 = \begin{bmatrix} \frac{3}{4} & \frac{1}{4} \\ \frac{7}{8} & \frac{1}{8} \end{bmatrix}, \qquad \mathbf{P}_2 = \begin{bmatrix} \frac{7}{8} & \frac{1}{8} \\ \frac{7}{8} & \frac{1}{8} \end{bmatrix}$$

where S_1 = running well, S_2 = not running well, and where $\mathbf{P}_1$ corresponds to no monthly service, and $\mathbf{P}_2$ corresponds to having monthly service.

(a) For each of the two policies, what is the equilibrium distribution of states?
(b) In equilibrium, what is the average monthly cost of each of the two policies?

15. *Simplified Blackjack.* A game between a "dealer" and a "customer" is played with a (very large) mixed deck of cards consisting of equal numbers of ones, twos, and threes. Two cards are initially dealt to each player. After looking at his cards the customer can elect to take additional cards one at a time until he signals that he will "stay." If the sum of the values of his cards exceeds six, he loses. Otherwise the dealer takes additional cards one at a time until his sum is five or more. If his sum exceeds six, the customer wins. Otherwise, the player with the highest sum (under seven) wins. Equal values under seven result in a draw.

    (a) Set up a Markov chain for the process of taking cards until a value of five or more is obtained. Identify the absorbing states. Find the probabilities of entering the various absorbing states for each initial sum. Find the probability of entering various absorbing states.
    (b) If the customer follows the strategy of taking cards until his value is five or more, at what rate will he lose?
    *(c) If the second card dealt to the dealer is face up, the customer can base his strategy on the value of that card. For each of the three possible cards showing, at what sum should the customer stay? What are the odds in this case?

16. *Periodic States.* Show that if one state in a given communicating class is periodic, then all states in that class are periodic.

17. For the following special cases of the infinite random walk, determine if (i) it is aperiodic, (ii) irreducible, and (iii) there is a solution to the eigenvector problem.

    (a) $r=0, p>q>0$.
    (b) $r>0, p=q>0$.

18. Suppose a (nonfair) coin is flipped successively with the probability of heads or tails on any trial being p and $1-p$, respectively. Define an infinite Markov chain where state S_j corresponds to a landing of the coin that represents a run of exactly j heads on the most recent flips. Show that this Markov chain is aperiodic and irreducible. Is it positive recurrent?

19. *Dynamics of Poverty and Wealth.* Income distribution is commonly represented by the distribution function $D(y)$, which measures the number of individuals with incomes exceeding y. It has been observed that with surprising regularity, in various

countries over long histories, the upper end of the distribution closely approximates the form

$$D(y) = Cy^{-\alpha}$$

for some parameters C and α. This is referred to as *Pareto's law.*

One theoretical explanation for the regularity encompassed by Pareto's law is based on a Markov chain model of individual income developed by Champernowne. In this model the income scale is divided into an infinity of income ranges. The income ranges are taken to be of uniform proportionate length; that is, they might be \$50 to \$100, \$100 to \$200, \$200 to \$400, and so on. At any one time a given individual's income falls within one of these ranges. At the next period (and periods might correspond to years), his or her income makes a transition, either upward or downward, according to a set of probabilities that are characteristic of the particular economy in which the person lives.

It is assumed that no income moves up by more than one or down by more than n income ranges in a period, where $n \geq 1$ is a fixed integer. Specifically, it is assumed that there are $n+2$ positive numbers, $p_{-n}, p_{-n+1}, \ldots, p_0, p_1$ such that

$$1 = \sum_{u=-n}^{1} p_u$$

Then, the transition probabilities for the process are defined as

$$p_{ij} = p_{j-i} \quad \text{for} \quad -n \leq j-i \leq 1, j \geq 1$$

$$p_{i0} = 1 - \sum_{r=0}^{i} p_{1-r} \quad \text{for} \quad 0 \leq i \leq n$$

$$p_{ij} = 0 \text{ otherwise}$$

The pattern of transition probabilities is illustrated in Fig. 7.8 for $n = 2$. (This model can be viewed as an extension of the infinite random walk.)

Finally, it is assumed that the average number of income ranges moved in one step is negative when at the upper levels. That is,

$$-np_{-n} + (-n+1)p_{-n+1} + \cdots + (-1)p_{-1} + 0 \cdot p_0 + p_1 < 0$$

(a) Verify that this Markov chain is irreducible and aperiodic.

(b) Let

$$F(\lambda) = -\lambda + \sum_{u=-n}^{1} p_u \lambda^{1-u}$$

and show that $F(1) = 0$, $F'(1) > 0$, and $F(0) = p_1 > 0$. Conclude that there is a root λ to the equation $F(\lambda) = 0$ in the range $0 < \lambda < 1$.

(c) Find an equilibrium probability vector.

(d) Show that the equilibrium income distribution satisfies Pareto's law.

i \ j	0	1	2	3	4	5	6	
0	$1-p_1$	p_1	0	0	0	0	0	⋯
1	$1-p_0-p_1$	p_0	p_1	0	0	0	0	⋯
2	$1-p_{-1}-p_0-p_1$	p_{-1}	p_0	p_1	0	0	0	⋯
3	0	p_{-2}	p_{-1}	p_0	p_1	0	0	⋯
4	0	0	p_{-2}	p_{-1}	p_0	p_1	0	⋯
5	0	0	0	p_{-2}	p_{-1}	p_0	p_1	⋯
6	0	0	0	0	p_{-2}	p_{-1}	p_0	⋯
7								
.								
.								

Figure 7.8. Champernowne model

NOTES AND REFERENCES

Sections 7.1–7.4. For general introductions to finite Markov chains and additional applications see Bhat [B8], Howard [H7], Kemeny and Snell [K11], or Clarke and Disney [C4], which is close to our level and order of presentation and considers some of the same examples. Bartholomew [B2] has several examples from the social sciences. The learning model is based on Estes [E1]. The landing probabilities for real game of *Monopoly* are available in tabulated form; see, for example, Walker and Lehman [W1].

Section 7.5. For more advanced treatments of Markov chains, including infinite chains, see Karlin [K8] and Feller [F1].

Section 7.6. Entropy of a Markov process (Problem 5) forms the basis of information theory as developed by Shannon; see Shannon and Weaver [S4]. For further discussion of the theory of positive matrices as presented in Problems 10 and 11, see Karlin [K7], Nikaido [N1], or Gantmacher [G3]. A full theory of Markov processes with reward (Problem 14) has been developed by Howard [H6]. The optimal strategy for real blackjack was found (by computer simulation) by Thorp [T2]. The model of Problem 19 together with various modifications is due to Champernowne [C2].

chapter 8.

Concepts of Control

The analysis of dynamic systems can certainly enhance our understanding of phenomena around us. But beyond analysis is the higher objective of influencing the behavior of a system by control or design. The field of *control theory* is directed toward this general objective.

8.1 INPUTS, OUTPUTS, AND INTERCONNECTIONS

The state space description of a linear dynamic system focuses on the behavior of the entire system—the evolution of all state variables as a result of prescribed initial conditions and input values. It is often the case, however, that certain variables, or combinations of variables, are of special interest, while others are of secondary interest. For example, an automobile might be represented as a high-order dynamic system. During operation, however, one is generally only interested in a few characterizing variables (such as vehicle position and velocity) rather than the whole assortment of internal variables (engine speed, etc.) required for a complete description of the state. It is useful to explicitly recognize the important variables as system *outputs*.

In a linear system the outputs are generally linear combinations of state variables and input variables. Thus, in the context of outputs, the definition of a linear discrete-time nth-order system is expanded to the general form

$$\mathbf{x}(k+1)=\mathbf{A}(k)\mathbf{x}(k)+\mathbf{B}(k)\mathbf{u}(k)$$

$$\mathbf{y}(k)=\mathbf{C}(k)\mathbf{x}(k)+\mathbf{D}(k)\mathbf{u}(k)$$

As usual, $\mathbf{x}(k)$ is an n-dimensional *state* vector, $\mathbf{u}(k)$ is an m-dimensional *input* vector, and $\mathbf{A}(k)$ and $\mathbf{B}(k)$ are $n\times n$ and $n\times m$ matrices, respectively. The vector $\mathbf{y}(k)$ is, say, a p-dimensional output vector, and accordingly $\mathbf{C}(k)$ and $\mathbf{D}(k)$ are $p\times n$ and $p\times m$ matrices, respectively.

Often only a single output variable $y(k)$ is specified. In this case $\mathbf{C}(k)$ is $1\times n$ and $\mathbf{D}(k)$ is $1\times m$, and hence one usually uses the lower-case notation $\mathbf{c}(k)^T$ and $\mathbf{d}(k)^T$. If the system has only a single input and a single output, it is expressed as

$$\mathbf{x}(k+1)=\mathbf{A}(k)\mathbf{x}(k)+\mathbf{b}(k)u(k)$$

$$y(k)=\mathbf{c}(k)^T\mathbf{x}(k)+d(k)u(k)$$

The choice of outputs is, from a mathematical viewpoint, quite arbitrary. However, in practice, variables are designated as outputs if they are available for measurement or if they have some special importance.

From the perspective of control, the structure of inputs and outputs is an integral component of the description of a dynamic system. The input structure determines the degree that system behavior can be modified, and the output structure governs the kind of information available for control. These structural components of a dynamic system interact, and are basic to the very objective of control.

If a system has only a single input and a single output, it is possible to deduce a single nth-order difference equation that governs the output variable. This is essentially the reverse of the procedure used in Sect. 4.1 to convert an nth-order difference equation to state space form. Thus, consideration of outputs leads us back to the study of linear difference equations. Therefore, the first few sections of this chapter are concerned with an alternative method of solution of such equations—the transform method.

If a system has several outputs and several inputs, it is often useful to partition the system into a number of interconnected subsystems each having a single input and single output. Each subsystem can be analyzed by the transform method, and the results can be appropriately combined. This combination process is an important product of the control (or input–output) viewpoint.

The sections of this chapter are divided into two groups. Those through Sect. 8.5 cover the *transform* approach to analysis discussed above. The transform material is a natural augmentation of methods developed in earlier chapters, but it is not essential for the second half of this chapter. Sections 8.6–8.10 are devoted more explicitly to control issues from a *state space* viewpoint. This material represents an introduction to the field of modern control theory.

TRANSFORM METHODS

8.2 z-TRANSFORMS

The z-transform is a mathematical operation that, when applied to a sequence of numbers, produces a function of a variable z. The formal definition is given below.

Definition. Given a sequence of numbers $y(0)$, $y(1)$, $y(2), \ldots,$ the *z-transform* of this sequence is the series (depending on z)

$$Y(z)=\sum_{k=0}^{\infty}\frac{y(k)}{z^k} \tag{8-1}$$

The series (8-1) can be looked at in two ways. First, and perhaps most naturally, it can be regarded as defining a function of the variable z. Thus, numerical values of z (either real or complex) yield numerical values for $Y(z)$. Viewed this way, one must examine the conditions under which the series (8-1) converges. If the values of $y(k)$ grow at most geometrically [that is, if $|y(k)|\leq c^k$ for some $c\geq 0$], then the series converges for $|z|$ sufficiently large [$|z|>c$]. In all our applications this condition is fulfilled.

Another approach is to treat this definition formally without concern for conditions of convergence. The variable z is regarded simply as a symbol, and the transform $Y(z)$ as a series that is never actually summed. This corresponds to the approach that one often uses in connection with polynomials. Polynomials in a symbol z can be defined and manipulated algebraically, even if it is never intended that the symbol z be assigned a numerical value. The same algebraic procedures can be applied to the z-transform series $Y(z)$. However, as shown below, there is often a shorthand representation of the series, which is closely related to the first, function viewpoint.

Example 1. Consider the constant sequence $y(k)=1$. This sequence is referred to as the unit step. The z-transform of this sequence is

$$Y(z)=\sum_{k=0}^{\infty}\frac{1}{z^k}$$

If z is a complex variable such that $|z|>1$, this sequence converges and has the value

$$Y(z)=\frac{1}{1-1/z}=\frac{z}{z-1}$$

Example 2. Consider the sequence $y(k)$ having the specific values 1, 2, 3, 0, $0, \ldots.$ The z-transform is

$$Y(z)=1+\frac{2}{z}+\frac{3}{z^2}=\frac{z^2+2z+3}{z^2}$$

Example 3. Consider the geometric sequence $y(k)=a^k$. Then

$$Y(z)=\sum_{k=0}^{\infty}\frac{a^k}{z^k}$$

For $|z|>|a|$, this series converges to

$$Y(z)=\frac{z}{z-a}$$

Rational Functions

A form of function that arises often in the study of linear dynamic equations, and particularly in the context of transforms, is that of a rational function. A function $F(z)$ of the variable z is said to be *rational* if it is a ratio of polynomials in z. Thus, a rational function has the form

$$F(z)=\frac{b_m z^m + b_{m-1} z^{m-1} + \cdots + b_0}{a_n z^n + a_{n-1} z^{n-1} + \cdots + a_0} \tag{8-2}$$

A rational function is called *proper* if the degree of the numerator polynomial is no greater than the degree of the denominator polynomial. It is *strictly proper* if the degree of the numerator polynomial is less than the degree of the denominator polynomial.

If the numerator and denominator polynomials of a rational function have no common factors, the function is said to be *reduced.* Clearly any rational function can be converted to a unique reduced form by cancelling out common factors. The *degree* of a rational function is equal to the degree of its denominator polynomial when the function is reduced.

In many cases, as in Examples 1, 2 and 3 above, the z-transform of a sequence converges (for large values of $|z|$) to a proper rational function. Thus, the rational function is a representation of the series defining the transform. If z is treated simply as a symbol rather than a number, the rational function can be used as a shorthand representation of the original series. If the denominator of the rational expression is divided into the numerator according to the standard rules of long division of polynomials, the original series is obtained. From this viewpoint the rational function acts as a *generator* for the series. Referring to Example 1, the rational function $z/(z-1)=1/(1-1/z)$ is a formal generator for the series $1+1/z+1/z^2+\cdots$. This formal viewpoint allows us to suppress considerations of series convergence, especially for sequences whose z-transforms have rational representations.

Fortunately, proper rational z-transforms are in direct correspondence with sequences generated by homogeneous linear, constant-coefficient, difference equations. This result, established below, thereby justifies the special attention we devote to rational forms.

Theorem 1. *A sequence $y(k)$, $k=0, 1, 2, \ldots$ has a z-transform $Y(z)$ that can be expressed as a reduced proper rational function of degree n if and only if* *there are n scalars $a_0, a_1, \ldots, a_{n-1}$ such that*

$$y(k+n)+a_{n-1}y(k+n-1)+\cdots+a_0y(k)=0 \tag{8-3}$$

for all $k>0$.

Proof. Suppose $y(k)$ has a reduced proper rational transform of degree n,

$$Y(z)=\frac{b_nz^n+b_{n-1}z^{n-1}+\cdots+b_0}{z^n+a_{n-1}z^{n-1}+\cdots+a_0}=y(0)+y(1)z^{-1}+y(2)z^{-2}+\cdots+\cdots \tag{8-4}$$

Then it follows that

$$b_nz^n+b_{n-1}z^{n-1}+\cdots+b_0=\{z^n+a_{n-1}z^{n-1}+\cdots+a_0\} \times\{y(0)+y(1)z^{-1}+\cdots\} \tag{8-5}$$

Equating coefficients of like powers of z gives

$$\begin{aligned} b_n &= y(0)\\ b_{n-1} &= a_{n-1}y(0)+y(1)\\ b_{n-2} &= a_{n-2}y(0)+a_{n-1}y(1)+y(2)\\ &\vdots\\ b_0 &= a_0y(0)+a_1y(1)+\cdots+a_{n-1}y(n-1)+y(n) \end{aligned} \tag{8-6}$$

and, since the coefficient of z^{-k} on the left side of (8-5) is zero for $k>0$, it follows that

$$0=a_0y(k)+a_1y(k+1)+\cdots+a_{n-1}y(k+n-1)+y(k+n) \tag{8-7}$$

for all $k>0$. This shows that if the transform is (8-4), the sequence must satisfy (8-3).

To prove the reverse implication, we start with (8-7), which is identical to (8-3), and then select $b_0, b_1, \ldots, b_n$ so that the equations (8-6) are satisfied. It is then possible to go in the reverse direction through the argument to show that $Y(z)$ is equal to (8-4). ∎

Transform Properties and Transform Pairs

Transforms of rather complicated sequences can be easily derived by knowing a few simple properties of transforms and a few simple transform pairs. Stated below are three useful properties.

Property 1. If $f(k)$ and $g(k)$ are sequences with z-transforms $F(z)$ and $G(z)$, respectively, then the transform of $f(k)+g(k)$ is $F(z)+G(z)$. The transform of $af(k)$ is $aF(z)$.

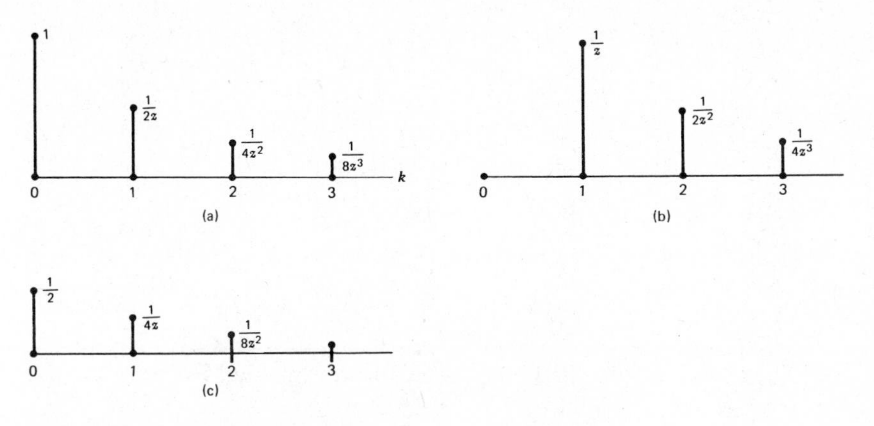

Figure 8.1. *z*-transform of $f(k)$, $f(k-1)$, and $f(k+1)$.

Proof. This property follows from the fact that the definition (8-1) is linear in the sequence values. ▮

Property 2. If the sequence $f(k)$ has the z-transform $F(z)$, then the unit delayed sequence

$$g(k)=\begin{cases} f(k-1) & k\geq 1 \\ 0 & k=0 \end{cases}$$

has z-transform

$$G(z)=z^{-1}F(z)$$

Proof. [Figure 8.1 represents the transform series corresponding to $f(k)=2^{-k}$ and its delayed and advanced versions. It can be seen that the sum obtained in (b) is just z^{-1} times that of (a).] The actual proof is obtained by writing

$$\begin{aligned} G(z) &= \sum_{k=0}^{\infty} z^{-k}g(k)=\sum_{k=1}^{\infty} z^{-k}f(k-1) \\ &= z^{-1}\sum_{k=1}^{\infty} z^{-(k-1)}f(k-1) \\ &= z^{-1}\sum_{j=0}^{\infty} z^{-j}f(j)=z^{-1}F(z). \end{aligned}$$ ▮

Property 3. If the sequence $f(k)$ has z-transform $F(z)$, then the unit advanced sequence $h(k)=f(k+1)$, $k\geq 0$ has the transform

$$H(z)=zF(z)-zf(0)$$

Proof. (See Fig. 8.1c.) We have

$$\begin{aligned} H(z) &= \sum_{k=0}^{\infty} z^{-k}h(k) = \sum_{k=0}^{\infty} z^{-k}f(k+1) \\ &= z\sum_{k=0}^{\infty} z^{-k-1}f(k+1) = z\sum_{j=1}^{\infty} z^{-j}f(j) \\ &= z\sum_{j=0}^{\infty} z^{-j}f(j) - zf(0) \\ &= zF(z) - zf(0). \blacksquare \end{aligned}$$

Let us illustrate how the rules can be used in conjunction with the known transform pairs to find the transform of other sequences.

Example 4. Consider a pulse of duration N defined by

$$f(k) = \begin{cases} 1 & 0 \le k < N \\ 0 & k \ge N \end{cases}$$

This can be regarded as the sum of a unit step (from Example 1) and the negative of a unit step delayed N time periods. Thus, using the result of Example 1 and N applications of Property 2, the transform of $f(k)$ is

$$\begin{aligned} F(z) &= \frac{z}{z-1} - \frac{z}{z-1}\cdot\frac{1}{z^N} \\ &= \frac{z}{z-1}\left(1 - \frac{1}{z^N}\right) \\ &= \frac{z^N - 1}{z^{N-1}(z-1)} \end{aligned}$$

Example 5. Consider the *geometric ramp* sequence

$$g(k) = ka^{k-1}$$

Let $f(k)$ be the geometric sequence $f(k) = a^k$. Then

$$g(k) = \frac{d}{da} f(k)$$

Using the result of Example 3,

$$G(z) = \frac{d}{da} F(z) = \frac{d}{da}\frac{z}{z-a}$$

$$G(z) = \frac{z}{(z-a)^2}$$

Table 8.1 lists some simple but frequently used z-transform pairs.

Table 8-1. z-Transform Pairs

$f(k)$	$F(z)$
unit impulse: $f(k)=\begin{cases}1, & k=0\\ 0, & k\geq 0\end{cases}$	$F(z)=1$
unit step: $f(k)=1$	$F(z)=\dfrac{z}{z-1}$
unit ramp: $f(k)=k$	$F(z)=\dfrac{z}{(z-1)^2}$
geometric series: $f(k)=a^k$	$F(z)=\dfrac{z}{z-a}$
delayed geometric series: $f(k)=\begin{cases}0, & k=0\\ a^{k-1}, & k>0\end{cases}$	$F(z)=\dfrac{1}{z-a}$
geometric ramp: $f(k)=ka^{k-1}$	$F(z)=\dfrac{z}{(z-a)^2}$
delayed geometric ramp: $f(k)=\begin{cases}0, & k=0\\ (k-1)a^{k-2}, & k>0\end{cases}$	$F(z)=\dfrac{1}{(z-a)^2}$

8.3 TRANSFORM SOLUTION OF DIFFERENCE EQUATIONS

The z-transform is the basis of a very effective method for solution of linear, constant-coefficient difference equations. It essentially automates the process of determining the coefficients of the various geometric sequences that comprise a solution.

The heart of the method is the fact that if the sequence $y(k)$ has transform $Y(z)$, then the advanced sequence $y(k+1)$ has transform $zY(z)-zy(0)$. (See Property 3 in Sect. 8.2.) By repeated application of this property one can deduce the successive correspondences

$$\begin{aligned} y(k) &\leftrightarrow Y(z)\\ y(k+1) &\leftrightarrow zY(z)-zy(0)\\ y(k+2) &\leftrightarrow z^2Y(z)-z^2y(0)-zy(1)\\ y(k+3) &\leftrightarrow z^3Y(z)-z^3y(0)-z^2y(1)-zy(2) \end{aligned} \tag{8-8}$$

and so forth.

Now consider the nth-order difference equation

$$y(k+n)+a_{n-1}y(k+n-1)+a_{n-2}y(k+n-2)+\cdots+a_0y(k)=g(k), \qquad k=0,1,2,\ldots \tag{8-9}$$

By taking the z-transform of both sides, an algebraic equation is obtained for the unknown transform $Y(z)$. Let us assume first that the initial conditions are zero; that is, $y(0)=y(1)=\cdots=y(n-1)=0$. In this case only the first terms in the right-hand side of the correspondences (8-8) are required. The transform of (8-9) becomes

$$z^n Y(z)+a_{n-1}z^{n-1}Y(z)+a_{n-2}z^{n-2}Y(z)+\cdots+a_0 Y(z)=G(z) \quad (8\text{-}10)$$

or

$$(z^n+a_{n-1}z^{n-1}+a_{n-2}z^{n-2}+\cdots+a_0)Y(z)=G(z) \quad (8\text{-}11)$$

Thus,

$$Y(z)=\frac{G(z)}{z^n+a_{n-1}z^{n-1}+a_{n-2}z^{n-2}+\cdots+a_0} \quad (8\text{-}12)$$

Once the z-transform of $g(k)$ is computed, the transform $Y(z)$ is determined. Finally, the solution $y(k)$ can be found by inverting the transform; that is, by finding the sequence that has this particular transform. (A method for this inversion is presented later in this section.)

We note that if the transform $G(z)$ is rational, then the transform $Y(z)$ is also rational. The polynomial appearing in the denominator of (8-12) is recognized as the characteristic polynomial of the original difference equation (8-9). To determine the basic geometric sequences comprising the solution $y(k)$, it is necessary to factor this polynomial.

The procedure is also applicable if the initial conditions are not all zero. In that case the additional terms in (8-8) must be incorporated. This, however, simply modifies the right-hand side of (8-11).

Let us illustrate the procedure with some simple examples. Then we shall consider a general procedure for inverting rational z-transforms.

Example 1. Consider the difference equation

$$y(k+1)-y(k)=0 \qquad k=0,1,2,\ldots$$

with initial condition $y(0)=1$. (One should be able to deduce the solution immediately, but let us work through the procedure.) The z-transform of the equation is

$$zY(z)-zy(0)-Y(z)=0$$

or, substituting the initial condition and solving,

$$Y(z)=\frac{z}{z-1}$$

From Table 8.1 of the previous section (or Example 1 of that section) we find that this transform corresponds to $y(k)=1$, $k=0, 1, 2, \ldots$, and this is clearly the correct solution.

Example 2. Consider the difference equation

$$y(k+1)+2y(k)=4^k \qquad k=0, 1, \ldots$$

with $y(0)=0$. Application of the z-transform yields

$$zY(z)+2Y(z)=\frac{z}{z-4}$$

Thus,

$$Y(z)=\frac{z}{(z+2)(z-4)}$$

One may verify that

$$\frac{z}{(z+2)(z-4)}=\frac{\frac{2}{3}}{z-4}+\frac{\frac{1}{3}}{z+2}$$

Therefore from Table 8.1 it follows that $y(k)$ is composed of two delayed geometric sequences: one with ratio 4 and the other with ratio -2. Specifically,

$$y(k)=\begin{cases} 0 & k=0 \\ \frac{2}{3}4^{k-1}+\frac{1}{3}(-2)^{k-1} & k\geq 1 \end{cases}$$

Inversion by Partial Fractions

Transform inversion is the process of going from a given z-transform, expressed as a function of z, to the corresponding original sequence. If the z-transform is given as a rational function, this inversion can always be accomplished directly (but tediously) by dividing the denominator polynomial into the numerator, recovering the series expansion that defines the transform in terms of the original sequence. There is, however, a simpler procedure in which the transform is first expressed in what is termed a partial fraction expansion. We shall outline the general technique and apply it to some examples.

Let $F(z)$ be a reduced strictly proper rational function

$$F(z)=\frac{b_{n-1}z^{n-1}+\cdots+b_0}{z^n+a_{n-1}z^{n-1}+\cdots+a_0} \tag{8-13}$$

The denominator can be factored in the form

$$z^n+a_{n-1}z^{n-1}+\cdots+a_0=(z-z_1)(z-z_2)\cdots(z-z_n)$$

Let us assume first that the roots $z_1, z_2, \ldots, z_n$ are distinct. Then the function

$F(z)$ can be written in *partial fraction* form

$$F(z)=\frac{c_1}{z-z_1}+\frac{c_2}{z-z_2}+\cdots+\frac{c_n}{z-z_n} \tag{8-14}$$

When the roots are distinct an expansion of the form (8-14) always exists. The n constants $c_1, c_2, \ldots, c_n$ must be determined so that when (8-14) is cross multiplied to convert it to the form (8-13), the n coefficients $b_0, b_1, \ldots, b_{n-1}$ are correct.

Once a transform is expressed in the partial fraction form (8-14), inversion is straightforward. Each term on the right-hand side of (8-14) corresponds to a delayed geometric series with ratio z_i. Thus, the inverse transform of $F(z)$ is the weighted sum of such sequences.

Specifically,

$$f(k)=\begin{cases} 0 & k=0 \\ c_1 z_1^{k-1}+c_2 z_2^{k-1}+\cdots+c_n z_n^{k-1} & k \geq 1 \end{cases}$$

If some of the z_i's are not distinct, then the partial fraction expansion must in general include higher-order terms. For instance, suppose a root, say z_i, is repeated m times. Then the expansion must include the terms

$$\frac{c_{i1}}{z-z_i}+\frac{c_{i2}}{(z-z_i)^2}+\cdots+\frac{c_{im}}{(z-z_i)^m}$$

Again, however, each of these can be inverted by extending the result of the last entry in Table 8.1. This leads to terms of the form $z_i^k, kz_i^{k-1}, k^2 z_i^{k-2}, \ldots, k^{(m-1)} z_i^{k-m+1}$.

Example 3. Consider the transform

$$F(z)=\frac{z-3}{z^2-3z+2}$$

Let us find its inverse.

Factoring the denominator we obtain

$$F(z)=\frac{z-3}{(z-1)(z-2)}$$

Therefore, we look for a partial fraction expansion of the form

$$F(z)=\frac{c_1}{z-1}+\frac{c_2}{z-2}$$

Cross multiplying and equating terms we find

$$F(z)=\frac{c_1(z-2)+c_2(z-1)}{(z-1)(z-2)}$$

Comparison with the original expression for $F(z)$ leads to the equations

$$c_1+c_2=1$$
$$-2c_1-c_2=-3$$

These have solution $c_1=2$, $c_2=-1$, and hence the partial fraction expansion is

$$F(z)=\frac{2}{z-1}-\frac{1}{z-2}$$

Therefore, using Table 8.1 we can deduce that

$$f(k)=\begin{cases}0 & k=0\\ 2-2^{k-1} & k\geq 1\end{cases}$$

Example 4. Let

$$F(z)=\frac{2z^2-7z+7}{(z-1)^2(z-2)}$$

To invert this we seek a partial fraction expansion of the form

$$F(z)=\frac{c_1}{z-1}+\frac{c_2}{(z-1)^2}+\frac{c_3}{z-2}$$

Cross multiplying and equating terms we find (using a procedure similar to that employed in Example 3)

$$F(z)=\frac{1}{z-1}-\frac{2}{(z-1)^2}+\frac{1}{z-2}$$

From Table 8.1 this is easily converted to

$$f(0)=\begin{cases}0, & k=0\\ 1-2(k-1)+2^{k-1}, & k\geq 1\end{cases}$$

Example 5. Let

$$F(z)=\frac{2z^2-3z}{z^2-3z+2}$$

This rational function is not strictly proper. However, z is a factor of the numerator. Thus,

$$\frac{F(z)}{z}=\frac{2z-3}{z^2-3z+2}$$

is strictly proper and can be expressed in partial fraction form

$$\frac{F(z)}{z}=\frac{1}{z-1}+\frac{1}{z-2}$$

This means that

$$F(z)=\frac{z}{z-1}+\frac{z}{z-2}$$

The first term corresponds to the unit step and the second corresponds to a geometric series with ratio 2. Therefore,

$$f(k)=1+2^k \qquad k\geq 0$$

Example 6. Consider

$$F(z)=\frac{z+2}{z-1}$$

This rational function is not strictly proper. However, it can be written as a constant plus a strictly rational function.

$$F(z)=1+\frac{3}{z-1}$$

Therefore,

$$f(k)=\begin{cases}1 & k=0\\ 3 & k\geq 1\end{cases}$$

8.4 STATE EQUATIONS AND TRANSFORMS

The transform technique can be applied to state equations, leading to new relations, new insights, and new analysis techniques.

Transfer Functions

Consider the following single-input, single-output system:

$$\mathbf{x}(k+1)=\mathbf{A}\mathbf{x}(k)+\mathbf{b}u(k) \tag{8-15a}$$

$$y(k)=\mathbf{c}^T\mathbf{x}(k)+du(k) \tag{8-15b}$$

The z-transform can be applied to these equations, yielding

$$z\mathbf{X}(z)-z\mathbf{x}(0)=\mathbf{A}\mathbf{X}(z)+\mathbf{b}U(z) \tag{8-16a}$$

$$Y(z)=\mathbf{c}^T\mathbf{X}(z)+dU(z) \tag{8-16b}$$

The transform $\mathbf{X}(z)$ is a vector with components being the transforms of the corresponding components of $\mathbf{x}(k)$. Since the z-transform is linear, it is possible to apply it to the linear equations (8-15) in the direct fashion indicated.

Let us assume $\mathbf{x}(0)=\mathbf{0}$. Then from (8-16a) one finds

$$\mathbf{X}(z)=[\mathbf{I}z-\mathbf{A}]^{-1}\mathbf{b}U(z) \tag{8-17}$$

Hence, substituting in (8-16b)

$$Y(z)=\mathbf{c}^T[\mathbf{I}z-\mathbf{A}]^{-1}\mathbf{b}U(z)+dU(z)$$

Therefore,

$$Y(z)=H(z)U(z) \tag{8-18}$$

where

$$H(z)=\mathbf{c}^T[\mathbf{I}z-\mathbf{A}]^{-1}\mathbf{b}+d \tag{8-19}$$

[Note that $\mathbf{c}^T[\mathbf{I}z-\mathbf{A}]^{-1}\mathbf{b}$ is 1×1.]

The transform $H(z)$ defined by (8-19) is called the *transfer function* of the system (8-15). It carries all the information necessary to determine the output in terms of the input. [Direct use of the transfer function always implicitly assumes $\mathbf{x}(0)=\mathbf{0}$.]

Since the state equations (8-15) are closely related to ordinary linear, constant-coefficient difference equations, it is perhaps not surprising that $H(z)$ can be expressed as a rational function of z. The proposition below verifies this fact.

Proposition. *The transfer function of a single-input, single-output nth-order linear system is a proper rational function of degree no greater than n.*

Proof. This result follows from the cofactor expression for the inverse of a matrix. We know that the determinant of $\mathbf{I}z-\mathbf{A}$ is a polynomial of degree n—the characteristic polynomial of $\mathbf{A}$. The elements of the inverse of $\mathbf{I}z-\mathbf{A}$ are, accordingly, each strictly proper rational functions with the characteristic polynomial as denominator. The transfer function $H(z)$ consists of weighted sums of these elements, plus the constant d. When expressed with a common denominator polynomial (which is the characteristic polynomial) the result is a proper rational function. The degree of the denominator polynomial is n. If it is possible to reduce the result, the degree is less than n. ∎

Impulse Response

Instead of employing the transform technique, let us solve (8-15) directly for $y(k)$. This is easily accomplished by using the general solution formula of Sect. 4.5. Assuming $\mathbf{x}(0)=\mathbf{0}$, the output $y(k)$ is given by the formula

$$y(k)=\sum_{l=0}^{k-1}\mathbf{c}^T\mathbf{A}^{k-l-1}\mathbf{b}u(l)+du(k) \tag{8-20}$$

Defining, for $k\geq 0$, the scalar-valued function

$$h(k)=\begin{cases} d & k=0 \\ \mathbf{c}^T\mathbf{A}^{k-1}\mathbf{b} & k\geq 1 \end{cases} \tag{8-21}$$

the expression for the output can be written

$$y(k)=\sum_{l=0}^{k} h(k-l)u(l) \tag{8-22}$$

The function $h(k)$ defined by (8-21) is the *impulse response function* of the system (8-15). The interpretation of the impulse response is, of course, that it is the output corresponding to a unit pulse at $k=0$. Clearly, in order to determine the output for any input sequence (under the assumption of zero initial state), it is sufficient to have computed the impulse response function, rather than all powers of $\mathbf{A}$. For this reason the impulse response function is a natural characterization of the system.

The impulse response and the transfer function are intimately related. Indeed, the transfer function is the z-transform of the impulse response. The easiest way to see this is to consider the output produced by application of a unit impulse at $k=0$. By definition, the output $y(k)$ is $h(k)$, the impulse response. On the other hand, in terms of z-transforms, the transform $y(z)$ of the output satisfies

$$Y(z)=H(z)U(z)$$

where $H(z)$ is the transfer function and $U(z)$ is the z-transform of the input. For an input equal to a unit impulse, one has $U(z)=1$ (see Table 8.1). Thus, $Y(z)=H(z)$. Since $Y(z)$ is the z-transform of the impulse response, so is $H(z)$. This result is summarized below.

Theorem. *For a linear, constant-coefficient, single-input, single-output system, the transfer function $H(z)$ is the z-transform of the impulse response $h(k)$.*

Example 1 (First-Order System). Consider again the first-order system

$$x(k+1)=ax(k)+bu(k)$$

$$y(k)=cx(k)$$

The transfer function is

$$H(z)=\frac{bc}{z-a}$$

In terms of a series expansion we have

$$H(z)=\frac{bc}{z}\left(1+\frac{a}{z}+\frac{a^2}{z^2}+\cdots\right)$$

Inverting this term by term we see that

$$h(k)=\begin{cases}0 & k=0\\ bca^{k-1} & k\geq 1\end{cases}$$

which agrees with the known form for the impulse response.

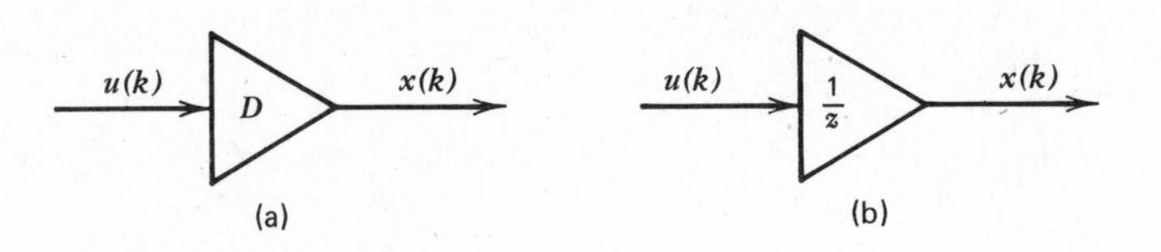

Figure 8.2. Unit delay.

The Unit Delay

Consider the most elementary dynamic system—a single delay. Its defining equations are

$$x(k+1) = u(k)$$
$$y(k) = x(k)$$

and we have often represented it diagrammatically, as in Fig. 8.2*a*. The transfer function of this elementary system is easily seen to be $H(z) = 1/z$. Therefore, the unit delay can be expressed diagrammatically as in Fig. 8.2*b*. Indeed we shall find it convenient to follow standard convention and refer to $1/z$ as the delay operation. It is, of course, the fundamental dynamic component for discrete-time systems. General systems are composed simply of a number of delays and various static components.

Combinations

An important characteristic of the transfer function, which is not shared by the impulse response function, is that the transfer function of a large structured system can be easily written in terms of the transfer functions of individual subsystems, thereby reducing a high-order calculation to a series of smaller-order calculations.

First consider a parallel arrangement as shown in Fig. 8.3. Suppose the two systems S_1 and S_2 have transfer functions $H_1(z)$ and $H_2(z)$, respectively.

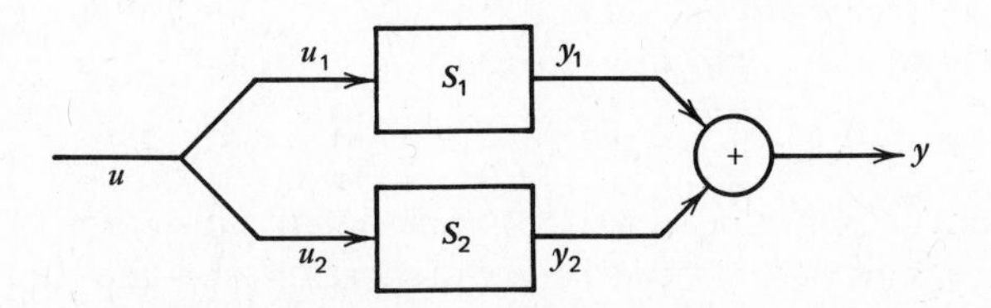

Figure 8.3. A parallel combination.

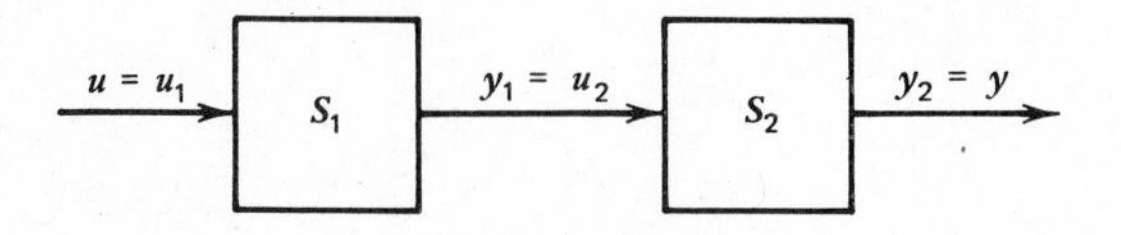

Figure 8.4. A series combination.

Then, assuming all state variables are initially zero, it follows that

$$Y_1(z) = H_1(z)U_1(z)$$
$$Y_2(z) = H_2(z)U_2(z)$$
$$U_1(z) = U_2(z) = U(z)$$
$$Y(z) = Y_1(z) + Y_2(z)$$

Therefore,

$$Y(z) = [H_1(z) + H_2(z)]U(z)$$

and hence the composite transfer function is the sum of the two individual transfer functions. Thus,

The transfer function of a parallel combination of systems is equal to the sum of the transfer functions of the individual subsystems.

The above property of addition for parallel combinations is important, but not particularly striking. This property is possessed by the impulse response function $h(k)$ as well. The unique property of the transfer function is how it decomposes in a *series* combination of systems.

Consider the system shown in Fig. 8.4. Suppose the two individual systems again have transfer functions $H_1(z)$ and $H_2(z)$, respectively. Then

$$Y_1(z) = H_1(z)U(z)$$
$$Y(z) = H_2(z)Y_1(z)$$

Thus,

$$Y(z) = H_2(z)H_1(z)U(z)$$

In general, therefore,

The transfer function of a series combination of systems is equal to the product of the transfer functions of the individual subsystems.

Feedback Structures

Systems composed of more complex arrangements of component subsystems can be treated by systematic application of the rules for parallel and series combinations. One important arrangement is the general feedback form illustrated in Fig. 8.5. In this figure $G_1(z)$ and $G_2(z)$ represent the transfer functions

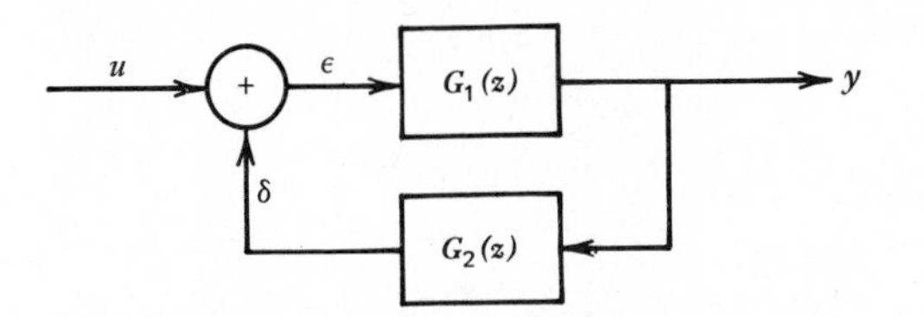

Figure 8.5. General feedback structure.

from the input to the output of the systems contained within their respective boxes. The overall transfer function $H(z)$ from u to y is to be determined.

The transfer function $H(z)$ can be determined by writing an equation that it must satisfy. If the overall transfer function from u to y is $H(z)$, then using the series rule the transfer function from u to δ is $H(z)G_2(z)$. Then, using the parallel rule, the transfer function from u to ε must be $1+H(z)G_2(z)$. Then again using the series rule the transfer function from u to y must be $G_1(z)\{1+H(z)G_2(z)\}$. We therefore obtain the equation

$$H(z)=G_1(z)\{1+H(z)G_2(z)\} \tag{8-23}$$

and thus we conclude that

$$H(z)=\frac{G_1(z)}{1-G_1(z)G_2(z)} \tag{8-24}$$

This formula for the transfer function of a feedback structure, together with the formulas for parallel and series combinations, generally enables one to quickly compute the transfer function of complex arrangements.

Example 2. A diagram of interconnected systems as shown in Fig. 8.6 might arise in the study of a control system for a mechanical or electrical system. This transfer function is readily found by application of the rules described in this section. The inner feedback loop can be regarded as a system with transfer function

$$\frac{G_1(z)}{1-G_1(z)H_1(z)}$$

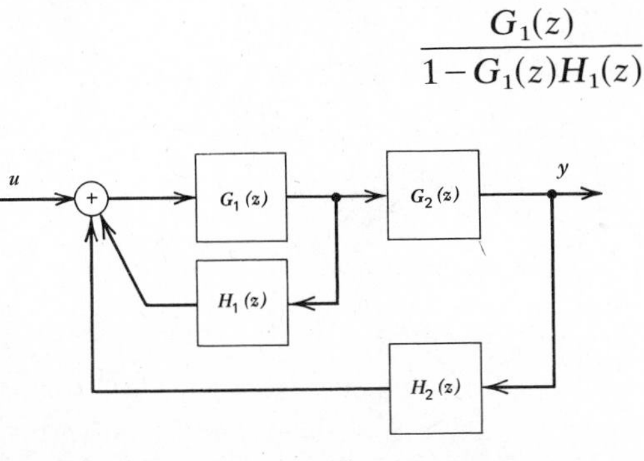

Figure 8.6. Interconnected systems.

This, combined with the system G_2, can then be regarded as the upper part of the outer feedback loop. The overall transfer function is

$$H(z)=\frac{\dfrac{G_1(z)G_2(z)}{1-G_1(z)H_1(z)}}{1-\dfrac{G_1(z)G_2(z)H_2(z)}{1-G_1(z)H_1(z)}}$$

$$H(z)=\frac{G_1(z)G_2(z)}{1-G_1(z)H_1(z)-G_2(z)G_1(z)H_2(z)}$$

Note that if the individual transfer functions are rational, the overall transfer function will also be rational.

8.5 LAPLACE TRANSFORMS

The foregoing developments can be paralleled for continuous-time systems through introduction of the Laplace transform.

Definition. Corresponding to a function $y(t)$, $t \geq 0$ the *Laplace transform* of this function is

$$Y(s)=\int_0^\infty y(t)e^{-st}\,dt \tag{8-25}$$

where s is an indeterminant variable.

There are certain convergence conditions associated with the integral in this definition. If $y(t)$ is a continuous function bounded by some exponential [say $|y(t)| \leq e^{Mt}$], then the integral exists for complex values of s having real part sufficiently large (larger than M). However, analogous to the z-transform situation, the integral can usually be considered as just a formalism; the transform itself being represented by a rational expression in the variable s. This is always possible when $y(t)$ is made up of various exponential terms.

Example 1. Consider the function $y(t)=e^{at}$. The corresponding Laplace transform is

$$Y(s)=\int_0^\infty e^{at}e^{-st}\,dt=\left.\frac{e^{(a-s)t}}{a-s}\right|_0^\infty$$

$$Y(s)=\frac{1}{s-a} \tag{8-26}$$

The integral converges if the real part of s is greater than the real part of a. However, one may sidestep the convergence issue and simply associate the rational expression (8-26) with the transform of e^{at}.

The Laplace transform enjoys a set of properties quite similar to those of the z-transform.

Property 1. If $f(t)$ and $g(t)$ are functions with Laplace transforms $F(s)$ and $G(s)$, respectively, then the transform of $f(t)+g(t)$ is $F(s)+G(s)$. The transform of $af(t)$ is $aF(s)$.

Proof. This follows from the linearity of the definition. ▮

Property 2. If $f(t)$ has Laplace transform $F(s)$, then the derivative function $g(t)=(d/dt)f(t)$ has transform $G(s)=sF(s)-f(0)$.

Proof. We have

$$G(s)=\int_0^\infty \frac{d}{dt} f(t)e^{-st}\,dt$$

Integration by parts yields

$$\begin{aligned} G(s) &= f(t)e^{-st}\Big|_0^\infty + s\int_0^\infty f(t)e^{-st}\,dt \\ &= sF(s)-f(0). \;▮ \end{aligned}$$

Property 3. If $f(t)$ has Laplace transform $F(s)$, then the integral function $g(t)=\int_0^t f(\tau)\,d\tau$ has transform

$$G(s)=\frac{1}{s}F(s)$$

Proof. This can be deduced from Property 2 above, since $(d/dt)g(t)=f(t)$. Thus,

$$F(s)=sG(s)-g(0)$$

Since $g(0)=0$, the result follows. ▮

Solution of Differential Equations

The Laplace transform provides a convenient mechanism for solving ordinary linear, constant-coefficient differential equations. The approach is exactly analogous to the z-transform procedure for difference equations. We illustrate the technique with a single example.

Example 2. Consider the differential equation

$$\frac{d^2y(t)}{dt^2}-3\frac{dy(t)}{dt}+2y(t)=1 \tag{8-27}$$

with the initial conditions

$$y(0)=1 \qquad \frac{dy(0)}{dt}=0$$

We note that by two applications of Property 2 the Laplace transform of (d^2y/dt^2) is $s^2Y(s)-sy(0)-(dy/dt)(0)$. Then, taking the transform of the differential equation (8-27) yields

$$(s^2-3s+2)Y(s)-sy(0)-\frac{dy(0)}{dt}+3y(0)=\frac{1}{s}$$

The $1/s$ term on the right is the transform of 1 (see Example 1, with $a=0$). Substituting the given initial conditions yields

$$(s^2-3s+2)Y(s)=\frac{1}{s}+s-3=\frac{s^2-3s+1}{s}$$

Therefore,

$$Y(s)=\frac{s^2-3s+1}{s(s^2-3s+2)}$$

This transform can be inverted by developing a partial function expansion. Thus,

$$Y(s)=\frac{1}{2s}+\frac{1}{s-1}-\frac{1}{2(s-2)}$$

Therefore, $y(t)$ can be expressed as the corresponding combination of exponential functions,

$$y(t)=\tfrac{1}{2}+e^t-\tfrac{1}{2}e^{2t}$$

State Space Equations

Let us now apply the Laplace transform to an nth-order, single-input, single-output system written in state space form

$$\dot{\mathbf{x}}(t)=\mathbf{A}\mathbf{x}(t)+\mathbf{b}u(t) \tag{8-28a}$$

$$y(t)=\mathbf{c}^T\mathbf{x}(t)+du(t) \tag{8-28b}$$

Assume that $\mathbf{x}(0)=\mathbf{0}$.

Application of the Laplace transform yields immediately

$$s\mathbf{X}(s)=\mathbf{A}\mathbf{X}(s)+\mathbf{b}U(s)$$

$$Y(s)=\mathbf{c}^T\mathbf{X}(s)+dU(s)$$

Elimination of $\mathbf{X}(s)$ produces the input–output relation

$$Y(s)=H(s)U(s)$$

where

$$H(s)=\mathbf{c}^T[s\mathbf{I}-\mathbf{A}]^{-1}\mathbf{b}+d \tag{8-29}$$

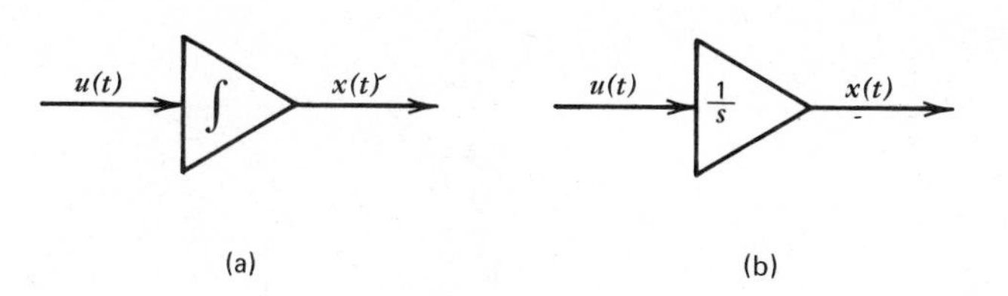

Figure 8.7. The pure integrator.

As before $H(s)$ is called the *transfer function* of the system (8-28). It is a proper rational function of degree n or less. The transfer function converts the complicated dynamic relation between input and output into a simple multiplicative operation of transforms. This greatly facilitates calculation, especially when a system has a natural decomposition as a collection of subsystems.

The Pure Integrator

The simplest continuous-time dynamic system is the system $\dot{x}(t)=u(t)$. The output $x(t)$ is just the integral of the input function. This is represented diagrammatically in Fig. 8.7*a*. The transfer function of this system is easily seen to be $1/s$. Thus, we often use $1/s$ instead of an integral sign in system diagrams, as illustrated in Fig. 8.7*b*.

Combinations

We conclude this section with an example that illustrates how the transfer function often leads to a more visual and more rapid approach to analysis. It works well in conjunction with dynamic diagrams.

Example 3. Suppose we are given the system

$$\begin{bmatrix} \dot{x}_1 \\ \dot{x}_2 \end{bmatrix} = \begin{bmatrix} 1 & 0 \\ 1 & 2 \end{bmatrix} \begin{bmatrix} x_1 \\ x_2 \end{bmatrix} + \begin{bmatrix} 1 \\ 0 \end{bmatrix} u(t)$$

$$y(t)=x_2(t)$$

The initial condition is $x_1(0)=x_2(0)=0$ and the input $u(t)=1$ is applied, for $t \geq 0$. What is the output?

The system is shown in diagram form in Fig. 8.8. From the diagram one sees immediately that the transfer function is

$$H(s)=\frac{1}{(s-1)(s-2)}$$

The Laplace transform of the constant input is, by Example 1 (with $a=0$),

$$U(s)=1/s$$

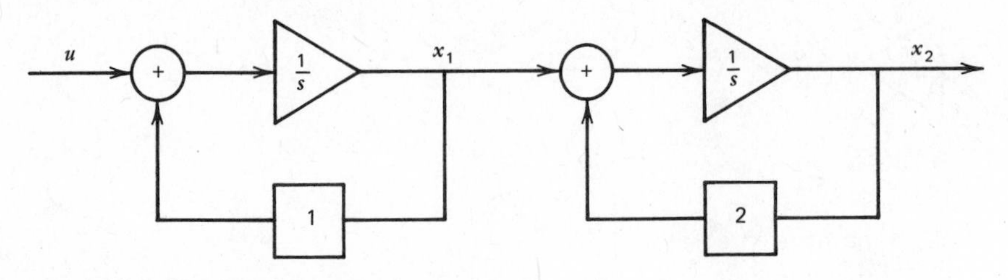

Figure 8.8. Diagram for Example 3.

Therefore, the output has Laplace transform $Y(s)=H(s)U(s)$. Thus,

$$Y(s)=\frac{1}{s(s-1)(s-2)}$$

This can be written in partial fraction form

$$Y(s)=\frac{1}{2s}-\frac{1}{s-1}+\frac{1}{2(s-2)}$$

It then follows that

$$y(t)=\tfrac{1}{2}-e^{t}+\tfrac{1}{2}e^{2t}$$

STATE SPACE METHODS

8.6 CONTROLLABILITY

The input and output structure of a system can significantly influence the available means for control. Two fundamental concepts characterizing the dynamic implications of input and output structure are the dual concepts of controllability and observability.

Controllability for Discrete-Time Systems

We begin by giving a general definition of controllability for linear discrete-time systems.

Definition. The nth order system

$$\mathbf{x}(k+1)=\mathbf{A}\mathbf{x}(k)+\mathbf{B}\mathbf{u}(k) \tag{8-30}$$

is said to be *completely controllable** if for $\mathbf{x}(0)=\mathbf{0}$ and any given n vector

* It is possible to define a notion of controllability with respect to a subset of the states that can be reached. The terminology *completely controllable* thus refers to the fact that all states can be reached. We shall have no need for anything other than the notion of complete controllability, and, often, for economy of language, we refer to this concept simply as *controllability*.

$\mathbf{x}_1$ there exists a finite index N and a sequence of inputs $\mathbf{u}(0), \mathbf{u}(1), \ldots,$ $\mathbf{u}(N-1)$ such that this input sequence, applied to (8-30), yields $\mathbf{x}(N) = \mathbf{x}_1$.

Thus, somewhat more loosely, a system is completely controllable if the state can be driven from the origin to any other state in a finite number of steps.† Actually, the choice of $\mathbf{x}(0) = \mathbf{0}$ is simply one of convenience. We will later show that if the system is completely controllable in the sense of the given definition, it is possible to drive the system from *any* initial state to any specified state within a finite number of steps. Thus, complete controllability corresponds directly to the intuitive notion of being able to control the system.

Example 1. In the system shown in Fig. 8.9 the state variable x_1 cannot be moved from zero by application of inputs $u(k)$. This system is *not* completely controllable.

It is possible to derive a simple set of conditions on the $n \times n$ matrix $\mathbf{A}$ and the $n \times m$ matrix $\mathbf{B}$ that are equivalent to controllability. This result is stated below.

Theorem 1. *A discrete-time system* (8-30) *is completely controllable if and only if the $n \times nm$ controllability matrix*

$$\mathbf{M} = [\mathbf{B}, \mathbf{AB}, \ldots, \mathbf{A}^{n-1}\mathbf{B}] \tag{8-31}$$

has rank n.

Before proving this theorem let us briefly study the structure of the controllability matrix. If the system has only a single, scalar input, the input matrix $\mathbf{B}$ reduces to a vector $\mathbf{b}$. In that case the controllability matrix is written

$$\mathbf{M} = [\mathbf{b}, \mathbf{Ab}, \ldots, \mathbf{A}^{n-1}\mathbf{b}]$$

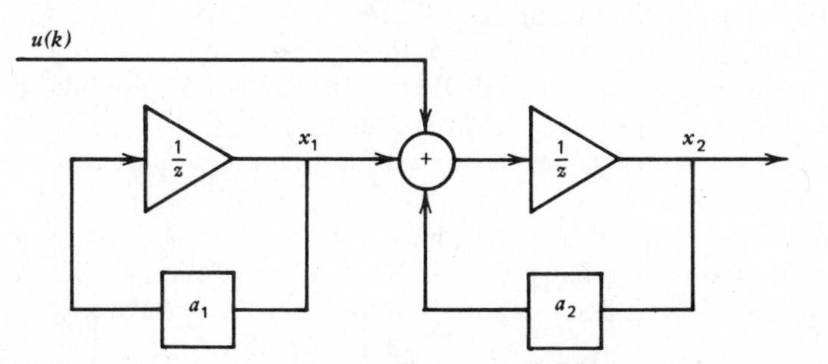

Figure 8.9. A system that is not completely controllable.

† Many authors term this property *complete reachability* and define a system as completely controllable if any initial state can be driven to zero in finite time by an appropriate input sequence. The two notions coincide in the discrete-time case if $\mathbf{A}$ is nonsingular, and they always coincide in the continuous-time case.

and it is a square $n \times n$ matrix. The condition that this matrix has rank n is equivalent, in this case, to its being nonsingular. In the general case where $\mathbf{B}$ consists of m columns, the controllability matrix can be regarded as being composed of m controllability matrices of dimension $n \times n$. As expressed in (8-31) they are interlaced with the first column of each group forming one of these matrices, the second forming another, and so forth. The condition for complete controllability, as stated by the theorem, is that from among the nm columns there are n that are linearly independent—so that the rank is n.

As an aid to the proof of the theorem, we first prove the following preliminary result. This result is used later in this section as well.

Lemma. *For any $N \geq n$, the rank of the matrix*

$$[\mathbf{B}, \mathbf{AB}, \ldots, \mathbf{A}^{N-1}\mathbf{B}]$$

is equal to the rank of the controllability matrix $\mathbf{M}$.

Proof. As k increases by one unit the rank of the matrix $\mathbf{M}_k = [\mathbf{B}, \mathbf{AB}, \ldots, \mathbf{A}^{k-1}\mathbf{B}]$ either increases (by at least 1) or remains constant. Suppose that k is an integer such that the rank of $\mathbf{M}_{k+1}$ is equal to the rank of $\mathbf{M}_k$. That means that the m columns comprising $\mathbf{A}^k\mathbf{B}$ are each linearly dependent on the (previous) columns in $\mathbf{M}_k$. That is, there is a relation of the form

$$\mathbf{A}^k\mathbf{B} = \mathbf{BD}_0 + \mathbf{ABD}_1 + \cdots + \mathbf{A}^{k-1}\mathbf{BD}_{k-1} \tag{8-32}$$

where each $\mathbf{D}_i$ is an $m \times m$ matrix. Now multiplication of this relation by $\mathbf{A}$ leads to the new relation

$$\mathbf{A}^{k+1}\mathbf{B} = \mathbf{ABD}_0 + \mathbf{A}^2\mathbf{BD}_1 + \cdots + \mathbf{A}^k\mathbf{BD}_{k-1} \tag{8-33}$$

which shows that the columns comprising $\mathbf{A}^{k+1}\mathbf{B}$ are linearly dependent on the columns in $\mathbf{M}_{k+1}$. Therefore, the rank of $\mathbf{M}_{k+2}$ is the same as the rank of $\mathbf{M}_{k+1}$. By continuing this argument, we see that for all $j > k$ the rank of $\mathbf{M}_j$ is equal to that of $\mathbf{M}_k$. Thus we have shown that, in the progression of $\mathbf{M}_k$'s, once the rank fails to increase, it will remain constant even as additional columns are adjoined.

In view of the above, the rank of $\mathbf{M}_k$ increases by at least 1 at each increment of k until it attains its maximum rank. Since the maximum rank is at most n, the maximum rank is attained within n steps (that is, by $\mathbf{M}_n$). ∎

Now we turn to the proof of the theorem.

Proof of Theorem 1. Suppose a sequence of inputs $\mathbf{u}(0), \mathbf{u}(1), \ldots, \mathbf{u}(N-1)$ is applied to the system (8-30), with $\mathbf{x}(0) = \mathbf{0}$. It follows that

$$\mathbf{x}(N) = \mathbf{A}^{N-1}\mathbf{Bu}(0) + \mathbf{A}^{N-2}\mathbf{Bu}(1) + \cdots + \mathbf{Bu}(N-1)$$

From this formula we see that points in state space can be reached if and only if they can be expressed as linear combinations of powers of **A** times **B**. Thus, the issue of complete controllability rests on whether the infinite sequence **B**, **AB**, $\mathbf{A}^2\mathbf{B}, \ldots$ has a finite number of columns that span the entire n-dimensional space. By the lemma, however, these span the full n-dimensional space if and only if **M** is of rank n. ▌

As a result of this theorem we see that the condition of complete controllability is slightly stronger than can be immediately inferred from the definition. The definition requires only that it be possible to transfer the system state from the origin to an arbitrary point in some finite number of steps. As a corollary of the theorem, however, we see that if such transfers are possible in a finite number of steps, the transfer can in fact be accomplished within n steps.

It also follows from the theorem that if a system is completely controllable in the sense of transference from the origin to an arbitrary point, it is in fact controllable in the stronger sense of being able to transfer the state between two arbitrary points within n steps. To show this, suppose $\mathbf{x}(0)$ and $\mathbf{x}(n)$ are specified arbitrarily. With zero input the system would move to $\mathbf{A}^n\mathbf{x}(0)$ at period n. Thus, the desired input sequence is the one that would transfer the state from the origin to $\mathbf{x}(n)-\mathbf{A}^n\mathbf{x}(0)$ at period n. We see that the modest definition of complete controllability actually implies rather strong and desirable control characteristics.

Example 2. The system in Fig. 8.9 is described by the equations

$$\begin{bmatrix} x_1(k+1) \\ x_2(k+1) \end{bmatrix} = \begin{bmatrix} a_1 & 0 \\ 1 & a_2 \end{bmatrix} \begin{bmatrix} x_1(k) \\ x_2(k) \end{bmatrix} + \begin{bmatrix} 0 \\ 1 \end{bmatrix} u(k)$$

Thus, the controllability matrix $\mathbf{M}=[\mathbf{b}, \mathbf{Ab}]$ is

$$\mathbf{M} = \begin{bmatrix} 0 & 0 \\ 1 & a_2 \end{bmatrix}$$

Its columns are linearly dependent, and, as deduced earlier, the system is not completely controllable.

Example 3. If the input of the system shown in Fig. 8.9 is shifted to a position entering the first stage rather than the second, the corresponding controllability matrix $\mathbf{M}=[\mathbf{b}, \mathbf{Ab}]$ is

$$\mathbf{M} = \begin{bmatrix} 1 & a_1 \\ 0 & 1 \end{bmatrix}$$

which has rank 2. Thus, this system is completely controllable.

Single-Input Systems

For single-input systems, there is a simple interpretation of complete controllability in terms of the diagonal representation of the system (assuming that the system can be diagonalized). Suppose the system

$$\mathbf{x}(k+1)=\mathbf{A}\mathbf{x}(k)+\mathbf{b}u(k) \tag{8-34}$$

is converted to diagonal form by the change of variable

$$\mathbf{x}(k)=\mathbf{P}\mathbf{z}(k) \tag{8-35}$$

That is, the matrix $\mathbf{P}^{-1}\mathbf{A}\mathbf{P}=\mathbf{\Lambda}$ is diagonal. In the new variables the system is described by the equations

$$\mathbf{z}(k+1)=\mathbf{\Lambda}\mathbf{z}(k)+\hat{\mathbf{b}}u(k) \tag{8-36}$$

where $\hat{\mathbf{b}}=\mathbf{P}^{-1}\mathbf{b}$. It should be clear that complete controllability of the system (8-34) is equivalent to complete controllability of (8-36), for if any point in state space for one system can be reached, so can any point in the other.

Complete controllability hinges on the nonsingularity of the matrix $[\hat{\mathbf{b}}, \mathbf{\Lambda}\hat{\mathbf{b}}, \ldots, \mathbf{\Lambda}^{n-1}\hat{\mathbf{b}}]$. This matrix is

$$\mathbf{M}=\begin{bmatrix} \hat{b}_1 & \lambda_1\hat{b}_1 & \lambda_1^2\hat{b}_1 & \cdots & \lambda_1^{n-1}\hat{b}_1 \\ \hat{b}_2 & \lambda_2\hat{b}_2 & \lambda_2^2\hat{b}_2 & \cdots & \lambda_2^{n-1}\hat{b}_2 \\ \cdot & \cdot & \cdot & & \cdot \\ \cdot & \cdot & \cdot & & \cdot \\ \cdot & \cdot & \cdot & & \cdot \\ \hat{b}_n & \lambda_n\hat{b}_n & \lambda_n^2\hat{b}_n & \cdots & \lambda_n^{n-1}\hat{b}_n \end{bmatrix} \tag{8-37}$$

To be nonsingular, it is clear that it is necessary that $\hat{b}_i \neq 0$ for $i=1, 2, \ldots, n$, for otherwise one row would be identically zero. It is also clear that there can be no repeated eigenvalues, for otherwise two rows would be proportional. These conditions are both necessary and sufficient for complete controllability of (8-36). (See Problem 14 for the nondiagonalizable case.)

The intuitive meaning of these results is made clear by Fig. 8.10. For the system to be completely controllable, there must be a nonzero connection from the input to each of the subsystems in its diagonal representation—otherwise that subsystem variable cannot be influenced by the input. Furthermore, the diagonal system cannot have repeated roots—otherwise the variables of the corresponding two subsystems will always be in fixed proportion.

Complete controllability of a system means that movement of the state can be directed by the input. This is possible only if the input is "fully connected" to the dynamics of the system as described above. The input must reach every individual first-order subsystem, and these subsystems must have different dynamic constants if they are to be independently controlled. Complete controllability is a general criteria for the "connectivity" of the input structure.

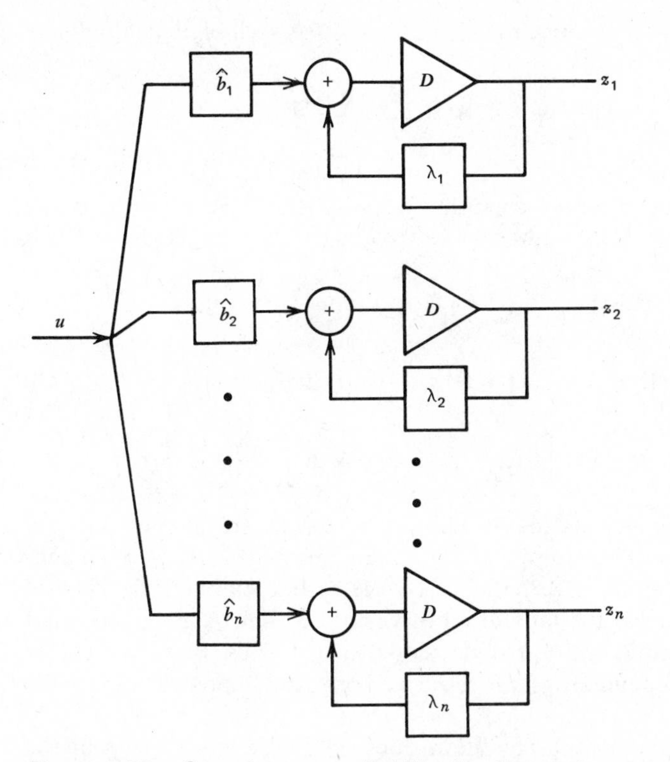

Figure 8.10. Complete controllability for diagonal systems.

Controllability for Continuous-Time Systems

The definition of complete controllability for continuous-time systems is a direct analog of the definition for discrete-time systems.

Definition. The system

$$\dot{\mathbf{x}}(t) = \mathbf{A}\mathbf{x}(t) + \mathbf{B}\mathbf{u}(t) \tag{8-38}$$

is said to be *completely controllable* if for $\mathbf{x}(0) = \mathbf{0}$ and any given state $\mathbf{x}_1$ there exists a finite time t_1 and a piecewise continuous input $\mathbf{u}(t)$, $0 \le t \le t_1$ such that $\mathbf{x}(t_1) = \mathbf{x}_1$.

The criterion for complete controllability in terms of the matrices **A** and **B** is also analogous to that for discrete-time systems, although the proof is somewhat different. In continuous time we must employ rather indirect arguments to translate the rank condition to specific input functions. As will be seen later, however, the interpretation of the condition is identical to that for discrete-time systems.

Theorem. *A continuous-time system* (8-38) *is completely controllable if and only if the controllability matrix*

$$\mathbf{M}=[\mathbf{B}, \mathbf{AB}, \mathbf{A}^2\mathbf{B}, \ldots, \mathbf{A}^{n-1}\mathbf{B}]$$

has rank n.

Proof. Suppose first that the rank condition does not hold. For any t_1 and any input function $\mathbf{u}(t)$, $0 \le t \le t_1$, we have

$$\mathbf{x}(t_1)=\int_0^{t_1} e^{\mathbf{A}(t_1-t)}\mathbf{B}\mathbf{u}(t)\,dt \tag{8-39}$$

$$\mathbf{x}(t_1)=\int_0^{t_1}\left\{\mathbf{I}+\mathbf{A}(t_1-t)+\frac{\mathbf{A}^2}{2!}(t_1-t)^2\cdots\right\}\mathbf{B}\mathbf{u}(t)\,dt \tag{8-40}$$

$$\mathbf{x}(t_1)=\mathbf{B}\int_0^{t_1}\mathbf{u}(t)\,dt+\mathbf{AB}\int_0^{t_1}(t_1-t)\mathbf{u}(t)\,dt+\mathbf{A}^2\mathbf{B}\int_0^{t_1}\frac{(t_1-t)^2}{2!}\mathbf{u}(t)\,dt+\cdots$$

When evaluated, the integrals in the above expression are simply constant m-dimensional vectors. Therefore, the expression shows that $\mathbf{x}(t_1)$ is a linear combination of the columns of $\mathbf{B}$, $\mathbf{AB}$, ... By the earlier lemma, if the rank of $\mathbf{M}$ is less than n, then even the infinite set of vectors $\mathbf{B}$, $\mathbf{AB}$, $\mathbf{A}^2\mathbf{B}$, ... does not contain a full basis for the entire n-dimensional space. Thus, there is a vector $\mathbf{x}_1$ that is linearly independent of all these vectors, and therefore cannot be attained.

Now suppose that the rank condition does hold. We will show that the system is completely controllable and that in fact the state can be transferred from zero to an arbitrary point $\mathbf{x}_1$ within an arbitrarily short period of time.

We first show that for any $t_1 > 0$ the matrix

$$\mathbf{K}=\int_0^{t_1} e^{-\mathbf{A}t}\mathbf{B}\mathbf{B}^T e^{-\mathbf{A}^T t}\,dt \tag{8-41}$$

is nonsingular. To prove this, suppose there is a vector $\mathbf{a}$ such that $\mathbf{Ka}=\mathbf{0}$. Then

$$\mathbf{a}^T\mathbf{K}\mathbf{a}=0 \tag{8-42}$$

or, more explicitly,

$$\int_0^{t_1}\mathbf{a}^T e^{-\mathbf{A}t}\mathbf{B}\mathbf{B}^T e^{-\mathbf{A}^T t}\mathbf{a}\,dt=0 \tag{8-43}$$

The integrand above has the form $\mathbf{c}(t)^T\mathbf{c}(t)$, where $\mathbf{c}(t)=\mathbf{B}^T e^{-\mathbf{A}^T t}\mathbf{a}$. It follows that the integrand is always nonnegative. For the integral (8-43) to vanish, it follows that the integrand must vanish identically for $0 \le t \le t_1$. Therefore,

$$\mathbf{a}^T e^{-\mathbf{A}t}\mathbf{B}=\mathbf{0}$$

for all t, $0 \le t \le t_1$. Evaluation of this expression, and its successive derivatives, with respect to t, at $t=0$ leads to the following sequence of equations:

$$\mathbf{a}^T\mathbf{B}=\mathbf{0}$$
$$\mathbf{a}^T\mathbf{A}\mathbf{B}=\mathbf{0}$$
$$\mathbf{a}^T\mathbf{A}^2\mathbf{B}=\mathbf{0}$$
$$\vdots$$
$$\mathbf{a}^T\mathbf{A}^{n-1}\mathbf{B}=\mathbf{0}$$

This means that the vector **a** must be orthogonal to all columns of the matrix **M**. Since it is assumed that this matrix has rank n, it must follow that $\mathbf{a}=\mathbf{0}$. Therefore, by the fundamental lemma of linear algebra, **K** is nonsingular.

Now, given $\mathbf{x}_1$, select any $t_1>0$ and set

$$\mathbf{u}(t)=\mathbf{B}^T e^{-\mathbf{A}^T t}\mathbf{K}^{-1}e^{-\mathbf{A}t_1}\mathbf{x}_1 \tag{8-44}$$

Then from (8-39)

$$\mathbf{x}(t_1)=\int_0^{t_1} e^{\mathbf{A}(t_1-t)}\mathbf{B}\mathbf{B}^T e^{-\mathbf{A}^T t}\mathbf{K}^{-1}e^{-\mathbf{A}t_1}\mathbf{x}_1\,dt$$
$$\mathbf{x}(t_1)=e^{\mathbf{A}t_1}\mathbf{K}\mathbf{K}^{-1}e^{-\mathbf{A}t_1}\mathbf{x}_1=\mathbf{x}_1$$

Therefore, the input (8-44) transfers the state from zero to $\mathbf{x}_1$, and the system is completely controllable. ∎

The interpretation of this result is the same as for the discrete-time case. If there is only a single input and the system has a complete set of eigenvectors, then again controllability corresponds to the conditions that the eigenvalues be distinct and that the connection between the input and each of the separate one-dimensional systems be nonzero.

Example 4 (Stick Balancing). There are several mechanical problems—including the maintenance of a satellite in proper orbit, the control of a helicopter, and the control of a rocket while being thrust upward—that have the character of complex balancing problems. As a simple version of a problem of this type let us consider the balancing of a stick on your hand, as illustrated in Fig. 8.11. We know from experience that this is possible—and thus the stick and hand system must be controllable. Let us verify that it is.

For simplicity we consider balancing a stick of length L all of whose mass M is concentrated at the top. From Newton's laws, it can be deduced that the system is governed by the equation

$$\ddot{u}(t)\cos\theta(t)+L\ddot{\theta}(t)=g\sin\theta(t) \tag{8-45}$$

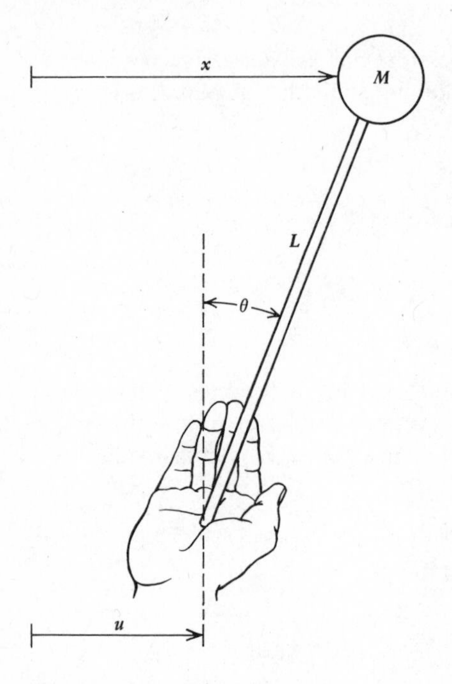

Figure 8.11. Stick balancing.

where g is the gravitational constant. We also have the relation

$$x(t) = u(t) + L \sin \theta(t) \tag{8-46}$$

Assuming that the stick is nearly at rest in the vertical position (with θ small), the two equations (8-45) and (8-46) can be written in terms of $x(t)$ as

$$\ddot{x}(t) = \frac{g}{L}[x(t) - u(t)]$$

For ease of notation let us set $L = 1$. Then defining the velocity $v(t) = \dot{x}(t)$, the system has the state space representation

$$\begin{bmatrix} \dot{x}(t) \\ \dot{v}(t) \end{bmatrix} = \begin{bmatrix} 0 & 1 \\ g & 0 \end{bmatrix} \begin{bmatrix} x(t) \\ v(t) \end{bmatrix} + g \begin{bmatrix} 0 \\ -1 \end{bmatrix} u(t)$$

The controllability matrix is

$$\mathbf{M} = g \begin{bmatrix} 0 & -1 \\ -1 & 0 \end{bmatrix}$$

Since **M** is nonsingular, the system is completely controllable. (This is easy; but

Figure 8.12. A controllability problem.

what about two sticks, one on top of the other as shown in Fig. 8.12!? You may wish to explore the controllability properties of this system without writing equations.)

8.7 OBSERVABILITY

Complete observability is a concept quite analogous to complete controllability. In fact, the two are often referred to as *dual* concepts in that results for one are the transpose of results for the other.

Observability for Discrete-Time Systems

We begin by stating the formal definition.

Definition. The discrete-time system

$$\mathbf{x}(k+1) = \mathbf{A}\mathbf{x}(k)$$
$$\mathbf{y}(k) = \mathbf{C}\mathbf{x}(k) \tag{8-47}$$

is *completely observable* if there is a finite index N such that knowledge of the outputs $\mathbf{y}(0), \mathbf{y}(1), \ldots, \mathbf{y}(N-1)$ is sufficient to determine the value of the initial state $\mathbf{x}(0)$.

In the above definition, it is assumed that the equations (8-47) governing the system and its outputs are known but that the initial state is unknown before the outputs become available. By watching the outputs, the value of the initial state can be inferred if the system is completely observable. In a sense the state is observed through the output structure.

The definition above easily extends to general time-invariant linear systems of the form

$$\mathbf{x}(k+1)=\mathbf{A}\mathbf{x}(k)+\mathbf{B}\mathbf{u}(k) \tag{8-48a}$$

$$\mathbf{y}(k)=\mathbf{C}\mathbf{x}(k)+\mathbf{D}\mathbf{u}(k) \tag{8-48b}$$

having both an input and an output structure. (As usual $\mathbf{A}$ is $n\times n$, $\mathbf{B}$ is $n\times m$, $\mathbf{C}$ is $p\times n$, and $\mathbf{D}$ is $p\times m$.) One simply imposes the requirement that all inputs $\mathbf{u}(k)$ be zero. This reduces the system to (8-47). Thus, both complete controllability and complete observability of (8-48) can be investigated.

Complete observability is an important concept in the context of system control. Control inputs are usually determined on the basis of observation of available outputs. If the output structure is deficient in that it does not eventually convey full information about the state vector, then it may not be possible to devise suitable control strategies. Thus, in general, good control requires both the ability to infer what the system is doing (observability) and the ability to change the behavior of the system (controllability).

Example 1. The system shown in Fig. 8.13 is not completely observable from the single output y. There is no way to infer the value of x_1.

The criterion for complete observability in terms of the specific system description is analogous to that of complete controllability.

Theorem 1. *The system* (8-47) *is completely observable if and only if the $pn\times n$ observability matrix*

$$\mathbf{S}=\begin{bmatrix}\mathbf{C}\\ \mathbf{CA}\\ \mathbf{CA}^2\\ \cdot\\ \cdot\\ \cdot\\ \mathbf{CA}^{n-1}\end{bmatrix}$$

has rank n.

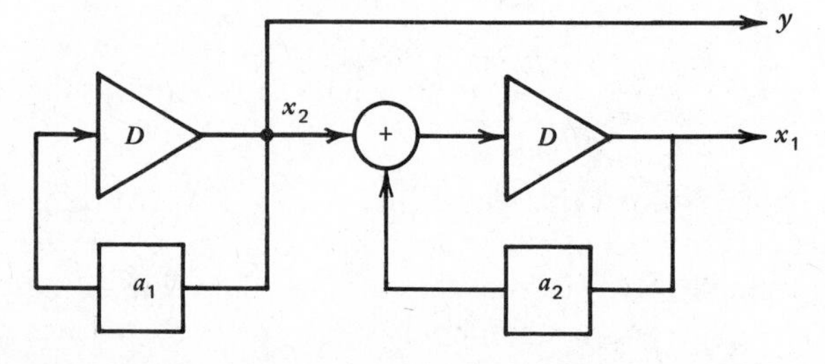

Figure 8.13. Not completely observable.

Proof. Suppose that $\mathbf{S}$ has rank n. We will show that we can take $N=n$ in the definition and that, accordingly, the initial state can be determined from knowledge of the n outputs $\mathbf{y}(0), \mathbf{y}(1), \ldots, \mathbf{y}(n-1)$.

Solution of the system equation (8-47) yields the set of equations

$$
\begin{aligned}
\mathbf{y}(0) &= \mathbf{C}\mathbf{x}(0) \\
\mathbf{y}(1) &= \mathbf{C}\mathbf{A}\mathbf{x}(0) \\
\mathbf{y}(2) &= \mathbf{C}\mathbf{A}^2\mathbf{x}(0) \\
&\;\;\vdots \\
\mathbf{y}(n-1) &= \mathbf{C}\mathbf{A}^{n-1}\mathbf{x}(0)
\end{aligned}
\tag{8-49}
$$

This is a set of pn equations. In view of the rank condition, n of these equations can be selected that are linearly independent, and they can be solved uniquely for $\mathbf{x}(0)$.

The proof of the converse follows a pattern similar to that for complete controllability. One can write

$$
\begin{bmatrix} \mathbf{C} \\ \mathbf{C}\mathbf{A} \\ \mathbf{C}\mathbf{A}^2 \\ \cdot \\ \cdot \\ \cdot \\ \mathbf{C}\mathbf{A}^{N-1} \end{bmatrix} = [\mathbf{C}^T, \mathbf{A}^T\mathbf{C}^T, (\mathbf{A}^2)^T\mathbf{C}^T, \ldots, (\mathbf{A}^{N-1})^T\mathbf{C}^T]^T
$$

Therefore, from the lemma of the previous section, it follows that for $N \geq n$ the matrix on the left is of rank n if and only if the observability matrix $\mathbf{S}$ is of rank n. Therefore, if an initial state can be uniquely determined from a finite number of output observations, it can be so determined from just $\mathbf{y}(0), \mathbf{y}(1), \ldots, \mathbf{y}(n-1)$. ∎

As a simple application, one can find that the observability matrix for the system of Fig. 8.13 is

$$
S = \begin{bmatrix} 0 & 1 \\ 0 & a_1 \end{bmatrix}
$$

Since this is not of full rank, it follows, as deduced earlier, that the system is not completely observable.

For systems with a single output and which have a diagonal form, the above result has a natural interpretation in terms of connections of the outputs to the individual subsystems. Referring to Fig. 8.14, the system is completely observable if and only if the λ_i's are distinct and the c_i's are all nonzero. This

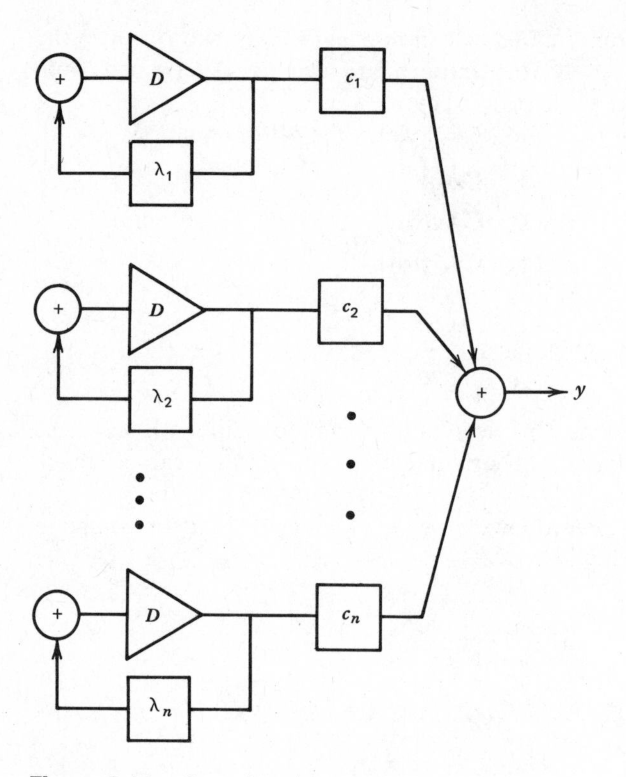

Figure 8.14. Complete observability for diagonal system.

means that for complete observability the output must be connected to each of the first-order subsystems (in the diagonal form) and it means that no two subsystems can be identical; for otherwise it would be impossible, on the basis of output observations alone, to distinguish responses of the two subsystems.

Observability for Continuous-Time Systems

For continuous-time systems the development exactly parallels what one would expect.

Definition. A system

$$\begin{aligned}\dot{\mathbf{x}}(t)&=\mathbf{A}\mathbf{x}(t)\\ \mathbf{y}(t)&=\mathbf{C}\mathbf{x}(t)\end{aligned} \tag{8-50}$$

is *completely observable* if there is a $t_1>0$ such that knowledge of $\mathbf{y}(t)$, for all t, $0\leq t\leq t_1$ is sufficient to determine $\mathbf{x}(0)$.

The statement of the following theorem should not be unexpected; we leave its proof to the reader.

Theorem 2. *The system* (8-50) *is completely observable if and only if the observability matrix*

$$\mathbf{S} = \begin{bmatrix} \mathbf{C} \\ \mathbf{CA} \\ \mathbf{CA}^2 \\ \cdot \\ \cdot \\ \cdot \\ \mathbf{CA}^{n-1} \end{bmatrix}$$

has rank n.

The concepts of complete controllability and complete observability should be viewed from both a mathematical and a practical perspective. In terms of the formal (mathematical) development of modern control theory, they are fundamental notions, and many techniques of control design rely on satisfaction of these conditions. The concepts are important from a practical viewpoint as well (because of their basis in theory), but it is rare that in specific practical applications controllability or observability represent cloudy issues that must be laboriously resolved. Usually, the context, or one's intuitive knowledge of the system, makes it clear whether the controllability and observability conditions are satisfied.

A good analogy is the concept of nonsingularity of a square matrix. The concept of nonsingularity is vitally important to the theory of linear algebra, and many analytical techniques are based on an assumption of nonsingularity. Nevertheless, in a given problem context, if a matrix is singular, there is usually a good reason for it, deriving from the general character of the problem; and the good analyst is not surprised. It is similar with controllability and observability. The good analyst rarely needs to go through the formal calculations.

8.8 CANONICAL FORMS

The concepts of controllability and observability provide a strong link between the state vector (matrix) description of linear systems and the (scalar) transfer function description, for it is these concepts that relate the input and output structure to the internal state mechanism. The linkage is displayed most directly by converting the state vector description to one of several available *canonical forms.*

These particular canonical forms are based on the *companion form* of matrices. Companion form matrices were used before, in Chapter 5, to convert high-order scalar difference equations to state variable form, and their use in this section is an extension of that original use.

The companion canonical forms provide valuable insight into the structural relations of a system; but, even beyond that, these canonical forms have significant utility. In the development of control theory it is often most convenient to work with systems expressed in companion form, just as in earlier chapters the diagonal form was often used. Moreover, because the companion canonical forms are relatively easy to compute (as compared to the diagonal form), they are often used for system representation in the course of a sophisticated analysis or control design.

The development of canonical forms applies identically to discrete-time and continuous-time systems since the canonical forms are really transformations on the matrix $\mathbf{A}$, the input matrix $\mathbf{B}$, and output matrix $\mathbf{C}$. For convenience we work with continuous-time systems. Also, for simplicity our development is restricted to systems with a single input and a single output.

There are essentially two classes of canonical forms. One class is based on complete controllability of the input, the other is based on complete observability of the output. The two classes are duals.

Controllability Canonical Forms

Suppose the system

$$\dot{\mathbf{x}}(t)=\mathbf{A}\mathbf{x}(t)+\mathbf{b}u(t) \tag{8-51}$$

is completely controllable. This is equivalent to nonsingularity of the controllability matrix

$$\mathbf{M}=[\mathbf{b}, \mathbf{A}\mathbf{b}, \mathbf{A}^2\mathbf{b}, \ldots, \mathbf{A}^{n-1}\mathbf{b}] \tag{8-52}$$

Therefore, given complete controllability, the controllability matrix can be used to define a change of variable in (8-51). Indeed let us set

$$\mathbf{x}(t)=\mathbf{M}\mathbf{z}(t) \tag{8-53}$$

This transforms (8-51) to the form

$$\dot{\mathbf{z}}(t)=\mathbf{M}^{-1}\mathbf{A}\mathbf{M}\mathbf{z}(t)+\mathbf{M}^{-1}\mathbf{b}u(t) \tag{8-54}$$

The system (8-54) has an especially simple structure, which we can easily derive. Let $\bar{\mathbf{A}}=\mathbf{M}^{-1}\mathbf{A}\mathbf{M}$, $\bar{\mathbf{b}}=\mathbf{M}^{-1}\mathbf{b}$. We find immediately that

$$\bar{\mathbf{b}}=\begin{bmatrix}1\\0\\ \cdot\\ \cdot\\ \cdot\\0\end{bmatrix}$$

because $\mathbf{M}^{-1}$ is a matrix whose rows are each orthogonal to $\mathbf{b}$; except the first row, and that row when multiplied by $\mathbf{b}$ yields 1. The matrix $\bar{\mathbf{A}}$ can be expressed as

$$\bar{\mathbf{A}} = \mathbf{M}^{-1}[\mathbf{Ab}, \mathbf{A}^2\mathbf{b}, \ldots, \mathbf{A}^n\mathbf{b}]$$

Each row of $\mathbf{M}^{-1}$ is orthogonal to all columns $\mathbf{A}^k\mathbf{b}$ in $\bar{\mathbf{A}}$, except for two. Consider, specifically, the ith row of $\mathbf{M}^{-1}$. It is orthogonal to $\mathbf{A}^k\mathbf{b}$ for all k, $0 \le k \le n$, except for $k = i-1$ and $k = n$. For $k = i-1$, the product is 1. For $k = n$, it is some, as yet unknown, value that we denote by $-a_{i-1}$. Therefore, we conclude that $\bar{\mathbf{A}}$ has the structure of a companion matrix, and the system (8-54) is in the canonical form

$$\dot{\mathbf{z}}(t) = \begin{bmatrix} 0 & 0 & 0 & \cdots & -a_0 \\ 1 & 0 & 0 & \cdots & -a_1 \\ 0 & 1 & 0 & \cdots & -a_2 \\ \cdot & & & & \\ \cdot & & & & \\ \cdot & & & & \\ 0 & 0 & 0 & \cdot\cdot 1 & -a_{n-1} \end{bmatrix} \mathbf{z}(t) + \begin{bmatrix} 1 \\ 0 \\ 0 \\ \cdot \\ \cdot \\ \cdot \\ 0 \end{bmatrix} u(t) \tag{8-55}$$

The constants $a_0, a_1, \ldots, a_{n-1}$ in (8-55) are the negatives of the values of the product of the individual rows of $\mathbf{M}^{-1}$ with the vector $\mathbf{A}^n\mathbf{b}$. However, although it might at first appear that these constants possibly depend on the vector $\mathbf{b}$, they actually do not. They are the coefficients of the characteristic polynomial of $\mathbf{A}$. That is, when the characteristic polynomial of $\mathbf{A}$ is written with its leading coefficient as $+1$, it is

$$p(\lambda) = \lambda^n + a_{n-1}\lambda^{n-1} + \cdots + a_1\lambda + a_0 \tag{8-56}$$

where the a_i's are those in the companion matrix $\bar{\mathbf{A}}$. This is most easily verified by simply calculating the characteristic polynomial of $\bar{\mathbf{A}}$, a calculation that has been carried out before (see Problem 3, Chapter 5), although the reader may wish to verify this again at this point. This shows that the characteristic polynomial of $\bar{\mathbf{A}}$ is given by (8–56). Then, since $\mathbf{A}$ and $\bar{\mathbf{A}}$ are similar, and therefore have the same characteristic polynomial (Problem 21, Chapter 3), it follows that the characteristic polynomial of $\mathbf{A}$ is given by (8-56).

If the system (8-51) also has a linear output structure, the output will have the representation

$$y(t) = \mathbf{c}^T\mathbf{z}(t) + du(t) \tag{8-57}$$

in the new coordinate system. The vector $\mathbf{c}^T$ in general does not possess any special structure in this canonical form. The diagram of a system in this canonical form is shown in Fig. 8.15. Any completely controllable single-input, single-output system can be put in the canonical form represented in Fig. 8.15 by a change of variables.

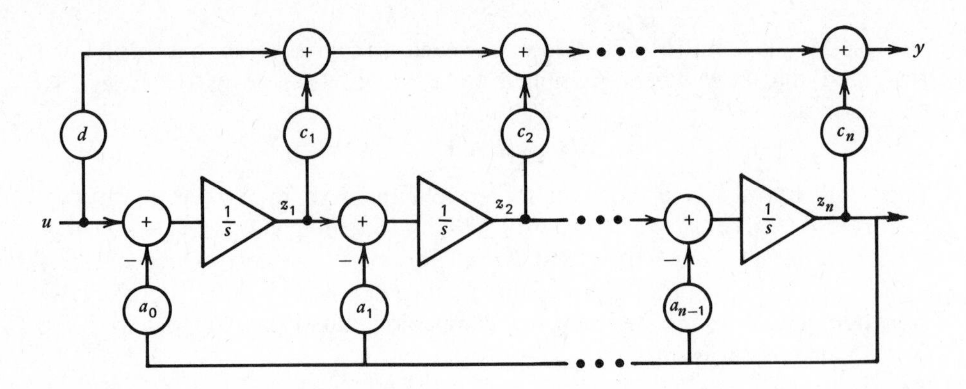

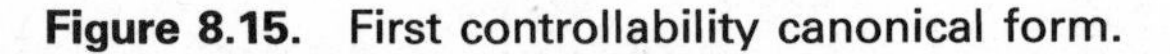

Figure 8.15. First controllability canonical form.

A converse result is also of interest. Given the structure of Fig. 8.15, suppose the coefficient values $a_0, a_1, \ldots, a_{n-1}$ and $c_1, c_2, \ldots, c_n$ are assigned arbitrary values. It can be easily shown that the resulting system is always completely controllable. Thus the structure shown in Fig. 8.15 is a complete characterization of completely controllable systems.

A second canonical form for a completely controllable system is obtained by selecting a different coordinate change. As before, corresponding to the system (8-51), let

$$\mathbf{M} = [\mathbf{b}, \mathbf{Ab}, \ldots, \mathbf{A}^{n-1}\mathbf{b}]$$

Suppose that $\mathbf{M}^{-1}$ is expressed in terms of its rows as

$$\mathbf{M}^{-1} = \begin{bmatrix} \mathbf{e}_1^T \\ \mathbf{e}_2^T \\ \cdot \\ \cdot \\ \cdot \\ \mathbf{e}_n^T \end{bmatrix} \tag{8-58}$$

We can show that the row vectors $\mathbf{e}_n^T, \mathbf{e}_n^T\mathbf{A}, \ldots, \mathbf{e}_n^T\mathbf{A}^{n-1}$ generated by the last row of $\mathbf{M}^{-1}$ are linearly independent. To prove this, suppose there were constants $\alpha_1, \alpha_2, \ldots, \alpha_n$ such that

$$\alpha_1\mathbf{e}_n^T + \alpha_2\mathbf{e}_n^T\mathbf{A} + \cdots + \alpha_n\mathbf{e}_n^T\mathbf{A}^{n-1} = \mathbf{0} \tag{8-59}$$

Multiplication on the right by $\mathbf{b}$ gives

$$\alpha_1\mathbf{e}_n^T\mathbf{b} + \alpha_2\mathbf{e}_n^T\mathbf{Ab} + \cdots + \alpha_n\mathbf{e}_n^T\mathbf{A}^{n-1}\mathbf{b} = 0 \tag{8-60}$$

However, in view of the definition of $\mathbf{e}_n^T$, the first $n-1$ terms of this equation vanish identically. The equation therefore implies that $\alpha_n = 0$. Now that it is known that $\alpha_n = 0$, multiplication of (8-59) by $\mathbf{Ab}$ yields $\alpha_{n-1} = 0$. Continuing in this way, we conclude that $\alpha_i = 0$ for all $i = 1, 2, \ldots, n$. Thus, the vectors are linearly independent.

Starting with the completely controllable combination $\mathbf{A}$, $\mathbf{b}$, we have deduced a corresponding row vector $\mathbf{e}_n^T$ such that the combination $\mathbf{e}_n^T$, $\mathbf{A}$ is completely observable. We now use this combination to define an appropriate change of coordinates, much the way that the $\mathbf{A}$, $\mathbf{b}$ combination was used above.

Let

$$\mathbf{P} = \begin{bmatrix} \mathbf{e}_n^T \\ \mathbf{e}_n^T\mathbf{A} \\ \cdot \\ \cdot \\ \cdot \\ \mathbf{e}_n^T\mathbf{A}^{n-1} \end{bmatrix} \tag{8-61}$$

and define the change of coordinates for the system (8-51) by

$$\mathbf{z}(t) = \mathbf{P}\mathbf{x}(t) \tag{8-62}$$

This transforms the original system to

$$\dot{\mathbf{z}}(t) = \mathbf{P}\mathbf{A}\mathbf{P}^{-1}\mathbf{z}(t) + \mathbf{P}\mathbf{b}u(t) \tag{8-63}$$

Now, following a procedure similar to that employed before, we find that (8-63) has the explicit structure

$$\dot{\mathbf{z}}(t) = \begin{bmatrix} 0 & 1 & 0 & \cdots & 0 \\ 0 & 0 & 1 & \cdots & 0 \\ \cdot & & & & \cdot \\ \cdot & & & & \cdot \\ \cdot & & & & \cdot \\ 0 & 0 & 0 & \cdots & 1 \\ -a_0 & -a_1 & -a_2 & \cdots & -a_{n-1} \end{bmatrix} \mathbf{z}(t) + \begin{bmatrix} 0 \\ 0 \\ \cdot \\ \cdot \\ \cdot \\ 0 \\ 1 \end{bmatrix} u(t) \tag{8-64}$$

$$y(t) = [c_1 \quad c_2 \quad c_3 \quad \cdots \quad c_n]\, \mathbf{z}(t) + du(t)$$

The values of the a_i's are the same as in the first form although the c_i's are not. The diagram of a system in this canonical form, having an arbitrary single output, is shown in Fig. 8.16.

Observability Canonical Forms

Exactly parallel procedures can be worked out for a completely observable single-output system

$$\begin{aligned} \dot{\mathbf{x}}(t) &= \mathbf{A}\mathbf{x}(t) + \mathbf{b}u(t) \\ y(t) &= \mathbf{c}^T\mathbf{x}(t) + du(t) \end{aligned} \tag{8-65}$$

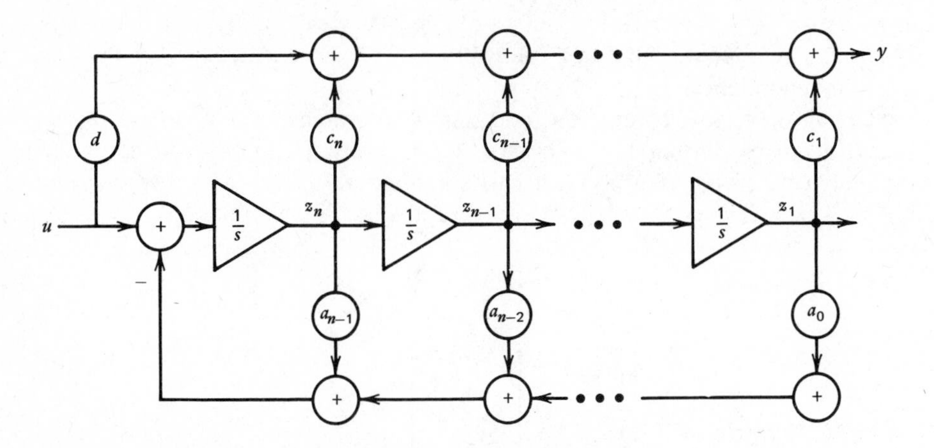

Figure 8.16. Second controllability canonical form.

These canonical forms and their underlying change of variables are based on the nonsingularity of the observability matrix

$$\mathbf{S} = \begin{bmatrix} \mathbf{c}^T \\ \mathbf{c}^T\mathbf{A} \\ \cdot \\ \cdot \\ \cdot \\ \mathbf{c}^T\mathbf{A}^{n-1} \end{bmatrix} \tag{8-66}$$

The derivations of these forms are entirely analogous to those for the controllability forms. The first observability canonical form is derived from the change of variable $\mathbf{z}(t) = \mathbf{S}\mathbf{x}(t)$. The second is based on the fact that the last column of $\mathbf{S}^{-1}$ together with $\mathbf{A}$ forms a completely controllable combination, and thus a nonsingular matrix $\mathbf{P}$ can be derived with columns equal to powers of $\mathbf{A}$ times this column. The resulting two canonical forms are, respectively,

$$\dot{\mathbf{z}}(t) = \begin{bmatrix} 0 & 1 & 0 & \dots & 0 \\ 0 & 0 & 1 & \dots & 0 \\ \cdot & & & & \cdot \\ \cdot & & & & \cdot \\ \cdot & & & & \cdot \\ 0 & 0 & 0 & \dots & 1 \\ -a_0 & -a_1 & -a_2 & \dots & -a_{n-1} \end{bmatrix} \mathbf{z}(t) + \begin{bmatrix} b_1 \\ b_2 \\ \cdot \\ \cdot \\ \cdot \\ b_{n-1} \\ b_n \end{bmatrix} u(t) \tag{8-67}$$

$$y(t) = [1 \quad 0 \quad 0 \quad \dots \quad 0]\ \mathbf{z}(t) + du(t)$$

and

$$\dot{\mathbf{z}}(t)=\begin{bmatrix} 0 & 0 & \dots & 0 & -a_0 \\ 1 & 0 & \dots & 0 & -a_1 \\ 0 & 1 & \dots & 0 & -a_2 \\ \cdot & & & & \cdot \\ \cdot & & & & \cdot \\ \cdot & & & & \cdot \\ 0 & 0 & \dots & 1 & -a_{n-1} \end{bmatrix}\mathbf{z}(t)+\begin{bmatrix} b_1 \\ b_2 \\ b_3 \\ \cdot \\ \cdot \\ \cdot \\ b_n \end{bmatrix}u(t) \tag{8-68}$$

$$y(t)=[0 \quad 0 \quad \dots \quad 0 \quad 1]\ \mathbf{z}(t)+du(t)$$

Again, the values of the a_i's are the same as in the earlier canonical forms. However, the b_i's are not necessarily the same as those in (8-65) and not the same in the two canonical forms.

Relation to Transfer Functions

The canonical forms presented in this section can be used to obtain state space representations of a given rational transfer function. This brings out the strong connection between transfer functions, controllability and observability, and canonical forms.

As an example, consider the system in controllability canonical form (8-64) represented in Fig. 8.16. The corresponding transfer function is

$$H(s)=\mathbf{c}^T[s\mathbf{I}-\mathbf{A}]^{-1}\mathbf{b}+d$$

The corresponding rational function can be derived by straightforward (but somewhat lengthy) substitution of the special forms for **A** and **b**. Another way, which we employ, is to apply the combination properties of transfer functions to the diagram of Fig. 8.16. For this purpose let us denote by $H_k(s)$ the transfer function from the input u to the state variable z_k. It is immediately clear that $H_{k-1}(s)=(1/s)H_k(s)$. Therefore,

$$s^{n-k}H_k(s)=H_n(s) \tag{8-69}$$

The transfer function $H_n(s)$ can be decomposed into the part that comes directly from u and the part that comes indirectly from the state variables. Thus,

$$H_n(s)=\frac{1}{s}\left(1-\sum_{k=1}^{n} a_{k-1}H_k(s)\right) \tag{8-70}$$

In view of (8-69) this becomes

$$H_n(s)=\frac{1}{s}\left(1-\sum_{k=1}^{n} \frac{a_{k-1}}{s^{n-k}}H_n(s)\right)$$

Multiplying by s^n leads to

$$[s^n + a_{n-1}s^{n-1} + \cdots + a_0]H_n(s) = s^{n-1}$$

Thus,

$$H_n(s) = \frac{s^{n-1}}{s^n + a_{n-1}s^{n-1} + \cdots + a_0} \tag{8-71}$$

Once this explicit form is known, the transfer function from u to y can be immediately deduced using (8-69). Thus,

$$H(s) = \frac{c_n s^{n-1} + c_{n-1}s^{n-2} + \cdots + c_1}{s^n + a_{n-1}s^{n-1} + \cdots + a_0} + d \tag{8-72}$$

From this result it follows that by appropriate choice of the parameters $a_0, a_1, \ldots, a_{n-1}$ and $c_1, c_2, \ldots, c_n$ the transfer function can be made equal to any proper rational function. Similar correspondences apply to the other canonical forms. Therefore, canonical forms provide a simple and direct solution of the problem of constructing a state vector representation of a system for which only the transfer function is specified.

8.9 FEEDBACK

In rough terms, the task of *control* is to manipulate the available inputs of a dynamic system to cause the system to behave in a manner more desirable than it otherwise would. Control can be designed for many purposes, however, and it can be applied in many ways. This section briefly discusses the difference between *open-loop control* and *feedback control* and illustrates some of the advantages of feedback control. The primary theoretical development in the section is the Eigenvalue Placement Theorem, which is an important achievement of modern control theory.

Open-loop Control

In *open-loop control* the input function is generated by some process external to the system itself, and then is applied to the system. The input might be generated by any of several means: by analysis, by some physical device, or by some random phenomenon. A specific input might be generated by analysis, for example, when designing a ten-year national immigration policy on the basis of a current census, or when developing the yearly production plan for a company. The control input might be repeatedly generated by a physical device when directing a physical process or machine. For example, in a heating system the furnace might be programmed to go on and off in cycles of fixed duration. In any case, the defining feature of open-loop control is that the input function is determined completely by an external process.

Closed-loop Control (Feedback)

In *closed-loop control* the input is determined on a continuing basis by the behavior of the system itself, as expressed by the behavior of the outputs. This kind of control is *feedback* control since the outputs are fed back (in perhaps modified form) to the input.

There are many common and familiar examples of feedback control. One is the thermostatic home heating system. The output of the heating system is the temperature in the house, which is measured by the thermostat. When the temperature falls below a certain level the furnace is turned on; and when the temperature rises above a given level the furnace is turned off. This operation represents feedback from the output to the input of the heating system. It contrasts sharply with the programmed furnace, since now the duration of the heating cycles depends on system behavior.

Perhaps one of the earliest examples of feedback control is a water clock believed to be built by Ktesibios in Alexandria in the third century B.C. The principle is still used today. A sketch of the water clock is shown in Fig. 8.17. In order to maintain a constant flow rate into the main tank of the clock, the water level in a regulating tank is held nearly constant. This constant level is achieved by a float valve, which is essentially a feedback mechanism. Water from an external supply enters the regulating tank through a small pipe. A ball floats on the surface of the water in the tank below the pipe opening. When the water level rises it forces the ball to tighten against the pipe opening, reducing the input supply rate. When the level drops, the input rate increases.

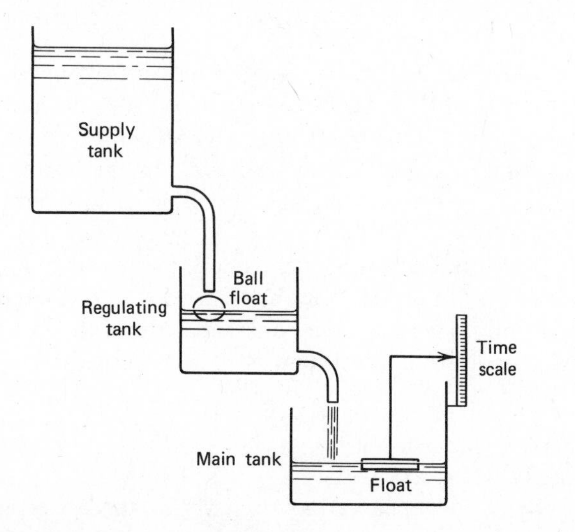

Figure 8.17. Water clock.

There are many reasons why feedback or closed-loop control is often preferable to open-loop control. One reason is that a feedback rule is often simple, while a comparable open-loop scheme might require a fair amount of calculation and complex implementation. One could, for example, conceive of a sophisticated open-loop furnace control that on a programmed basis varied the cycle lengths during a 24-hour period to account for expected outside temperature variations. A thermostat, however, would be simpler and more effective. In general, a feedback control rule does the required computation as it goes along, and immediately implements the results.

Beyond the issue of implementation, feedback is often superior to open-loop control from a performance standpoint. Feedback can automatically adjust to unforeseen system changes or to unanticipated disturbance inputs. The heating system with thermostatic control adjusts automatically to varying external temperatures, the water clock regulating tank adjusts to changes in incoming water pressure, and feedback introduced in planning can correct for unforeseen events. This desirable characteristic, which we intuitively attribute to feedback control, is analyzed mathematically in terms of the general concepts of dynamic systems. In particular, the fact that feedback can rapidly adjust to changes is a manifestation of the fact that feedback can increase the stability of a system. Indeed a basic feature of feedback is that it can influence the characteristic polynomial of a system.

Example 1. Let us consider the effect of feedback applied to the very simple system

$$\dot{x}(t) = u(t)$$

This might, for example, represent the equation governing the water level in a closed tank, with $u(t)$ denoting the inflow. Suppose it is desired to bring the value of x from some initial value to a value x_0. This can be done, of course, by open-loop control using a suitable choice of $u(t)$. In this form, however, the system is only marginally stable, and any error or disturbance will be perpetuated indefinitely.

Now suppose that the input is controlled by feedback. In particular, suppose $u(t) = \alpha[x_0 - x(t)]$. For $x(t) < x_0$, but close to x_0, this might correspond to the flow through the float valve, the constant α depending on the pipe size and incoming water pressure. The system is then governed by

$$\dot{x}(t) = \alpha[x_0 - x(t)]$$

This system certainly yields $x(t) \rightarrow x_0$.

We note that by defining the variable $y = x - x_0$, the closed-loop system is governed by the simple first-order equation

$$\dot{y}(t) = -\alpha y(t)$$

The variable $y(t)$ is the *error*. Feedback has converted the original marginally stable system to one which is asymptotically stable.

Eigenvalue Placement Theorem

As stated above, an important objective of feedback (either implicitly or explicitly) is often to make the system somewhat more stable than it would be otherwise. It is natural, therefore, to ask how much influence feedback can really have on the eigenvalues of a system. A major result in this direction is the Eigenvalue Placement Theorem. This result states that if a system is completely controllable and if *all* state variables are available (as outputs), then by suitable direct feedback it is possible for the closed-loop system to have any desired characteristic polynomial. In a sense this provides another characterization of complete controllability. If the system is completely controllable, the input is connected to the internal dynamic structure in sufficient diversity so that feedback to the input can arbitrarily change the characteristic polynomial.

Eigenvalue Placement Theorem. *Let* $(\mathbf{A}, \mathbf{B})$ *be a completely controllable pair of real matrices; that is, the rank of* $[\mathbf{B}, \mathbf{AB}, \ldots, \mathbf{A}^{n-1}\mathbf{B}]$ *is n. Then given any nth-order real polynomial* $p(\lambda) = \lambda^n + a_{n-1}\lambda^{n-1} + \cdots + a_0$ *there is a real matrix* $\mathbf{C}$ *such that* $\mathbf{A} + \mathbf{BC}$ *has* $p(\lambda)$ *as its characteristic polynomial.*

Proof. We prove this result only for the case where $m = 1$, corresponding to a single-input system. In that case we put $\mathbf{B} = \mathbf{b}$ and seek an n-vector $\mathbf{c}^T$.

Since the pair $(\mathbf{A}, \mathbf{b})$ is completely controllable, we may, without loss of generality, assume that they are in a form corresponding to the second controllability canonical form of the previous section. Specifically,

$$\mathbf{A} = \begin{bmatrix} 0 & 1 & 0 & \ldots & 0 \\ 0 & 0 & 1 & \ldots & 0 \\ \cdot & & & & \\ \cdot & & & & \\ \cdot & & & & \\ 0 & 0 & 0 & \ldots & 1 \\ -a_0 & -a_1 & -a_2 & \ldots & -a_{n-1} \end{bmatrix} \qquad \mathbf{b} = \begin{bmatrix} 0 \\ 0 \\ \cdot \\ \cdot \\ \cdot \\ 0 \\ 1 \end{bmatrix}$$

Selection of $\mathbf{c}^T = [c_1 \quad c_2 \quad \cdots \quad c_n]$ gives

$$\mathbf{bc}^T = \begin{bmatrix} 0 & 0 & 0 & \ldots & 0 \\ 0 & 0 & 0 & \ldots & 0 \\ \cdot & & & & \\ \cdot & & & & \\ \cdot & & & & \\ 0 & 0 & 0 & \ldots & 0 \\ c_1 & c_2 & c_3 & \ldots & c_n \end{bmatrix}$$

Therefore, the matrix $\mathbf{A}+\mathbf{b}\mathbf{c}^T$ will have a characteristic polynomial with coefficients $a_{i-1}-c_i$, and these can be made arbitrary by appropriate choice of the c_i's. ▮

Example 2 (Stick Balancing). The stick balancing system of Example 4, Sect. 6, is unstable without control, but since it is completely controllable, it must be possible to select a feedback strategy that will stabilize it.

As a first attempt, one might try a control strategy based solely on $x(t)$; that is, one might set $u(t)=cx(t)$ for some values of c. However, this leads to the closed-loop system $\ddot{x}=g(1-c)x$ and the characteristic equation $\lambda^2=g(1-c)$. Hence, at best this system will be marginally stable.

To attain a suitable feedback strategy it is necessary to use both $x(t)$ and $v(t)$. Thus, setting $u(t)=c_1x(t)+c_2v(t)$, we obtain the closed-loop characteristic equation

$$\lambda^2+gc_2\lambda-g(1-c_1)=0$$

Clearly, as predicted by the Eigenvalue Placement Theorem, we have complete flexibility in the choice of the characteristic polynomial through selection of the two coefficients c_1 and c_2. If we decide to place both roots at -1, we select $c_1=1+1/g$, $c_2=2/g$. The resulting system is then asympototically stable.

8.10 OBSERVERS

Many sophisticated analytical procedures for control design are based on the assumption that the full state vector is available for measurement. These procedures specify the current input value as a function of the current value of the state vector—that is, the control is a static function of the state. Mathematically, of course, there is very good reason for this kind of control specification. The system evolves according to its state vector equations, and thus intelligent control, influencing future behavior, should be based on the current value of the state. A simple but important example of such a control procedure is embodied by the eigenvalue placement theorem presented in Sect. 8.9. In that case, the complete flexibility in specification of the characteristic polynomial assumes that all state variables can be measured.

In many systems of practical importance, however, the entire state vector is not available for measurement. In many physical systems, for example, measurements require the use of costly measurement devices and it may be unreasonable to measure all state variables. In large social or economic systems, measurements may require extensive surveys or complex record-keeping procedures. And, in some systems, certain components of the state vector correspond to inaccessible internal variables, which cannot be measured. In all these situations, control strategies must be based on the values of a subset of the state variables.

When faced with this rather common difficulty, there are two avenues of approach. The first is to look directly for new procedures that require fewer measurements—either restricting the choice of static feedback functions or developing more complex (dynamic) feedback processing procedures. The second (simpler) approach is to construct an approximation to the full state vector on the basis of available measurements. Any of the earlier static control procedures can then be implemented using this approximate state in place of the actual state. In this way the relatively simple and effective control procedures, which assume that the state is available, are applicable to more general situations.

We recall that a system is completely observable if by observation of the system outputs the value of the initial state can be deduced within a finite time period. In our earlier discussion of observability, the required calculation was treated rather indirectly. Within the present context, however, it is apparent that such calculations become a matter of practical significance. Effective control can be dependent on the results.

It is shown in this section that the state (or an approximation to it) can be conveniently computed by a device known as an *observer*. The observer is itself a linear dynamic system. Its input values are the values of measured outputs from the original system, and its state vector generates missing information about the state of the original system. The observer can be regarded as a dynamic device that, when connected to the available system outputs, generates the entire state.

A Trivial Observer

A trivial solution to the problem of estimating the state of a system is to build a copy of the original system. If, for example, the original system is

$$\dot{\mathbf{x}}(t) = \mathbf{A}\mathbf{x}(t) + \mathbf{B}\mathbf{u}(t)$$

The observer would be

$$\dot{\mathbf{z}}(t) = \mathbf{A}\mathbf{z}(t) + \mathbf{B}\mathbf{u}(t)$$

The inputs $\mathbf{u}(t)$ to the original system are controls that we supply, and hence they can be applied to the copy as well. Also, since the second system is a model, its state $\mathbf{z}(t)$ can be measured. If $\mathbf{z}(0) = \mathbf{x}(0)$, the model will follow the original system exactly. (See Fig. 8.18.)

The trouble with this technique is that errors do not die out quickly. Indeed it is easy to see that

$$[\dot{\mathbf{z}}(t) - \dot{\mathbf{x}}(t)] = \mathbf{A}[\mathbf{z}(t) - \mathbf{x}(t)]$$

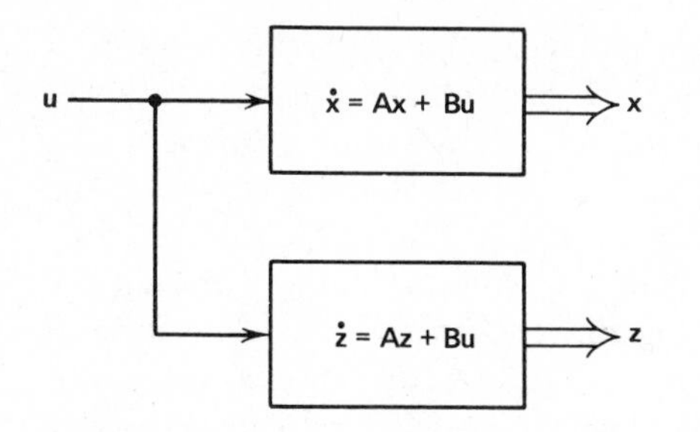

Figure 8.18. A system copy used as observer.

and hence the error in the estimate tends to zero only if the original system is stable—and then only at a speed determined by the eigenvalues of the original system. This is a serious limitation, which is overcome by the more general approach to observers.

Identity Observer

To begin the development of more general observers, consider the following system:

$$\begin{aligned}\dot{\mathbf{x}}(t) &= \mathbf{A}\mathbf{x}(t)+\mathbf{B}\mathbf{u}(t)\\ \mathbf{y}(t) &= \mathbf{C}\mathbf{x}(t)\end{aligned} \tag{8-73}$$

The system is assumed to be n dimensional, and the output vector is p dimensional.* The system is assumed to be completely observable.

We construct an observer for (8-73) of the form

$$\dot{\mathbf{z}}(t) = \mathbf{A}\mathbf{z}(t)+\mathbf{E}[\mathbf{y}(t)-\mathbf{C}\mathbf{z}(t)]+\mathbf{B}\mathbf{u}(t) \tag{8-74}$$

where the $n \times p$ matrix $\mathbf{E}$ is yet to be specified (see Fig. 8.19). The observer is an n-dimensional system with state vector $\mathbf{z}(t)$. The inputs to the observer are of two types. The first set consists of the measurements $\mathbf{y}(t)$ available from the original system. The second set is a copy of the inputs to the original system. These inputs are the control inputs, so they are available to us. Note that this observer is a generalization of the trivial observer discussed above. If $\mathbf{z}(t) = \mathbf{x}(t)$, then, since $\mathbf{y}(t) = \mathbf{C}\mathbf{x}(t)$, the observer (8-74) would collapse to a copy of the original system.

Using $\mathbf{y}(t) = \mathbf{C}\mathbf{x}(t)$ and subtracting (8-73) from (8-74), we find in general that

$$\dot{\mathbf{z}}(t)-\dot{\mathbf{x}}(t) = [\mathbf{A}-\mathbf{E}\mathbf{C}][\mathbf{z}(t)-\mathbf{x}(t)]$$

If the observer is initiated such that $\mathbf{z}(0) = \mathbf{x}(0)$, then it follows that $\mathbf{z}(t) = \mathbf{x}(t)$

* More generally, the output will have the form $\mathbf{y}(t) = \mathbf{C}\mathbf{x}(t)+\mathbf{D}\mathbf{u}(t)$. The additional term can be easily incorporated into our development.

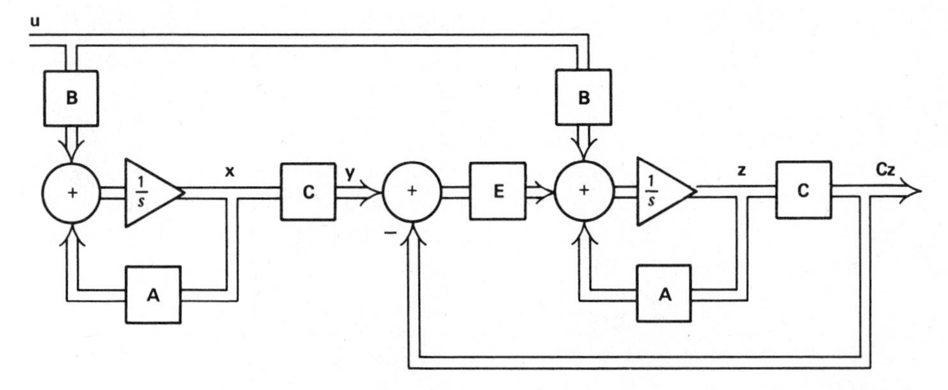

Figure 8.19. An identity observer.

for all $t>0$. The state of the observer tracks the state of the original system. An observer of this type is an *identity* observer since it tracks the entire state.

If $\mathbf{z}(0)\neq\mathbf{x}(0)$, then the *error vector* $\boldsymbol{\varepsilon}(t)=\mathbf{z}(t)-\mathbf{x}(t)$ is governed by the homogeneous system

$$\dot{\boldsymbol{\varepsilon}}(t)=[\mathbf{A}-\mathbf{EC}]\boldsymbol{\varepsilon}(t)$$

If the system matrix $\mathbf{A}-\mathbf{EC}$ is asymptotically stable, the error vector tends to zero. Indeed, the error tends to zero at a rate determined by the dominant eigenvalue of $\mathbf{A}-\mathbf{EC}$. The eigenvalues of this matrix are controlled by our choice of the matrix $\mathbf{E}$. A significant result is the following theorem.

Identity Observer Theorem. *Given a completely observable system* (8-73), *an identity observer of the form* (8-74) *can be constructed, and the coefficients of the characteristic polynomial of the observer can be selected arbitrarily.*

Proof. This follows from the Eigenvalue Placement Theorem. The pair $(\mathbf{A}^T, \mathbf{C}^T)$ is completely controllable. Thus, $\mathbf{E}$ can be selected so that $\mathbf{A}^T-\mathbf{C}^T\mathbf{E}^T$ has an arbitrary characteristic polynomial, and this is the same as the characteristic polynomial of $\mathbf{A}-\mathbf{EC}$. ∎

Example 1 (A First-Order System). Consider the first-order system

$$\dot{x}(t)=ax(t)$$

with $x(t)$ as an available output. There is, of course, no need to construct an observer for this system, since the (one-dimensional) state is assumed to be an output. However, consideration of this simple system should help relate the above observer result to the basic principles of dynamic systems.

Suppose that we decide to construct an observer having a root of $b\neq a$. The general form of the observer must be

$$\dot{z}(t)=az(t)+e[x(t)-z(t)]$$

In order that the root of the observer be equal to b, we set $e=a-b$. Thus, the observer is defined by

$$\dot{z}(t)=bz(t)+(a-b)x(t) \tag{8-75}$$

If $z(0)=x(0)$, then $z(t)=x(t)$ for all $t>0$.

Now let us interpret this result in terms of elementary properties of differential equations. We know explicitly that $x(t)=x(0)e^{at}$. When this is substituted into the observer (8-75), the resulting solution $z(t)$ will consist of two exponential terms—one with e^{at} and the other with e^{bt}. Explicitly

$$z(t)=x(0)e^{at}+[z(0)-x(0)]e^{bt}$$

Thus, if $z(0)$ is properly chosen, only the exponential associated with the input to the observer will appear in the solution.

This first-order example illustrates the general mechanism of observers. With zero input, the outputs of the original system consist of various exponential terms (perhaps multiplied by a power of t.) These exponentials serve as inputs to the observer. The state variables of the observer will, accordingly, each consist of a sum of exponential terms; some generated internally and some being passed through from the input. By properly arranging the structure and the initial condition of the observer, it can be made to follow the exponential terms of the original system. If the original system has a nonzero input, its effect on the observer is cancelled out by suitably applying this same input to the observer.

Reduced-Order Observers

The identity observer described above possesses a certain degree of redundancy. It reconstructs all n state variables of the original system even though p of these variables—the output variables—are already known precisely. It seems plausible that by eliminating this redundancy an observer could be devised with order equal to $n-p$ rather than n; the full state of the original system being obtained from the $n-p$ state variables of the observer together with the p measurements. This is indeed possible, as explained below.

Let us again consider the completely observable system

$$\begin{aligned}\dot{\mathbf{x}}(t)&=\mathbf{A}\mathbf{x}(t)+\mathbf{B}\mathbf{u}(t)\\ \mathbf{y}(t)&=\mathbf{C}\mathbf{x}(t)\end{aligned} \tag{8-76}$$

We assume now that the $p\times n$ matrix $\mathbf{C}$ has rank p. This corresponds to the condition that the p measurements are linearly independent.

It is possible to transform (8-76) so that the output structure is particularly simple. Let $\mathbf{V}$ be any $(n-p)\times n$ matrix such that the matrix

$$\mathbf{P}=\begin{bmatrix}\mathbf{V}\\ \mathbf{C}\end{bmatrix} \tag{8-77}$$

is nonsingular. It is possible to find such a $\mathbf{V}$ since $\mathbf{C}$ has rank p. Now introduce the change of variable

$$\bar{\mathbf{x}}(t)=\mathbf{P}\mathbf{x}(t) \tag{8-78}$$

The new state vector $\bar{\mathbf{x}}(t)$ can be partitioned as

$$\bar{\mathbf{x}}(t)=\begin{bmatrix}\mathbf{w}(t)\\ \mathbf{y}(t)\end{bmatrix} \tag{8-79}$$

where $\mathbf{w}(t)$ is $n-p$ dimensional and $\mathbf{y}(t)$ is the p-dimensional vector of outputs. In this form the output variables are equal to the last p state variables.

We assume that the system has been put in the special form indicated above. Specifically, after transforming the system matrices to reflect the change of variable (8-78), the system can itself be written in partitioned form

$$\begin{bmatrix}\dot{\mathbf{w}}(t)\\ \dot{\mathbf{y}}(t)\end{bmatrix}=\begin{bmatrix}\mathbf{A}_{11} & \mathbf{A}_{12}\\ \mathbf{A}_{21} & \mathbf{A}_{22}\end{bmatrix}\begin{bmatrix}\mathbf{w}(t)\\ \mathbf{y}(t)\end{bmatrix}+\begin{bmatrix}\mathbf{B}_1\\ \mathbf{B}_2\end{bmatrix}\mathbf{u}(t) \tag{8-80}$$

We can extract from this system a subsystem of order $n-p$ that has as inputs the known quantities $\mathbf{u}(t)$ and $\mathbf{y}(t)$. Furthermore, we shall show that a subsystem with arbitrary characteristic polynomial can be selected.

Multiply the bottom part of (8-80) by an arbitrary $(n-p)\times p$ matrix $\mathbf{E}$ and subtract from the top part. This leads to

$$\begin{aligned}\dot{\mathbf{w}}(t)-\mathbf{E}\dot{\mathbf{y}}(t)&=(\mathbf{A}_{11}-\mathbf{E}\mathbf{A}_{21})\mathbf{w}(t)\\ &\quad+(\mathbf{A}_{12}-\mathbf{E}\mathbf{A}_{22})\mathbf{y}(t)+(\mathbf{B}_1-\mathbf{E}\mathbf{B}_2)\mathbf{u}(t)\end{aligned} \tag{8-81}$$

This can be rewritten as

$$\begin{aligned}\dot{\mathbf{w}}(t)-\mathbf{E}\dot{\mathbf{y}}(t)&=(\mathbf{A}_{11}-\mathbf{E}\mathbf{A}_{21})[\mathbf{w}(t)-\mathbf{E}\mathbf{y}(t)]\\ &\quad+[\mathbf{A}_{11}\mathbf{E}-\mathbf{E}\mathbf{A}_{21}\mathbf{E}+\mathbf{A}_{12}-\mathbf{E}\mathbf{A}_{22}]\mathbf{y}(t)\\ &\quad+(\mathbf{B}_1-\mathbf{E}\mathbf{B}_2)\mathbf{u}(t)\end{aligned} \tag{8-82}$$

Letting $\mathbf{v}(t)=\mathbf{w}(t)-\mathbf{E}\mathbf{y}(t)$, we have

$$\begin{aligned}\dot{\mathbf{v}}(t)&=(\mathbf{A}_{11}-\mathbf{E}\mathbf{A}_{21})\mathbf{v}(t)\\ &\quad+[\mathbf{A}_{11}\mathbf{E}-\mathbf{E}\mathbf{A}_{21}\mathbf{E}+\mathbf{A}_{12}-\mathbf{E}\mathbf{A}_{22}]\mathbf{y}(t)\\ &\quad+(\mathbf{B}_1-\mathbf{E}\mathbf{B}_2)\mathbf{u}(t)\end{aligned} \tag{8-83}$$

In this equation $\mathbf{v}(t)$ is unknown, while $\mathbf{y}(t)$ and $\mathbf{u}(t)$ serve as known inputs. We have no observations of the $(n-p)$-dimensional vector $\mathbf{v}(t)$. Thus, the observer is formed by merely copying the system (8-83). It is,

$$\begin{aligned}\dot{\mathbf{z}}(t)&=(\mathbf{A}_{11}-\mathbf{E}\mathbf{A}_{21})\mathbf{z}(t)\\ &\quad+[\mathbf{A}_{11}\mathbf{E}-\mathbf{E}\mathbf{A}_{21}\mathbf{E}+\mathbf{A}_{12}-\mathbf{E}\mathbf{A}_{22}]\mathbf{y}(t)\\ &\quad+(\mathbf{B}_1-\mathbf{E}\mathbf{B}_2)\mathbf{u}(t)\end{aligned} \tag{8-84}$$

This serves as an observer for (8-80). By subtracting (8-83) from (8-84), it follows immediately that

$$\dot{\mathbf{z}}(t)-\dot{\mathbf{v}}(t)=(\mathbf{A}_{11}-\mathbf{E}\mathbf{A}_{21})[\mathbf{z}(t)-\mathbf{v}(t)] \tag{8-85}$$

Thus, the state $\mathbf{z}(t)$ of the observer tends toward $\mathbf{v}(t)$ at a speed determined by the eigenvalues of the matrix $\mathbf{A}_{11}-\mathbf{E}\mathbf{A}_{21}$.

From the state vector $\mathbf{z}(t)$ of the observer, the original state of (8-80) is estimated by $\hat{\mathbf{w}}(t)$ and $\hat{\mathbf{y}}(t)$ determined by

$$\begin{aligned}\hat{\mathbf{w}}(t)&=\mathbf{z}(t)+\mathbf{E}\mathbf{y}(t)\\ \hat{\mathbf{y}}(t)&=\mathbf{y}(t)\end{aligned} \tag{8-86}$$

The state vector $\mathbf{z}(t)$ together with the original measurements $\mathbf{y}(t)$ provide enough information to construct an approximation to the state of the original system.

The effectiveness of the observer depends to a great extent on the location of the eigenvalues of the observer system matrix $\mathbf{A}_{11}-\mathbf{E}\mathbf{A}_{21}$. It can be shown (see Problem 16) that if the original system (8-80) [or equivalently (8-76)] is completely observable, the pair $\mathbf{A}_{21}$, $\mathbf{A}_{11}$ is completely observable. By the Eigenvalue Placement Theorem the coefficients of the characteristic polynomial of $\mathbf{A}_{11}-\mathbf{E}\mathbf{A}_{21}$ can be selected arbitrarily by proper selection of $\mathbf{E}$. Thus, the reduced-order observer can be constructed with the same kind of dynamic flexibility as the identity observer. We summarize by the following

Observer Theorem. *Given an nth-order completely observable system* (8-76) *with p linearly independent outputs, an observer of order $n-p$ can be constructed and the coefficients of the characteristic polynomial of the observer can be selected arbitrarily.*

Example 2 (Stick Balancing). Consider once again the stick balancing problem originally posed in Sect. 8.6. Let us suppose that only position can be measured (as might be the case in complex mechanical balancing problems). We shall construct a reduced-order observer to produce an approximation to the velocity.

The system is

$$\begin{bmatrix}\dot{v}(t)\\ \dot{x}(t)\end{bmatrix}=\begin{bmatrix}0 & g\\ 1 & 0\end{bmatrix}\begin{bmatrix}v(t)\\ x(t)\end{bmatrix}+g\begin{bmatrix}-1\\ 0\end{bmatrix}u(t) \tag{8-87}$$

where, to be consistent with the above development, we have reordered the state variables so that the measured variable $x(t)$ is the bottom part of the state vector. Applying (8-84) and (8-86), the observer has the general form

$$\dot{z}(t)=-ez(t)+(g-e^2)x(t)-gu(t) \tag{8-88a}$$

$$\hat{v}(t)=z(t)+ex(t) \tag{8-88b}$$

An estimate of the velocity can be obtained from the observer, which is first-order. The position is available by direct measurement, and hence an approximation to the whole state is available by use of the first-order observer. The eigenvalue of the observer is $-e$, which can be selected arbitrarily.

Eigenvalue Separation Theorem

There is an important result which shows that in some sense it is meaningful to separate the problem of observer design from that of control design. Suppose one has the system

$$\dot{\mathbf{x}}(t)=\mathbf{A}\mathbf{x}(t)+\mathbf{B}\mathbf{u}(t)$$
$$\mathbf{y}(t)=\mathbf{C}\mathbf{x}(t)$$

which is both completely controllable and completely observable. A feedback control law might be first designed under the assumption that the entire state were available for direct measurement. This would entail setting $\mathbf{u}(t)=\mathbf{K}\mathbf{x}(t)$ resulting in the closed-loop system

$$\dot{\mathbf{x}}(t)=(\mathbf{A}+\mathbf{B}\mathbf{K})\mathbf{x}(t)$$

The characteristic polynomial of $\mathbf{A}+\mathbf{B}\mathbf{K}$ can be selected arbitrarily.

Next an observer (either identity or reduced-order) can be designed that will generate an approximation $\hat{\mathbf{x}}(t)$ of the state $\mathbf{x}(t)$. The characteristic polynomial of the observer can also be selected arbitrarily.

Finally, the two design can be combined by setting the control input equal to $\mathbf{u}(t)=\mathbf{K}\hat{\mathbf{x}}(t)$—that is, using the observer generated state in place of the actual state. The resulting closed-loop system containing the observer is a somewhat complicated dynamic system. It can be shown, however, that the characteristic polynomial of this composite system is the product of the characteristic polynomial of the feedback system matrix $\mathbf{A}+\mathbf{B}\mathbf{K}$ and the characteristic polynomial of the observer. This means that the eigenvalues of the composite system are those of the feedback system together with those of the observer. In other words, insertion of an observer in a feedback system to replace unavailable measurements does not affect the eigenvalues of the feedback system; it merely adjoins its own eigenvalues.

We prove this eigenvalue separation theorem for the special case of an identity observer.

Eigenvalue Separation Theorem. *Consider the system*

$$\dot{\mathbf{x}}(t)=\mathbf{A}\mathbf{x}(t)+\mathbf{B}\mathbf{u}(t)$$
$$\mathbf{y}(t)=\mathbf{C}\mathbf{x}(t)$$

the identity observer

$$\dot{\mathbf{z}}(t)=(\mathbf{A}-\mathbf{E}\mathbf{C})\mathbf{z}(t)+\mathbf{E}\mathbf{y}(t)+\mathbf{B}\mathbf{u}(t)$$

and the control law

$$\mathbf{u}(t)=\mathbf{K}\mathbf{z}(t)$$

The characteristic polynomial of this composite is the product of the characteristic polynomials of $\mathbf{A}+\mathbf{BK}$ *and* $\mathbf{A}-\mathbf{EC}$.

Proof. Substituting the control law $\mathbf{u}(t)=\mathbf{K}\mathbf{z}(t)$ and the output equation $\mathbf{y}(t)=\mathbf{C}\mathbf{x}(t)$ yields the two differential equations

$$\dot{\mathbf{x}}(t)=\mathbf{A}\mathbf{x}(t)+\mathbf{BK}\mathbf{z}(t)$$
$$\dot{\mathbf{z}}(t)=(\mathbf{A}-\mathbf{EC})\mathbf{z}(t)+\mathbf{EC}\mathbf{x}(t)+\mathbf{BK}\mathbf{z}(t)$$

Defining $\boldsymbol{\varepsilon}(t)=\mathbf{z}(t)-\mathbf{x}(t)$ and subtracting the first equation from the second yields

$$\dot{\mathbf{x}}(t)=(\mathbf{A}+\mathbf{BK})\mathbf{x}(t)+\mathbf{BK}\boldsymbol{\varepsilon}(t)$$
$$\dot{\boldsymbol{\varepsilon}}(t)=(\mathbf{A}-\mathbf{EC})\boldsymbol{\varepsilon}(t)$$

The corresponding system matrix in partitioned form is

$$\begin{bmatrix} \mathbf{A}+\mathbf{BK} & \mathbf{BK} \\ \mathbf{0} & \mathbf{A}-\mathbf{EC} \end{bmatrix}$$

The result then follows from the fact that the characteristic polynomial of a matrix of this type is the product of the characteristic polynomials of the two blocks on the diagonal. (See Problem 22.) ∎

Example 3. Consider once again the stick balancing problem. In Example 2, Sect. 8.9, a feedback control scheme was devised that placed both closed-loop eigenvalues at -1. This scheme required both position and velocity measurements. In Example 2 of this section an observer was derived to estimate velocity when only position can be measured. Let us select $e=2$ so that the single eigenvalue of the observer is -2, somewhat more negative than the roots of the feedback control system designed earlier. We can then combine these designs and check whether the eigenvalue separation theorem holds using the reduced-order observer.

The corresponding complete control system, consisting of the original system, the control law, and the observer, is governed by the equations

$$\dot{x}(t)=v(t)$$
$$\dot{v}(t)=gx(t)-gu(t)$$
$$\dot{z}(t)=-2z(t)+(g-4)x(t)-gu(t)$$
$$gu(t)=(g+1)x(t)+2\hat{v}(t)$$
$$\hat{v}(t)=z(t)+2x(t)$$

Eliminating the static equations for $gu(t)$ and $\hat{v}(t)$ shows that the closed-loop system is governed by

$$\dot{x}(t) = v(t)$$

$$\dot{v}(t) = gx(t) - (g+1)x(t) - 2z(t) - 4x(t)$$

$$\dot{z}(t) = -2z(t) + (g-4)x(t) - (g+1)x(t) - 2z(t) - 4x(t)$$

or finally

$$\begin{bmatrix} \dot{x}(t) \\ \dot{v}(t) \\ \dot{z}(t) \end{bmatrix} = \begin{bmatrix} 0 & 1 & 0 \\ -5 & 0 & -2 \\ -9 & 0 & -4 \end{bmatrix} \begin{bmatrix} x(t) \\ v(t) \\ z(t) \end{bmatrix} \tag{8-89}$$

This system has characteristic equation

$$-\lambda^2(\lambda+4) + 18 - 5(\lambda+4) = 0$$

or, equivalently,

$$\lambda^3 + 4\lambda^2 + 5\lambda + 2 = 0$$

This can be factored as

$$(\lambda+1)^2(\lambda+2) = 0$$

Therefore, the roots of the overall system are -1, -1, -2 in agreement with the eigenvalue separation theorem.

8.11 PROBLEMS

1. Find a rational expression for the z-transform of the sequence $1, 4, 9, 16, \ldots$.

2. If $f(k)$ has a rational z-transform $F(z)$, find the z-transform of the sequence $g(k)$ defined by

$$g(k) = \sum_{j=0}^{k} f(j)$$

3. Show that the z-transforms of the sequences $f(k) = b^k \sin ak$ and $g(k) = b^k \cos ak$ are:

$$F(z) = \frac{bz \sin a}{z^2 - 2bz \cos a + b^2}$$

$$G(z) = \frac{z^2 - bz \cos a}{z^2 - 2bz \cos a + b^2}$$

4. *Partial Fractions.* The coefficients of a partial fraction expansion can be found in a simple straightforward way as follows. Suppose $F(z)$ is a reduced strictly proper rational function of degree n.

(a) Suppose the roots of the denominator of $F(z)$ are distinct. Then

$$F(z)=\frac{c_1}{z-z_1}+\frac{c_2}{z-z_2}+\cdots+\frac{c_n}{z-z_n}$$

Show that c_i can be evaluated as

$$c_i=\lim_{z\to z_i}(z-z_i)F(z)$$

(b) Suppose z_i is a root that is repeated m times. Then the partial fraction of $F(z)$ includes terms of the form

$$\frac{c_{i1}}{z-z_i}+\frac{c_{i2}}{(z-z_i)^2}+\cdots+\frac{c_{im}}{(z-z_i)^m}$$

Show that c_{ip}, $1\leq p\leq m$ can be evaluated as

$$c_{ip}=\lim_{z\to z_i}\frac{1}{(m-p)!}\frac{d^{m-p}}{dz^{m-p}}[(z-z_i)^m F(z)]$$

5. Find the inverse z-transform of

$$\frac{z(z-1)(z-2)}{(z^2+2z+2)(z^2+4z+4)}$$

(*Hint:* Use the results of Problems 3 and 4.)

6. A firm's inventory is governed by the equation

$$I(k+1)=\beta I(k)+P(k)-S(k)$$

where $I(k)$ = inventory at time k, $P(k)$ = production during time period k, $S(k)$ = sales during time period k, and $1-\beta$ = spoilage rate. Suppose the firm decides to produce as much during a given time period as was sold during the previous time period; that is, $P(k)=S(k-1)$. This gives

$$I(k+1)=\beta I(k)+S(k-1)-S(k)$$

(a) Using z-transforms, find $I(k)$ if $S(k)=\alpha^k$, $\alpha>1$, $k\geq 0$. [Assume $I(0)=2$, $S(-1)=0$.]

(b) Using z-transforms, find $I(k)$ if

$$S(k)=\begin{cases}1 & k \text{ even}\\ \frac{1}{2} & k \text{ odd}\end{cases} \qquad \text{where } k\geq 0$$

7. *Transfer Gain.* The transfer function $H(z)$ of a single-input, single-output discrete-time system can be viewed as a gain function. Assume that a is a number such that $H(a)$ is finite. Show that if the geometric sequence a^k is applied as input to the system, the same geometric series appears in the output multiplied by the gain $H(a)$ [that is, $H(a)a^k$ appears as an output term].

8. *Diagram Manipulation.* Find the transfer function from u to y for the systems represented by the diagrams of Fig. 8.20.

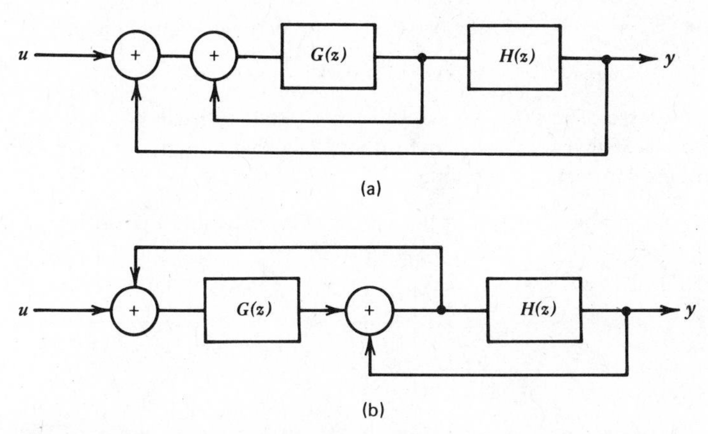

Figure 8.20. Interconnected systems. (a) Double feedback. (b) Interlaced feedback.

9. *State Space and Series Connections.* Consider the series arrangement of two single-input, single-output systems

$$\mathbf{x}_1(k+1) = \mathbf{A}_1\mathbf{x}_1(k) + \mathbf{b}_1 u(k)$$
$$v(k) = \mathbf{c}_1^T\mathbf{x}_1(k) + d_1 u(k)$$
$$\mathbf{x}_2(k+1) = \mathbf{A}_2\mathbf{x}_2(k) + \mathbf{b}_2 v(k)$$
$$y(k) = \mathbf{c}_2^T\mathbf{x}_2(k) + d_2 v(k)$$

Let $H_i(z) = \mathbf{c}_i^T(z\mathbf{I} - \mathbf{A}_i)^{-1}\mathbf{b}_i + d_i$, for $i = 1, 2$.

(a) Eliminate $v(k)$ and find the state space representation of the composite system.
(b) Show directly, by matrix manipulation, that the transfer function of the composite system is $H(z) = H_1(z)H_2(z)$.

10. Use Laplace transforms to solve each of the following differential equations, with initial conditions $y(0) = 0$, $\dot{y}(0) = 1$.

(a) $\dfrac{d^2y}{dt^2} + \dfrac{dy}{dt} - 6y = 1$

(b) $\dfrac{d^2y}{dt^2} - y = e^{-t}$

(c) $\dfrac{d^2y}{dt^2} + \dfrac{dy}{dt} + y = te^{-t}$

11. Show that two sticks (with masses concentrated at their tops) placed side by side a little distance apart on one hand can be simultaneously balanced in most cases. What conditions on the lengths and masses render the simultaneous balancing feat impossible?

*12. *The Finite-Settling-Time Problem.* Suppose the nth-order single-input system

$$\mathbf{x}(k+1)=\mathbf{A}\mathbf{x}(k)+\mathbf{b}u(k)$$

is completely controllable. Then it is known that any initial state can be driven to zero within n steps. We shall seek a linear feedback law that automatically drives the state to zero in the shortest possible time.

(a) Let $\mathbf{M}$ be the controllability matrix of the system. Let $\mathbf{e}_i^T$ denote the ith row of $\mathbf{M}^{-1}$; that is,

$$\mathbf{M}^{-1}=\begin{bmatrix}\mathbf{e}_1^T\\ \mathbf{e}_2^T\\ \cdot\\ \cdot\\ \cdot\\ \mathbf{e}_n^T\end{bmatrix}$$

Show that setting

$$u(k)=-\mathbf{e}_{n-k}^T\mathbf{A}^n\mathbf{x}(0) \qquad k=0,1,2,\ldots,n-1$$

will drive the state to zero in the shortest time.

(b) Define $\mathbf{c}^T=-\mathbf{e}_n^T\mathbf{A}^n$. Show that the feedback rule $u(k)=\mathbf{c}^T\mathbf{x}(k)$ will yield the same result as part (a).

(c) The closed-loop system is governed by

$$\mathbf{x}(k+1)=(\mathbf{A}+\mathbf{b}\mathbf{c}^T)\mathbf{x}(k)$$

Conclude that all eigenvalues of $\mathbf{A}+\mathbf{b}\mathbf{c}^T$ are zero. What is the Jordan form of $\mathbf{A}+\mathbf{b}\mathbf{c}^T$?

13. *Controllability.* For the discrete-time system

$$\mathbf{x}(k+1)=\mathbf{A}\mathbf{x}(k)+\mathbf{b}u(k)$$

where

$$\mathbf{A}=\begin{bmatrix}2 & 1\\ 0 & 2\end{bmatrix}$$

let

$$\mathbf{b}_1=\begin{bmatrix}0\\ 1\end{bmatrix} \qquad \mathbf{b}_2=\begin{bmatrix}1\\ 0\end{bmatrix}$$

(a) For each $\mathbf{b}$, determine if the system is controllable.

(b) For each $\mathbf{b}$ that results in a completely controllable system, find the shortest input sequence that drives the state to zero if

$$\mathbf{x}(0)=\begin{bmatrix}2\\ 1\end{bmatrix}$$

14. *Multiple Eigenvalues and Controllability.* Suppose the system

$$\dot{\mathbf{x}}(t)=\mathbf{A}\mathbf{x}(t)+\mathbf{b}u(t)$$

has all eigenvalues equal to λ. Show that this system is completely controllable if and only if the Jordan form of $\mathbf{A}$ has only a single chain and (in the same coordinate system) $\mathbf{b}$ has a nonzero component in the last component. Give a dynamic diagram interpretation of this result.

15. *Controllability and Feedback.* Show that feedback does not destroy complete controllability. Specifically, show that if the system

$$\mathbf{x}(k+1)=\mathbf{A}\mathbf{x}(k)+\mathbf{B}\mathbf{u}(k)$$

is completely controllable, then the system

$$\mathbf{x}(k+1)=\mathbf{A}\mathbf{x}(k)+\mathbf{B}\mathbf{C}\mathbf{x}(k)+\mathbf{B}\mathbf{u}(k)$$

is also completely controllable.
(*Hint:* Rather than checking the rank condition, apply the original definition of complete controllability.)

16. *Observability.* Suppose that the partitioned system

$$\begin{bmatrix}\dot{\mathbf{w}}(t)\\ \dot{\mathbf{y}}(t)\end{bmatrix}=\begin{bmatrix}\mathbf{A}_{11} & \mathbf{A}_{12}\\ \mathbf{A}_{21} & \mathbf{A}_{22}\end{bmatrix}\begin{bmatrix}\mathbf{w}(t)\\ \mathbf{y}(t)\end{bmatrix}$$

with output $\mathbf{y}(t)$ is completely observable. Show that the combination of $\mathbf{A}_{11}$ as system and $\mathbf{A}_{21}$ as output matrix is a completely observable pair. (*Hint:* You may find it simpler to prove complete controllability of the transposed combination.)

17. *Structure of Canonical Forms.* Show that the controllability canonical forms are completely controllable and the observability canonical forms are completely observable for all values of the parameters $a_0, a_1, \ldots, a_{n-1}$.

18. *Observability Canonical Forms.* Work out the detailed derivation of the two observability canonical forms.

19. Consider the two dynamic systems

S_1	S_2
$\dot{x}_1=x_2+u$	$\dot{x}_3=x_3+w$
$\dot{x}_2=-2x_1-3x_2$	$z=x_3$
$y=\alpha x_1+x_2$	

S_1 has state (x_1, x_2), control u, and output y. S_2 has state x_3, control w, and output z.

(a) Determine whether each system is controllable, observable, stable. (Note α is a parameter.)
(b) These two systems are connected in series, with $w=y$. The resulting system is called S_3. Determine whether it is controllable, observable, stable.
(c) The systems are now connected in a feedback configuration as shown in Fig. 8.21 to produce S_4. Determine whether S_4 is controllable, observable.

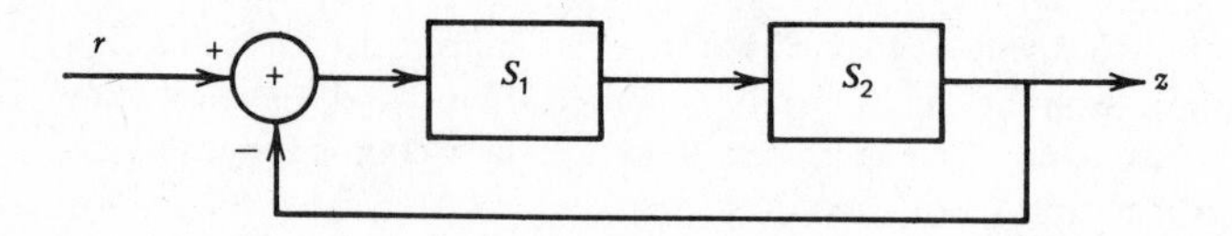

Figure 8.21. System for Problem 19.

20. *Observer Design.* Design an observer for the system shown in Fig. 8.22. The observer should be of second order, with both eigenvalues equal to -3.

21. *Uncontrollability of Observers.* Let S_1 be a linear time-invariant system with input $\mathbf{u}(t)$ and output $\mathbf{y}(t)$. An observer S_2 is connected appropriately to S_1. Show that the resulting composite system is always uncontrollable from the input $\mathbf{u}(t)$ (that is, it is not completely controllable.)

22. Assume that $\mathbf{A}$ and $\mathbf{C}$ are square matrices. Use Laplace's expansion to show that

$$\begin{vmatrix} \mathbf{I} & \mathbf{0} \\ \mathbf{0} & \mathbf{C} \end{vmatrix} = |\mathbf{C}|, \qquad \begin{vmatrix} \mathbf{A} & \mathbf{B} \\ \mathbf{0} & \mathbf{I} \end{vmatrix} = |\mathbf{A}|$$

Use this result to show that

$$\begin{vmatrix} \mathbf{A} & \mathbf{B} \\ \mathbf{0} & \mathbf{C} \end{vmatrix} = |\mathbf{A}||\mathbf{C}|.$$

23. Prove the Eigenvalue Separation Theorem for reduced-order observers.

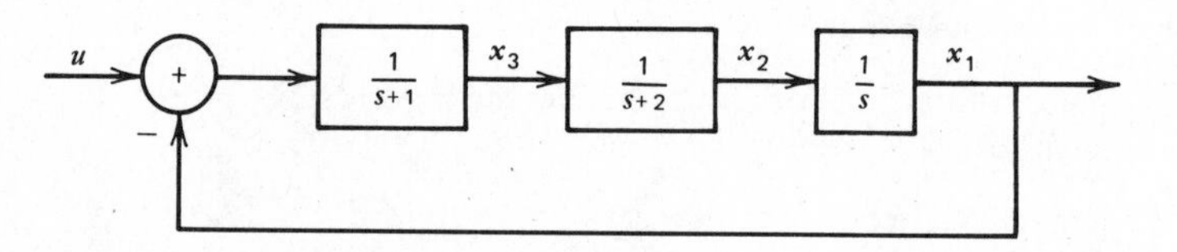

Figure 8.22 System for Problem 20.

NOTES AND REFERENCES

Sections 8.2–8.5. Transform methods have long been used to solve linear, constant-coefficient differential and difference equations. Actually, Laplace transform methods were developed first, since differential equations were generally used to represent dynamic phenomena. The books by Gardner and Barnes [G4] and Carslaw and Jaeger [C1] published in the early 1940s served as the classic references for many years. With the advent of digital techniques in the 1950s the z-transform also attained popularity. See Aseltine [A4]. A good deal of attention typically was devoted to issues of convergence in transform methods. The recent trend, however, is to view them more algebraically in terms of an indeterminant symbol rather than as a branch of complex variable theory. An example of this is Rugh [R7].

Sections 8.6–8.7. The concepts of controllability and observability are due to Kalman. (See, for example, [K1] which was highly instrumental in development of modern control theory.) Also see Gilbert [G6] for discussion of the relation of controllability to diagonal representations. For early work on canonical forms, see Kalman [K3].

Section 8.8. For multivariable (many inputs and outputs) versions of the canonical forms, see Luenberger [L8].

Section 8.9. We have only been able to hint at the general area of feedback and its many applications and ramifications. See, for example, Clark [C3]. The eigenvalue placement theorem for single-input systems follows directly from the single-input canonical forms. The more general result is closely tied to more general canonical forms. See Luenberger [L7], [L8] or, for a more explicit statement, Wonham [W6].

Section 8.10. For the original development of reduced-order observers refer to Luenberger [L6], [L7], and [L10]. The particular presentation at the end of this section (and the result of Problem 16) is based on a construction due to Gopinath [G10].

chapter 9.

Analysis of Nonlinear Systems

9.1 INTRODUCTION

The analysis of nonlinear systems is in some respects similar to that of linear systems, but in other respects it is quite different. The similarity derives from the fact that one of the principal techniques for analysis of nonlinear systems is to approximate or bound them by appropriate linear systems, and then use linear theory. The differences derive from the fact that entirely new types of behavior can arise in nonlinear systems, compared to linear systems, as is illustrated below. The analysis is also different because explicit solutions are rarely available for nonlinear systems, and thus behavioral characteristics must be inferred by more subtle methods. Nevertheless, accounting for both the similarities and differences with linear system analysis, there is a set of useful general principles for nonlinear systems analysis that provides coherence to this important topic.

We focus attention on nonlinear systems defined by either a set of difference equations or differential equations, as described in the beginning of Chapter 4. An nth-order discrete-time system has the following general form:

$$\begin{aligned}
x_1(k+1) &= f_1(x_1(k), x_2(k), \ldots, x_n(k), k) \\
x_2(k+1) &= f_2(x_1(k), x_2(k), \ldots, x_n(k), k) \\
&\vdots \\
x_n(k+1) &= f_n(x_1(k), x_2(k), \ldots, x_n(k), k)
\end{aligned} \tag{9-1}$$

for $k=0, 1, 2, \ldots$. Of course, we often find it convenient to express (9-1) in the vector form

$$\mathbf{x}(k+1)=\mathbf{f}(\mathbf{x}(k), k) \tag{9-2}$$

where $\mathbf{x}(k)$ is the n-dimensional *state* vector, and $\mathbf{f}$ is the n-dimensional vector function whose components are the f_i's.

Similarly, an nth-order continuous-time system has the following form:

$$\begin{aligned}
\dot{x}_1(t)&=f_1(x_1(t), x_2(t), \ldots, x_n(t), t)\\
\dot{x}_2(t)&=f_2(x_1(t), x_2(t), \ldots, x_n(t), t)\\
&\vdots\\
\dot{x}_n(t)&=f_n(x_1(t), x_2(t), \ldots, x_n(t), t)
\end{aligned} \tag{9-3}$$

which is often expressed in the vector form

$$\dot{\mathbf{x}}(t)=\mathbf{f}(\mathbf{x}(t), t) \tag{9-4}$$

The functions defining the systems (9-2) and (9-4) [or equivalently (9-1) and (9-3)] depend on both the state $\mathbf{x}$ and on time, k or t. These systems are thus, in general, time-varying. If the functions do not depend explicitly on time, the system is said to be *time-invariant.* Our attention is devoted mainly to such systems; that is, to systems that can be written as

$$\mathbf{x}(k+1)=\mathbf{f}(\mathbf{x}(k)) \tag{9-5}$$

or

$$\dot{\mathbf{x}}(t)=\mathbf{f}(\mathbf{x}(t)) \tag{9-6}$$

New Forms of Behavior

Certain even first-order nonlinear systems exhibit forms of behavior that are either somewhat different or drastically different than that obtainable in linear systems. The following two examples are classic.

Example 1 (The Logistic Curve). The standard first-order linear differential equation defining exponential growth is often modified to reflect the fact that, due to crowding, limited resources, or a variety of other reasons, growth cannot continue indefinitely. There are invariably restraining factors whose influence eventually dominates. A standard modified equation that accounts for the restraining influence is

$$\dot{x}(t)=a[1-x(t)/c]x(t)$$

where $a>0$, $c>0$. The term $a[1-x(t)/c]$ can be interpreted as the instantaneous growth rate. This rate decreases as the growth variable $x(t)$ increases toward its maximum possible level c.

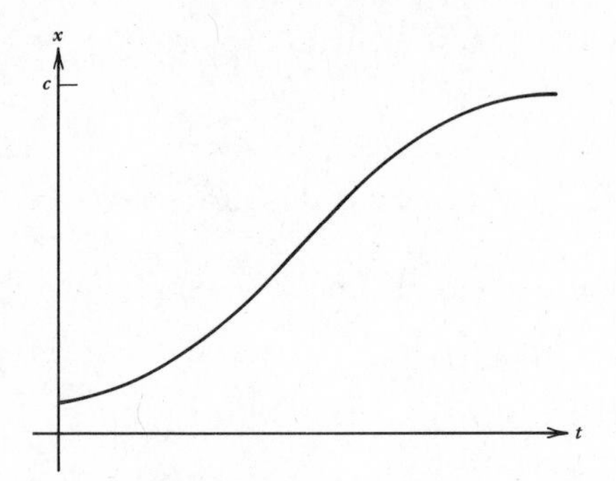

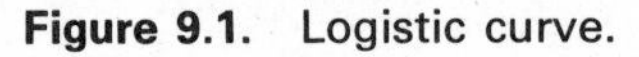
Figure 9.1. Logistic curve.

The solution to this equation (which is easy to obtain; see Problem 1) is

$$x(t)=\frac{c}{1+be^{-at}}$$

where $b>0$ is determined by the initial condition $x(0)<c$. The corresponding curve, shown in Fig. 9.1, is the *logistic curve.* It has a characteristic *S* shape that approximates exponential growth at the low end and saturation at the upper end.

This solution does not represent particularly unusual behavior. Indeed, the point of this example is that nonlinear terms can be meaningfully introduced into system equations in order to better represent natural phenomena. The general character of the resulting solution should be consistent with intuitive expectations. In this particular example, the linear model is modified to account for growth limitations, and the solution does exhibit the limitation property. Nonlinear modifications, then, are not necessarily obtuse, but in fact they may be quite consistent with one's intuitive understanding of both the system structure and the solution pattern.

Example 2 (Finite Escape Time). As an example of an entirely new form of behavior, not exhibited by linear systems, we consider a growth model where the growth rate *increases* with size. Specifically, we consider the differential equation

$$\dot{x}(t)=a(1+x(t)/c)x(t)$$

where $a>0$, $c>0$. If $x(0)>0$, the solution for $t>0$ is

$$x(t)=\frac{c}{be^{-at}-1}$$

where $b>1$. This solution is illustrated in Fig. 9.2. Its primary characteristic is that the variable $x(t)$ not only tends to infinity (as would be expected) but it gets there in finite time!

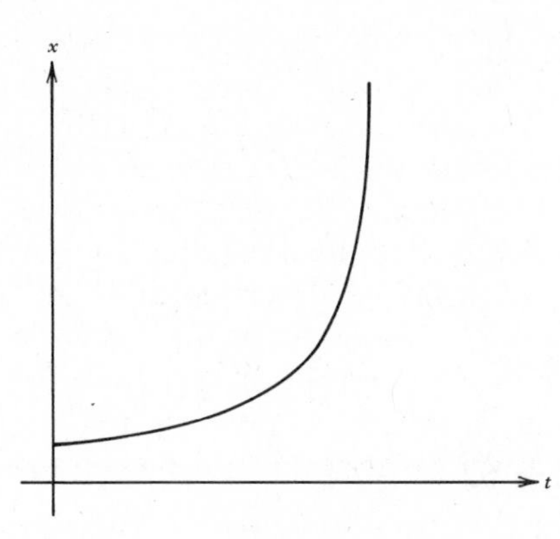

Figure 9.2. Finite escape time.

The Main Tool for Analysis of Nonlinear Systems

The objectives of analysis for nonlinear systems are similar to the objectives pursued when investigating complex linear systems. In general, one does not seek detailed solutions either in numerical or analytical form, but rather one seeks to characterize some aspect of system behavior. For example, one might ask whether there are equilibrium points, and whether they are stable. One seeks estimates of growth rates, estimates of behavior resulting from perturbations, and characterizations of limiting behavior. In nonlinear systems, one might also look for finite escape time phenomena, saturation effects, or threshold effects. The role of analysis, therefore, is to characterize in broad outline the critical aspects of system behavior—not the details.

By far the most useful device or tool for nonlinear system analysis is the *summarizing function.* The idea is a simple one, but one of great power and utility. The very concept of a summarizing function is a reflection of the general objectives of analysis. It summarizes behavior, suppressing detail.

In formal terms, a summarizing function is just some function of the system state vector. As the system evolves in time, the summarizing function takes on various values conveying some information. It is often possible, however, and this is a key requirement for analysis, to write an approximate first-order difference or differential equation that describes the behavior of the summarizing function. An analysis of that first-order equation is then in some sense a summary analysis of the entire system.

This idea was first systematically introduced by Liapunov for the study of stability of nonlinear systems. The special summarizing functions used for this purpose are, accordingly, referred to as *Liapunov functions.* It is now recognized that this idea in its generalized form is perhaps the most powerful device for the analysis of nonlinear systems—its power being derived from its simplicity, its generality, and its flexibility. It is this idea that is developed in this chapter and applied to several examples in both this chapter and the next.

This chapter contains most of the general theoretical development of the summarizing function concept, emphasizing the Liapunov theory. The theory is built up in stages, and one does not have to go through it all in order to begin to apply it. Some examples are interleaved with the development in this chapter, but other more extended examples are presented in the following chapter. It is suggested that the reader essentially read this chapter and the next in parallel. The examples in the next chapter themselves progress from the relatively simple to the more complex, so it will be helpful to refer back and forth between the two chapters.

9.2 EQUILIBRIUM POINTS

The concept of an equilibrium point, which was used extensively in earlier chapters in connection with linear dynamic systems, carries over directly to nonlinear dynamic systems. The general definition is repeated here.

Definition. A vector $\bar{\mathbf{x}}$ is an *equilibrium point* for a dynamic system if once the state vector is equal to $\bar{\mathbf{x}}$ it remains equal to $\bar{\mathbf{x}}$ for all future time.

In particular, if a system is described by a set of difference equations

$$\mathbf{x}(k+1)=\mathbf{f}(\mathbf{x}(k), k) \tag{9-7}$$

an equilibrium point is a state $\bar{\mathbf{x}}$ satisfying

$$\bar{\mathbf{x}}=\mathbf{f}(\bar{\mathbf{x}}, k) \tag{9-8}$$

for all k. Similarly, for a continuous-time system

$$\dot{\mathbf{x}}(t)=\mathbf{f}(\mathbf{x}(t), t) \tag{9-9}$$

an equilibrium point is a state $\bar{\mathbf{x}}$ satisfying

$$\mathbf{f}(\bar{\mathbf{x}}, t)=\mathbf{0} \tag{9-10}$$

for all t.

In most situations of practical interest the system is time-invariant, in which case the equilibrium points $\bar{\mathbf{x}}$ are solutions of an n-dimensional system of algebraic equations. Specifically,

$$\bar{\mathbf{x}}=\mathbf{f}(\bar{\mathbf{x}}) \tag{9-11}$$

in discrete time, or

$$\mathbf{0}=\mathbf{f}(\bar{\mathbf{x}}) \tag{9-12}$$

in continuous time.

An analysis of a nonlinear dynamic system may devote considerable attention to the characterization of equilibrium points. This contrasts with a typical analysis of linear systems where equilibrium points are basically solutions to linear equations, and hence are treated in a rather routine manner. The nonlinear case is different in two essential respects. First, since equilibrium points are solutions, in this case, to nonlinear equations, finding such solutions is somewhat more of an accomplishment than in the linear case (from a purely technical viewpoint). Thus, a description of equilibrium points often constitutes significant information. Second, and perhaps more fundamentally, the equilibrium point distribution is potentially more complex in the nonlinear case than in the linear case. A system may have none, one, any finite number, or an infinity of equilibrium points in virtually any spacial pattern in state space. Thus, characterization of equilibrium points is not only technically more difficult, it is a much broader question. Ultimately, however, as in the linear case, interest centers not just on the existence of equilibria but also on their stability properties.

Example 1. Consider again the equation for the logistic curve

$$\dot{x}(t) = a(1 - x(t)/c)x(t)$$

A value $\bar{x}$ is an equilibrium point of this first order equation if it satisfies the algebraic equation

$$0 = a[1 - \bar{x}/c]\bar{x}$$

There are two solutions: $\bar{x} = 0$ and $\bar{x} = c$.

In terms of the population system that is modeled by this system, a population level of zero or of c represents a level that once attained will not change.

Example 2 (A Discrete-time System). Consider the system

$$x_1(k+1) = \alpha x_1(k) + x_2(k)^2$$
$$x_2(k+1) = x_1(k) + \beta x_2(k)$$

an equilibrium point is a two-dimensional vector $\bar{\mathbf{x}} = (\bar{x}_1, \bar{x}_2)$ satisfying

$$\bar{x}_1 = \alpha \bar{x}_1 + \bar{x}_2^{\,2}$$
$$\bar{x}_2 = \bar{x}_1 + \beta \bar{x}_2$$

The second equation can be solved for $\bar{x}_1$ in terms of $\bar{x}_2$. This can then be substituted in the first equation yielding

$$(1-\alpha)(1-\beta)\bar{x}_2 = \bar{x}_2^{\,2}$$

Clearly, there are two equilibrium points

$$\bar{\mathbf{x}} = (0, 0) \quad \text{and} \quad \bar{\mathbf{x}} = ((1-\alpha)(1-\beta)^2, (1-\alpha)(1-\beta)).$$

9.3 STABILITY

Stability properties characterize how a system behaves if its state is initiated close to, but not precisely at, a given equilibrium point. If a system is initiated with the state exactly equal to an equilibrium point, then by definition it will never move. When initiated close by, however, the state may remain close by, or it may move away. Roughly speaking, an equilibrium point is stable if whenever the system state is initiated near that point, the state remains near it, perhaps even tending toward the equilibrium point as time increases.

Suppose that $\bar{\mathbf{x}}$ is an equilibrium point of a time-invariant system. That is, $\bar{\mathbf{x}}$ is an equilibrium point of either

$$\mathbf{x}(k+1)=\mathbf{f}(\mathbf{x}(k)) \tag{9-13}$$

or of

$$\dot{\mathbf{x}}(t)=\mathbf{f}(\mathbf{x}(t)) \tag{9-14}$$

For a precise definition of stability, it is convenient to introduce the notation $S(\bar{\mathbf{x}}, R)$ to denote the spherical region* in the state space with center at $\bar{\mathbf{x}}$ and radius R. Using this notation we then can state four important definitions related to stability. These definitions might at first appear somewhat obscure because of their somewhat mathematically involuted character. For this reason they require careful study. We state them all as a unit, and then interpret them verbally and geometrically.

Definition.

(1) An equilibrium point $\bar{\mathbf{x}}$ is *stable* if there is an $R_0>0$ for which the following is true: For every $R<R_0$, there is an r, $0<r<R$, such that if $\mathbf{x}(0)$ is inside $S(\bar{\mathbf{x}}, r)$, then $\mathbf{x}(t)$ is inside $S(\bar{\mathbf{x}}, R)$ for all $t>0$.
(2) An equilibrium point $\bar{\mathbf{x}}$ is *asymptotically stable* whenever it is stable and in addition there is an $\bar{R}_0>0$ such that whenever the state is initiated inside $S(\bar{\mathbf{x}}, \bar{R}_0)$, it tends to $\bar{\mathbf{x}}$ as time increases.
(3) An equilibrium point $\bar{\mathbf{x}}$ is *marginally stable* if it is stable but not asymptotically stable.
(4) An equilibrium point $\bar{\mathbf{x}}$ is *unstable* if it is not stable. Equivalently, $\bar{\mathbf{x}}$ is unstable if for some $R>0$ and any $r>0$ there is a point in the spherical region $S(\bar{\mathbf{x}}, r)$ such that if initiated there, the system state will eventually move outside of $S(\bar{\mathbf{x}}, R)$.

* Specifically, we define the distance between two points $\mathbf{x}$ and $\mathbf{y}$ in the n-dimensional state space by $\|\mathbf{x}-\mathbf{y}\|=(\sum_{i=1}^{n}(x_i-y_i)^2)^{1/2}$. This is called the *Euclidean distance*. The region $S(\bar{\mathbf{x}}, R)$ is the set of vectors $\mathbf{x}$ satisfying $\|\mathbf{x}-\bar{\mathbf{x}}\|<R$; that is, it is the set of all points whose Euclidean distance from $\bar{\mathbf{x}}$ is less than R.

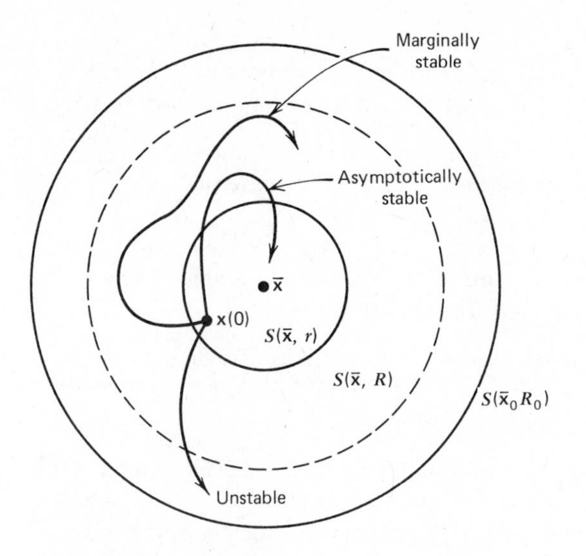

Figure 9.3. Definitions.

These definitions are best understood in terms of their geometric interpretations. Let us refer to the path traced by the state of a system as the state *trajectory*. Any initial point defines a corresponding trajectory emanating from it. In this terminology, the first definition says that an equilibrium point $\bar{\mathbf{x}}$ is stable if it is possible to confine the trajectory of the system to within an arbitrary distance from $\bar{\mathbf{x}}$ by restricting the initial state to within some other (perhaps smaller) distance of $\bar{\mathbf{x}}$. According to the definition, one first selects an $R>0$, and then to assure that the state trajectory remains within $S(\bar{\mathbf{x}}, R)$, a (smaller) $r>0$ is found specifying the allowable region $S(\bar{\mathbf{x}}, r)$ for the initial state. This is the formalization of the intuitive notion that stability means that if the state is initiated near $\bar{\mathbf{x}}$, it remains near $\bar{\mathbf{x}}$.

The other three definitions are, of course, based upon the first. Asymptotic stability requires that, in addition to simply remaining near $\bar{\mathbf{x}}$, the state trajectory, when initiated close to $\bar{\mathbf{x}}$, should tend toward $\bar{\mathbf{x}}$. (It can get there in either finite or infinite time.) Asymptotic stability is the strongest of the stability properties and the one which in most instances is considered most desirable. The definition of marginal stability is introduced primarily for convenience of discussion. It distinguishes stability from asymptotic stability. Instability implies that there are trajectories that start close to $\bar{\mathbf{x}}$ but eventually move far away.

The definitions are illustrated in Fig. 9.3. This figure should be studied in conjunction with the formal definitions. Note, however, that the definitions of stability in this section apply both to discrete-time and continuous-time systems. For purposes of discussion (and following tradition) it is often convenient and most picturesque to talk as if the trajectories are continuous. Nevertheless, such discussions, and figures such as Fig. 9.3, must be understood as having direct discrete-time analogs.

Example 1 (First-order Linear Equation). The differential equation

$$\dot{x}(t) = ax(t)$$

has the origin as an equilibrium point. This point is (1) stable if $a \leq 0$, (2) asymptotically stable if $a < 0$, and (3) unstable if $a > 0$.

Example 2 (The Logistic Equation Again). The equation for logistic population growth

$$\dot{x}(t) = a(1 - x(t)/c)x(t)$$

with $a > 0$, $c > 0$, has the equilibrium points $\bar{x} = 0$ and $\bar{x} = c$. The point $\bar{x} = 0$ is unstable since any small, but positive, initial population level will increase monotonically toward $x = c$. The point $\bar{x} = c$ is asymptotically stable since if slightly displaced from that point, either upwards or downwards, the population level will tend toward it again.

Example 3. Consider the discrete-time system

$$x(k+1) = \frac{x(k)}{1 + x(k)}$$

which arises in genetics. (See Chapter 10.) An equilibrium point $\bar{x}$ of this system must satisfy the equation

$$\bar{x} = \frac{\bar{x}}{1 + \bar{x}}$$

The only solution is $\bar{x} = 0$.

Any positive initial state will produce a trajectory tending toward zero as $k \to \infty$. Small negative initial states, however, lead to movement away from zero, at least to the point -1. Thus, in terms of the general definition, the equilibrium point $\bar{x} = 0$ is unstable.

9.4 LINEARIZATION AND STABILITY

According to the basic definitions, stability properties depend only on the nature of the system near the equilibrium point. Therefore, to conduct an analysis of stability it is often theoretically legitimate and mathematically convenient to replace the full nonlinear description by a simpler description that approximates the true system near the equilibrium point. Often a linear approximation is sufficient to reveal the stability properties. This idea of

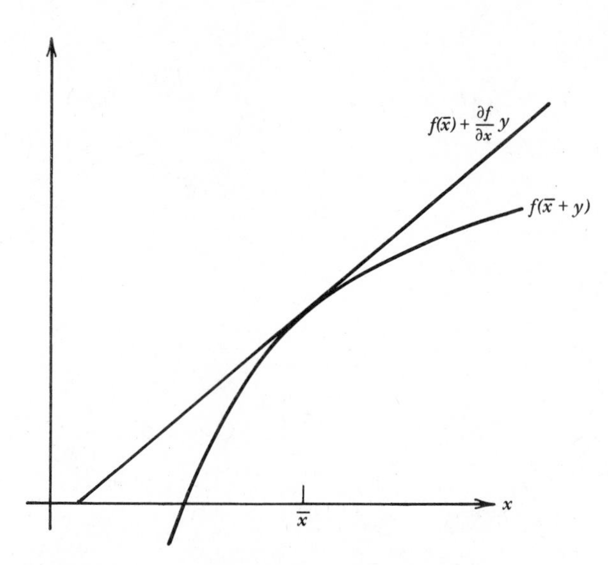

Figure 9.4. Linear approximation.

checking stability by examination of a linearized version of the system is referred to as *Liapunov's first method*, or sometimes as *Liapunov's indirect method*. It is a simple and powerful technique, and is usually the first step in the analysis of any equilibrium point.

The linearization of a nonlinear system is based on linearization of the nonlinear function **f** in its description. For a first-order system, defined by a single function $f(x)$ of a single variable, the procedure is to approximate f near $\bar{x}$ by

$$f(\bar{x}+y)=f(\bar{x})+\frac{d}{dx}f(\bar{x})y \tag{9-15}$$

This is illustrated in Fig. 9.4.

An nth-order system is defined by n functions, each of which depends on n variables. In this case each function is approximated by the relation

$$\begin{aligned}
&f_i(\bar{x}_1+y_1, \bar{x}_2+y_2, \ldots, \bar{x}_n+y_n)\\
&\qquad\simeq f_i(\bar{x}_1, \bar{x}_2, \ldots, \bar{x}_n)+\frac{\partial}{\partial x_1}f_i(\bar{x}_1, \bar{x}_2, \ldots, \bar{x}_n)y_1\\
&\qquad\quad+\frac{\partial}{\partial x_2}f_i(\bar{x}_1, \bar{x}_2, \ldots, \bar{x}_n)y_2+\cdots+\frac{\partial}{\partial x_n}f_i(\bar{x}_1, \bar{x}_2, \ldots, \bar{x}_n)y_n
\end{aligned}$$

The linear approximation for the vector $\mathbf{f}(\mathbf{x})$ is made up of the n separate approximations for each component function. The complete result is expressed compactly in vector notation as

$$\mathbf{f}(\bar{\mathbf{x}}+\mathbf{y}) \simeq \mathbf{f}(\bar{\mathbf{x}})+\mathbf{F}\mathbf{y} \tag{9-16}$$

In this expression $\mathbf{F}$ is the $n \times n$ matrix

$$\mathbf{F}=\begin{bmatrix} \frac{\partial f_1}{\partial x_1} & \frac{\partial f_1}{\partial x_2} & \cdots & \frac{\partial f_1}{\partial x_n} \\ \frac{\partial f_2}{\partial x_1} & \frac{\partial f_2}{\partial x_2} & \cdots & \frac{\partial f_2}{\partial x_n} \\ \vdots & & & \\ \frac{\partial f_n}{\partial x_1} & \frac{\partial f_n}{\partial x_2} & \cdots & \frac{\partial f_n}{\partial x_n} \end{bmatrix}_{\mathbf{x}=\bar{\mathbf{x}}} \tag{9-17}$$

The matrix $\mathbf{F}$ is referred to as the *Jacobian matrix* of $\mathbf{f}$.

Now let us apply this procedure to derive the linearized versions of discrete- and continuous-time systems. Suppose first that $\bar{\mathbf{x}}$ is an equilibrium point of the discrete-time system

$$\mathbf{x}(k+1)=\mathbf{f}(\mathbf{x}(k)) \tag{9-18}$$

Then substituting $\mathbf{x}(k)=\bar{\mathbf{x}}+\mathbf{y}(k)$ in (9-18) and using (9-16), we obtain

$$\bar{\mathbf{x}}+\mathbf{y}(k+1) \simeq \mathbf{f}(\bar{\mathbf{x}})+\mathbf{F}\mathbf{y}(k) \tag{9-19}$$

However, the equilibrium point $\bar{\mathbf{x}}$ satisfies $\bar{\mathbf{x}}=\mathbf{f}(\bar{\mathbf{x}})$. It follows therefore that (approximately)

$$\mathbf{y}(k+1)=\mathbf{F}\mathbf{y}(k) \tag{9-20}$$

This is the linear approximation valid for small deviations $\mathbf{y}(k)$ from the equilibrium point $\bar{\mathbf{x}}$.

Next suppose that $\bar{\mathbf{x}}$ is an equilibrium point of the continuous-time system

$$\dot{\mathbf{x}}(t)=\mathbf{f}(\mathbf{x}(t)) \tag{9-21}$$

Setting $\mathbf{x}(t)=\bar{\mathbf{x}}+\mathbf{y}(t)$ and using the approximation (9-16) leads in a similar way to the linear approximation

$$\dot{\mathbf{y}}(t)=\mathbf{F}\mathbf{y}(t) \tag{9-22}$$

Thus, in either discrete or continuous time, the linear approximation of a nonlinear system has $\mathbf{F}$ as its system matrix. The state vector of the approximation is the deviation of the original state from the equilibrium point.

As we know, the stability properties of a linear system are determined by the location (in the complex plane) of the eigenvalues of the system matrix, and the stability properties of the linearized version of a nonlinear system can be determined that way. Then, stability properties of the original system can be inferred from the linearized system using the following general results:

(1) If all eigenvalues of $\mathbf{F}$ are strictly inside the unit circle for discrete-time systems (strictly in the left half-plane for continuous-time systems), then $\bar{\mathbf{x}}$ is asymptotically stable for the nonlinear system.
(2) If at least one eigenvalue of $\mathbf{F}$ has absolute value greater than one for discrete-time systems (or has a positive real part for continuous-time systems), then $\bar{\mathbf{x}}$ is unstable for the nonlinear system.
(3) If the eigenvalues of $\mathbf{F}$ are all inside the unit circle, but at least one is on the boundary for discrete-time systems (or all in the left half-plane, but at least one has a zero real part in continuous-time systems), then $\bar{\mathbf{x}}$ may be either stable, asymptotically stable, or unstable for the nonlinear system.

The essence of these rules is that, except for the boundary situation, the eigenvalues of the linearized system completely reveal the stability properties of an equilibrium point of a nonlinear system. The reason is that, for small deviations from the equilibrium point, the performance of the system is approximately governed by the linear terms. These terms dominate and thus determine stability—provided that the linear terms do not vanish. If there are boundary eigenvalues, a separate analysis is required.

Example 1 (First-order Quadratic System). Consider the system

$$\dot{x}(t) = ax(t) + cx(t)^2$$

The origin $x = 0$ is an equilibrium point for any parameter values a and c. The linearized version of the system, linearized about the point $x = 0$, is

$$\dot{y}(t) = ay(t)$$

Based on the general principles above, we can deduce the following relations between parameters and stability.

(1) $a < 0$: asymptotically stable
(2) $a > 0$: unstable
(3) $a = 0$: cannot tell from linearization.

In the third case, $a = 0$, it is not possible to infer stability characteristics without an analysis of higher-order terms. For this case the system reduces to

$$\dot{x}(t) = cx(t)^2$$

If $c = 0$, it is clear that the origin is stable. If $c \neq 0$, then it is easy to see that it is unstable. [For example, if $c > 0$, then any $x(0) > 0$ will lead to ever-increasing $x(t)$.]

Example 2. Consider the discrete-time system

$$x_1(k+1)=\alpha x_1(k)+x_2(k)^2$$
$$x_2(k+1)=x_1(k)+\beta x_2(k)$$

In Example 2, Sect. 9.2, it was found that this system has the two equilibrium points $\bar{\mathbf{x}}=(0,0)$ and $\bar{\mathbf{x}}=((1-\alpha)(1-\beta)^2,\ (1-\alpha)(1-\beta))$. Let us attempt to analyze the stability of each of these equilibrium points. We assume that $0<\alpha<1$, $0<\beta<1$.

For $\bar{\mathbf{x}}=(0,0)$ we find that

$$\mathbf{F}=\begin{bmatrix}\alpha & 0\\ 1 & \beta\end{bmatrix}$$

and thus the corresponding linearized system is

$$y_1(k+1)=\alpha y_1(k)$$
$$y_2(k+1)=y_1(k)+\beta y_2(k)$$

The eigenvalues of this lower triangular system are α and β; hence under our assumptions on α and β, we can conclude that the equilibrium point is asymptotically stable.

For $\bar{\mathbf{x}}=((1-\alpha)(1-\beta)^2,\ (1-\alpha)(1-\beta))$, we find that

$$\mathbf{F}=\begin{bmatrix}\alpha & 2(1-\alpha)(1-\beta)\\ 1 & \beta\end{bmatrix}$$

The characteristic equation of this matrix is $(\lambda-\alpha)(\lambda-\beta)=2(1-\alpha)(1-\beta)$. The left side of this equation increases with λ, and is smaller than the right side at $\lambda=1$. It is clear, therefore, that there is a root λ with $\lambda>1$. Thus, this second equilibrium point is unstable.

9.5 EXAMPLE: THE PRINCIPLE OF COMPETITIVE EXCLUSION

The principle of competitive exclusion in biology states that it is unlikely for two or more similar species to coexist in a common environment. The competitive struggle for food and other resources results in extinction of all but the most fit. When similar species do coexist over a long period of time, they generally evolve distinct differences in their food and habits. Each of the species tends to occupy a unique ecological *niche* so it does not directly compete with other species. A version of the principle of competitive exclusion can be demonstrated mathematically by developing a model of interaction between species. The model presented here was originally developed by Volterra.

Suppose that a number of different species share the resources of a common environment. There is no predation among them, and indeed the only interaction between species that influences growth is the indirect fact that they share the common environment. The growth rate of each of the different species is slowed as the overall community population level increases because of crowding, deterioration of the environment, and lack of food. The effect is an aggregate one—due to all species, and influencing each of them.

The starting point for a corresponding mathematical description is the logistic curve, which is a commonly accepted model of growth in a crowded environment. Let us denote the population of the various species by x_i for $i=1, 2, \ldots, n$. Each of the species imposes somewhat different burdens on the environment (due to differences in average size, etc.). We assume that the aggregate burden is a linear combination

$$F(\mathbf{x}) = \sum_{i=1}^{n} \alpha_i x_i \tag{9-23}$$

where $\alpha_i > 0$, $i=1, 2, \ldots, n$. Then, as an extension of the usual logistic model, it is reasonable to hypothesize that population growth is governed by the set of equations

$$\begin{aligned} \dot{x}_1(t) &= [\beta_1 - \gamma_1 F(\mathbf{x}(t))]x_1(t) \\ \dot{x}_2(t) &= [\beta_2 - \gamma_2 F(\mathbf{x}(t))]x_2(t) \\ &\vdots \\ \dot{x}_n(t) &= [\beta_n - \gamma_n F(\mathbf{x}(t))]x_n(t) \end{aligned} \tag{9-24}$$

where $\beta_i > 0$, $\gamma_i > 0$ for $i=1, 2, \ldots, n$. The β_i's represent the natural growth rates in the absence of crowding effects. The γ_i's represent the sensitivities of the growth rates to the aggregate crowding effect. For technical reasons (which shall soon be apparent) we assume that

$$\frac{\beta_i}{\gamma_i} \neq \frac{\beta_j}{\gamma_j} \tag{9-25}$$

for all $i \neq j$.

Let us look for the (nonnegative) equilibrium points of the system. These, of course, are solutions to the set of equations

$$0 = [\beta_i - \gamma_i F(\bar{\mathbf{x}})]\bar{x}_i \qquad i=1, 2, \ldots, n$$

For each such equation, we distinguish the cases $\bar{x}_i > 0$ and $\bar{x}_i = 0$. If $\bar{x}_i > 0$ for some i, it follows that

$$\beta_i - \gamma_i F(\bar{\mathbf{x}}) = 0$$

or, equivalently,

$$F(\bar{\mathbf{x}}) = \beta_i/\gamma_i \tag{9-26}$$

In view of our assumption (9-25), however, such an expression for $F(\mathbf{x})$ can hold for at most one index i since otherwise $F(\mathbf{x})$ would supposedly have two different values. Therefore, at most one component can be positive in an equilibrium point. In other words, a group of species cannot coexist in equilibrium in this model; at most one can have nonzero population.

Suppose $\bar{x}_i$ is the single positive equilibrium component. Its value can be found from (9-23) and (9-26), yielding

$$\bar{x}_i = \beta_i/(\alpha_i\gamma_i) \tag{9-27}$$

This system has a total of $n+1$ equilibrium points; the zero point and the n equilibria corresponding to a single positive population level, as given by (9-27).

The next question, of course, is whether these various equilibria are stable. Or, in the terms of the biological setting, if one of the species dominates the environment, is its position secure, or can it be driven out of existence by the least perturbation? A simple analysis of this question can be conducted by the linearization technique of the previous section.

Let us consider the equilibrium point $\bar{\mathbf{x}}$ corresponding to

$$\bar{x}_1 = \beta_1/(\alpha_1\gamma_1) \tag{9-28a}$$

$$\bar{x}_i = 0, \qquad i > 1 \tag{9-28b}$$

This is really quite general, since the species can always be renumbered so that the one under consideration is the first.

The linearized system is found by differentiating the original system at the equilibrium point. The required operations fall into a number of cases, corresponding to the equation index i and the variable index j.

(a) $i=1$, $j=1$.

$$\begin{aligned}\frac{\partial}{\partial x_1}[\beta_1 - \gamma_1 F(\mathbf{x})]x_1 \bigg|_{\bar{\mathbf{x}}} &= \beta_1 - \gamma_1\frac{\alpha_1\beta_1}{\alpha_1\gamma_1} - \gamma_1\frac{\alpha_1\beta_1}{\alpha_1\gamma_1} \\ &= -\beta_1\end{aligned}$$

(b) $i=1$, $j>1$.

$$\begin{aligned}\frac{\partial}{\partial x_j}[\beta_1 - \gamma_1 F(\mathbf{x})]x_1 \bigg|_{\bar{\mathbf{x}}} &= -\gamma_1\alpha_j\frac{\beta_1}{\gamma_1\alpha_1} \\ &= -\alpha_j\frac{\beta_1}{\alpha_1}\end{aligned}$$

(c) $i>1,\ j=i$.

$$\frac{\partial}{\partial x_i}[\beta_i-\gamma_i F(\mathbf{x})]x_i\bigg|_{\bar{x}}=\beta_i-\gamma_i\frac{\beta_1}{\gamma_1}$$

(d) $i>1,\ j\neq i$.

$$\frac{\partial}{\partial x_j}[\beta_i-\gamma_i F(\mathbf{x})]x_i\bigg|_{\bar{x}}=0$$

As a result of these tedious but elementary calculations, it is found that the linearized system has the form

$$\dot{\mathbf{y}}(t)=\begin{bmatrix}-\beta_1 & \varepsilon_2 & \varepsilon_3 & \cdots & \varepsilon_n\\ 0 & \Delta_2 & 0 & \cdots & 0\\ 0 & 0 & \Delta_3 & \cdots & 0\\ & & & & \cdot\\ & & & & \cdot\\ & & & & \cdot\\ 0 & 0 & 0 & \cdots & \Delta_n\end{bmatrix}\mathbf{y}(t) \tag{9-29}$$

where

$$\Delta_i=\beta_i-\frac{\gamma_i}{\gamma_1}\beta_1 \tag{9-30a}$$

$$\varepsilon_j=-\frac{\alpha_j\beta_1}{\alpha_1} \tag{9-30b}$$

It follows, because of its upper triangular structure, that the eigenvalues of the linearized system are equal to the diagonal elements of the matrix in (9-29). The equilibrium point, corresponding to having only the first population nonzero, will be asymptotically stable if $\Delta_i<0$ for all i. If any $\Delta_i>0$, the equilibrium point is unstable. The condition $\Delta_i<0$ corresponds to

$$\frac{\beta_1}{\gamma_1}>\frac{\beta_i}{\gamma_i}$$

Therefore, in order for this equilibrium point to be stable, the ratio β_i/γ_i must be maximized by $i=1$.

We have now reached the point where we can summarize the results of our analysis. First, under our assumptions, it is *not* possible for a group of species to live together in equilibrium—an equilibrium can have a nonzero population level of only one of the species. If the *fitness factors* β_i/γ_i, $i=1,2,\ldots,n$ (the ratios of natural growth rate to crowding sensitivity) are associated with the species, the largest fitness factor determines the one, among the many species, for which the corresponding equilibrium is stable.

9.6 LIAPUNOV FUNCTIONS

The second method of Liapunov, often referred to as the *direct method*, works explicitly with the nonlinear system rather than the linearized version. This has the advantage, first, of being applicable in marginal situations, and, second, of enabling the analysis to extend beyond only a small region near the equilibrium point.

The basic idea of the direct method for verifying stability is to seek an aggregate summarizing function that continually decreases toward a minimum as the system evolves. The classic example for mechanical systems (which is treated in detail later) is that of *energy*. The *energy* of a free mechanical system with friction always decreases unless the system is at rest; and this fact can be used to establish stability. In general, a function of this type, which allows one to deduce stability, is termed a Liapunov function.

General Requirements

Since the Liapunov function concept applies to both discrete-time and continuous-time systems, with slightly different forms, we first outline the general considerations. Later we present the separate results for discrete-time and continuous-time systems.

Suppose that $\bar{\mathbf{x}}$ is an equilibrium point of a given dynamic system. A *Liapunov function* for the system and the equilibrium point $\bar{\mathbf{x}}$ is a real-valued function V, which is defined over a region Ω of the state space that contains $\bar{\mathbf{x}}$, and satisfies the three requirements:

(1) V is continuous.
(2) $V(\mathbf{x})$ has a unique minimum at $\bar{\mathbf{x}}$ with respect to all other points in Ω.
(3) Along any trajectory of the system contained in Ω, the value of V never increases.

Let us go over these three requirements in order to bring out their full meaning. The function V can be conveniently visualized in two ways. The first is to imagine its graph constructed over the state space, as illustrated in Fig. 9.5*a*. The first requirement, that of continuity, simply means that the graph is connected without breaks. The second requirement, that $V(\mathbf{x})$ is minimized at $\bar{\mathbf{x}}$, means that the graph has its lowest point at $\bar{\mathbf{x}}$. Liapunov and many researchers after him required in addition that this minimum value be zero (as in the figure), but this is neither necessary nor always convenient. The important property is simply that $\bar{\mathbf{x}}$ be the unique minimum point.

The third requirement is perhaps the most difficult to visualize, at least at first. It is this condition, however, which relates the function V to the system. Let us consider the successive values of $V(\mathbf{x})$ taken as the point $\mathbf{x}(t)$ moves along a path defined by the system. (We use continuous-time notation simply

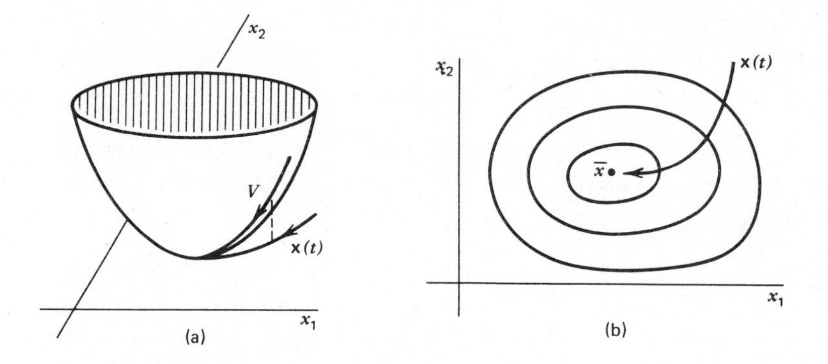

Figure 9.5. Graph and contours.

for convenience. The discussion applies to both discrete and continuous time.) As the state moves in state space according to the laws of motion, we associate with each time point t the corresponding value of the function $V(\mathbf{x}(t))$. This is the summarizing idea discussed before. The time behavior of the function V is a summary of the behavior of $\mathbf{x}(t)$. The third requirement on a Liapunov function is that the value of V, associated with the state vector, never increases with time as the system evolves. In the geometric terms of the figure this requirement means that the curves on the graph corresponding to system motion must run downhill—never uphill.

A second way to visualize a Liapunov function is in terms of its contours in the state space, as illustrated in Fig. 9.5*b*. The center point is the equilibrium point, which corresponds to the minimum value of the Liapunov function. The closed curves in the figure are contours; that is, loci of points where V is constant. The value of V increases as one moves to contours further distant from the equilibrium point. The condition that V does not increase for movement along trajectories can be interpreted as meaning that the trajectories must always cross contours in the direction toward the center—never outward.

The important point to remember is that the Liapunov function is just a function on the state space; that is, given $\mathbf{x}$ in Ω there is a corresponding value $V(\mathbf{x})$. The function is not defined by motion of the system; rather, as the system evolves, the state moves and the corresponding value of V changes.

Example 1. Consider the system

$$x_1(k+1)=\frac{x_2(k)}{1+x_2(k)^2}$$

$$x_2(k+1)=\frac{x_1(k)}{1+x_2(k)^2}$$

which has $\mathbf{x}=(0,0)$ as an equilibrium point.

Let us define the function

$$V(x_1, x_2)=x_1^{\,2}+x_2^{\,2}$$

This V is continuous and has a unique minimum at the equilibrium point. To check whether V increases along a trajectory, we consider any two vectors $\mathbf{x}(k)$ and $\mathbf{x}(k+1)$ related by the system equation. We find

$$\begin{aligned} V(\mathbf{x}(k+1)) &= x_1(k+1)^2+x_2(k+1)^2 \\ &= \frac{x_2(k)^2}{[1+x_2(k)^2]^2}+\frac{x_1(k)^2}{[1+x_2(k)^2]^2} \\ &= \frac{x_1(k)^2+x_2(k)^2}{[1+x_2(k)^2]^2} \\ &= \frac{V(\mathbf{x}(k))}{[1+x_2(k)^2]^2} \le V(\mathbf{x}(k)) \end{aligned}$$

Thus, this V is a Liapunov function for this system and this equilibrium point. There are, of course, many other functions that satisfy the first two requirements but fail to satisfy the third. One must often search hard for a function satisfying all three requirements.

Liapunov Theorem For Discrete Case

We now focus specifically on a discrete-time system

$$\mathbf{x}(k+1)=\mathbf{f}(\mathbf{x}(k)) \tag{9-31}$$

together with a given equilibrium point $\bar{\mathbf{x}}$. We assume that the function $\mathbf{f}(\mathbf{x})$ is continuous. The requirement on a Liapunov function that it never increases along a trajectory can then be translated into a specific mathematical relation.

If at any time k the state of the system (9-31) is equal to $\mathbf{x}$, then at the next time instant $k+1$ the state will be $\mathbf{f}(\mathbf{x})$. The values of the Liapunov function at these points are, accordingly, $V(\mathbf{x})$ and $V(\mathbf{f}(\mathbf{x}))$. Therefore, the change in value is

$$\Delta V(\mathbf{x})=V(\mathbf{f}(\mathbf{x}))-V(\mathbf{x}) \tag{9-32}$$

If V is a Liapunov function on Ω, this change is less than or equal to zero for all possible states $\mathbf{x}$ in Ω. In other words, the requirement that the Liapunov function not increase along a trajectory translates into the relation

$$\Delta V(\mathbf{x})\equiv V(\mathbf{f}(\mathbf{x}))-V(\mathbf{x})\le 0 \tag{9-33}$$

for all $\mathbf{x}$ in Ω. It is this form that is used in the formal definition for discrete-time systems.

Definition. A function V defined on a region Ω of the state space of (9-31) and containing $\bar{\mathbf{x}}$ is a *Liapunov function* for the discrete-time system (9-31) if it satisfies the following three requirements:

(1) V is continuous.
(2) $V(\mathbf{x})$ has a unique minimum at $\bar{\mathbf{x}}$ with respect to all other points in Ω.
(3) The function $\Delta V(\mathbf{x}) = V(\mathbf{f}(\mathbf{x})) - V(\mathbf{x})$ satisfies

$$\Delta V(\mathbf{x}) \leq 0$$

for all $\mathbf{x}$ in Ω.

The geometric interpretation of a Liapunov function makes it almost immediately clear that if a Liapunov function exists the corresponding equilibrium point must be stable. The general idea is that if V can only decrease with time as the system evolves, it must tend toward its minimum value. Accordingly $\mathbf{x}$ must tend to $\bar{\mathbf{x}}$. The precise statement is the following theorem.

Theorem 1 (Liapunov Theorem—Discrete Time). *If there exists a Liapunov function $V(\mathbf{x})$ in a spherical region $S(\bar{\mathbf{x}}, R_0)$ with center $\bar{\mathbf{x}}$, then the equilibrium point $\bar{\mathbf{x}}$ is stable. If, furthermore, the function $\Delta V(\mathbf{x})$ is strictly negative at every point (except $\bar{\mathbf{x}}$), then the stability is asymptotic.*

Proof. The proof is based on geometric relations illustrated in Fig. 9.6. Suppose $V(\mathbf{x})$ exists within the spherical region $S(\bar{\mathbf{x}}, R_0)$. Let R be arbitrary with $0 < R < R_0$. Let $R_1 < R$ be selected so that if $\mathbf{x} \in S(\bar{\mathbf{x}}, R_1)$ then $\mathbf{f}(\mathbf{x}) \in S(\bar{\mathbf{x}}, R_0)$. Such an R_1 exists because $\mathbf{f}$ is continuous. With this choice, if the state vector lies inside $S(\bar{\mathbf{x}}, R_1)$, it will not jump out of the larger sphere $S(\bar{\mathbf{x}}, R_0)$ in one step.

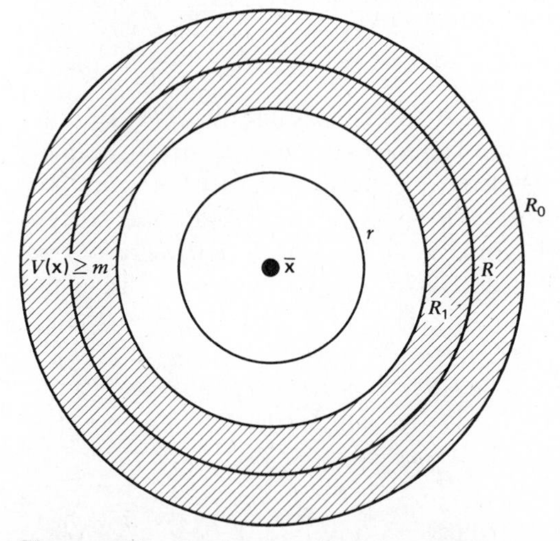

Figure. 9.6. Regions defined in proof.

Now let m be the minimum value of $V(\mathbf{x})$ over the region defined by $R_1 \leq \|\mathbf{x}-\bar{\mathbf{x}}\| \leq R_0$. This minimum value exists because $V(\mathbf{x})$ is assumed to be continuous. Also we have $m > V(\bar{\mathbf{x}})$ since V has a unique minimum at $\bar{\mathbf{x}}$.

Now again since $V(\mathbf{x})$ is continuous, it is possible to select an r, $0<r<R_1$ such that for $\mathbf{x}$ in $S(\bar{\mathbf{x}}, r)$ there holds $V(\mathbf{x})<m$. This is because near $\bar{\mathbf{x}}$ the function V must take values close to $V(\bar{\mathbf{x}})$.

Now suppose $\mathbf{x}(0)$ is taken as an arbitrary point in $S(\bar{\mathbf{x}}, r)$. Then $V(\mathbf{x}(0))< m$. Since $\Delta V(\mathbf{x}) \leq 0$ the value of V cannot increase with time. Therefore, the trajectory can never go outside $S(\mathbf{x}, R_1)$, and consequently it can never go outside $S(\bar{\mathbf{x}}, R)$. Thus, for this arbitrary $R>0$ we have found $r>0$, corresponding to the requirements of the definition of stability.

If in addition $\Delta V(\mathbf{x})<0$ for every point except $\bar{\mathbf{x}}$, then $V(\mathbf{x})$ must actually decrease continually (either for all k, or until the state reaches $\bar{\mathbf{x}}$ if this happens in finite time). Thus V converges to some limiting value $\bar{m}$. The only question is whether it is possible for $\bar{m} > V(\bar{\mathbf{x}})$. This is not possible because, since $V(\mathbf{x}(k))$ converges to $\bar{m}$, it must be true that $\Delta V(\mathbf{x}(k))$ converges to zero. But $\Delta V(\mathbf{x})$ is strictly negative everywhere except at $\bar{\mathbf{x}}$. Thus, since $\Delta V(\mathbf{x})$ is continuous, $\mathbf{x}(k)$ must converge to $\bar{\mathbf{x}}$ and $V(\mathbf{x}(k))$ must converge to $V(\bar{\mathbf{x}})$. This is asymptotic stability. ∎

Example 1 (continued). Since there is a Liapunov function for Example 1, it can be concluded that the equilibrium point is stable. Note that this conclusion cannot be obtained by examination of the linearized system, since the eigenvalues of this system are $\lambda = \pm 1$.

Liapunov Theorem for Continuous Case

We now consider the continuous-time system

$$\dot{\mathbf{x}}(t) = \mathbf{f}(\mathbf{x}(t)) \tag{9-34}$$

together with a given equilibrium point $\bar{\mathbf{x}}$. Again we assume that $\mathbf{f}$ is continuous. In the continuous-time case the requirement that the value of a Liapunov function never increases along a trajectory is expressed in terms of the time derivative. Suppose $\mathbf{x}(t)$ is a trajectory. Then $V(\mathbf{x}(t))$ represents the corresponding values of V along the trajectory. In order that V not increase, we require that $\dot{V}(\mathbf{x}(t)) \leq 0$ for all t. This derivative can be expressed, using the chain rule for differentiation, as

$$\dot{V}(\mathbf{x}(t)) = \frac{\partial V}{\partial x_1}\dot{x}_1(t) + \frac{\partial V}{\partial x_2}\dot{x}_2(t) + \cdots + \frac{\partial V}{\partial x_n}\dot{x}_n(t) \tag{9-35}$$

Then using the original system equation (9-34) this becomes

$$\dot{V}(\mathbf{x}(t)) = \frac{\partial V}{\partial x_1} f_1(\mathbf{x}(t)) + \frac{\partial V}{\partial x_2} f_2(\mathbf{x}(t)) + \cdots + \frac{\partial V}{\partial x_n} f_n(\mathbf{x}(t)) \tag{9-36}$$

Defining the *gradient (row) vector*

$$\nabla V(\mathbf{x}) = \left[\frac{\partial V(\mathbf{x})}{\partial x_1}, \frac{\partial V(\mathbf{x})}{\partial x_2}, \ldots, \frac{\partial V(\mathbf{x})}{\partial x_n}\right] \tag{9-37}$$

(9-36) can be written as

$$\dot{V}(\mathbf{x}(t)) = \nabla V(\mathbf{x}(t))\mathbf{f}(\mathbf{x}(t)) \tag{9-38}$$

Therefore, the requirement that V not increase along any trajectory of the system translates into the requirement that $\dot{V}(\mathbf{x}) \equiv \nabla V(\mathbf{x})\mathbf{f}(\mathbf{x}) \leq 0$ for all $\mathbf{x}$ in Ω. It is this form that is used in continuous time. The definition of Liapunov function and the theorem are given below.

Definition. A function V defined on a region Ω of the state space of (9-34) and containing $\bar{\mathbf{x}}$ is a *Liapunov function* if it satisfies the following three requirements:

(1) V is continuous and has continuous first partial derivatives.
(2) $V(\mathbf{x})$ has a unique minimum at $\bar{\mathbf{x}}$ with respect to all other points in Ω.
(3) The function $\dot{V}(\mathbf{x}) \equiv \nabla V(\mathbf{x})\mathbf{f}(\mathbf{x})$ satisfies $\dot{V}(\mathbf{x}) \leq 0$ for all $\mathbf{x}$ in Ω.

Theorem 2 (Liapunov Theorem—Continuous Time). *If there exists a Liapunov function $V(\mathbf{x})$ in a spherical region $S(\bar{\mathbf{x}}, R_0)$ with center $\bar{\mathbf{x}}$, then the equilibrium point $\bar{\mathbf{x}}$ is stable. If, furthermore, the function $\dot{V}(\mathbf{x})$ is strictly negative at every point (except $\bar{\mathbf{x}}$), then the stability is asymptotic.*

Proof. The proof is similar to that for the discrete-time case, but simpler because it is not possible for the trajectory to jump outside of a region under consideration. We leave the details to the reader. ∎

Example 2. Consider the system

$$\dot{x}_1(t) = x_2(t)$$
$$\dot{x}_2(t) = -x_1(t) - x_2(t)$$

which has an equilibrium point at $x_1 = x_2 = 0$. Define the function

$$V(x_1, x_2) = x_1{}^2 + x_2{}^2$$

This function is certainly continuous with continuous first derivatives, and it is clearly minimized at the origin (which is also the equilibrium point). This function satisfies the first two requirements of a Liapunov function. To check the final requirement, we write

$$\begin{aligned}\dot{V}(x_1, x_2) &= 2x_1\dot{x}_1 + 2x_2\dot{x}_2 \\ &= 2x_1x_2 + 2x_2(-x_1 - x_2) \\ &= -2x_2^{\,2} \leq 0\end{aligned}$$

where we have substituted the system equations for the time derivatives of the state variables. Thus, V is a Liapunov function, and the system is stable. We cannot infer asymptotic stability, however, since $\dot{V}$ is not strictly negative at every nonzero point.

Constants of Motion

An important special situation is when a Liapunov function can be found that does not change value along a trajectory; that is, it neither increases nor decreases. This corresponds to $\Delta V(\mathbf{x}) \equiv 0$ or $\dot{V}(\mathbf{x}) \equiv 0$ in discrete and continuous time, respectively. In this case, V is constant along a trajectory—and the function V is said to be a *constant of motion.* If such a function can be determined, it conveys a good deal of information because we then know that any trajectory must lie on a contour of the function V. In an n-dimensional space a contour usually defines an $(n-1)$-dimensional surface. If $n = 2$, for example, knowledge of a constant of motion yields curves along which trajectories must lie.

Example 3. Consider the system

$$\begin{aligned}\dot{x}_1(t) &= x_2(t) \\ \dot{x}_2(t) &= -x_1(t)\end{aligned}$$

Let $V(\mathbf{x}) = x_1^{\,2} + x_2^{\,2}$. Then it is seen that $\dot{V}(\mathbf{x}) \equiv 0$. The function V is a constant of motion. Therefore, the system trajectories are restricted to contours of V. In this case the contours are circles centered at the origin. Thus any trajectory must travel around such a circle. The particular circle followed depends on the initial condition.

Extent of Stability

It should be clear from the proof of the Liapunov stability theorems that if a Liapunov function is defined over a large region Ω, we can say more than if it is defined only in a small region. In fact, it is clear that if the initial point $\mathbf{x}(0)$ is selected with, say, $V(\mathbf{x}(0)) = q$, then the subsequent trajectory never goes outside of the region $V(\mathbf{x}) \leq q$. Therefore, the region over which the Liapunov function is defined delineates a region throughout which system performance is easily related to the equilibrium point. This kind of information, concerning the *extent of stability*, cannot be obtained by the simple linearization technique, and for this reason a Liapunov function analysis often follows a linearization analysis even if stability has been established.

An important special case of asymptotic stability is when the extent includes the entire state space; that is, when an equilibrium point $\bar{\mathbf{x}}$ is asymptotically stable, and, in addition, when initiated at any point in the entire state space the system state tends to $\bar{\mathbf{x}}$. This property is referred to as *asymptotic stability in the large*. It is often a most desirable property. The following theorem, stated without proof, gives conditions guaranteeing asymptotic stability in the large.

Theorem 3. *Suppose V is a Liapunov function for a dynamic system and an equilibrium point* $\bar{\mathbf{x}}$. *Suppose in addition that*

(i) *V is defined on the entire state space.*
(ii) $\Delta V(\mathbf{x})<0$ [*or* $\dot{V}(\mathbf{x})<0$] *for all* $\mathbf{x}\neq\bar{\mathbf{x}}$.
(iii) $V(\mathbf{x})$ *goes to infinity as any component of* $\mathbf{x}$ *gets arbitrarily large in magnitude.*

Then $\bar{\mathbf{x}}$ *is asymptotically stable in the large.*

9.7 EXAMPLES

This section presents three examples illustrating the construction and use of Liapunov functions. From these examples, the reader should begin to recognize that the construction of a suitable Liapunov function generally springs from the context, or original motivation, of the system equations. This theme is elaborated throughout this chapter and the next. Once it is appreciated, the concept of a Liapunov function becomes much more than an abstract mathematical tool. It becomes an integral component of overall system description.

Example 1 (Iterative Procedure for Calculating Square Roots). Successive approximation techniques or other iterative procedures can be formulated as dynamic processes. Convergence of a procedure can be guaranteed if a Liapunov function is found. As an example, consider the problem of finding the square root of a positive number a. If we start with an estimate of the square root, we might square it to see how close the result is to a and then try to improve our estimate. By repeating this process, we develop a successive approximations procedure.

In algebraic terms, we seek a solution to the equation

$$x^2-a=0 \tag{9-39}$$

This can be written as

$$x=x+a-x^2$$

Successive approximations can then be generated by the dynamic system

$$x(k+1)=x(k)+a-x(k)^2 \tag{9-40}$$

This seems a reasonable, if not terribly inspired, approach to the calculation. We modify the current estimate by adding to it the difference between a and the square of the estimate. Clearly, if the current estimate is too small, the process will increase it, and vice versa. Will it work?

As a first step of the analysis it is simplest to use the linearization method. The linearized version (9–40) at the point $\bar{x}=\sqrt{a}$ is

$$y(k+1)=(1-2\sqrt{a})y(k) \tag{9-41}$$

where, as usual, $y(k)=x(k)-\bar{x}$. The condition for asymptotic stability is therefore

$$|1-2\sqrt{a}|<1$$

or, equivalently,

$$0<a<1$$

Already we can conclude that this method can work only for a limited range of a values.

The linear analysis tells us that for $0<a<1$ the method will work provided we start close enough. But it does not tell us how close the initial estimate must be. A Liapunov function would give us more information.

It is natural to define

$$V(x)=|a-x^2| \tag{9-42}$$

This function is continuous and has a minimum at $\bar{x}$. Denoting the right-hand side of (9-40) by $f(x(k))$, it follows that

$$\begin{aligned} V(f(x)) &= |a-(x+a-x^2)^2| \\ &= |a-x^2-2x(a-x^2)-(a-x^2)^2| \\ &= |(a-x^2)[1-2x-(a-x^2)]| \\ &= |a-x^2|\,|(1-x)^2-a| \end{aligned} \tag{9-43}$$

Therefore,

$$\Delta V(x)\equiv V(f(x))-V(x)=V(x)[|(1-x)^2-a|-1] \tag{9-44}$$

The condition for $\Delta V(x)<0$ is thus

$$|(1-x)^2-a|<1 \tag{9-45}$$

For $0<a<1$, this is equivalent to the requirement that

$$1-\sqrt{1+a}<x<1+\sqrt{1+a}$$

In particular, the requirement is satisfied for all x, $0 \le x \le 1$. Thus, the Liapunov results show us that this simple procedure converges to $\sqrt{a}$, $0 < a < 1$, provided only that the initial estimate x satisfies $0 \le x \le 1$.

As a special instance, suppose we set $a = \frac{1}{4}$ and try to calculate its square root (which we already know is $x = \frac{1}{2}$) using this method and starting at $x = 1$. We obtain the following successive estimates:

	x
0	1.0000000
1	.25000000
2	.43750000
3	.49609375
4	.49998474
5	.49999999
6	.50000000

However, let us try a number close to unity, say $a = .98$. Then we obtain:

0	1.0000	20	.9966471659
1	.980	30	.9954176614
2	.9996	40	.9944139252
3	.98039984	50	.9935944066
4	.9992159937	51	.9863645618
5	.9807833916	75	.9877494455
6	.9988473304	76	.9921004784
7	.981151341	100	.9912712204
8	.9984933871	101	.988652588
9	.981504343	125	.989153085
10	.9981535677	126	.9907292594
11	.981843023	150	.9904285409
12	.9978273012	151	.9894798463

The exact result is $x = .9899494936$

This slow rate of convergence is, of course, not unexpected in view of the linearized version. The eigenvalue of the linearized system is $-.979899$, which indicates extremely slow convergence.

We now know the range of a values for which the procedure will work, have found a range of acceptable starting values, and have an estimate of the speed of convergence. Through a combination of linear analysis and a Liapunov function, we have been able to obtain a fairly complete characterization of the properties of the iteration procedure (9-40).

Example 2 (Swinging Pendulum). This is an example of the type that apparently originally motivated the invention of the Liapunov function concept. A simple swinging pendulum has an equilibrium point when hanging straight

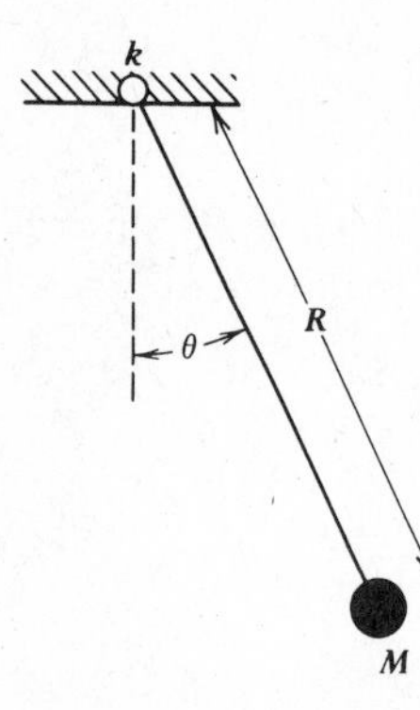

Figure 9.7. Swinging pendulum.

down. This point is stable, as is intuitively clear. Furthermore, if it is assumed that there is some friction in the bearing, then, as is equally clear, the point is asymptotically stable. The equations of motion come from Newton's laws, but they are nonlinear in this case. How can one mathematically establish the strong stability properties that one feels must hold for this simple system? The answer is to look at the energy of the system. Since energy is constantly dissipated by the system, mechanical energy serves as a suitable Liapunov function.

We assume that the pendulum is of length R and has point mass M concentrated at the end. The position of the pendulum at any time is described by the angle θ. We assume that the frictional force is proportional to the speed of the pendulum. (See Fig. 9.7.) To write the equations governing the pendulum, we equate mass times acceleration to total force in the direction perpendicular to the pendulum arm. Mass times acceleration is $MR\ddot{\theta}(t)$. The force is the component of gravitational force in this direction $-\mathrm{Mg}\sin\theta(t)$ plus the retarding force due to friction $-Mk\dot{\theta}(t)$. Here $g>0$ is the gravitational constant and $k>0$ is a friction coefficient. Thus, we have

$$MR\ddot{\theta}(t) = -Mg\sin\theta(t) - Mk\dot{\theta}(t) \tag{9-46}$$

In state variable form this becomes

$$\begin{aligned} \dot{\theta}(t) &= \omega(t) \\ \dot{\omega}(t) &= -\frac{g}{R}\sin\theta(t) - \frac{k}{R}\,\omega(t) \end{aligned} \tag{9-47}$$

The second state variable $\omega(t)$ is the angular velocity of the pendulum.

We now define the function

$$V(\theta, \omega) = \tfrac{1}{2}MR^2\omega^2 + MgR(1-\cos\theta) \tag{9-48}$$

This function V is the mechanical energy of the system (which is the sum of the kinetic and the potential energy). Although the physical significance of V is a major conceptual tool, it is for our present purposes irrelevant. Our main interest is that this V is a Liapunov function.

To verify that V is a Liapunov function, we note first that V is continuous and has continuous partial derivatives. It is positive everywhere except at $\theta=0$, $\omega=0$, where it is zero. Thus, V is minimized at the equilibrium point. Finally, we compute $\dot{V}$:

$$\begin{aligned}\dot{V}(\theta,\omega) &= MR^2\omega\dot{\omega}+MgR\dot{\theta}\sin\theta \\ &= -MRg\omega\sin\theta - kMR\omega^2 + MgR\omega\sin\theta \\ &= -kMR\omega^2 \leq 0 \end{aligned} \tag{9-49}$$

Thus, V is a Liapunov function, and we can immediately conclude that the equilibrium is stable.

Example 3 (A Pursuit Problem). Suppose a hound is chasing a rabbit. The rabbit runs a straight course along the x-axis at constant velocity R. The hound runs at a constant velocity H, but in such a way as to always point directly toward the rabbit. Let us write the differential equations describing the motion of the hound and the rabbit. (See Fig. 9.8.)

Let $x_r(t)$, $y_r(t)$ and $x_h(t)$, $y_h(t)$ denote the x and y coordinates of the rabbit and hound, respectively. Then

$$\dot{x}_r = R \tag{9-50a}$$

$$\dot{y}_r = y_r = 0 \tag{9-50b}$$

Figure 9.8. Hound and rabbit.

The fact that the velocity of the hound is H means that

$$\dot{x}_h^2 + \dot{y}_h^2 = H^2 \tag{9-51}$$

The fact that the velocity vector of the hound always points toward the rabbit means that

$$\dot{x}_h = -k(x_h - x_r) \tag{9-52a}$$

$$\dot{y}_h = -k(y_h - y_r) \tag{9-52b}$$

for some positive constant k.

Using (9-52a) and (9-51) one may determine k for (9-52b), thereby obtaining the equations

$$\dot{x}_h = \frac{-(x_h - x_r)H}{\sqrt{(x_h - x_r)^2 + y_h^2}} \tag{9-53a}$$

$$\dot{y}_h = \frac{-y_h H}{\sqrt{(x_h - x_r)^2 + y_h^2}} \tag{9-53b}$$

The system is perhaps more meaningful when expressed in terms of relative coordinates—the coordinates of the difference in position of the hound and rabbit. Defining

$$x = x_h - x_r$$

$$y = y_h$$

there results

$$\dot{x} = \frac{-xH}{\sqrt{x^2 + y^2}} - R \tag{9-54a}$$

$$\dot{y} = \frac{-yH}{\sqrt{x^2 + y^2}} \tag{9-54b}$$

It is this final system that we examine.

We ask whether the hound will always catch the rabbit. With respect to the system (9-54) this is equivalent to asking whether a trajectory with an arbitrary initial condition $x(0)$, $y(0)$ will eventually get to the origin, where the relative coordinates are zero. This particular system is indeterminant at the origin, but it is well-defined everywhere else. Clearly for our purpose we can consider the origin as an equilibrium point. To establish the desired conclusion it is natural to seek a Liapunov function for (9-54). But how can we find a suitable Liapunov function for such a complicated highly nonlinear system? The answer is found most easily by recalling the original source of the equations. The hound is trying to catch the rabbit, and his movement at every instant directly contributes to satisfying that objective. It is natural therefore to suppose that distance (or distance squared) from the origin might serve as a Liapunov function, since it is the separating distance that the hound seeks to diminish.

Indeed setting

$$V(x, y) = x^2 + y^2 \tag{9-55}$$

one finds

$$\dot{V}(x, y) = -2H\sqrt{x^2+y^2} - 2Rx \tag{9-56}$$

We can show that if $H > R$, then $\dot{V}(x, y)$ is negative for any point other than the origin. If $x = 0$, $y \neq 0$, this is clear. If $x \neq 0$, then $-H\sqrt{x^2+y^2} - Rx < -(H-R)|x| < 0$. Thus, $\dot{V}(x, y) < 0$ for all x, y except the origin. It follows that if the hound runs faster than the rabbit, he always catches the rabbit.

*9.8 INVARIANT SETS

The Liapunov function concept and the stability theorems can be generalized in several directions to treat special circumstances. One generalization, based on the idea of an invariant set, is particularly useful for two sorts of common situations. The first situation is where a Liapunov function is found, and $\Delta V(\mathbf{x})$ [or $\dot{V}(\mathbf{x})$] is strictly less than zero for some values of $\mathbf{x}$ but not for all $\mathbf{x}$. The original Liapunov theorem only assures stability in this case. By employing the invariant set concept, however, one can often establish asymptotic stability with the same Liapunov function.

The second place where the invariant set concept is useful is for systems that do not have equilibrium points, but in which the state vector does tend to follow a fixed pattern as time increases. For example, in two dimensions the state may tend toward a trajectory that endlessly travels clockwise around the unit circle. The Liapunov function concept can be extended to handle such situations by use of the invariant set concept.

Definition. A set G is an *invariant set* for a dynamic system if whenever a point $\mathbf{x}$ on a system trajectory is in G, the trajectory remains in G.

An equilibrium point is perhaps the simplest example of an invariant set. Once the system reaches such a point, it never leaves. Also, if a system has several equilibrium points, the collection G of these points is an invariant set. Here is a somewhat different example.

Example 1 (A Limit Cycle). Consider the two-dimensional system.

$$\dot{x} = y + x[1 - x^2 - y^2]$$
$$\dot{y} = -x + y[1 - x^2 - y^2]$$

The origin is an equilibrium point, and it can be easily shown that this system has no other equilibrium points. However, once the system is on the unit circle $x^2+y^2=1$, it will stay there. The unit circle is an invariant set of the system. To verify this we simply note that $x\dot{x}+y\dot{y}=(x^2+y^2)[1-x^2-y^2]$, and hence the velocity vector is orthogonal to the state vector on the unit circle. Thus, the unit circle is an invariant set.

To obtain the useful generalized Liapunov result, the concept of invariant sets is combined with another key idea that emerged (at least briefly) in the course of the proof of the Liapunov stability theorem. When $\Delta V(\mathbf{x})\leq 0$ for all $\mathbf{x}$ [or $\dot{V}(\mathbf{x})\leq 0$ for all $\mathbf{x}$], then certainly V must always decrease—moreover, and this is the important observation, $\Delta V(\mathbf{x})$ [or $\dot{V}(\mathbf{x})$] must tend to zero if V has a lower limit. So in some sense it is more relevant to look at the places where $\Delta V(\mathbf{x})=0$ [or where $\dot{V}(\mathbf{x})=0$] than where V is minimized (although the latter includes the former). The following theorem combines the two ideas. Essentially it states that if a V is defined such that $\Delta V(\mathbf{x})\leq 0$, then the state must go both to an invariant set and to a place where $\Delta V(\mathbf{x})=0$.

Theorem (Invariant Set Theorem). *Let $V(\mathbf{x})$ be a scalar function with continuous first partial derivatives. Let Ω_s denote the region where $V(\mathbf{x})<s$. Assume that Ω_s is bounded and that $\Delta V(\mathbf{x})\leq 0$ [or $\dot{V}(\mathbf{x})\leq 0$ in continuous time] within Ω_s. Let S be the set of points within Ω_s where $\Delta V(\mathbf{x})=0$ [or $\dot{V}(\mathbf{x})=0$], and let G be the largest invariant set within S. Then every trajectory in Ω_s tends to G as time increases.*

Proof. The conditions on $\Delta V(\mathbf{x})$ imply that V is a nonincreasing function of time. Therefore, any solution initiated within Ω_s does not leave Ω_s. Furthermore, since V must be bounded from below (because Ω_s is bounded), it follows that $V(\mathbf{x})$ tends to a finite limiting value, and accordingly $\Delta V(\mathbf{x})$ tends to zero. Again since Ω_s is bounded, the trajectory must tend to the set S.

Define the limiting set Γ as the set of all points to which the trajectory tends as time increases. It can be shown (using techniques beyond the scope of this book) that this set contains at least one point and is an invariant set. This set must be contained in the set S, and therefore it must be part of the largest invariant set G within S. ▌

This one theorem is an extremely powerful tool for system analysis. It contains the original Liapunov stability theorem as a special case, but can often supply additional results.

Example 2 (The Pendulum). Let us return to the pendulum example that was treated in the previous section. The energy Liapunov function V had

$$\dot{V}(\theta,\omega)=-kMR\omega^2$$

Since $\dot{V}\leq 0$, the original Liapunov theorem only establishes marginal stability.

However, using the invariant set concept we can go somewhat further. The set S is in this case the set where $\omega=0$. An invariant set with $\omega=0$ must be a rest position, and the largest invariant set within S consists of the two rest positions; at the bottom and the top. The invariant set theorem enables us to conclude that solutions must tend to one of the two equilibrium points.

Example 3 (The Limit Cycle). For the system

$$\dot{x} = y + x[1 - x^2 - y^2]$$
$$\dot{y} = -x + y[1 - x^2 - y^2]$$

define the function

$$V(x, y) = (1 - x^2 - y^2)^2$$

Then

$$\begin{aligned}\dot{V}(x, y) &= -2(1 - x^2 - y^2)(2x\dot{x} + 2y\dot{y}) \\ &= -4(x^2 + y^2)(1 - x^2 - y^2)^2\end{aligned}$$

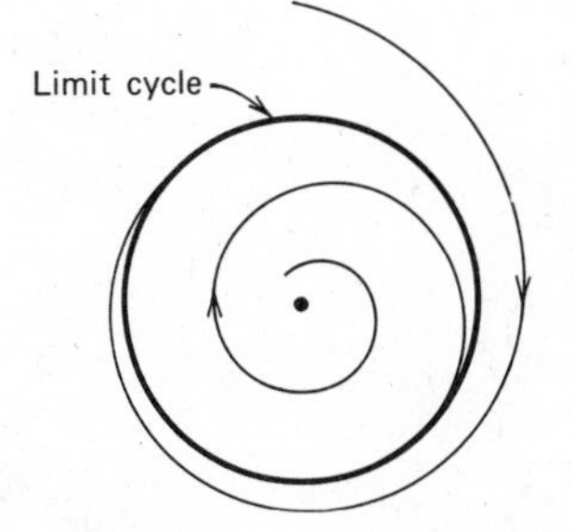

Figure 9.9. A limit cycle.

Thus, $\dot{V} \leq 0$ for all points in the state space. The set S consists of the origin and the unit circle. Since we have seen that S is also an invariant set, every solution tends to S. Furthermore, the origin is not included in Ω_1. It easily follows that no solution tends to the origin, except the solution that starts there. The origin is therefore unstable, and all other solutions tend toward the unit circle. The form of the solutions are illustrated in Fig. 9.9.

9.9 A LINEAR LIAPUNOV FUNCTION FOR POSITIVE SYSTEMS

In general there is no easy way to find a Liapunov function for a stable nonlinear system. Like modeling itself, one must rely on experience, a spark of insight, and familiarity with what has worked in the past (which in this case is other Liapunov functions). There are, however, general types of Liapunov functions that work for certain classes of systems. Some of these are presented in the following four sections, and they can be used as suitable references in other situations.

The basic results for positive linear systems can be interpreted in terms of Liapunov functions (or more generally in terms of summarizing functions). This interpretation adds a bit of insight and clarification to both subjects.

For simplicity let us first consider a discrete-time strictly positive homogeneous system

$$\mathbf{x}(k+1)=\mathbf{A}\mathbf{x}(k) \tag{9-57}$$

Consistent with the general motivation of such systems, we only consider nonnegative solutions. Likewise, the Liapunov function need only be defined for nonnegative vectors in this case.

Suppose that the system is asymptotically stable. The eigenvalue of greatest absolute value is the Frobenius–Perron eigenvalue λ_0, with $0 \leq \lambda_0 < 1$. Let $\mathbf{f}_0^T$ be the corresponding strictly positive *left* eigenvector, and define

$$V(\mathbf{x})=\mathbf{f}_0^T\mathbf{x}$$

For this function we have (since $\mathbf{f}_0^T > \mathbf{0}$)

$$\begin{aligned} &V(\mathbf{x})>0 \quad \text{for all } \mathbf{x}\geq\mathbf{0} \\ &V(\mathbf{0})=0 \end{aligned} \tag{9-58}$$

Thus, V has a minimum point (with respect to nonnegative vectors) at $\mathbf{x}=\mathbf{0}$. Furthermore, for any $\mathbf{x}(k) \geq \mathbf{0}$, we have

$$\begin{aligned} V(\mathbf{x}(k+1)) &= \mathbf{f}_0^T\mathbf{A}\mathbf{x}(k) \\ &= \lambda_0\mathbf{f}_0^T\mathbf{x}(k) \\ &= \lambda_0 V(\mathbf{x}(k)) < V(\mathbf{x}(k)) \end{aligned} \tag{9-59}$$

Since V strictly decreases as k increases, V is a Liapunov function, and it explicitly demonstrates the asymptotic stability of the origin.

This idea can be extended to include nonhomogeneous positive systems and to allow for arbitrary state vectors by slight modification. The resulting Liapunov function employs absolute values and hence is no longer linear, but it is still of first degree. Consider the system

$$\mathbf{x}(k+1)=\mathbf{A}\mathbf{x}(k)+\mathbf{b} \tag{9-60}$$

where $\mathbf{A}>\mathbf{0}$. Suppose that there is a unique equilibrium point

$$\bar{\mathbf{x}}=[\mathbf{I}-\mathbf{A}]^{-1}\mathbf{b} \tag{9-61}$$

which is asymptotically stable.

We now define

$$V(\mathbf{x}) = \mathbf{f}_0^T |\mathbf{x} - \bar{\mathbf{x}}| \tag{9-62}$$

where here $|\mathbf{x} - \bar{\mathbf{x}}|$ is the vector with components equal to the absolute values of the corresponding components of $\mathbf{x} - \bar{\mathbf{x}}$.

Clearly $V(\mathbf{x})$ is minimized at $\mathbf{x} = \bar{\mathbf{x}}$. Also we have

$$\begin{aligned} V(\mathbf{x}(k+1)) &= \mathbf{f}_0^T |\mathbf{x}(k+1) - \bar{\mathbf{x}}| \\ &= \mathbf{f}_0^T |\mathbf{A}\mathbf{x}(k) - \mathbf{A}\bar{\mathbf{x}}| \\ &\leq \mathbf{f}_0^T \mathbf{A} |\mathbf{x}(k) - \bar{\mathbf{x}}| = \lambda_0 \mathbf{f}_0^T |\mathbf{x}(k) - \bar{\mathbf{x}}| \\ &< V(\mathbf{x}(k)) \end{aligned} \tag{9-63}$$

for $\mathbf{x}(k) \neq \bar{\mathbf{x}}$. Again V is a Liapunov function.

9.10 AN INTEGRAL LIAPUNOV FUNCTION

Another important Liapunov function is constructed by integrating the right-hand side of the system equation. Although this construction, by itself, is applicable only to scalar systems, the idea can be combined with other forms of Liapunov functions in some high-order systems.

Consider the system

$$\dot{x}(t) = f(x(t)) \tag{9-64}$$

Here $x(t)$ is just scalar-valued. The function $f(x)$ is assumed to satisfy the following properties:

(1) $f(x)$ is continuous
(2) $xf(x) < 0$ for $x \neq 0$
(3) $-\int_0^\infty f(x)\,dx = \infty$, $\int_{-\infty}^0 f(x)\,dx = \infty$

The general form of $f(x)$ is shown in Fig. 9.10. These properties can be regarded as a rather general extension of what one would require if $f(x)$ were linear. If, say, $f(x) = ax$, then the requirement $a < 0$ would imply the three requirements above. Overall, the conditions on f are very modest.

The origin is clearly an equilibrium point (the only one) since $f(x) = 0$ only for $x = 0$. We can prove that the origin is asymptotically stable in the large. To do so we let

$$V(x) = -\int_0^x f(\sigma)\,d\sigma$$

Clearly $V(x)$ is continuous and has a continuous derivative. In view of the sign

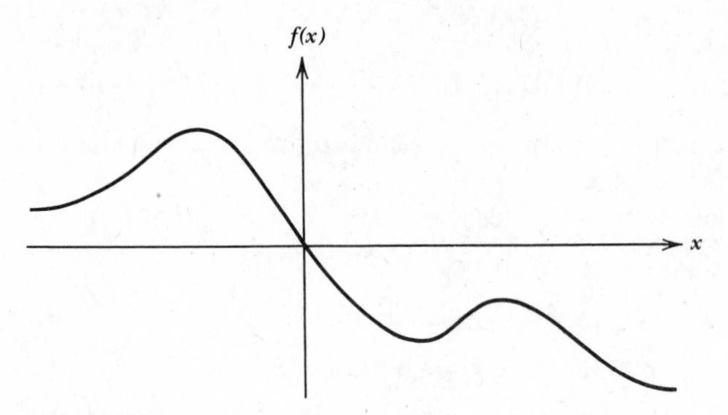

Figure 9.10. General form of $f(x)$.

assumption on f, it is easily seen that $V(x)>0$ for $x \neq 0$, $V(0)=0$. Also

$$\dot{V}(x) = \frac{dV(x)}{dx} f(x)$$
$$= -f(x)^2 < 0 \qquad (9\text{-}65)$$

for all $x \neq 0$. Finally, $V(x) \to \infty$ as $|x| \to \infty$ because of the assumption of the integral of $f(x)$. Thus, all conditions for asymptotic stability in the large are satisfied, according to Theorem 3 of Sect. 6.

As an example, this result establishes asymptotic stability in the large of a system such as

$$\dot{x} = \tanh(x) - x$$

which when linearized about $x=0$ gives $\dot{x}=0$. The linearization technique is inconclusive, while the integral Liapunov function yields extremely strong results.

*9.11 A QUADRATIC LIAPUNOV FUNCTION FOR LINEAR SYSTEMS

A very important form for a Liapunov function is a general quadratic function. It is used in numerous special cases. It has a special role in linear systems, since in that case the time derivative of the function is also quadratic.

Quadratic Forms

By a quadratic form in the variables $x_1, x_2, \ldots, x_n$ we mean a function of the form

$$V(\mathbf{x}) = p_{11}x_1^2 + p_{12}x_1x_2 + p_{13}x_1x_3 + \cdots + p_{1n}x_1x_n$$
$$+ p_{21}x_2x_1 + \cdots + p_{2n}x_2x_n + \cdots + p_{nn}x_n^2 \qquad (9\text{-}66)$$

It is a function where each term is a constant times the product of two variables. When written out in length this way, each pair of distinct variables may enter two different terms. For example, there is a term of the form $p_{12}x_1x_2$ and another of the form $p_{21}x_2x_1$. Only one of these is really necessary, but it is convenient to divide the term involving the product of two variables into two equal parts. Then $p_{ij} = p_{ji}$ for all i and j.

The quadratic function (9-66) can be expressed in matrix notation as

$$V(\mathbf{x}) = \mathbf{x}^T\mathbf{P}\mathbf{x} \tag{9-67}$$

where $\mathbf{P}$ is an $n \times n$ matrix whose entries are the coefficients in the long expression. By selecting the coefficients as described above, the matrix $\mathbf{P}$ will be symmetric. Here are some examples of quadratic forms.

(1) $V(x_1, x_2) = x_1{}^2 + x_2{}^2$ is equivalent to

$$V(\mathbf{x}) = [x_1\ x_2]\begin{bmatrix} 1 & 0 \\ 0 & 1 \end{bmatrix}\begin{bmatrix} x_1 \\ x_2 \end{bmatrix}$$

$$= \mathbf{x}^T\mathbf{I}\mathbf{x}$$

(2) $V(x_1, x_2) = (x_1 + x_2)^2 = x_1{}^2 + 2x_1x_2 + x_2{}^2$

$= x_1{}^2 + x_1x_2 + x_2x_1 + x_2{}^2$

This is equivalent to

$$V(\mathbf{x}) = [x_1\ x_2]\begin{bmatrix} 1 & 1 \\ 1 & 1 \end{bmatrix}\begin{bmatrix} x_1 \\ x_2 \end{bmatrix}$$

(3) $V(x_1, x_2) = x_1{}^2 - x_2{}^2 + 4x_1x_2$ is equivalent to

$$V(\mathbf{x}) = [x_1\ x_2]\begin{bmatrix} 1 & 2 \\ 2 & -1 \end{bmatrix}\begin{bmatrix} x_1 \\ x_2 \end{bmatrix}$$

A quadratic form $\mathbf{x}^T\mathbf{P}\mathbf{x}$ is said to be *positive semidefinite* if $\mathbf{x}^T\mathbf{P}\mathbf{x} \geq 0$ for every $\mathbf{x}$. It is said to be *positive definite* if $\mathbf{x}^T\mathbf{P}\mathbf{x} > 0$ for every $\mathbf{x} \neq \mathbf{0}$. The form $x_1{}^2 + x_2{}^2$ is positive definite. The form $(x_1 - x_2)^2$ is positive semidefinite. The form $x_1{}^2 - x_2{}^2 + 4x_1x_2$ is not positive semidefinite. Since a quadratic form is defined completely by its associated symmetric matrix $\mathbf{P}$, we can apply these same definitions to the matrix itself. Thus, a symmetric matrix $\mathbf{P}$ is said to be *positive semidefinite* if its associated quadratic form is positive semidefinite—and similarly for the other definitions.

It is, of course, important to be able to determine whether a given symmetric matrix $\mathbf{P}$ is positive definite without directly verifying that $\mathbf{x}^T\mathbf{P}\mathbf{x} > 0$ for all $\mathbf{x}$. Fortunately, this is not too difficult. One procecure is to examine the eigenvalues of $\mathbf{P}$. Since $\mathbf{P}$ is symmetric the eigenvalues are all real. (See Problem 19, Chapter 3.) The symmetric matrix $\mathbf{P}$ is positive semidefinite if and only if all its eigenvalues are nonnegative. It is positive definite if and only if the eigenvalues are strictly positive.

Another test is in terms of the principal minors of **P**. Specifically, the matrix **P** is positive definite if and only if

$$p_{11}>0, \quad \begin{vmatrix} p_{11} & p_{12} \\ p_{21} & p_{22} \end{vmatrix}>0, \ldots \quad |\mathbf{P}|>0$$

The reader should be able to quickly apply either of these tests to the three examples given above.

Liapunov Functions

Consider the linear homogeneous continuous-time system

$$\dot{\mathbf{x}}(t)=\mathbf{A}\mathbf{x}(t)$$

Let us associate with this system and the equilibrium point $\bar{\mathbf{x}}=\mathbf{0}$ the quadratic function

$$V(\mathbf{x})=\mathbf{x}^T\mathbf{P}\mathbf{x} \tag{9-68}$$

where **P** is symmetric and positive definite. This V is continuous and has continuous first partial derivatives. Furthermore, since **P** is positive definite, the origin is the unique minimum point of V. Thus in terms of general characteristics, such a positive definite quadratic form is a suitable candidate for a Liapunov function. It remains, of course, to determine how $\dot{V}(\mathbf{x})$ is influenced by the dynamics of the system.

We have*

$$\begin{aligned} \dot{V}(\mathbf{x}) &= \frac{d}{dt}\mathbf{x}^T\mathbf{P}\mathbf{x} \\ &= \dot{\mathbf{x}}^T\mathbf{P}\mathbf{x}+\mathbf{x}^T\mathbf{P}\dot{\mathbf{x}} \\ &= \mathbf{x}^T\mathbf{A}^T\mathbf{P}\mathbf{x}+\mathbf{x}^T\mathbf{P}\mathbf{A}\mathbf{x} \\ &= \mathbf{x}^T(\mathbf{A}^T\mathbf{P}+\mathbf{P}\mathbf{A})\mathbf{x} \end{aligned} \tag{9-69}$$

Therefore, defining the symmetric matrix

$$-\mathbf{Q}=\mathbf{A}^T\mathbf{P}+\mathbf{P}\mathbf{A} \tag{9-70}$$

we have

$$\dot{V}(\mathbf{x})=-\mathbf{x}^T\mathbf{Q}\mathbf{x} \tag{9-71}$$

* The earlier formula $\dot{V}(\mathbf{x})=\nabla V(\mathbf{x})f(\mathbf{x})$ yields $\dot{V}(\mathbf{x})=2\mathbf{x}^T\mathbf{P}\mathbf{A}\mathbf{x}$. However, $2\mathbf{x}^T\mathbf{P}\mathbf{A}\mathbf{x}=\mathbf{x}^T(\mathbf{P}\mathbf{A}+\mathbf{A}^T\mathbf{P})\mathbf{x}$, giving the same result as (9-69). The latter form is preferred since it expresses the result as a *symmetric* quadratic form.

We see that the function $\dot{V}(\mathbf{x})$ is also a quadratic form. The function V will be a Liapunov function if the matrix $\mathbf{Q}$ is positive semidefinite. If, in fact, $\mathbf{Q}$ is positive definite (rather than just semidefinite), we can infer that the system is asymptotically stable.

Example 1. Consider the system

$$\dot{\mathbf{x}}(t) = \begin{bmatrix} 0 & 1 \\ -2 & -3 \end{bmatrix} \mathbf{x}(t) \tag{9-72}$$

This system is asymptotically stable. Indeed, its eigenvalues are $\lambda = -1$, $\lambda = -2$. However, using $\mathbf{P} = \mathbf{I}$, corresponding to the positive definite function $V(\mathbf{x}) = x_1^2 + x_2^2$, does not yield a Liapunov function because in this case

$$\mathbf{Q} = -\mathbf{A} - \mathbf{A}^T = \begin{bmatrix} 0 & -1 \\ 2 & 3 \end{bmatrix} + \begin{bmatrix} 0 & 2 \\ -1 & 3 \end{bmatrix} = \begin{bmatrix} 0 & 1 \\ 1 & 6 \end{bmatrix} \tag{9-73}$$

which is not positive semidefinite. [Note, for instance, that $|\mathbf{Q}| = -1$.]

Example 2. For the system (9-72) of Example 1, let us use

$$\mathbf{P} = \begin{bmatrix} 5 & 1 \\ 1 & 1 \end{bmatrix}$$

This $\mathbf{P}$ is positive definite. The corresponding $\mathbf{Q}$ is

$$\begin{aligned} \mathbf{Q} &= -\begin{bmatrix} 5 & 1 \\ 1 & 1 \end{bmatrix} \begin{bmatrix} 0 & 1 \\ -2 & -3 \end{bmatrix} - \begin{bmatrix} 0 & -2 \\ 1 & -3 \end{bmatrix} \begin{bmatrix} 5 & 1 \\ 1 & 1 \end{bmatrix} \\ &= \begin{bmatrix} 4 & 0 \\ 0 & 4 \end{bmatrix} \end{aligned}$$

which is positive definite. Thus, the function $V(\mathbf{x}) = \mathbf{x}^T \mathbf{P} \mathbf{x}$ is a Liapunov function that explicitly demonstrates the asymptotic stability of (9-72). This illustrates that in general only certain positive definite quadratic forms can serve as a Liapunov function for a given asymptotically stable system. Nevertheless, if the system is asymptotically stable, it is always possible to find a suitable $\mathbf{P}$. (See Problem 17.)

9.12 COMBINED LIAPUNOV FUNCTIONS

When faced with a new system structure, it is sometimes possible to combine two or more of the simple forms presented in the last few sections in order to construct an appropriate Liapunov function. As an example, consider a nonlinear oscillatory system defined by

$$\ddot{x} + k\dot{x} + g(x) = 0 \tag{9-74}$$

where $g(x)$ satisfies $xg(x)>0$ for $x\neq 0$ and where $k>0$. This might represent the equation of a mass and nonlinear spring, subject to friction.

In state variable form the system becomes

$$\dot{x} = y$$
$$\dot{y} = -ky - g(x) \tag{9-75}$$

We shall show that a suitable Liapunov function is

$$V(x, y) = \tfrac{1}{2}y^2 + \int_0^x g(\sigma)\, d\sigma \tag{9-76}$$

This is a combination of a quadratic Liapunov function and an integral Liapunov function. It is easily verified that this function satisfies the first two requirements for a Liapunov function. To verify that the third requirement is satisfied, we calculate

$$\begin{aligned}\dot{V}(x, y) &= y\dot{y} + g(x)\dot{x} \\ &= -ky^2 - yg(x) + g(x)y \\ &= -ky^2 \le 0\end{aligned}$$

This establishes the stability of this general system. By using the invariant set stability theorem, asymptotic stability can be established.

As a final note, we point out that although systematic trial and error of analytical forms can often successfully lead to a suitable Liapunov function (as it apparently did in the above example), generally a Liapunov function has some intuitive significance within the context of the system itself—beyond simple mathematics. As an illustration, one should look again at the pendulum example of Sect. 9.7. It will be found to be a special case of the example of this section with $g(\theta)=(g/R)\sin\theta$. The Liapunov function used here is the same as that used in this example: namely, the energy of the system. Thus, in this case, as in many others, the appropriate mathematical construct has great physical significance.

9.13 GENERAL SUMMARIZING FUNCTIONS

As discussed in the beginning of this chapter, the Liapunov function can be regarded as a special case (a most important special case) of the concept of a summarizing function. The general underlying idea is to simplify the analysis of a complex high-order dynamic system by considering a single scalar-valued function whose time behavior can be estimated. In the case of a Liapunov function one concludes that the V function goes to a minimum. It follows that the state must go to the equilibrium point—although we do not know its precise path. This idea can often be used to summarize the general nature of the system even if it is unstable. The following two examples illustrate the general approach.

Example 1. Consider the system

$$\dot{x}_1 = x_1 x_2 \tag{9-77a}$$

$$\dot{x}_2 = x_2 \tag{9-77b}$$

Define $V(\mathbf{x}) = x_1^2 + x_2^2$, which is the square of the distance from the origin. We then have

$$\begin{aligned} \dot{V}(\mathbf{x}) &= 2x_1\dot{x}_1 + 2x_2\dot{x}_2 \\ &= 2(x_1^2 x_2 + x_2^2) \end{aligned} \tag{9-78}$$

For $x_2 \geq 1$ we can write

$$\dot{V}(\mathbf{x}) \geq 2V(\mathbf{x}) \tag{9-79}$$

From (9-77b) it is easy to see that if $x_2(0) > 1$, then $x_2(t) > 1$ for all $t > 0$. Thus, the inequality (9-79) will be valid for all $t > 0$ provided only that $x_2(0) > 1$. We can conclude that

$$V(\mathbf{x}(t)) \geq V(\mathbf{x}(0))e^{2t} \tag{9-80}$$

We conclude that the square of the length of the state vector grows at least as fast as e^{2t}. The length itself, the distance from the origin, increases at least as fast as e^t. This general qualitative information is obtained without detailed knowledge of the solution.

Example 2. The summarizing concept is sometimes valuable even in connection with linear systems. By selecting the summarizing function as an aggregate of several variables, a simple approximation can sometimes be deduced.

As a simple illustration of this idea, consider the positive system

$$\begin{bmatrix} x_1(k+1) \\ x_2(k+1) \\ x_3(k+1) \end{bmatrix} = \begin{bmatrix} 1 & 2 & 3 \\ 2 & 1 & 1 \\ 0 & 1 & 1 \end{bmatrix} \begin{bmatrix} x_1(k) \\ x_2(k) \\ x_3(k) \end{bmatrix} \tag{9-81}$$

where each $x_i(k)$ is nonnegative. Define the summarizing function

$$V(\mathbf{x}) = x_1 + x_2 + x_3 \tag{9-82}$$

By application of (9-81) we find

$$\begin{aligned} V(\mathbf{x}(k+1)) &= x_1(k+1) + x_2(k+1) + x_3(k+1) \\ &= 3x_1(k) + 4x_2(k) + 5x_3(k) \end{aligned}$$

Recalling that each $x_i(k) \geq 0$, we may write from the above

$$V(\mathbf{x}(k+1)) \leq 5[x_1(k) + x_2(k) + x_3(k)] = 5V(\mathbf{x}(k))$$

Likewise,

$$V(\mathbf{x}(k+1)) \geq 3[x_1(k)+x_2(k)+x_3(k)] = 3V(\mathbf{x}(k))$$

Thus, we have obtained the two inequalities

$$V(\mathbf{x}(k+1)) \leq 5V(\mathbf{x}(k)) \tag{9-83a}$$

$$V(\mathbf{x}(k+1)) \geq 3V(\mathbf{x}(k)) \tag{9-83b}$$

Thus, we can easily deduce that

$$3^k V(\mathbf{x}(0)) \leq V(\mathbf{x}(k)) \leq 5^k V(\mathbf{x}(0)) \tag{9-84}$$

Therefore, we have both upper and lower bounds on the summarizing function, and these bounds were quite easily found.

9.14 PROBLEMS

1. By a suitable change of variable, convert the logistic equation

$$\dot{x}(t) = a\left(1 - \frac{x(t)}{c}\right)x(t)$$

(with $a>0$, $c>0$) to a linear first-order differential equation. Show that if $0< x(0)<c$, then

$$x(t) = \frac{c}{1+be^{-at}}$$

and $b>0$.

2. Verify that the solution to the equation

$$\dot{x}(t) = a\left[1 + \frac{x(t)}{c}\right]x(t)$$

where $a>0$, $c>0$, $x(0)>0$ is

$$x(t) = \frac{c}{be^{-at}-1}$$

for some $b>0$.

3. For the system

$$\dot{x}(t) = \sin[x(t)+y(t)]$$

$$\dot{y}(t) = e^{x(t)} - 1$$

determine *all* the equilibrium points, and using Liapunov's first method, classify each equilibrium point as stable or unstable.

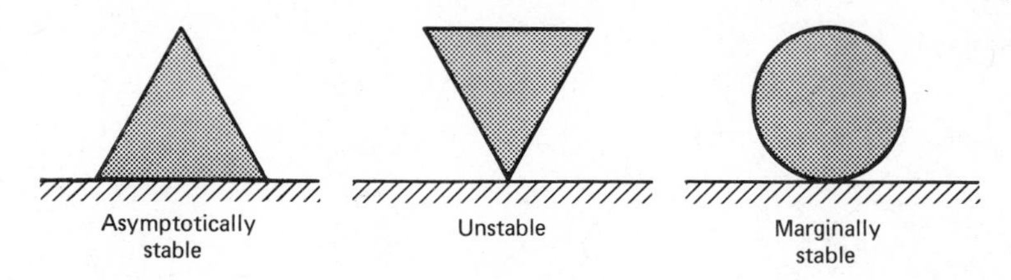

Figure 9.11. Which label is incorrect?

4. A classic "textbook" illustration of stability is that of Fig. 9.11. (Only pivoting motion is considered for the first two objects, and only rolling for the third.) Each object when at rest is in a state of mechanical equilibrium, but a small displacement of *angular* position leads in the three cases, respectively, to return to equilibrium, a large displacement, and no further change. Actually, although the above statements are correct, one of the labels is incorrect if the objects are considered as dynamic systems governed by Newton's laws. Which one, and why? (You do not need to write any equations.)

5. Using Liapunov's first method, determine whether the origin is a stable equilibrium point for each of the following systems:

   (a) $\dot{x}_1 = -x_1 + x_2^2$

   $\dot{x}_2 = -x_2(x_1 + 1)$

   (b) $\dot{x}_1 = x_1^3 + x_2$

   $\dot{x}_2 = x_1 - x_2$

   (c) $\dot{x}_1 = -x_1 + x_2$

   $\dot{x}_2 = -x_2 + x_1^2$

   (d) $x_1(k+1) = 2x_1(k) + x_2(k)^2$

   $x_2(k+1) = x_1(k) + x_2(k)$

   (e) $x_1(k+1) = 1 - e^{x_1(k)x_2(k)}$

   $x_2(k+1) = x_1(k) + 2x_2(k)$

6. *Model of Bacterial Growth.* An industrial plant's effluent waste is fed into a pool of bacteria that transforms the waste to nonpolluting forms. Maintaining the bacterial concentrations at effective levels is a critical problem. If the pool's oxygen supply, temperature, and pH are kept within acceptable limits, then the bacteria's ability to grow is primarily dependent on the supply of some nourishing organic substrate (for example, glucose or the waste itself).

   A simple mathematical model of growth can be derived from a few basic assumptions that were deduced from batch culture experiments by Monod. One observation is that the rate of growth of bacteria in the culture is approximately

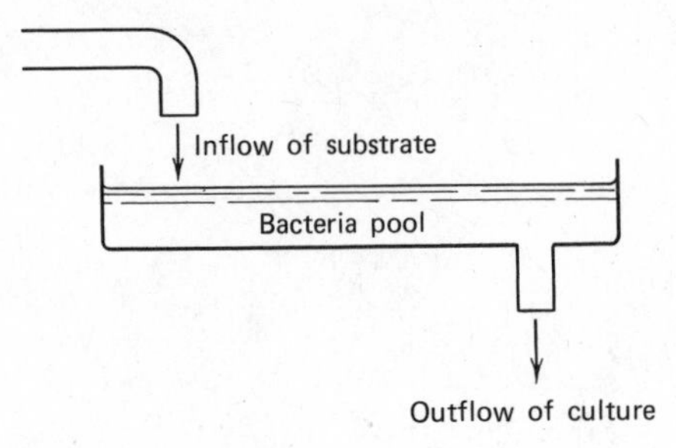

Figure 9.12. Bacteria pool.

proportional to the substrate concentration when the substrate is scarce, but tends towards a constant saturation level as the substrate becomes more plentiful. If we let x_1 equal the concentration of bacteria and x_2 equal the concentration of substrate, then we can represent the result by the equation

$$\dot{x}_1 = a\frac{x_2}{x_2+K}x_1$$

where a and K are positive constants.

A second observation is that the weight ratio of bacteria formed to substrate used is a constant. This gives a rate of change equation for the substrate

$$\dot{x}_2 = -b\frac{x_2}{x_2+K}x_1$$

where b is a positive constant.

In the pollution control application, there is a continual flow of nourishing substrate into the pool that is matched by a flow of bacterial culture out of the pool. (See Fig. 9.12.) Also, the rate of flow is controlled to be proportional to the volume of culture present in the pool. Adding this information to the equations above, setting a and b equal 1, and normalizing so that the concentration of substrate in the inflow stream is 1, yields the following model:

$$\dot{x}_1 = \frac{x_2}{x_2+K}x_1 - Dx_1$$

$$\dot{x}_2 = \frac{-x_2}{x_2+K}x_1 + D[1-x_2]$$

A stability analysis will reveal whether in the face of slight disturbances the culture will continue to be effective. In the following, assume $K>0$ and $0<D<1$.

(a) Determine the equilibrium points. What condition on D and K is required to insure that all equilibrium points are nonnegative?

(b) For each equilibrium point determine what conditions on D and K are required to insure asymptotic stability. Compare with part (a).

*7. *Liapunov Instability Theorem.* Prove the following: Let $\bar{\mathbf{x}}$ be an equilibrium point of the dynamic system $\mathbf{x}(k+1)=\mathbf{f}(\mathbf{x}(k))$. Assume there is a function $V(\mathbf{x})$ such that

$V(\bar{\mathbf{x}})=0$, $V(\mathbf{x})$ assumes positive values arbitrarily near $\bar{\mathbf{x}}$, and $\Delta V(\mathbf{x})>0$ for all $\mathbf{x}\neq\bar{\mathbf{x}}$. Then $\bar{\mathbf{x}}$ is unstable. (You should also state and prove the continuous-time version.)

8. *No Marginal Stability.* Consider the system

$$\dot{x}_1 = x_1[a^2 - x_1{}^2 - x_2{}^2] + x_2[a^2 + x_1{}^2 + x_2{}^2]$$
$$\dot{x}_2 = -x_1[a^2 + x_1{}^2 + x_2{}^2] + x_2[a^2 - x_1{}^2 - x_2{}^2]$$

and the function $V(\mathbf{x}) = x_1^2 + x_2^2$. Show that the system is asymptotically stable for $a = 0$ and unstable for $a \neq 0$. (Use Problem 7.) Thus, the transition from asymptotic stability to instability does not pass through a point of marginal stability.

*9. *Alternative System Formulations.* In many situations the choice between a discrete-time and a continuous-time representation is somewhat arbitrary. It is, of course, possible to approximate one version by another. Suppose that the system

$$\dot{\mathbf{x}}(t) = \mathbf{f}(\mathbf{x}(t)) \tag{0}$$

is given. By selecting a step size of Δ this can be approximated by a discrete-time system by either of two methods:

$$\frac{\mathbf{x}(k+1)-\mathbf{x}(k)}{\Delta} = \mathbf{f}(\mathbf{x}(k)) \tag{A}$$

$$\frac{\mathbf{x}(k+1)-\mathbf{x}(k)}{\Delta} = \mathbf{f}(\mathbf{x}(k+1)) \tag{B}$$

Suppose that the origin is stable for the original system (0) and that $V(\mathbf{x})$ is a corresponding Liapunov function. Suppose also that $V(\mathbf{x})$ is quasi-convex; that is, for any c if $V(\mathbf{x}) \leq c$ and $V(\mathbf{y}) \leq c$, then $V(\alpha\mathbf{x} + (1-\alpha)\mathbf{y}) \leq c$ for all α, $0 \leq \alpha \leq 1$. [The contours of $V(\mathbf{x})$ are convex.] Finally, assume that $\nabla V(\mathbf{x}) \neq \mathbf{0}$ for all $\mathbf{x} \neq \mathbf{0}$.

(a) Show that the corresponding discrete-time system defined by method A is not necessarily stable.

(b) Show that the corresponding discrete-time system defined by method B is stable.

(c) Suppose the linear system

$$\dot{\mathbf{x}}(t) = \mathbf{A}\mathbf{x}(t)$$

is asymptotically stable. Show directly, in terms of eigenvalues, that the corresponding discrete-time system defined by method B is also asymptotically stable.

10. Prove that the origin is stable for each of the systems below using Liapunov's second method. [In parts (a) and (b) find a suitable Liapunov function. In part (c) try the suggested function.]

(a) $\dot{x} = y$
$\dot{y} = -x^3$

(b) $\dot{x} = -x^3 - y^2$
$\dot{y} = xy - y^3$

(c) $\dot{x} = y(1-x)$
$\dot{y} = -x(1-y)$
$V = -x - \log(1-x) - y - \log(1-y)$

11. For a discrete-time linear system

$$\mathbf{x}(k+1) = \mathbf{A}\mathbf{x}(k)$$

let $V(\mathbf{x}) = \mathbf{x}^T\mathbf{P}\mathbf{x}$, where $\mathbf{P}$ is a symmetric positive definite matrix. What condition will insure that $V(\mathbf{x})$ is a Liapunov function. (Express the answer in terms of the positive semidefiniteness of some other symmetric matrix.)

12. *Margin of Safety.* Show that if there are symmetric, positive definite matrices $\mathbf{P}$ and $\mathbf{Q}$ such that

$$\mathbf{P}\mathbf{A} + \mathbf{A}^T\mathbf{P} + 2\lambda\mathbf{P} = -\mathbf{Q}$$

then all eigenvalues of $\mathbf{A}$ have a real part that is less than $-\lambda$.

13. *Steepest Descent.* Suppose that a function $f(x)$ is to be minimized with respect to the n-dimensional vector $\mathbf{x}$. A numerical procedure for finding the minimum is the method of steepest descent, which starts from an initial estimate $\mathbf{x}_0$ and successively computes new estimates according to

$$\mathbf{x}_{k+1} = \mathbf{x}_k - \alpha_k \mathbf{g}(\mathbf{x}_k)$$

where $\mathbf{g}(\mathbf{x}_k)$ is (the transpose of) the gradient of f at $\mathbf{x}_k$. The scalar α_k is chosen so as to minimize the function $f(\mathbf{x}_k - \alpha\mathbf{g}(\mathbf{x}_k))$ with respect to α. (This last operation involves a one-dimensional numerical search.)

Assume that the function f satisfies the following properties:

(1) f has a unique minimum at the point $\bar{\mathbf{x}}$.
(2) f has continuous partial derivatives, and the gradient of f vanishes only at $\bar{\mathbf{x}}$.
(3) $f(\mathbf{x}) \to \infty$ as the magnitude of any component $\mathbf{x}$ goes to infinity.
(4) $\alpha_k = \alpha(x_k)$ is continuous in $\mathbf{x}_k$.

Show that the iterative procedure converges to the solution $\bar{\mathbf{x}}$ from any starting point.

14. *Musicians, Jugglers, and Biological Clocks.* How is it that a group of musicians, playing together without a leader, are able to stay together? How do a pair of jugglers, whose performance depends on precise timing, keep synchronized? Or, in a biological context, why do the muscle elements that compose the heart contract at the same rate, and how do certain organisms become synchronized to the daily light cycle?

In general we can conjecture that synchronization is possible because an individual responds to the general performance of others, speeding up if the others are ahead, and slowing down if they are behind. To construct a model along these lines, we define the variable x_i as the *position* of the ith member of the group. The position is a general measure of the total phase transversed from some reference point (e.g., in music, x_i is the position in the musical score, which should increase linearly with time). We assume that each member i has a notion as to the proper speed A_i, which he would adopt if he were alone. In the presence of others, however, he modifies his speed if he deviates from the group average. Thus letting

$\bar{x}=\sum_{k=1}^{n} x_k/n$ we postulate

$$\dot{x}_1(t)=A_1-f_1(x_1(t)-\bar{x}(t))$$
$$\dot{x}_2(t)=A_2-f_2(x_2(t)-\bar{x}(t))$$
$$\vdots$$
$$\dot{x}_n(t)=A_n-f_n(x_n(t)-\bar{x}(t))$$

We assume that the response function f has a continuous derivative and satisfies

(1) $f_i(0)=0, \quad i=1,2,\ldots,n$

(2) $f_i'(y)>0$ all y and $i=1,2,\ldots,n$

(3) $|f_i(y)|\to\infty$ as $|y|\to\infty$ $i=1,2,\ldots,n$

(a) In this example one does not seek an equilibrium point in the usual sense, but rather a trajectory that moves according to $\dot{x}_i(t)=A$ for all i, corresponding to a situation where the *relative* positions of all individuals are constant. To show that such a condition is sustainable, assume there is a unique A and a set of numbers δ_i, $i=1,2,\ldots,n$ with $\Sigma\,\delta_i=0$ such that $A_i-f_i(\delta_i)=A$ for all i. Find a corresponding synchronized trajectory, and show that it is an invariant set of the system.

(b) For the case $n=2$, convert the two equations to a single equation in the variable $z(t)=x_1(t)-\bar{x}(t)$ and show that the equilibrium point is asymptotically stable in the large using an integral Liapunov function.

15. *Observability and Stability.* Consider the nth-order system

$$\dot{\mathbf{x}}(t)=\mathbf{A}\mathbf{x}(t)$$

Suppose that there is a symmetric positive definite matrix $\mathbf{P}$ such that $\mathbf{PA}+\mathbf{A}^T\mathbf{P}=-\mathbf{c}\mathbf{c}^T$, where $\mathbf{c}^T$ is an n-dimensional vector. Suppose also that the pair $\mathbf{A}$, $\mathbf{c}^T$ is completely observable. Show that the system is *asymptotically* stable. Interpret this result.

16. *The van der Pol Equation.* The equation

$$\ddot{x}+\varepsilon[x^2-1]\dot{x}+x=0$$

arises in the study of vacuum tubes. Show that if $\varepsilon<0$, the origin is asymptotically stable. (An important but deep result is that for $\varepsilon>0$ there is a limit cycle that is approached by all other trajectories.)

17. *Liapunov Equation.* The quadratic Liapunov function for linear systems is actually completely general. We can show: If $\mathbf{A}$ is an $n\times n$ matrix with all eigenvalues in the left half of the complex plane, and if $\mathbf{Q}$ is any symmetric positive definite $n\times n$ matrix, then there is a positive definite $\mathbf{P}$ such that $\mathbf{PA}+\mathbf{A}^T\mathbf{P}=-\mathbf{Q}$. To prove this define

$$\mathbf{P}=\int_0^\infty e^{\mathbf{A}^Tt}\mathbf{Q}e^{\mathbf{A}t}\,dt$$

(a) Show that the integral exists and that $\mathbf{P}$ is symmetric and positive definite.
(b) Show that

$$-\mathbf{Q}=\int_0^\infty \frac{d}{dt}(e^{\mathbf{A}^Tt}\mathbf{Q}e^{\mathbf{A}t})\,dt=\mathbf{PA}+\mathbf{A}^T\mathbf{P}$$

18. *A Cubic System.* Given the system

$$\dot{x}_1=x_2$$
$$\dot{x}_2=x_3$$
$$\dot{x}_3=-(x_1+cx_2)^3-bx_3$$

consider the function

$$V(x_1,x_2,x_3)=\frac{b}{4}x_1^4+\frac{1}{4c}(x_1+cx_2)^4$$
$$-\frac{1}{4c}x_1^4+\frac{b^2}{2}x_2^2+bx_2x_3+\tfrac{1}{2}x_3^2$$

(a) What condition on $b>0$, $c>0$ insures that $V(x_1, x_2, x_3)>0$ for $(x_1, x_2, x_3)\neq 0$?
(b) Show that $\dot{V}$ has the form

$$\dot{V}(x_1,x_2,x_3)=\gamma x_2^2[(\tfrac{3}{2}x_1+cx_2)^2+\tfrac{3}{4}x_1^2]$$

and determine the constant γ.
(c) Is V always a Liapunov function if b and c satisfy the conditions of part (a)?
(d) Is the origin asymptotically stable under these conditions?

19. *Krasovskii's Method.* Consider the system $\dot{\mathbf{x}}=\mathbf{f}(\mathbf{x})$. Assume that $\mathbf{f}(\mathbf{x})=\mathbf{0}$ if and only if $\mathbf{x}=\mathbf{0}$, and that $\mathbf{F}$, the Jacobian, exists in the region of interest. A trial Liapunov function is chosen to be the Euclidean norm of $\dot{\mathbf{x}}$ squared

$$V(\mathbf{x})=\|\dot{\mathbf{x}}\|^2=\mathbf{f}(\mathbf{x})^T\mathbf{f}(\mathbf{x})$$

(a) Find *sufficient* conditions for V to be a Liapunov function (with respect to $\bar{\mathbf{x}}=\mathbf{0}$). Express the answer in terms of the positive semidefiniteness of a symmetric matrix. Note that if the matrix is positive definite the origin is asymptotically stable.
(b) Consider the control system given by (assume $g_1(0)=g_2(0)=0$)

$$\dot{x}_1=g_1(x_1)+g_2(x_2)$$
$$\dot{x}_2=x_1-ax_2$$

Use the results of part (a) to establish conditions for asymptotic stability.
(c) Now suppose for the system of part (b)

$$g_1(x_1)=-x_1^3-x_1$$
$$g_2(x_2)=\tfrac{1}{2}x_2^2$$
$$a=1$$

Apply the results of part (b) to this system.

NOTES AND REFERENCES

Sections 9.1–9.13. Most of the foundation for stability analysis was laid by Liapunov [L14]. A very readable introduction is LaSalle and Lefschetz [L2]. Also see Kalman and Bertram [K4], [K5]. The principle of competitive exclusion example is due to Volterra [V1], who actually considered more general forms for $F(\mathbf{x})$. The quadratic Liapunov function for linear systems was proposed in Liapunov's original investigation. The combined integral and quadratic function is referred to as the Luré form.

Section 9.14. The result of Problem 9 appears to be new. For details on the bacteria culture problem see Monod [M6].

chapter 10.

Some Important Dynamic Systems

As emphasized in Chapter 9, one of the most useful principles for analysis of nonlinear systems is that of the Liapunov function—or, more generally, the summarizing function. To apply this principle, however, it is necessary to construct a special function suitable for the particular situation at hand; and such a construction for an unfamiliar set of equations is rarely easy. Knowledge of particular examples can be helpful, for, sometimes, a suitable function can be found by combining or modifying functions that work in other situations. Nevertheless, as a purely mathematical venture, discovery of a suitable summarizing function is far from routine. Indeed, from this viewpoint, the summarizing function principle might be dismissed as elegant in concept, but not readily useful. From a broader viewpoint, however, the principle has great utility, for a suitable function often has significance within the physical or social context of the system. We observed this earlier by noting that a Liapunov function might correspond to energy in the pendulum example, or distance in a pursuit problem. Most Liapunov functions have similar intuitive or instructive interpretations.

This theme is expanded in this chapter. It is argued that the summarizing function concept is almost a fundamental principle of scientific advance—at least in connection with phenomena described in terms of nonlinear dynamics. Many sciences were finally considered to have attained a state of maturity only when the underlying dynamic laws possessed the degree of simplicity and order represented by the discovery of a suitable summarizing function. Indeed, in some cases the summarizing function is regarded as perhaps more important than the dynamic equations of motion. And in nearly every case, the summarizing function has important scientific or intuitive meaning and is considered to be an integral part of the discipline. Thus, the summarizing function is more than a fragile mathematical concept; it is a concept that links together various scientific fields.

This chapter explores mechanics, thermodynamics, population ecology, epidemics, economics, and population genetics to illustrate how the summarizing function concept relates to the underlying scientific laws. The corresponding summarizing functions include the widely known constructs of energy, entropy, and fitness. The sections are essentially independent and some are simply brief outlines of broad fields, but together this collection forms a fascinating set of important dynamic systems.

10.1 ENERGY IN MECHANICS

The dynamic behavior of a mechanical system is governed by *Newton's Second Law of Motion*, which for a single particle of fixed mass m is

$$\mathbf{f} = m\frac{d\mathbf{v}}{dt} \tag{10-1}$$

where $\mathbf{f}$ is the total *force vector* acting on the particle and $\mathbf{v}$ is the velocity vector of the particle. To apply this equation to a given situation, it is necessary to have a clear understanding of what constitutes force. To a great extent this is clarified through the introduction of energy.

We define the work done by a force $\mathbf{f}$ acting on a particle going from point 1 to point 2 to be the integral

$$W_{12} = \int_1^2 \mathbf{f}^T\, d\mathbf{s}$$

That is, incremental work is the scalar product of the force vector with the vector of incremental movement. When this definition is applied to a particle of mass m, Eq. (10-1) yields*

$$W_{12} = \int_1^2 m\frac{d\mathbf{v}^T}{dt}\frac{d\mathbf{s}}{dt}\,dt = \int_1^2 m\frac{d\mathbf{v}^T}{dt}\mathbf{v}\,dt = \tfrac{1}{2}m\int_1^2 \frac{dv^2}{dt}\,dt$$

and therefore

$$W_{12} = \tfrac{1}{2}m(v_2{}^2 - v_1{}^2) \tag{10-2}$$

The quantity

$$T = \tfrac{1}{2}mv^2 \tag{10-3}$$

* In this section a symbol such as v denotes the magnitude of the corresponding vector $\mathbf{v}$.

is termed the *kinetic energy* of the particle, and, accordingly, the change in kinetic energy is equal to the work done.

If the force $\mathbf{f}$ has the property that the total work done around any closed path is zero, then the force is said to be *conservative.* For example, gravitational forces are conservative, but frictional forces are nonconservative. When the force is conservative, it is possible to define a *potential energy* function V on the coordinate space. Setting the potential at some reference point equal to zero, the potential of any other point is defined to be the negative of the work done in moving to it from the reference point. This value is unique, independent of the path chosen between the two points, because the work around any closed path is zero. The potential function is often simpler to deal with than the force because potential is scalar-valued while force is vector-valued. However, the force vector can be easily recovered from the potential energy by the relation

$$\mathbf{f} = -\nabla V \tag{10-4}$$

That is, the force is the negative of the gradient of the potential energy function.

It is now possible to formulate the law of conservation of energy. For a conservative force, the work done in moving from point 1 to point 2 is

$$W_{12} = V_1 - V_2$$

However, from (10-2) and (10-3)

$$W_{12} = T_2 - T_1$$

By subtracting these two equations we obtain the relation

$$T_1 + V_1 = T_2 + V_2 \tag{10-5}$$

This result can be formulated as the law of conservation of mechanical energy: *If the forces acting on a particle are conservative, then the total mechanical energy, $T + V$, is conserved.*

The above development for motion of a single particle can be extended to more complex mechanical systems consisting of several interacting bodies. Again, if the external forces are conservative, total mechanical energy is conserved. In many situations, of course, a system is subjected to frictional and other dissipative forces that generate heat. (The pendulum example of Sect. 9.7 is a good illustration.) These systems are not mechanically conservative—the total mechanical energy *decreases* with time. Because of this decreasing property, it is clear that the mechanical energy can serve as a Liapunov function for dissipative as well as conservative mechanical systems. It is important to recognize that this general property can be applied to any mechanical system, even if the associated differential equations appear complex. One simply expresses the total mechanical energy in terms of the system variables. Thus, mechanical energy serves both as a fundamental physical concept, which adds clarification and unity to Newton's laws, and as a Liapunov function for stability analysis.

Example (Planetary Motion). Newton's explanation of planetary motion, as described by Kepler's laws, is that planets are subjected to a conservative force derived from a potential of the form

$$V = -k/r$$

where r is the distance from the sun and k is a constant (depending on the masses of the planet and the sun). The associated force on the planet is

$$\mathbf{f} = -\nabla V$$

The force has magnitude

$$f = -k/r^2$$

and is directed toward the sun. The force can therefore be regarded as a (gravitational) attraction of the planet toward the sun. The fact that the force is derived from a potential guarantees that the force is conservative and that total energy is conserved. Thus, even without writing and solving the specific system of differential equations governing planetary motion, we can conclude that periodic orbits are sustainable.

10.2 ENTROPY IN THERMODYNAMICS

Entropy in thermodynamics represents one of the most significant scientific laws having a Liapunov character, and it is therefore an important example supporting the general theme of the chapter. Our discussion of thermodynamics is, however, brief, for it is merely intended to illustrate the general nature of the field and the role of entropy. A more complete discussion would require a thorough study of background material.

Thermodynamics is concerned with processes involving heat exchanges. In fact, the science of thermodynamics began in about 1760 with the recognition by Joseph Black of the distinction between heat and temperature. Different substances of the same weight and temperature may contain different amounts of heat. The *first law of thermodynamics* states that heat is a form of energy and that, when account is taken of this equivalence, the energy of an isolated system is conserved. This is, of course, a generalization of the result for conservative mechanical systems.

Another fundamental principle is the *second law of thermodynamics.* The second law is expressed in terms of *entropy*, and the law, stating that entropy of a thermodynamic system cannot decrease, can be interpreted in our terms as stating that entropy is a Liapunov function. Unlike our previous examples, however, the second law is stated as a universal law governing a broad class of dynamic processes, even without an explicit statement of the underlying dynamic equations. Whatever the dynamic processes might be, they must obey the second law.

A given substance constituting a thermodynamic system (an amount of liquid, gas, or solid) is described by its state—its temperature, pressure, volume, and so forth. The substance may undergo changes in state by the addition of heat, the performance of work, or by simply moving toward equilibrium. Among all these processes of change, there is distinguished an idealized set of *reversible processes*, defining paths of movement in the state space that ideally could be traversed in either direction. Such processes are only hypothetical, for to achieve them would require that the substance be held completely homogeneous (with respect to temperature, pressure, etc.) throughout the change. In practice the ideal sometimes can be approximated by conducting the process very slowly.

The first part of the second law of thermodynamics says that there is a function S, called entropy, that is a function of the state of a thermodynamic system. If two states 1 and 2 are connected by a reversible path, then the difference in entropy is

$$S_2 - S_1 = \int_1^2 \frac{dQ}{T} \tag{10-6}$$

where Q is the instantaneous heat added to the substance, and T is the temperature. If a value of entropy is assigned at a reference state, the value at another state can be found by devising a reversible process from one to the other, and evaluating the integral.

The second part of the second law of thermodynamics states that for real (irreversible) processes, (10–6) is replaced by

$$S_2 - S_1 \geq \int_1^2 \frac{dQ}{T} \tag{10-7}$$

Thus, an entropy change is greater for an irreversible process than for a reversible one. As a consequence of the second law, the entropy of an isolated system (one in which no heat or work is exchanged with the external environment) can never decrease. Therefore, thermodynamic processes of isolated systems follow paths of nondecreasing entropy, and accordingly entropy acts as a (negative) Liapunov function, which assures us that natural thermodynamic processes tend toward equilibrium.

Example. Suppose two identical bricks of material are initially at different temperatures T_1 and T_2. They are placed together so that through the process of heat conduction, heat may pass from one to the other. We assume that the heat contained in such a material is proportional to its temperature; that is, $Q = cT$.

If no work is done and no additional heat is added or none is lost from the system, the first law states that the energy in the two bricks must not change during the conduction process. Thus, the heat added to one brick must equal the heat subtracted from the other. It is easy to see that the condition of both bricks having the common temperature of $T_F = (T_1 + T_2)/2$ is consistent with this energy requirement. However, it is not possible, using the first law alone, to deduce that the two-brick system actually tends to this common temperature configuration. That is, it does not establish that this configuration is actually a stable equilibrium point, since there are many other configurations with the same total heat content.

Now let us consider the change in entropy of the two brick system if it moves from the initial condition to the configuration where the two temperatures are equal. For each brick we have $Q = cT$ for some positive constant c. The change in entropy of the first brick, going from temperature T_1 to T_F, is therefore

$$\Delta S_1 = \int_{T_1}^{T_F} \frac{dQ}{T} = \int_{T_1}^{T_F} c\frac{dT}{T} = c \log \frac{T_F}{T_1}$$

Likewise for the second brick

$$\Delta S_2 = c \log \frac{T_F}{T_2}$$

The total entropy change is therefore

$$\Delta S = \Delta S_1 + \Delta S_2 = c \log \frac{T_F^2}{T_1 T_2}$$

Using $T_F = \frac{1}{2}(T_1 + T_2)$, we find

$$\begin{aligned} \Delta S &= c \log \frac{(T_1 + T_2)^2}{4T_1 T_2} \\ &= c \log \tfrac{1}{4}\left(\frac{T_1}{T_2} + 2 + \frac{T_2}{T_1}\right) \end{aligned} \tag{10-8}$$

This last term is always nonnegative (being zero only if $T_1 = T_2$). Thus, the entropy is greater for equal brick temperatures than for the original configuration.

One can show directly that the state corresponding to equal temperatures of T_F represents maximization of entropy consistent with the given amount of

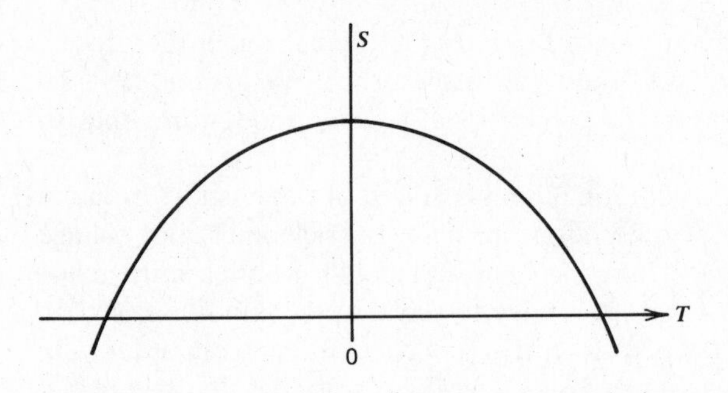

Figure 10.1. Entropy vs. temperature difference.

heat. If the bricks share heat of $2cT_F$, this heat can be divided to have one brick at temperature $T_F - T$ and the other at $T_F + T$, where T is arbitrary (but with $0 \leq |T| \leq T_F$). The entropy of the combination for various T is shown in Fig. 10.1 and is maximized at $T = 0$, corresponding to equality of temperature. Thus, even without reference to a specific set of equations governing conduction, the Liapunov character of entropy assures us that the equal temperature configuration is a stable equilibrium.

10.3 INTERACTING POPULATIONS

In this section we investigate the rather rich theory of interacting populations. The theory can be considered as the study of a certain class of nonlinear differential equations characterized by quadratic terms. The theory has endured, and has been borrowed by other fields as well, partly because the quadratic term has a natural probabilistic interpretation, and partly because a simple explicit Liapunov function is available for a large class of these equations. The Lotka–Volterra equations are essential items in one's catalog of examples.

The Predator–Prey Model

A classical model for a predator–prey system of two interacting populations (say wolves and goats) is given by the Lotka–Volterra equations

$$\begin{aligned} \dot{N}_1 &= aN_1 - bN_1N_2 \\ \dot{N}_2 &= -cN_2 + dN_1N_2 \end{aligned} \tag{10-9}$$

In these equations N_1 and N_2 represent, respectively, the prey and predator populations. The constants a, b, c, and d are all positive.

The model is based on the assumption that in the absence of predators the prey population will increase exponentially with a growth rate factor a.

Similarly, in the absence of prey, the predator population will diminish at a rate c. When both populations are present, the frequency of "encounters" is assumed to be proportional to the product of the two populations. The encounters directly decrease the number of prey and increase the number of predators; the associated coefficients being b and d, respectively. Of course, these equations are highly simplified and do not account for a number of external factors. Other factors that may either influence the coefficient values or add additional terms include the general environmental conditions (temperature, rainfall, etc.), the supply of other food for both predators and prey, and migration of the populations.

Volterra originally developed this model in order to explain the periodic oscillation in the composition of fish catches in the Adriatic Sea. The model has been used, however, in a variety of contexts to explain or predict fluctuating populations. An important application of a model of this type is to the study and control of pests that feed on agricultural crops. The pest population is often controlled by the introduction of predators, and the predator–prey model often forms a foundation for the design of important programs of ecological intervention.

The nonlinear dynamic equations (10-9) cannot be solved analytically in terms of elementary functions. It is, however, easy to see that there are equilibrium points. They are found by setting $\dot{N}_1 = \dot{N}_2 = 0$. This produces

$$0 = aN_1 - bN_1N_2$$
$$0 = -cN_2 + dN_1N_2$$

Thus, there is one equilibrium point at $N_1 = N_2 = 0$ and another at

$$N_1 = c/d \qquad N_2 = a/b$$

It is convenient to normalize variables so as to eliminate the need to carry along four parameters. Let

$$x_1 = \frac{d}{c} N_1 \qquad x_2 = \frac{b}{a} N_2$$

In terms of these variables the dynamic equations are

$$\begin{aligned} \dot{x}_1 &= ax_1(1 - x_2) \\ \dot{x}_2 &= -cx_2(1 - x_1) \end{aligned} \tag{10-10}$$

with the nonzero equilibrium point at $x_1 = 1$, $x_2 = 1$.

Let us investigate the stability of the two equilibrium points $(0, 0)$ and $(1, 1)$. First it is clear that $(0, 0)$ is unstable, for if x_1 is increased slightly it will grow exponentially. The point $(1, 1)$, however, requires a more detailed analysis. A linearization of the system, in terms of displacements Δx_1, Δx_2 from the equilibrium point (1, 1) is found, by evaluation of the first partial derivatives of (10-10) at (1, 1), to be

$$(\Delta\dot{x}_1) = -a(\Delta x_2)$$
$$(\Delta\dot{x}_2) = c(\Delta x_1)$$

or, in matrix form,

$$\begin{bmatrix}\Delta\dot{x}_1\\ \Delta\dot{x}_2\end{bmatrix} = \begin{bmatrix}0 & -a\\ c & 0\end{bmatrix}\begin{bmatrix}\Delta x_1\\ \Delta x_2\end{bmatrix}$$

The linearized system has eigenvalues $\pm i\sqrt{ac}$ representing a marginally stable system. From a linear analysis (the first method of Liapunov) it is impossible to infer whether in fact the equilibrium point is stable or unstable. It is necessary, therefore, to study the nonlinearities more explicitly.

We derive a function V that is constant along solutions. From (10-10) we can write

$$\frac{\dot{x}_2}{\dot{x}_1} = \frac{-cx_2(1-x_1)}{ax_1(1-x_2)}$$

Rearranging this so as to collect x_1 terms together and x_2 terms together leads to

$$c\dot{x}_1 - c\frac{\dot{x}_1}{x_1} + a\dot{x}_2 - a\frac{\dot{x}_2}{x_2} = 0$$

Each term can be integrated separately, producing

$$cx_1 - c\log x_1 + ax_2 - a\log x_2 = \log k$$

where k is a constant.

In view of the above, let us define the function, for $x_1 > 0$, $x_2 > 0$:

$$V(x_1, x_2) = cx_1 - c\log x_1 + ax_2 - a\log x_2 \tag{10-11}$$

We can conclude that V is a constant of motion, since its time derivative is zero. Therefore, the trajectory of population distribution lies on a fixed curve defined by $V = k$. Figure 10.2 shows what the curve might look like for various values of k.

From this analysis we see that the trajectories cycle around the equilibrium point. Hence the equilibrium is stable, but not asymptotically stable. The function V is easily shown to achieve a minimum at the equilibrium point (1, 1). Thus, V serves as a Liapunov function for the predator–prey system, and establishes stability. This function is a natural summarizing function associated with the system of interacting populations, and as shown below it plays an important role even in cases where it is not constant along trajectories. We refer to this function as the *ecological Liapunov function.*

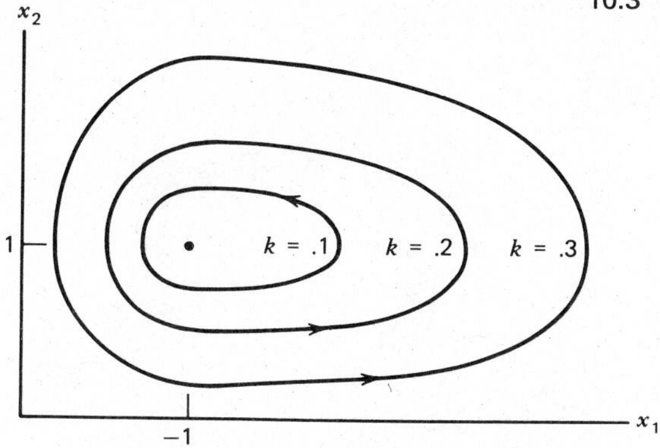

Figure 10.2. Predator–prey cycles.

The Effect of Crowding

As an illustration of how the basic ecological Liapunov function for the pure predator–prey system can be used to study modifications of the predator–prey model, let us introduce a term representing retardation of prey growth due to crowding. In particular, we consider the equations

$$\begin{aligned} \dot{N}_1 &= aN_1 - bN_1N_2 - eN_1^2 \\ \dot{N}_2 &= -cN_2 + dN_1N_2 \end{aligned} \tag{10-12}$$

The interpretation of these equations is essentially the same as before, except that now in the absence of predators growth of the prey population is governed by a standard logistic equation. The constants a, b, c, d, and e are again assumed to be positive.

One equilibrium point is again $N_1 = N_2 = 0$. Another is $N_1 = a/e$, $N_2 = 0$, corresponding to the equilibrium of the logistic growth of the prey in the absence of predators. Any equilibrium point with nonzero values of both N_1 and N_2 must satisfy the equations

$$\begin{aligned} 0 &= a - bN_2 - eN_1 \\ 0 &= -c + dN_1 \end{aligned}$$

This set of equations has the unique solution

$$\bar{N}_1 = c/d \quad \bar{N}_2 = \frac{da - ec}{bd} \tag{10-13}$$

Therefore, there is a positive equilibrium only if $a/e > c/d$.

Following the earlier procedure, we introduce the change of variables

$$x_1 = \frac{d}{c} N_1 \quad x_2 = \frac{bd}{da - ec} N_2 \tag{10-14}$$

which converts (10-12) to the simpler form

$$\begin{aligned}\dot{x}_1 &= \alpha x_1(1-x_2)+\beta x_1(1-x_1)\\ \dot{x}_2 &= -cx_2(1-x_1)\end{aligned} \tag{10-15}$$

where

$$\alpha = a - \frac{ec}{d}, \qquad \beta = \frac{ec}{d}$$

Now let us define, as before, for $x_1 > 0$, $x_2 > 0$:

$$V(x_1, x_2) = cx_1 - c\log x_1 + \alpha x_2 - \alpha \log x_2 \tag{10-16}$$

This function has its minimum at the equilibrium point (1, 1). Furthermore, by direct calculation we find

$$\begin{aligned}\dot{V}(x_1, x_2) &= c\dot{x}_1 - c\frac{\dot{x}_1}{x_1} + \alpha\dot{x}_2 - \alpha\frac{\dot{x}_2}{x_2}\\ &= c\alpha(1-x_2)(x_1-1) - c\beta(1-x_1)^2\\ &\quad - c\alpha(1-x_1)(x_2-1)\end{aligned}$$

Thus,

$$\dot{V}(x_1, x_2) = -c\beta(1-x_1)^2 \leq 0$$

Therefore, V is a Liapunov function of the system. Using the invariant theory of Sect. 9.8 it can be shown that (1, 1) is asymptotically stable (over the interior of the positive quadrant).

The *n*-species Case

Let us consider the general n-species model of population interaction:

$$\dot{x}_i = k_i x_i + b_i^{-1} \sum_{j=1}^{n} a_{ij} x_i x_j \qquad i = 1, 2, \ldots, n \tag{10-17}$$

In this system k_i is the linear growth constant, which can be either positive or negative. The term a_{ii} represents the quadratic growth term, which is usually zero or negative. The a_{ij}'s, $j \neq i$ represent the species interaction terms arising from predation, competition for resources, and so forth. These terms may have any value. Finally, the b_i's are positive normalizing factors that are used simply to provide some flexibility in defining the matrix $\mathbf{A}$ of coefficients a_{ij}.

An important special case is the case where each $a_{ii} = 0$ and the interaction coefficients a_{ij} are due to predation. An encounter of species i and j results in an increase of one and a decrease of the other in their individual growth rates. In that case a_{ij} and a_{ji} have opposite signs. It may be possible to select b_i's so that in every case b_i^{-1}/b_j^{-1} is the ratio of the respective increase and decrease. In this case $a_{ij} = -a_{ji}$ for all i, j and this antisymmetric case is referred to as a *Volterra ecology*.

Let us assume that the system (10-17) has an equilibrium point in which all species have nonzero population. In this case the equilibrium population levels $\bar{x}_i$, $i = 1, 2, \ldots, n$ must satisfy the system of linear equations

$$k_i b_i + \sum_{j=1}^{n} a_{ij}\bar{x}_j = 0 \tag{10-18}$$

Motivated by the two-dimensional case, we seek a generalization of the ecological Liapunov function. We define

$$V(\mathbf{x}) = \sum_{i=1}^{n} b_i(x_i - \bar{x}_i \log x_i) \tag{10-19}$$

This function is minimized at the equilibrium point. It remains to calculate $\dot{V}(\mathbf{x})$. We have

$$\begin{aligned} \dot{V}(\mathbf{x}) &= \frac{d}{dt}\sum_{i=1}^{n} b_i(x_i - \bar{x}_i \log x_i) \\ &= \sum_{i=1}^{n} b_i\left(1 - \frac{\bar{x}_i}{x_i}\right)\dot{x}_i \\ &= \sum_{i=1}^{n} b_i\left(1 - \frac{\bar{x}_i}{x_i}\right)\left(k_i x_i + b_i^{-1}\sum_{j=1}^{n} a_{ij}x_i x_j\right) \\ &= \sum_{i=1}^{n} b_i(x_i - \bar{x}_i)\left(k_i + b_i^{-1}\sum_{j=1}^{n} a_{ij}x_j\right) \end{aligned}$$

Using (10-18)

$$\dot{V}(\mathbf{x}) = \sum_{i=1}^{n} (x_i - \bar{x}_i) \sum_{j=1}^{n} a_{ij}(x_j - \bar{x}_j)$$

Finally,

$$\dot{V}(\mathbf{x}) = (\mathbf{x} - \bar{\mathbf{x}})^T \mathbf{A}(\mathbf{x} - \bar{\mathbf{x}}) \tag{10-20}$$

Therefore, $\dot{V}(\mathbf{x})$ is a quadratic form, and stability of the ecology can be easily inferred under appropriate conditions on the matrix $\mathbf{A}$.

In the case of a Volterra ecology, $\mathbf{A}$ is antisymmetric and it follows that $(\mathbf{x} - \bar{\mathbf{x}})^T \mathbf{A}(\mathbf{x} - \bar{\mathbf{x}}) = 0$ for all $\mathbf{x}$. Thus, as in the simplest two-dimensional predator–prey system, the function V is a constant of motion and the trajectories follow closed paths.

In general, if $\mathbf{A} + \mathbf{A}^T$ is negative semidefinite, the equilibrium of the n-species model is stable. This general result provides a simple and effective basis for the analysis of many cases.

10.4 EPIDEMICS

Epidemics of disease in human or other populations represent large-scale, important, and often tragic examples of dynamic phenomena. From a small initial group of infected individuals, a disease can spread to enormous proportions. The bubonic plague, smallpox, typhus, and other diseases have in some cases killed a substantial portion of the existing population of a country. While it is often possible to cure affected individuals directly, the most effective means of health maintenance has been preventative in nature, essentially altering the dynamic mechanism by which the epidemic spreads. Mathematical analysis has provided the necessary basis for these control programs.

A Simple Deterministic Model

A simple model for an epidemic is the basis for the famous *threshold effect* stated in 1927 by Kermac and McKendrick. This model, which as we shall see is a degenerate form of the predator–prey equations, captures much of the essence of the epidemic process and provides a solid starting point for analysis. Many generalizations are, of course, possible.

Let us consider a large population of n individuals, and a disease in which infection spreads by contact between individuals. Individuals who are once infected eventually either die, are isolated, or recover and are immune. Thus, at any one time the population is comprised of x susceptible individuals, y infected and circulating individuals, and z individuals who either have been removed (by death or isolation) or are immune. We have $x+y+z=n$ for all t. We assume that the population is subject to some form of homogeneous mixing, and that the rate of contact between susceptibles and infectives is proportional to the product xy. The rate of generation of new infectives is therefore βxy, where β is an infection-rate constant. Infectives are assumed to be removed (or become immune) at a rate proportional to their number with an associated removal constant γ. The governing differential equations are therefore

$$\frac{dx}{dt}=-\beta xy \tag{10-21a}$$

$$\frac{dy}{dt}=\beta xy-\gamma y \tag{10-21b}$$

$$\frac{dz}{dt}=\gamma y \tag{10-21c}$$

It is easy to verify that

$$\frac{d}{dt}(x+y+z)=0$$

so that $x+y+z=n$ for all t in this model.

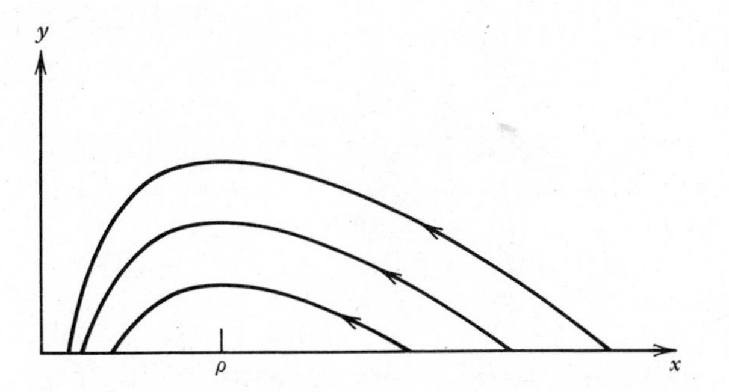

Figure 10.3. Epidemic trajectories.

It is sufficient to consider the first two equations, since z does not appear in them. It is recognized that they represent a degenerate form of the predator–prey equations. In this case, however, it is clear that every point on the x-axis (that is, any point where $y=0$) is an equilibrium point. There are no other equilibrium points. There is, however, a constant of motion for the system that reveals the qualitative nature of the solutions to these equations.

Dividing the first equation by the second, we obtain

$$\frac{\dot{x}}{\dot{y}}=\frac{-\beta x}{\beta x-\gamma}$$

or upon rearrangement

$$\dot{x}-\frac{\rho\dot{x}}{x}+\dot{y}=0$$

where $\rho=\gamma/\beta$. Therefore we conclude that the function

$$V(x, y)=x-\rho\log x+y \tag{10-22}$$

is a constant of motion. It follows then that along a trajectory

$$x-\rho\log x+y=x_0-\rho\log x_0+y_0$$

where x_0, y_0 are the initial values of x and y. We may thus solve for y in terms of x along the trajectory as

$$y(x)=y_0+x_0-x+\rho\log(x/x_0) \tag{10-23}$$

The family of trajectories is shown in Fig. 10.3.

Several interesting qualitative conclusions follow from these curves, as determined by (10-23).

(1) The Threshold Effect. It is easily verified that the maximum value of y, the number of infectives, occurs at the point $x=\rho$. Suppose, then, that a

number of infectives y_0 is introduced into a population x_0 of susceptibles. If $x_0 < \rho$, the level of infectives will decrease monotonically to zero, while if $x_0 > \rho$, the level of infectives will increase until the number of susceptibles is reduced to ρ and will decrease thereafter. Thus, ρ represents a threshold value of susceptibles for the phenomena of an epidemic to occur. Recalling that $\rho = \gamma/\beta$, it follows that within a given community, epidemic activity is increased as β is increased (by high population density, for example) or as γ is decreased (by not isolating infectives).

(2) The Escape Effect. Since in (10-23) $y(0) = -\infty$, it follows that y must vanish at some positive value of x. This means that, as shown in Fig. 10.3, the trajectories terminate on the x-axis at a positive value. Therefore, the epidemic terminates for lack of infectives rather than lack of susceptibles, and some individuals escape the disease entirely.

(3) Symmetry Effect. For the case where $x_0 > \rho$ but $x_0 - \rho$ is small, the epidemic curves are nearly symmetric with respect to the point $x = \rho$. This means that during the course of the epidemic the number of susceptibles is ultimately reduced to a level about as far below the critical value ρ as it was initially above this value.

10.5 STABILITY OF COMPETITIVE ECONOMIC EQUILIBRIA

Consider an economy in which n different commodities are produced and consumed. In order to facilitate orderly exchange of the commodities in a market, the n commodities are each given a price. A set of prices is an equilibrium set if at those prices the amount of each commodity demanded by consumers is exactly equal to the amount supplied by producers. Under rather general assumptions it is possible to prove that such equilibrium prices do exist. However, an important fundamental issue in economic theory is concerned with the construction and analysis of a market adjustment mechanism by which the set of equilibrium prices can be determined. That is, starting with an arbitrary set of prices, what market mechanism will adjust these prices toward equilibrium? To consider this question, economic theorists impose a dynamic structure representing market adjustments on top of the purely static framework of equilibrium theory, and then seek conditions guaranteeing that the equilibrium is stable with respect to this dynamic structure.

The Tâtonnement Process

The Swiss economist Walras, who at the end of the nineteenth century laid much of the foundation for present day equilibrium theory, dealt with the issue of stability by introducing a price adjustment mechanism referred to as *tâtonnement.* The process assumes the existence of a "referee" who initiates the process by announcing a set of prices for the n commodities to the members of the economy. Then, following that announcement, each member submits a list to the referee showing the amount of each commodity that he intends to demand or supply at those prices. If, as finally determined by these lists, the aggregate supply equals aggregate demand for each commodity, the announced prices are equilibrium prices and trading takes place at these prices. If, on the other hand, there is a mismatch between supply and demand for some commodity, the referee adjusts that price—increasing the price if demand exceeds supply and decreasing the price if supply exceeds demand. No trading is allowed at a nonequilibrium set of prices. Instead, after the referee makes the price adjustments, the participants submit new lists and the process is continued until equilibrium is attained, at which time trading takes place. Assuming an equilibrium exists, it is considered (asymptotically) stable if this adjustment mechanism converges to it.

The above description of the tâtonnement process serves essentially as motivation for considering a more explicit dynamic mathematical process where the prices are governed by a system of differential equations. Let $\mathbf{p}=(p_1, p_2, \ldots, p_n)$ be the vector of (announced) prices of the n commodities. The aggregate demand for the ith commodity, given these prices, is given by a function $x_i(\mathbf{p})$. Similarly, the aggregate supply is given by a function $y_i(\mathbf{p})$. Thus, the *excess demand* of the ith commodity is $f_i(\mathbf{p})=x_i(\mathbf{p})-y_i(\mathbf{p})$. Accordingly, following the spirit of the description of the price adjustment mechanism, the mathematical version of the tâtonnement process is defined by the system of equations:

$$\begin{aligned} \dot{p}_1 &= d_1 f_1(\mathbf{p}) \\ \dot{p}_2 &= d_2 f_2(\mathbf{p}) \\ &\cdot \\ &\cdot \\ &\cdot \\ \dot{p}_n &= d_n f_n(\mathbf{p}) \end{aligned} \tag{10-24}$$

In this system, the constants $d_1, d_2, \ldots, d_n$ are arbitrary positive adjustment factors, which reflect the possibility of adjusting various commodity prices at different rates.

D-Stability

A special case is, of course, the case where the excess demand functions f_i, $i=1, 2, \ldots, n$ are linear with respect to the price vector. Then we may assume, without loss of generality, that the system (10-24) has the form

$$\dot{\mathbf{p}}(t)=\mathbf{DA}[\mathbf{p}(t)-\bar{\mathbf{p}}] \tag{10-25}$$

where $\mathbf{p}(t)$ is the current price vector, $\bar{\mathbf{p}}$ is the equilibrium price vector, $\mathbf{A}$ is an $n \times n$ matrix defining the n excess demand functions, and $\mathbf{D}$ is an $n \times n$ diagonal matrix whose diagonal entries are the arbitrary positive adjustment factors $d_1, d_2, \ldots, d_n$.

Equation (10-25) is equivalent to

$$\frac{d}{dt}(\mathbf{p}-\bar{\mathbf{p}}) = \mathbf{DA}(\mathbf{p}-\mathbf{p}) \tag{10-26}$$

and therefore, as is always the case for linear systems, stability is determined entirely by the coefficient matrix, which in this case is $\mathbf{DA}$. Within the context of the tâtonnement process, however, the positive diagonal of $\mathbf{D}$ is considered arbitrary. It is therefore natural to seek conditions for which (10-25) is asymptotically stable for all choices of the adjustment parameters. This is the motivation for the following definition.

Definition. An $n \times n$ matrix $\mathbf{A}$ is said to be $\mathbf{D}$-*stable* if the matrix $\mathbf{DA}$ is asymptotically stable for all diagonal matrices $\mathbf{D}$ with positive diagonal entries.

It is possible to derive a simple condition guaranteeing that a matrix is $\mathbf{D}$-stable.

***Theorem* (Arrow–McManus).** *If there exists a diagonal matrix $\mathbf{C}$ with positive diagonal entries such that $\mathbf{CA}+\mathbf{A}^T\mathbf{C}$ is negative definite, then $\mathbf{A}$ is $\mathbf{D}$-stable.*

Proof. Consider $\mathbf{DA}$ where $\mathbf{D}$ is diagonal with arbitrary positive diagonal entries. To show that $\mathbf{DA}$ is asymptotically stable, it is sufficient to find a positive definite matrix $\mathbf{P}$ such that $\mathbf{P}(\mathbf{DA})+(\mathbf{DA})^T\mathbf{P}$ is negative definite. (See Sect. 9.11.) However, the matrix $\mathbf{P}=\mathbf{CD}^{-1}$ satisfies this requirement; it is itself diagonal with positive diagonal entries, and $\mathbf{P}(\mathbf{DA})+(\mathbf{DA})^T\mathbf{P}=\mathbf{CA}+\mathbf{A}^T\mathbf{C}$, which is negative definite by assumption. Thus, $\mathbf{A}$ is $\mathbf{D}$-stable. ▌

A special case of the criterion of this theorem is where $\mathbf{C}=\mathbf{I}$. This yields $\mathbf{A}+\mathbf{A}^T$ negative definite as a sufficient condition for $\mathbf{D}$-stability. This condition can in turn be expressed as the requirement that $\mathbf{x}^T\mathbf{Ax}<0$ for all $\mathbf{x}\neq\mathbf{0}$, which can be interpreted as a negative definiteness condition when $\mathbf{A}$ is not necessarily symmetric. It follows, as the reader may easily show, that under this assumption the quadratic form $V(\mathbf{p})=(\mathbf{p}-\bar{\mathbf{p}})^T\mathbf{D}^{-1}(\mathbf{p}-\bar{\mathbf{p}})$ serves as a Liapunov function for the system.

From economic considerations, it is clear that the diagonal elements of $\mathbf{A}$ are usually negative, so that an increase in price of a commodity reduces the excess demand for that commodity. Also, it is frequently true that the off-diagonal elements of $\mathbf{A}$ are positive; that is, an increase in the price of one commodity tends to increase the excess demand for other goods. In this situation **A** is a Metzler matrix, and all of the strong stability results for such matrices apply to the adjustment process.

The above analysis of **D**-stability is really only a slight extension of earlier results for stability of linear systems. To take a more significant step, we must relate the requirement of negative definiteness to natural economic assumptions. That is, we must show that the quadratic form is a Liapunov function under reasonable economic hypotheses. This is the task to which we next turn, while simultaneously generalizing to the nonlinear case.

Nonlinear Theory

Suppose now that the excess demand functions are nonlinear. In order to establish global stability of the tâtonnement process, it is necessary to present some additional economics and to introduce some assumptions on the behavior of the members of the economy.

First, we make explicit the assumption that we are dealing with a *closed* economy. All supply and all consumption of the n commodities is restricted to a fixed set of individuals. Furthermore, since there is only a single trading period, individuals in the closed economy can purchase commodities only to the extent that they obtain cash through supply of some other commodity.

Based on this assumption of a closed economy, it is assumed that the excess demand functions satisfy *Walras' law*

$$\sum_{i=1}^{n} p_i f_i(\mathbf{p}) = 0 \tag{10-27}$$

This result is essentially an accounting identity, under the assumption that everyone will spend for consumption all the income they derive from supply. Given a set of announced prices $\mathbf{p}$ with associated demand vector $\mathbf{x}(\mathbf{p})$ and supply vector $\mathbf{y}(\mathbf{p})$, the aggregate income that is derived from sales is $\mathbf{p}^T\mathbf{y}(\mathbf{p})$. Likewise the total expenditure in the form of demand is $\mathbf{p}^T\mathbf{x}(\mathbf{p})$. These two must be equal, leading to $\mathbf{p}^T\mathbf{x}(\mathbf{p}) - \mathbf{p}^T\mathbf{y}(\mathbf{p}) \equiv \mathbf{p}^T\mathbf{f}(\mathbf{p}) = 0$. Thus, Walras' law is applicable.

A further economic assumption is based on the *weak axiom of revealed preference.* Consider two price vectors $\mathbf{p}_a$ and $\mathbf{p}_b$ with associated excess demands $\mathbf{f}(\mathbf{p}_a) = \mathbf{z}_a$ and $\mathbf{f}(\mathbf{p}_b) = \mathbf{z}_b$, with $\mathbf{z}_a \neq \mathbf{z}_b$. If $\mathbf{p}_a^T\mathbf{z}_b \leq \mathbf{p}_a^T\mathbf{z}_a$, it follows that at the prices $\mathbf{p}_a$, the vector $\mathbf{z}_b$ is no more costly than the vector $\mathbf{z}_a$. Since $\mathbf{z}_a$ was actually selected even though $\mathbf{z}_b$ could have been, we say that $\mathbf{z}_a$ is revealed preferred to $\mathbf{z}_b$. The weak axiom of revealed preference asserts that if $\mathbf{z}_a \neq \mathbf{z}_b$ it is not possible for both $\mathbf{z}_a$ to be revealed preferred to $\mathbf{z}_b$ and $\mathbf{z}_b$ to be revealed preferred to $\mathbf{z}_a$. That is, $\mathbf{p}_a^T\mathbf{z}_b \leq \mathbf{p}_a^T\mathbf{z}_a$ implies that $\mathbf{p}_b^T\mathbf{z}_a > \mathbf{p}_b^T\mathbf{z}_b$. This axiom is always assumed to hold for individuals, and in some cases it will hold for the aggregate excess demand functions as well. This is our second economic assumption.

It is now possible to formulate an important global stability result for economic equilibria. We state the result for the dynamic system

$$\dot{\mathbf{p}} = \mathbf{f}(\mathbf{p}) \tag{10-28}$$

leaving the generalization to arbitrary adjustment rates to a problem.

Theorem. *If Walras' law is satisfied, if the aggregate excess demand functions satisfy the weak axiom of revealed preference, and if there is a unique equilibrium price* $\bar{\mathbf{p}}$, *then the system* (10-28) *is asymptotically stable in the large and** $V(\mathbf{p}) = \frac{1}{2}\|\mathbf{p} - \bar{\mathbf{p}}\|^2$ *is a Liapunov function.*

Proof. Let $\mathbf{V}(\mathbf{p}) = \frac{1}{2}\|\mathbf{p} - \bar{\mathbf{p}}\|^2$. Then for any $\mathbf{p} \neq \bar{\mathbf{p}}$, $\dot{V}(\mathbf{p}) = (\mathbf{p} - \bar{\mathbf{p}})^T\mathbf{f}(\mathbf{p}) = \mathbf{p}^T\mathbf{f}(\mathbf{p}) - \bar{\mathbf{p}}^T\mathbf{f}(\mathbf{p})$. By Walras' law the first term is zero, so $\dot{V}(\mathbf{p}) = -\bar{\mathbf{p}}^T\mathbf{f}(\mathbf{p})$.

Now by definition of $\bar{\mathbf{p}}$ it follows that $\mathbf{p}^T\mathbf{f}(\bar{\mathbf{p}}) = 0$. By Walras' law $\mathbf{p}^T\mathbf{f}(\mathbf{p}) = 0$. Therefore,

$$0 = \mathbf{p}^T\mathbf{f}(\bar{\mathbf{p}}) \leq \mathbf{p}^T\mathbf{f}(\mathbf{p}) = 0$$

(actually equality holds). Using the fact that, by uniqueness, $\mathbf{f}(\mathbf{p}) \neq \mathbf{f}(\bar{\mathbf{p}})$, it follows by the weak axiom of revealed preference of the aggregate that

$$\bar{\mathbf{p}}^T\mathbf{f}(\mathbf{p}) > \bar{\mathbf{p}}^T\mathbf{f}(\bar{\mathbf{p}}) = 0$$

Therefore $-\bar{\mathbf{p}}^T\mathbf{f}(\mathbf{p}) < 0$ and, accordingly, $\dot{V}(\mathbf{p}) < 0$ for all $\mathbf{p} \neq \bar{\mathbf{p}}$. This shows that $V(\mathbf{p})$ is a Liapunov function. ▮

10.6 GENETICS

Genetic evolution is the basis for perhaps one of the most profound dynamic processes. Its subtle action repeated over generations shapes the composition of life, providing both diversity and viability. The most famous concept in this area, of course, is Darwin's principle of evolution, based on survival of the fittest. This principle can be interpreted as postulating that average fitness is a Liapunov-type function, which tends to increase from generation to generation. Darwin enunciated the principle on the basis of aggregate observation without reference to an explicit dynamic mechanism. The genetic theory of evolution, on the other hand, provides a specific dynamic mechanism in the form of a system of nonlinear difference equations. It is natural, then, to attempt to reconcile the two theories by investigating whether average fitness is a Liapunov function for the nonlinear system. This is indeed the case, at least for the simplest genetic mechanism, and provides a profound example of how the search for summarizing functions can be regarded as a fundamental component

* We employ the "norm" notation $\|\mathbf{p} - \bar{\mathbf{p}}\|^2 = (\mathbf{p} - \bar{\mathbf{p}})^T(\mathbf{p} - \bar{\mathbf{p}})$.

of scientific investigation, and how the analyst must probe for functions with strong contextual meaning.

Background

Genetic information in living matter is stored in *chromosomes*, which are thread-shaped bodies occurring in cell nuclei. Along the chromosome structure there is a linear progression of specific *locations*, each location being occupied by exactly one of a set of possible *genes*. Two or more genes that are each capable of occupying a specific location are called *alleles*. Thus, the gene at a particular location represents a specific choice from a set of possible alleles. Human beings, and many other forms of life (including laboratory mice), are *diploid*, meaning that chromosomes occur in homologous pairs, but with perhaps different alleles at the two corresponding locations.

Individual inherited physical characteristics are generally traceable to specific combinations of genes. Some characteristics, such as ABO blood type or eye pigmentation in humans, are determined by the combination of genes on a pair of chromosomes at a single location. Other characteristics, however, involve genes at two or more locations. We shall focus on single location characteristics, for which the theory is simplest and most complete. In this case if there are two alternative genes (alleles), say A and a, individuals are characterized as being one of the three genotypes AA, Aa, or aa.

In diploid reproduction, one set of chromosomes is obtained from each of two parents to form the diploid structure of the offspring. Thus, in the case of one locus and two alleles, each parent, depending on its genotype, may contribute either A or a.

The Hardy–Weinberg Proportions

Our specific interest here is that of population genetics, which is the study of the evolution of the genetic composition of a population. We assume distinct generations and a *random mating* system in which any member of one sex is equally likely to mate with any member of the opposite sex in the same generation. It is then of interest to calculate the way in which the distribution of genotypes evolves from generation to generation. Also, as stated above, we concentrate on the one-locus two-allele case.

Under the random mating assumption, the relative proportions of the three genotypes AA, Aa, and aa in an offspring population is determined directly by the relative proportions of the alleles A and a in the parent population. The resulting genotype proportions are referred to as the *Hardy–Weinberg proportions*. Specifically, let p and q (with $p+q=1$) denote the portion of A and a alleles, respectively, in a parent population. During a generation of random mating, these alleles form a gene pool from which pairs

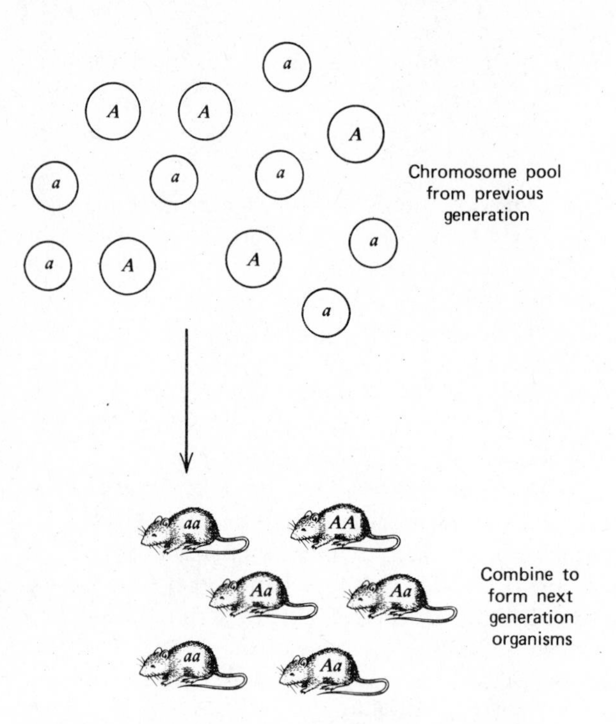

Figure 10.4. Hardy–Weinberg proportions.

are drawn to form the offspring genotypes. The proportions of AA, Aa, and aa genotypes formed will be p^2, $2pq$, and q^2, respectively. These are the Hardy-Weinberg proportions. The process of recombination is illustrated in Fig. 10.4. It is the random mating assumption that allows us to deduce the offspring genotype proportions by knowledge only of the parent allele proportions rather than the parent genotype proportions.

Example 1 (Eye Color). Blue eyes in humans occur when both genes at a certain location are the recessive gene that does not produce the brown pigment, melanin, in the iris of the eyes. When the dominant allele is present, some pigment is present, but its amount and distribution are controlled by genes at other locations. With no pigment in the outer layer of the iris, the eyes appear blue, with a little they appear green, and with more they appear hazel, light brown, dark brown, and finally black.

Suppose that in a random mating population ten percent of the people have blue eyes. What percent of the people have at least one recessive gene at the location that controls the production of melanin?

Let the alleles be A and a, respectively, with a being the gene that does not produce pigment. Assume that these genes occur in the proportions p and

q, respectively. Then in equilibrium the genotype *aa* that yields blue eyes has frequency

$$q^2 = .10$$

This yields $q = .32$. And accordingly $p = .68$. The frequency of the genotype *Aa* is $2pq = .43$. Altogether, the proportion of those individuals who have at least one *a* gene is $.43 + .10 = .53$, and they therefore constitute over half of the population.

Fitness and Selection

Evolutionary behavior is introduced into the genetic process through the phenomena of natural selection based on genotype. Individuals of a certain genotype may be disadvantaged, as compared with others, in terms of their likelihood to reproduce. (This might be due to a lower likelihood to survive, as in the case of sickle cell anemia, or to a lower reproductive rate, as in the case of a gene that reduces fertility.) Such genotypes, accordingly, contribute relatively less to the gene pool from which the next generation is composed than do other genotypes. This selective difference mechanism leads to a gradual evolution in the proportions of the various alleles—and consequently in the proportions of the various genotypes.

Again for the one-locus two-allele case, suppose that on an average basis an individual of genotype *AA*, *Aa*, or *aa* contributes genes to form individuals in the next generation with relative frequency w_{11}, w_{12}, or w_{22}, respectively. The numbers w_{11}, w_{12}, and w_{22} are the relative *fitness factors* of the genotypes.

Suppose that in generation k the proportions of alleles *A* and *a* are p and q, respectively. The genotypes *AA*, *Aa*, and *aa* therefore occur in the proportions p^2, $2pq$, and q^2, respectively. However, in terms of contribution to the gene pool for the next generation they have effective proportions $w_{11}p^2$, $2w_{12}pq$, and $w_{22}q^2$. (Here the proportions do not sum to one, but we shall normalize shortly.) The ratio of *A* alleles to *a* alleles in the next generation is, accordingly,

$$\frac{w_{11}p^2 + w_{12}pq}{w_{12}pq + w_{22}q^2}$$

Denoting the new proportions of *A* and *a* by $p(k+1)$ and $q(k+1)$, and the old proportions by $p(k)$ and $q(k)$, respectively, we deduce the recursive form

$$p(k+1) = \frac{w_{11}p(k)^2 + w_{12}p(k)q(k)}{w_{11}p(k)^2 + 2w_{12}p(k)q(k) + w_{22}q(k)^2} \tag{10-29a}$$

$$q(k+1) = \frac{w_{12}p(k)q(k) + w_{22}q(k)^2}{w_{11}p(k)^2 + 2w_{12}p(k)q(k) + w_{22}q(k)^2} \tag{10-29b}$$

This is the nonlinear system that governs the process of natural selection.

As a special case, consider $w_{11}=w_{12}=w_{22}$, corresponding to no genotype differentiation. In that case $p(k+1)=p(k)$, $q(k+1)=q(k)$, for all k. In the general case, however, the process is more complex.

Fitness

A casual examination of the difference equations (10-29) tends to be rather discouraging. How would one deduce a suitable Liapunov function for this highly nonlinear system? The answer, of course, is found by explicit consideration of the mean fitness of the population.

Given the allele proportions p and q (with $p+q=1$) the corresponding mean fitness is

$$\begin{aligned} w &= w_{11}p^2+2w_{12}pq+w_{22}q^2 \\ &= w_{11}p^2+2w_{12}p(1-p)+w_{22}(1-p)^2 \\ &= w_1p+w_2(1-p) \end{aligned} \tag{10-30}$$

where

$$w_1=pw_{11}+qw_{12}$$

$$w_2=pw_{12}+qw_{22}$$

We show first that the mean fitness never decreases. Then we examine the various equilibrium points of the system.

The gene frequencies at the next generation (denoted here by p' and q') can be expressed as

$$p'=p\frac{w_1}{w} \tag{10-31a}$$

$$q'=q\frac{w_2}{w} \tag{10–31b}$$

Thus,

$$\begin{aligned} \Delta p \equiv p'-p &= \frac{w_1-w}{w}p=\frac{w_1-w_1p-w_2q}{w}p \\ &= pq\frac{w_1-w_2}{w} \end{aligned} \tag{10-32}$$

Since mean fitness is a quadratic function of p, its value at p' can be expressed in the form

$$w'=w+\frac{dw}{dp}\Delta p+\frac{1}{2}\frac{d^2w}{dp^2}(\Delta p)^2 \tag{10-33}$$

The first derivative is evaluated at p, and the second derivative will be a constant. We find

$$\begin{aligned}\frac{dw}{dp} &= 2[w_{11}p + w_{12} - 2w_{12}p - w_{22}(1-p)] \\ &= 2[w_{11}p + w_{12}(1-p) - w_{12}p - w_{22}(1-p)] \\ &= 2(w_1 - w_2) \end{aligned} \tag{10-34}$$

Also,

$$\frac{d^2w}{dp^2} = 2(w_{11} - 2w_{12} + w_{22}) \tag{10-35}$$

Therefore, combining (10-32), (10-33), (10-34), and (10-35)

$$\begin{aligned}\Delta w &= w' - w \\ &= 2\frac{(w_1 - w_2)^2}{w} pq + (w_{11} - 2w_{12} + w_{22})\frac{(w_1 - w_2)^2}{w^2} p^2q^2 \\ &= pq\frac{(w_1 - w_2)^2}{w^2}[2w + (w_{11} - 2w_{12} + w_{22})pq] \\ &= pq\frac{(w_1 - w_2)^2}{w^2}(w + w_{11}p + w_{22}q) \geq 0 \end{aligned} \tag{10-36}$$

Thus, mean fitness never decreases.

The dynamic system (10-29) describing the evolution of gene frequencies has, in general, three distinct equilibrium points. The first two are the degenerate proportions $p = 0$ and $p = 1$. These points correspond to absence of one of the alleles. The third equilibrium point is found by solving

$$p = \frac{w_{11}p^2 + w_{12}p(1-p)}{w_{11}p^2 + 2w_{12}p(1-p) + w_{22}(1-p)^2}$$

This equation can be solved by multiplying through by the denominator of the right-hand side. The resulting cubic equation can then be reduced to a linear equation by dividing by p and by $(1-p)$ corresponding to the known solutions $p = 0$, $p = 1$. This leads to the equilibrium point

$$\bar{p} = \frac{w_{12} - w_{22}}{2w_{12} - w_{11} - w_{22}} \tag{10-37}$$

On the other hand, the point where $\Delta w = 0$ is found by setting $w_1 = w_2$, where as before

$$w_1 = pw_{11} + (1-p)w_{12}$$

$$w_2 = pw_{12} + (1-p)w_{22}$$

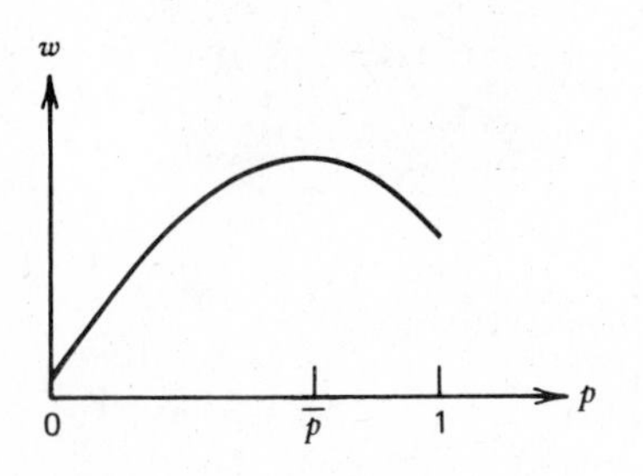

Figure 10.5. Fitness.

This yields exactly the same value of p as in (10-37). Thus, the equilibrium points are exactly the same as the points where $\Delta w = 0$. We can conclude therefore that mean fitness always increases unless the gene frequencies are in equilibrium.

Depending on the values of w_{11}, w_{12}, and w_{22} and on the initial gene frequencies, the process may converge toward any one of the three equilibrium points. Of greatest interest, however, are conditions under which the equilibrium $\bar{p}$ is stable. Accordingly, let us assume that $0 < \bar{p} < 1$. We recall that the mean fitness function w is quadratic in p, with second derivative equal to $w_{11} - 2w_{12} + w_{22}$. If this derivative is negative, then the function is shaped as in Fig. 10.5 and has a maximum at the point $\bar{p}$. In this case $\bar{p}$ is the only stable equilibrium point. If, on the other hand, the second derivative is positive, the curve is shaped the opposite way, and $\bar{p}$ is unstable.

A more explicit condition can be derived for asymptotic stability of a nondegenerate equilibrium point. As noted above, for asymptotic stability we must have

$$2w_{12} > w_{11} + w_{22} \tag{10-38}$$

Then from (10-37) it follows that for $\bar{p} > 0$ we must have

$$w_{12} > w_{22} \tag{10-39}$$

Similarly for $\bar{p} < 1$ we must have

$$w_{12} > w_{11} \tag{10-40}$$

The inequalities (10-39) and (10-40) of course imply (10-38). This leads to the important qualitative conclusion that in order for a population to have a stable mixed distribution of genotypes, it is necessary that the fitness of genotype *Aa* be greater than the fitness of each of the genotypes *AA* and *aa*.

The above analysis, showing that the mean fitness is always increasing at nonequilibrium points, can be generalized to the case of multiple alleles at a single location. The analysis, however, cannot be extended to the case where traits are governed by genes at several locations. Maximal fitness no longer always corresponds to equilibria, and fitness does not necessarily increase. This leaves open the possibility of determining an alternative simple governing principle underlying the complex process of inheritance.

Example 2. Many congenital diseases can be explained as the result of both genes at a single location being the same recessive gene. Suppose that due to almost certain death or due to voluntary nonreproduction, individuals with the disease do not have any offspring. Suppose that the other two genotypes have identical fitness factors. In this case the frequency of the recessive gene will decrease, but at a very slow rate.

Let $w_{11}=w_{12}=1$, $w_{22}=0$. Then the frequency of the recessive gene in generation $k+1$ is

$$q(k+1)=\frac{[1-q(k)]q(k)}{[1-q(k)]^2+2[1-q(k)]q(k)}$$

$$=\frac{q(k)}{1+q(k)}$$

It is easily verified that this converges to zero, but very slowly. This explains why even deadly genetic diseases can remain active for hundreds of generations.

10.7 PROBLEMS

1. *Conservation of Angular Momentum.* Show that in planetary motion, not only is energy conserved, but so is angular momentum $A = mr\dot{\theta}$.

2. *Escape Velocity.* A rocket near the Earth is subjected to a conservative gravitational force defined by a potential of the form

$$V=-k/r$$

where r is the distance to the center of the Earth. The rocket is to be launched by giving it an initial upward velocity at the Earth's surface r_0, but supplying no additional thrust thereafter. What is the minimum initial velocity such that the rocket will not fall back to Earth?

3. *Thermal Efficiency.* A difference in temperature of two heat reservoirs (such as a furnace and the outside atmosphere) can be used as the basis for an engine to produce work. During a complete cycle of the engine (which might consist of a gas-filled chamber and piston) an amount of heat Q_h is taken from the higher-temperature reservoir and an amount Q_L is added to the lower-temperature reservoir. The net heat lost $Q = Q_h - Q_L$ will (ideally) be W, the amount of work generated. The thermal efficiency of the engine is defined to be

$$\eta = W/Q_h$$

Using the second law of thermodynamics, show that the maximum possible efficiency is

$$\eta = 1 - T_L/T_h$$

where T_L and T_h are the temperatures of the low- and high-temperature reservoirs, respectively.

4. Show that the predator–prey model with crowding is a special case of the general n-species theory at the end of Sect. 10.3. What is the $\mathbf{A}$ matrix in this case?

5. *Epidemics in an Evolving Society.* In the epidemic model of Sect. 10.4 it is tacitly assumed that the dynamic behavior is fast compared with the general population turnover. Suppose that one accounts for the turnover. For a small but extended time period epidemic it is reasonable to modify the equation for $\dot{x}$ to

$$\dot{x} = \alpha x - \beta xy$$

for some $\alpha > 0$. This reflects the fact that new susceptibles are continually being introduced into the population. Discuss the qualitative nature of the resulting solutions. Does this provide a possible explanation of the observation that some diseases, such as measles, have recurred in periodic outbreaks?

6. It is often the case that the $\mathbf{A}$ matrix defining excess demand functions is a Metzler matrix. Show that a Metzler matrix is $\mathbf{D}$-stable if and only if it is (asymptotically) stable.

7. *Arbitrary Adjustment Rates.* Replace Eq. (10-28) by

$$\dot{\mathbf{p}} = \mathbf{D}\mathbf{f}(\mathbf{p})$$

where $\mathbf{D}$ is an $n \times n$ diagonal matrix with positive diagonal terms. Find an analog for this case of the theorem that applies to (10-28).

*8. *A Barter Process.* In the tâtonnement process no exchange of goods is allowed to take place until after the equilibrium prices are obtained. More advanced models allow for the possibility of exchange during the price adjustment process.

As a simple example, consider two individuals A and B and two commodities 1 and 2. Each individual has a utility function that depends on the amounts of the two commodities he possesses. Thus if A has amounts x_A and y_A of the two commodities, he has utility $U^A(x_A, y_A)$—and similarly for B. At each point in the process there is a price p for commodity B in terms of commodity A, which is the current exchange rate between the two commodities. From these assumptions it follows that

$$\dot{x}_B = -\dot{x}_A, \quad \dot{y}_B = -\dot{y}_A$$

$$\dot{x}_A + p\dot{y}_A = 0, \quad \dot{x}_B + p\dot{y}_B = 0$$

We assume that individual B essentially governs the price. Specifically, for all t,

$$p(t) = \frac{U_y^B(x_B(t), y_B(t))}{U_x^B(x_B(t), y_B(t))}$$

The level of exchange is governed by individual A according to the equation

$$\dot{x}_A(t) = U_x^A(x_A(t), y_A(t)) - U_y^A(x_A(t), y_A(t))/p(t)$$

Consider the function $V = U^A$. Show that $\dot{V} \geq 0$. Show that $\dot{V} = 0$ implies $\dot{x}_A = 0$. What is the interpretation of the final value of p?

9. *Albinism.* Albinism in humans occurs when an individual is of genotype *aa* for a certain gene. If all albinos voluntarily agreed to avoid having children, how many generations would it take to reduce the incidence of albinism in a society from one in ten thousand to one in one hundred thousand?

10. *Sickle Cell Anemia.* Sickle cell anemia, a deficiency of the red blood cells, is controlled by a single allele (Hb^s). Individuals of genotype (Hb^sHb^s) suffer severe sickle cell anemia and usually die in childhood. Individuals of genotype (Hb^sHb^A) have the sickle cell trait only mildly, and suffer no serious ill-effects. And of course individuals of genotype (Hb^AHb^A) do not show the trait.

    It has been determined that individuals of the genotype (Hb^sHb^A) are decidedly more resistant to malaria than those of genotype (Hb^AHb^A). This suggests that in regions with high malaria incidence, $w_{12} > w_{11} + w_{22}$ for this gene location. This is of course the condition for stability of a population of mixed genotypes. If $w_{22} = 0$, $w_{11} = .8$, and $w_{12} = 1$, what is the equilibrium distribution of genotypes?

NOTES AND REFERENCES

There is, of course, a vast selection of literature, including introductory textbooks, on each of the subjects included within this chapter. We cannot hope to cite even a representative sample of the most important references. Instead, we settle for a scattered selection—including the ones found to be most helpful in preparing this chapter.

Section 10.1. For a general introduction to mechanics see Housner and Hudson [H5], or Goldstein [G9].

Section 10.2. For introductory treatments of thermodynamics see Prigogine [P6] or Weinreich [W3].

Section 10.3. The classic references for interacting populations are Volterra [V1] and Lotka [L5] for the theory, and Gause [G5] for experimental verification. A modern (somewhat advanced) treatment is Goel, Maitra, and Montroll [G7]. Also see Watt [W2] and Pielou [P4].

Section 10.4. The original mathematical paper on epidemic theory is Kermac and McKendrick [K12]. The standard book on epidemics, including stochastic as well as deterministic theory, is Bailey [B1]. For stochastic models of epidemics see also Bartlett [B3].

Section 10.5. An excellent survey of the dynamic theory of economic equilibria is contained in Quirk and Saposnik [Q1]. The Liapunov function of the tâtonnement process, the squared Euclidean distance of the current price from the equilibrium price, does not appear to have a strong economic interpretation. The study of more advanced processes, however, leads one to consider other Liapunov-type functions directly related to fundamental economic notions. See Problem 8, and for a more advanced treatment see Arrow and Hahn [A2] or Takayama [T1].

Section 10.6. A very readable introduction to population genetics and various other subjects that overlap with some of the examples in this chapter is Wilson and Bossert [W5]. For more advanced work on population genetics see Ewens [E2] and Karlin [K9].

chapter 11.

Optimal Control

Underlying a serious study of a specific dynamic system is often a motivation to improve system behavior. When this motivation surfaces in explicit form, the subject of optimal control provides a natural framework for problem definition.

The general structure of an optimal control problem is straightforward. In the simplest version, there is a given dynamic system (linear or nonlinear, discrete-time or continuous-time) for which input functions can be specified. There is also an objective function whose value is determined by system behavior, and is in some sense a measure of the quality of that behavior. The optimal control problem is that of selecting the input function so as to optimize (maximize or minimize) the objective function. For example, the dynamic system might be a space vehicle (as often it was in some of the first modern applications) with inputs corresponding to rocket thrust. The objective might then be to reach the moon with minimum expenditure of fuel. As another example, the system might represent the dynamics of an individual's accumulation of wealth, with controls corresponding to yearly work effort and expenditure levels. The problem might then correspond to planning the lifetime pattern of work and expenditure in order to maximize enjoyment. Finally, as a third example, the system might be the nation's economy, with controls corresponding to government monetary and fiscal policy. The objective might be to minimize the aggregate deviations of unemployment and interest rates from fixed target values.

There is a diversity of mathematical issues associated with optimal control—and these form the subject material for *optimal control theory*. There is, first of all, the question of characterizing an optimal solution. That is, how can a particular control function be recognized as being optimal or not. This question is treated in the first several sections of the chapter. There is also the issue of open-loop versus closed-loop control, as introduced in Chapter 8. This is treated in the later sections of the chapter. Finally, there is the issue of computation. That is, how does one actually find the optimal control function—either in analytical form (if possible) or numerically? The analytic approach is emphasized in this chapter. Overall, the subject of optimal control theory has a long and deep history, an expanding range of applications, and a large assortment of numerical techniques. This chapter (including the assortment of applications treated in the problems at the end) presents a sampling of this important subject.

11.1 THE BASIC OPTIMAL CONTROL PROBLEM

Optimal control problems can be formulated for both discrete-time and continuous-time systems. In keeping with the spirit of the rest of the book, both types are discussed in this chapter, but in this instance greater attention is devoted to the continuous-time case. The reasons for this emphasis are that continuous-time problems are notationally simpler; most illustrative examples are more natural in continuous time; and, most importantly, the Pontryagin maximum principle results are stronger in continuous time than in discrete time. Naturally, however, discrete-time formulations also arise frequently, especially in large-scale systems treated by digital computers. Much of the theory is directly applicable to both discrete- and continuous-time systems.

This section formulates a basic continuous-time optimal control problem and develops the associated Pontryagin maximum principle. This section is relatively long, and the development is quite different in character than that of earlier chapters. A good strategy for the reader, on first encountering this material, might be to read this first subsection where the problem is formulated, and then skip to the end of the section where the final result is presented. Next the examples in Sect. 11.2 should be studied. Finally, this whole first section should be studied on a second reading.

The *basic optimal control problem* in continuous time is formulated as follows: one is given a system, defined on a fixed time interval $0 \leq t \leq T$,

$$\dot{\mathbf{x}}(t) = \mathbf{f}(\mathbf{x}(t), \mathbf{u}(t)) \tag{11-1a}$$

a (fixed) initial condition

$$\mathbf{x}(0) = \mathbf{x}_0 \tag{11-1b}$$

a set of allowable controls

$$\mathbf{u}(t) \in U \tag{11-1c}$$

and an objective function

$$J=\psi(\mathbf{x}(T))+\int_0^T l(\mathbf{x}(t),\mathbf{u}(t))\,dt \tag{11-1d}$$

to be maximized.*

Let us examine the elements of the problem (11-1). The structure of system equation (11-1a) together with the initial condition (11-1b) is by this point quite familiar. Generally, we assume that the state $\mathbf{x}(t)$ is n dimensional and the control (or input) $\mathbf{u}(t)$ is m dimensional. The function $\mathbf{f}$ is composed of n separate component functions that are all well behaved so that the system (11-1a) with the initial condition (11-1b) has a unique solution, once the control is specified.

The set of allowable controls U is an important component of the optimal control problem. In some cases, this set is the whole of m-dimensional space—in which case there is no real constraint on $\mathbf{u}(t)$. In other cases, however, this set takes the form of inequalities on the components of $\mathbf{u}(t)$. For example, if $u(t)$ is one dimensional, there may be a limitation of the form $u(t)\geq 0$, or perhaps $0\leq u(t)\leq 1$. Such constraints reflect the fact that, in some systems, control, to be physically meaningful, must be positive or must not exceed a certain bound. For example, in a planning problem the control variable might represent the fraction of current profit that is reinvested (and thus must lie between 0 and 1).

In the objective function (11-1d), both ψ and l are real-valued functions of their respective arguments. The term $\psi(\mathbf{x}(T))$ is the contribution to the objective of the final state. For example, this form of objective arises if it is desired to control an object so as to attain maximum velocity in a given time, or to plan resource allocations over time so as to obtain as much as possible at the end, and so forth. The integral term represents a contribution that accumulates over time. Such a term arises, for example, if the objective is to minimize total fuel expenditure of a machine or to maximize total production in a production facility. A specific problem may, of course, have either ψ or l identically equal to zero (but not both).

The interpretation of the optimal control problem (11-1) is straightforward, but worth emphasizing. The unknown is the control function $\mathbf{u}(t)$ on $0\leq t\leq T$. Once this function is specified, it determines, in conjunction with the system equation (11-1a) and the initial condition (11-1b), a unique state trajectory $\mathbf{x}(t)$, $0\leq t\leq T$. This trajectory and the control function then determine a value of J according to (11-1d). The problem is to find the control function $\mathbf{u}(t)$, $0\leq t\leq T$, satisfying the constraint (11-1c), which leads to the largest possible value of J.

* Almost all components of this problem ($\mathbf{f}$, l, and U) can depend explicitly on time without changing the nature of the results that follow.

The Modified Objective and the Hamiltonian

In many respects optimal control is a natural extension of the maximization of a function of a single variable, as considered in the study of calculus. The conditions for maximization are derived by considering the effect of small changes near the maximum point. This similarity is explicitly recognized by the term *calculus of variations*, which is the name of that branch of mathematics that first extended the "variational" idea to problems where the unknown is a function rather than a scalar. This general "variational" approach is followed here.

To characterize an optimal control, we shall trace out the effect of an arbitrary small change in $\mathbf{u}(t)$ and require that it be nonimproving for the objective. That is, we start with the assumption that the control function $\mathbf{u}(t)$ is optimal; we then make a small change in $\mathbf{u}(t)$ and determine the corresponding change in the objective J. This change should be negative (nonimproving) if the original $\mathbf{u}(t)$ is optimal. However, because a change in $\mathbf{u}(t)$ also changes $\mathbf{x}(t)$, it is difficult to carry out this plan and directly determine the net influence on the value of the objective. Therefore, a somewhat indirect approach is helpful. The "trick" that is used is to adjoin to J some additional terms, which sum to zero. In particular, we form the *modified objective function*

$$\bar{J} = J - \int_0^T \boldsymbol{\lambda}(t)^T[\dot{\mathbf{x}}(t) - \mathbf{f}(\mathbf{x}(t), \mathbf{u}(t))]\, dt \tag{11-2}$$

The term in brackets is zero for any trajectory. The coefficient n-vector $\boldsymbol{\lambda}(t)$ is at this point arbitrary. It is clear, however, that for any choice of $\boldsymbol{\lambda}(t)$ the value of $\bar{J}$ is the same as that of J for any $\mathbf{x}(t)$ and $\mathbf{u}(t)$ satisfying (11-1a). We can therefore consider the problem of maximizing $\bar{J}$ rather than J. The flexibility in the choice of $\boldsymbol{\lambda}(t)$ can then be used to make the problem as simple as possible.

For convenience we define the *Hamiltonian function*

$$H(\boldsymbol{\lambda}, \mathbf{x}, \mathbf{u}) = \boldsymbol{\lambda}^T\mathbf{f}(\mathbf{x}, \mathbf{u}) + l(\mathbf{x}, \mathbf{u}) \tag{11-3}$$

In terms of the Hamiltonian, the modified objective takes the explicit form

$$\bar{J} = \psi(\mathbf{x}(T)) + \int_0^T \{H(\boldsymbol{\lambda}(t), \mathbf{x}(t), \mathbf{u}(t)) - \boldsymbol{\lambda}(t)^T\dot{\mathbf{x}}(t)\}\, dt \tag{11-4}$$

The Hamiltonian is therefore fundamental for consideration of this modified objective.

Suppose now that a nominal control function $\mathbf{u}(t)$, satisfying the constraint $\mathbf{u}(t) \in U$, is specified. This determines a corresponding state trajectory $\mathbf{x}(t)$. Now we consider a "small" change in the control function to a new function $\mathbf{v}(t) \in U$. This change is "small" in the sense that the integral of absolute value of the

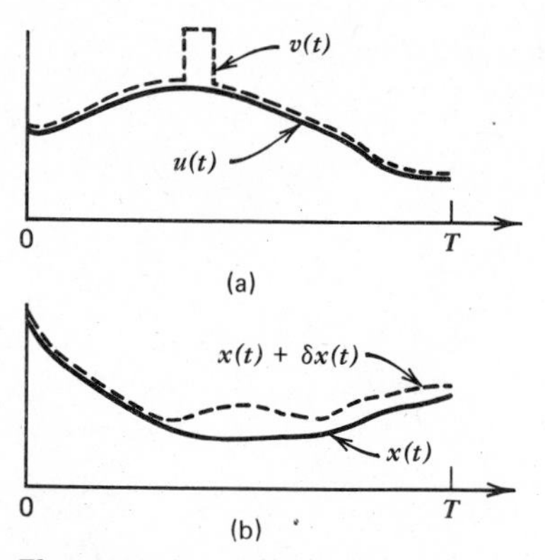

Figure 11.1. Effect of a small change.

difference is small (for each component of the control); that is,

$$\int_0^T |u_i(t) - v_i(t)|\, dt < \varepsilon \tag{11-5}$$

for each i, where ε is small. Therefore, the actual change can be large over a very short interval of time. (See Fig. 11.1*a*.)

This new control leads to a new state trajectory, which we write as $\mathbf{x}(t) + \delta\mathbf{x}(t)$. The change $\delta\mathbf{x}(t)$ is small for all t because the state depends (essentially) on the integral of the control function. (See Fig. 11.1*b*.)

If we define $\delta\bar{J}$ as the corresponding change in the modified objective, we have*

$$\delta\bar{J} = \psi(\mathbf{x}(T) + \delta\mathbf{x}(T)) - \psi(\mathbf{x}(T)) + \int_0^T [H(\boldsymbol{\lambda}, \mathbf{x} + \delta\mathbf{x}, \mathbf{v}) - H(\boldsymbol{\lambda}, \mathbf{x}, \mathbf{u}) - \boldsymbol{\lambda}^T \delta\dot{\mathbf{x}}]\, dt \tag{11-6}$$

An integration by parts yields

$$\int_0^T \boldsymbol{\lambda}^T \delta\dot{\mathbf{x}}\, dt = \boldsymbol{\lambda}(T)^T \delta\mathbf{x}(T) - \boldsymbol{\lambda}(0)^T \delta\mathbf{x}(0) - \int_0^T \dot{\boldsymbol{\lambda}}^T \delta\mathbf{x}\, dt \tag{11-7}$$

Therefore, we have

$$\delta\bar{J} = \psi(\mathbf{x}(T) + \delta\mathbf{x}(T)) - \psi(\mathbf{x}(T)) - \boldsymbol{\lambda}(T)^T \delta\mathbf{x}(T) + \boldsymbol{\lambda}(0)^T \delta\mathbf{x}(0) + \int_0^T [H(\boldsymbol{\lambda}, \mathbf{x} + \delta\mathbf{x}, \mathbf{v}) - H(\boldsymbol{\lambda}, \mathbf{x}, \mathbf{u}) + \dot{\boldsymbol{\lambda}}^T \delta\mathbf{x}]\, dt \tag{11-8}$$

We now approximate this expression (11-8) to first order (that is, to the order of ε) by using differential expressions for small differences.

* We drop the t arguments for simplicity and write $\mathbf{x}$ for $\mathbf{x}(t)$, and so forth.

We use the multidimension version of Taylor's Theorem to write*

$$\int_0^T [H(\boldsymbol{\lambda}, \mathbf{x}+\delta\mathbf{x}, \mathbf{v}) - H(\boldsymbol{\lambda}, \mathbf{x}, \mathbf{u})]\, dt$$

$$= \int_0^T [H(\boldsymbol{\lambda}, \mathbf{x}+\delta\mathbf{x}, \mathbf{v}) - H(\boldsymbol{\lambda}, \mathbf{x}, \mathbf{v}) + H(\boldsymbol{\lambda}, \mathbf{x}, \mathbf{v}) - H(\boldsymbol{\lambda}, \mathbf{x}, \mathbf{u})]\, dt$$

$$\simeq \int_0^T [H_{\mathbf{x}}(\boldsymbol{\lambda}, \mathbf{x}, \mathbf{v})\delta\mathbf{x} + H(\boldsymbol{\lambda}, \mathbf{x}, \mathbf{v}) - H(\boldsymbol{\lambda}, \mathbf{x}, \mathbf{u})]\, dt$$

$$= \int_0^T [H_{\mathbf{x}}(\boldsymbol{\lambda}, \mathbf{x}, \mathbf{u})\delta\mathbf{x} + (H_{\mathbf{x}}(\boldsymbol{\lambda}, \mathbf{x}, \mathbf{v}) - H_{\mathbf{x}}(\boldsymbol{\lambda}, \mathbf{x}, \mathbf{u}))\delta\mathbf{x} + H(\boldsymbol{\lambda}, \mathbf{x}, \mathbf{v}) - H(\boldsymbol{\lambda}, \mathbf{x}, \mathbf{u})]\, dt$$

$$\simeq \int_0^T [H_{\mathbf{x}}(\boldsymbol{\lambda}, \mathbf{x}, \mathbf{u})\delta\mathbf{x} + H(\boldsymbol{\lambda}, \mathbf{x}, \mathbf{v}) - H(\boldsymbol{\lambda}, \mathbf{x}, \mathbf{u})]\, dt \qquad (11\text{-}9)$$

where at each stage $\simeq$ denotes "equal to within the order of ε." (The last line follows by noting that both $\delta\mathbf{x}$ and the integral of $H_{\mathbf{x}}(\boldsymbol{\lambda}, \mathbf{x}, \mathbf{v}) - H_{\mathbf{x}}(\boldsymbol{\lambda}, \mathbf{x}, \mathbf{u})$ are of order ε; hence the product is of order ε^2.)

Substituting (11–9) into (11-8) and using a differential approximation to the first two terms in (11-8) yields

$$\delta\bar{J} = [\psi_{\mathbf{x}}(\mathbf{x}(T)) - \boldsymbol{\lambda}(T)^T]\delta\mathbf{x}(T) + \boldsymbol{\lambda}(0)^T\delta\mathbf{x}(0) + \int_0^T [H_{\mathbf{x}}(\boldsymbol{\lambda}, \mathbf{x}, \mathbf{u}) + \dot{\boldsymbol{\lambda}}^T]\delta\mathbf{x}\, dt + \int_0^T [H(\boldsymbol{\lambda}, \mathbf{x}, \mathbf{v}) - H(\boldsymbol{\lambda}, \mathbf{x}, \mathbf{u})]\, dt + p(\varepsilon) \qquad (11\text{-}10)$$

where $p(\varepsilon)$ denotes terms that are of smaller order than ε. This then is the general expression for the change in $\bar{J}$ resulting from an arbitrary change in $\mathbf{u}(t)$. We next simplify this expression by proper selection of the function $\boldsymbol{\lambda}(t)$.

The Adjoint Equation

Note that $\delta\mathbf{x}(0) = \mathbf{0}$, since a change in the control function does not change the

* Throughout this chapter the subscript notation is used for partial derivatives—scalar, vector, and matrix—as appropriate. Thus, for a function $f(x_1, x_2, \ldots, x_n) = f(\mathbf{x})$ we write $f_{x_1}(x_1, x_2, \ldots, x_n)$ for $(\partial f/\partial x_1)(x_1, x_2, \ldots, x_n)$. The gradient of f is denoted $\nabla f(\mathbf{x})$ or $f_{\mathbf{x}}(\mathbf{x})$ and is the row vector $[\partial f/\partial x_1, \partial f/\partial x_2, \ldots, \partial f/\partial x_n]$. If $\mathbf{f}(\mathbf{x})$ is m dimensional, then $\nabla \mathbf{f}(\mathbf{x}) = \mathbf{f}_{\mathbf{x}}(\mathbf{x})$ is the $m \times n$ Jacobian matrix $[(\partial f_i/\partial x_j)(\mathbf{x})]$. For a function $\mathbf{f}(\mathbf{x}, \mathbf{u})$ the notation $\mathbf{f}_{\mathbf{x}}(\mathbf{x}, \mathbf{u})$ similarly represents the matrix of partial derivatives of $\mathbf{f}$ with respect to the x_i's; that is,

$$\mathbf{f}_{\mathbf{x}}(\mathbf{x}, \mathbf{u}) = \left(\frac{\partial}{\partial x_j} f_i(\mathbf{x}, \mathbf{u})\right)$$

initial state. Thus, the second term on the right-hand side of (11-10) is always zero. We shall select $\boldsymbol{\lambda}(t)$ to make all other terms vanish except for the last integral. This is accomplished by selecting $\boldsymbol{\lambda}(t)$ as the solution to the *adjoint differential equation*

$$-\dot{\boldsymbol{\lambda}}(t)^T = H_{\mathbf{x}}(\boldsymbol{\lambda}(t), \mathbf{x}(t), \mathbf{u}(t)) \tag{11-11}$$

or, more explicitly,

$$-\dot{\boldsymbol{\lambda}}(t)^T = \boldsymbol{\lambda}(t)^T \mathbf{f}_{\mathbf{x}}(\mathbf{x}(t), \mathbf{u}(t)) + l_{\mathbf{x}}(\mathbf{x}(t), \mathbf{u}(t)) \tag{11-12}$$

with *final condition*

$$\boldsymbol{\lambda}(T)^T = \psi_{\mathbf{x}}(\mathbf{x}(T)) \tag{11-13}$$

Let us see what is involved here. First, we recall that $\mathbf{f}_{\mathbf{x}}(\mathbf{x}(t), \mathbf{u}(t))$ is an $n \times n$ matrix. It is, in general, time varying; but for a particular nominal $\mathbf{u}(t)$ and $\mathbf{x}(t)$ it is known. Likewise $l_{\mathbf{x}}(\mathbf{x}(t), \mathbf{u}(t))$ is a known (time-varying) n-dimensional row vector. Therefore (11-12) is a linear (time-varying) differential equation in the unknown (row) vector $\boldsymbol{\lambda}(t)^T$. Associated with this system is a *final* condition on $\boldsymbol{\lambda}(T)^T$. Thus, one can consider solving the adjoint equation by moving backward in time from T to 0. This determines a unique solution $\boldsymbol{\lambda}(t)$.

With this particular $\boldsymbol{\lambda}(t)$, the expression (11-10) for $\delta\bar{J}$ becomes simply

$$\delta\bar{J} = \int_0^T [H(\boldsymbol{\lambda}(t), \mathbf{x}(t), \mathbf{v}(t)) - H(\boldsymbol{\lambda}(t), \mathbf{x}(t), \mathbf{u}(t))]\, dt + p(\varepsilon) \tag{11-14}$$

Since $\boldsymbol{\lambda}(t)$, $\mathbf{x}(t)$, and $\mathbf{u}(t)$ are known and are independent of $\mathbf{v}(t)$, this expression gives a direct simple way to calculate the approximate consequence of a change to a new control function $\mathbf{v}(t)$. We can use this expression to deduce the conditions for optimality.

If the original control function $\mathbf{u}$ is optimal, it follows that for any t

$$H(\boldsymbol{\lambda}(t), \mathbf{x}(t), \mathbf{v}) \leq H(\boldsymbol{\lambda}(t), \mathbf{x}(t), \mathbf{u}(t))$$

for all $\mathbf{v} \in U$. [Here t is fixed, $\mathbf{u}(t)$ is the value of the optimal control at t, while $\mathbf{v}$ is arbitrary in U; $\mathbf{v}$ is not a time function.] To verify this inequality, suppose that for some t there were a $\mathbf{v} \in U$ with

$$H(\boldsymbol{\lambda}(t), \mathbf{x}(t), \mathbf{v}) > H(\boldsymbol{\lambda}(t), \mathbf{x}(t), \mathbf{u}(t))$$

Then we could change the function $\mathbf{u}$ as indicated in Fig. 11.1a so as to make the integrand in (11-14) positive over a small interval (say of width ε) containing this t. The integral itself would be positive (and of order ε). Thus, $\delta\bar{J}$ would be positive, contradicting the fact that the function $\mathbf{u}$ produces the maximal $\bar{J}$.

This result means that at every t the particular value $\mathbf{u}(t)$ in an optimal control has the property that it maximizes the Hamiltonian. This result is the Pontryagin maximum principle for this problem.

The Maximum Principle

We now summarize the complete set of conditions for optimality.

In the original problem one seeks $\mathbf{u}(t)$ and $\mathbf{x}(t)$ satisfying the system equation (11-1a), with the initial condition (11-1b), and the constraint (11-1c), while maximizing the objective function (11-1d). The necessary conditions serve as a test for a given $\mathbf{u}(t)$ and $\mathbf{x}(t)$ that must be satisfied if they are optimal. These conditions are stated in terms of an adjoint vector function $\boldsymbol{\lambda}(t)$.

Theorem (Maximum Principle). *Suppose $\mathbf{u}(t) \in U$ and $\mathbf{x}(t)$ represent the optimal control and state trajectory for the optimal control problem (11-1). Then there is an adjoint trajectory $\boldsymbol{\lambda}(t)$ such that together $\mathbf{u}(t)$, $\mathbf{x}(t)$, and $\boldsymbol{\lambda}(t)$ satisfy*

$$\dot{\mathbf{x}}(t) = \mathbf{f}(\mathbf{x}(t), \mathbf{u}(t)) \quad \text{(system equation)} \tag{11-15a}$$

$$\mathbf{x}(0) = \mathbf{x}_0 \quad \text{(initial state condition)} \tag{11-15b}$$

$$-\dot{\boldsymbol{\lambda}}(t)^T = \boldsymbol{\lambda}(t)^T \mathbf{f}_{\mathbf{x}}(\mathbf{x}(t), \mathbf{u}(t)) + l_{\mathbf{x}}(\mathbf{x}(t), \mathbf{u}(t)) \quad \text{(adjoint equation)} \tag{11-15c}$$

$$\boldsymbol{\lambda}(T)^T = \psi_{\mathbf{x}}(\mathbf{x}(T)) \quad \text{(adjoint final condition)} \tag{11-15d}$$

For all t, $0 \leq t \leq T$, and all $\mathbf{v} \in U$

$$H(\boldsymbol{\lambda}(t), \mathbf{x}(t), \mathbf{v}) \leq H(\boldsymbol{\lambda}(t), \mathbf{x}(t), \mathbf{u}(t)) \quad \text{(maximum condition)} \tag{11-15e}$$

where H is the Hamiltonian

$$H(\boldsymbol{\lambda}, \mathbf{x}, \mathbf{u}) = \boldsymbol{\lambda}^T \mathbf{f}(\mathbf{x}, \mathbf{u}) + l(\mathbf{x}, \mathbf{u})$$

This set of conditions can be regarded as a set of equations. The last condition, the maximum condition, is essentially a set of m static equations. One way to see this is to suppose that $\mathbf{u}(t)$ is interior to U (that is, not on the boundary of an inequality). Then the condition that H has a maximum at $\mathbf{u}(t)$ means that the derivatives of H with respect to each component of $\mathbf{u}$ must vanish. This gives m equations for each t.

The set of conditions (11-15) is *complete* in the sense that there are as many equations as unknowns. The unknowns are $\mathbf{u}(t)$, $\mathbf{x}(t)$, and $\boldsymbol{\lambda}(t)$—a total of $2n+m$ functions. The necessary conditions consist of $2n$ differential equations with $2n$ end conditions and m static equations (depending on t)—the total being sufficient to determine $2n+m$ functions. Thus, barring possible singular situations, these conditions can be used to find the optimal solution.

Before turning to some examples, one final point should be noted. As expressed in (11-15c) the adjoint equation is written in terms of $\boldsymbol{\lambda}(t)^T$. This is a natural consequence of the development. In practice, however, it is often convenient to write it in column-vector form in terms of $\boldsymbol{\lambda}(t)$. The result is

$$-\dot{\boldsymbol{\lambda}} = \mathbf{f}_{\mathbf{x}}(\mathbf{x}, \mathbf{u})^T \boldsymbol{\lambda} + l_{\mathbf{x}}(\mathbf{x}, \mathbf{u})^T \tag{11-15c$'$}$$

The important point to note is that the adjoint equation has system matrix $(\mathbf{f_x})^T$ rather than $\mathbf{f_x}$.

11.2 EXAMPLES

This section presents three examples of the use of the maximum principle. They illustrate how the necessary conditions can be used to find a solution. The particular method employed, however, depends on the structure of the problem.

Example 1 (The Triangle). It is desired to draw a curve $x(t)$, $0 \leq t \leq T$, starting at $x(0)=0$, whose slope at each point is no greater than 1 and that attains maximum height at T. This is a simple problem, for it is clear that the solution is to select $x(t)$ as the straight line with slope equal to 1. (See Fig. 11.2.) However, it is instructive to go through the mechanics of the maximum principle.

We may formulate the problem as having the components

$$\dot{x}(t) = u(t)$$
$$x(0) = 0$$
$$u(t) \leq 1$$
$$J = x(T)$$

In this problem, both $f_x(x, u) = 0$ and $l(x, u) = 0$. Therefore, the adjoint equation is

$$-\dot{\lambda}(t) = 0$$

The final condition is $\lambda(T) = 1$, since $\psi(x) = x$. Hence, we conclude immediately that $\lambda(t) \equiv 1$. The optimal control must maximize the Hamiltonian, which in this case reduces to

$$H = \lambda u = u$$

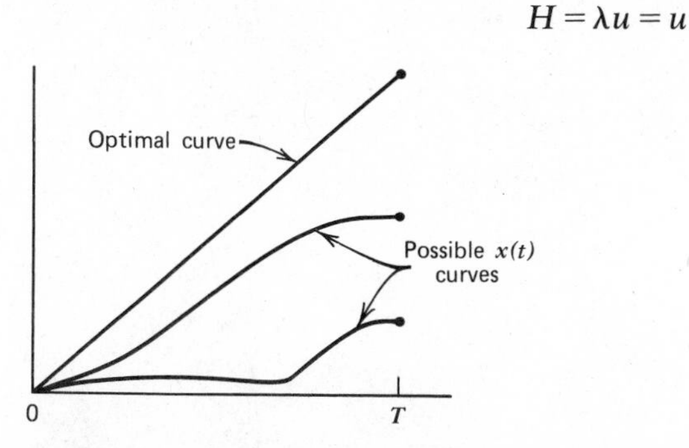

Figure 11.2. Triangle problem.

Thus, $u(t)$ must be as large as possible, subject to its constraint. We find therefore $u(t)=1$, which agrees with our intuitively derived solution.

Example 2 (Pushcart). A problem involving a second-order system is the problem of accelerating a cart in such a way as to maximize the total distance traveled in a given time, minus the total effort. The system is

$$\ddot{x}(t)=u(t)$$
$$x(0)=0$$
$$\dot{x}(0)=0$$

where x is the horizontal position and u is the applied force. The objective is

$$J=x(T)-\frac{1}{2}\int_0^T u(t)^2\,dt$$

where the integral term represents a penalty for effort. There are no control constraints.

Defining the state variables $x_1=x$, $x_2=\dot{x}$, the problem takes the standard form

$$\begin{bmatrix}\dot{x}_1\\ \dot{x}_2\end{bmatrix}=\begin{bmatrix}0 & 1\\ 0 & 0\end{bmatrix}\begin{bmatrix}x_1\\ x_2\end{bmatrix}+\begin{bmatrix}0\\ 1\end{bmatrix}u$$
$$x_1(0)=x_2(0)=0$$
$$J=x_1(T)-\frac{1}{2}\int_0^T u(t)^2\,dt$$

The adjoint system equation is*

$$-\dot{\lambda}_1(t)=0$$
$$-\dot{\lambda}_2(t)=\lambda_1(t)$$

The final conditions on the adjoint equations are (since $\psi=x_1(T)$)

$$\lambda_1(T)=1$$
$$\lambda_2(T)=0$$

The adjoint equations can be solved to yield

$$\lambda_1(t)=1$$
$$\lambda_2(t)=T-t$$

The Hamiltonian is

$$H(\boldsymbol{\lambda},\mathbf{x},\mathbf{u})=\lambda_1x_2+\lambda_2u-\tfrac{1}{2}u^2$$

* For a linear system with system matrix $\mathbf{A}$, the system matrix for the adjoint equation is $\mathbf{A}^T$.

Since u is unconstrained, the Hamiltonian can be maximized by setting its derivative with respect to u equal to zero. This yields

$$u(t)=\lambda_2(t)$$

as the condition for maximization. Thus, the final result is

$$u(t)=T-t$$

We conclude that the applied force should decrease linearly with time, reaching zero at the final time.

Example 3 (Insects as Optimizers). Many insects, such as wasps (including hornets and yellowjackets), live in colonies and have an annual life cycle. Their population consists of two castes: *workers* and *reproductives* (the latter comprised of *queens* and *males*). At the end of each summer all members of the colony die out except for the young queens who may start new colonies in early spring. From an evolutionary perspective, it is clear that a colony, in order to best perpetuate itself, should attempt to program its production of reproductives and workers so as to maximize the number of reproductives at the end of the season—in this way they maximize the number of colonies established the following year.

In reality, of course, this programming is not deduced consciously, but is determined by complex genetic characteristics of the insects. We may hypothesize, however, that those colonies that adopt nearly optimal policies of production will have an advantage over their competitors who do not. Thus, it is expected that through continued natural selection, existing colonies should be nearly optimal. We shall formulate and solve a simple version of the insects' optimal control problem to test this hypothesis.

Let $w(t)$ and $q(t)$ denote, respectively, the worker and reproductive population levels in the colony. At any time t, $0\le t\le T$, in the season the colony can devote a fraction $u(t)$ of its effort to enlarging the worker force and the remaining fraction $1-u(t)$ to producing reproductives. Accordingly, we assume that the two populations are governed by the equations:

$$\dot{w}(t)=bu(t)w(t)-\mu w(t)$$

$$\dot{q}(t)=c(1-u(t))w(t)$$

These equations assume that only workers gather resources. The positive constants b and c depend on the environment and represent the availability of resources and the efficiency with which these resources are converted into new workers and new reproductives. The per capita mortality rate of workers is μ, and for simplicity the small mortality rate of the reproductives is neglected. For the colony to be productive during the season it is assumed that $b>\mu$. The problem of the colony is to maximize

$$J=q(T)$$

subject to the constraint $0 \le u(t) \le 1$, and starting from the initial conditions

$$w(0)=1 \qquad q(0)=0$$

(The founding queen is counted as a worker since she, unlike subsequent reproductives, forages to feed the first brood.)

To apply the maximum principle to this problem we write the adjoint equations and terminal conditions, which, as the reader should verify, are

$$-\dot{\lambda}_1(t) = bu(t)\lambda_1(t) - \mu\lambda_1(t) + c(1-u(t))\lambda_2(t)$$
$$-\dot{\lambda}_2(t) = 0$$
$$\lambda_1(T)=0, \qquad \lambda_2(T)=1$$

In this case, unlike the previous two examples, the adjoint equations cannot be solved directly, since they depend on the unknown function $u(t)$. The other necessary conditions must be used in conjunction with the adjoint equations to determine the adjoint variables.

The Hamiltonian of the problem is

$$H(\lambda_1, \lambda_2, w, q, u) = \lambda_1(bu-\mu)w + \lambda_2 c(1-u)w$$
$$= w(\lambda_1 b - \lambda_2 c)u + (\lambda_2 c - \lambda_1 \mu)w$$

Since this Hamiltonian is linear in u, and since $w>0$, it follows that it is maximized with respect to $0 \le u \le 1$ by either $u=0$ or $u=1$, depending on whether $\lambda_1 b - \lambda_2 c$ is negative or positive, respectively.

It is now possible to solve the adjoint equations and determine the optimal u by moving *backward* in time from the terminal point T. In view of the known conditions on $\lambda_1(t)$ and $\lambda_2(t)$ at T, we find $\lambda_1(T)b - \lambda_2(T)c = -c < 0$, and hence the condition for maximization of the Hamiltonian yields $u(T)=0$. Also, it is clear, from the second adjoint equation, that $\lambda_2(t)=1$ for all t. Therefore near the terminal time T the first adjoint equation becomes

$$-\dot{\lambda}_1(t) = -\mu\lambda_1(t) + c$$

which has the solution

$$\lambda_1(t) = \frac{c}{\mu}(1 - e^{\mu(t-T)})$$

Viewed *backward* in time it follows that $\lambda_1(t)$ increases from its terminal value of 0. When it reaches a point $t_s < T$ where $\lambda_1(t_s) = c/b$, the value of u switches from 0 to 1. At that point the first adjoint equation becomes

$$-\dot{\lambda}_1(t) = (b-\mu)\lambda_1(t)$$

which, in view of the assumption that $b > \mu$, implies that, moving backward in time, $\lambda_1(t)$ continues to increase. Thus, there is no additional switch in u. The

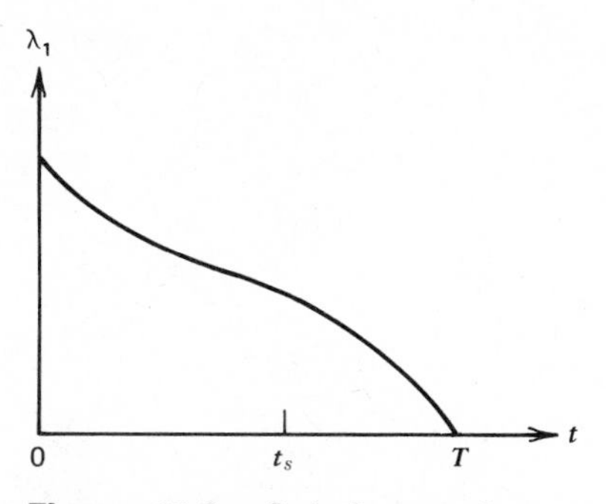

Figure 11.3. Solution to insect adjoint equation.

solution for $\lambda_1(t)$ is shown in Fig. 11.3. The point t_s is easily found to be

$$t_s = T + \frac{1}{\mu} \ln\left(1 - \frac{\mu}{b}\right)$$

In terms of the original colony problem, it is seen that the optimal solution is for the colony to produce only workers for the first part of the season, and then beyond a certain critical time to produce only reproductives. Social insect colonies do in fact closely follow this policy, and experimental evidence indicates that they adopt a switch time that is nearly optimal for their natural environment.

11.3 PROBLEMS WITH TERMINAL CONSTRAINTS

The basic optimal control problem discussed in Sect. 11.1 is referred to as a free endpoint problem since the final value of the state vector is completely arbitrary. There are, however, many problems in which the terminal value of the state vector is constrained in some manner. A simple example is the problem of finding the shortest path between two specified points. Another is the problem of economic planning to reach a given goal at minimum cost.

The maximum principle can be extended to such problems. This extension is simple and very natural in terms of the form of the final result. A rigorous proof, however, is exceedingly complex. Indeed, this extended result represents a major mathematical achievement, and is the capstone of the calculus of variations approach to optimal control. Our objective here is simply to state the result so that we can apply it, and to give a plausibility argument.

In an *optimal control problem with terminal constraints* one is given a time interval $0 \leq t \leq T$, a system

$$\dot{\mathbf{x}}(t) = \mathbf{f}(\mathbf{x}(t), \mathbf{u}(t)) \tag{11-16a}$$

an initial condition

$$\mathbf{x}(0) = \mathbf{x}_0 \tag{11-16b}$$

a control constraint

$$\mathbf{u}(t) \in U \tag{11-16c}$$

a set of terminal constraints

$$x_i(T) = \bar{x}_i \qquad i = 1, 2, \ldots, r \tag{11-16d}$$

and an objective function

$$J = \psi(\mathbf{x}(T)) + \int_0^T l(\mathbf{x}(t), \mathbf{u}(t))\, dt \tag{11-16e}$$

The terminal constraints take the form of specification of the final values of some of the state variables. The number of variables specified can vary from 1 to n (the dimension of the state). All other components of the problem are the same as in the free endpoint problem of Sect. 11.1.

When terminal constraints are present, there is a possibility that the problem has a certain degree of degeneracy. One new possibility, of course, is that there is no feasible control function and state trajectory satisfying the terminal constraint. No solution exists in this case. Another possibility is that there is only one trajectory satisfying the terminal constraint. In that case there is a solution, but it is unaffected by the particular objective function. The maximum principle must account for this type of degeneracy. A well-formulated problem, however, will not have such anomalies; rather, there will be a complete family of competing solutions.*

To deduce the conditions satisfied by an optimal solution for this problem, we go through the same procedure as in Sect. 11.1. Thus, we form the modified objective function and consider the change induced by a change in control from $\mathbf{u}(t)$ to a new control $\mathbf{v}(t)$ also satisfying all constraints. This leads to (11-10) of Sect. 11.1, which is repeated below:

$$\begin{aligned}\delta\bar{J} = {} & [\psi_{\mathbf{x}}(\mathbf{x}(T)) - \boldsymbol{\lambda}(T)^T]\delta\mathbf{x}(T) + \boldsymbol{\lambda}(0)^T\delta\mathbf{x}(0) \\ & + \int_0^T [H_{\mathbf{x}}(\boldsymbol{\lambda}, \mathbf{x}, \mathbf{u}) + \dot{\boldsymbol{\lambda}}^T]\delta\mathbf{x}\, dt \\ & + \int_0^T [H(\boldsymbol{\lambda}, \mathbf{x}, \mathbf{v}) - H(\boldsymbol{\lambda}, \mathbf{x}, \mathbf{u})]\, dt + p(\varepsilon)\end{aligned} \tag{11-17}$$

Again we select $\boldsymbol{\lambda}(t)$ to make all terms on the right-hand side of (11-17) vanish, except the last integral. As before $\delta\mathbf{x}(0)$ is always zero, so the second term on the right is zero. Also as before we require $\boldsymbol{\lambda}(t)$ to satisfy the adjoint equation

$$-\dot{\boldsymbol{\lambda}}(t)^T = H_{\mathbf{x}}(\boldsymbol{\lambda}(t), \mathbf{x}(t), \mathbf{u}(t)) \tag{11-18}$$

* The mathematical condition for a well-formulated problem is closely related to a controllability criterion in that a full range of terminal positions can be achieved by various control functions.

However, specification of boundary conditions is somewhat different than before. Since $x_i(T)$ is constrained for $i=1, 2, \ldots, r$, it follows that $\delta x_i(T)=0, i=1, 2, \ldots, r$ for all possible competing trajectories. Therefore, it is not necessary to specify all components of $\boldsymbol{\lambda}(T)$ in order to guarantee that the first term on the right-hand side of (11-17) vanishes. Only components corresponding to possibly nonzero components of $\delta\mathbf{x}(T)$ must be specified. Accordingly, it is necessary only to require

$$\lambda_i(T)=\psi_{\mathbf{x}}(\mathbf{x}(T))_i \tag{11-19}$$

for $i=r+1, r+2, \ldots, n$. This guarantees that all component products in the first term of (11-17) vanish. Thus, the general rule is: If $x_i(T)$ is constrained, then $\lambda_i(T)$ is free; if $x_i(T)$ is free, then $\lambda_i(T)$ is constrained.

The complete statement of the maximum principle for this problem is given below.

Theorem (Maximum Principle). *Let $\mathbf{x}(t)$, $\mathbf{u}(t)\in U$ be an optimal solution to the problem with terminal constraints (11-16). Then there is an adjoint trajectory $\boldsymbol{\lambda}(t)$ and a constant $\lambda_0\geq 0$ [with $(\lambda_0, \boldsymbol{\lambda}(t))\neq\mathbf{0}$] such that together $\mathbf{u}(t)$, $\mathbf{x}(t)$, $\boldsymbol{\lambda}(t)$, and λ_0 satisfy*

$$\dot{\mathbf{x}}(t)=\mathbf{f}(\mathbf{x}(t), \mathbf{u}(t)) \quad \text{(system equation)} \tag{11-20a}$$

$$\mathbf{x}(0)=\mathbf{x}_0 \quad \text{(initial state condition)} \tag{11-20b}$$

$$x_i(T)=\bar{x}_i, \qquad i=1, 2, \ldots, r \quad \text{(terminal state conditions)} \tag{11-20c}$$

$$-\dot{\boldsymbol{\lambda}}(t)^T=\boldsymbol{\lambda}(t)^T\mathbf{f}_{\mathbf{x}}(\mathbf{x}(t), \mathbf{u}(t))+\lambda_0 l_{\mathbf{x}}(\mathbf{x}(t), \mathbf{u}(t)) \quad \text{(adjoint equation)} \tag{11-20d}$$

$$\lambda_i(T)=\lambda_0\psi_{\mathbf{x}}(x(T))_i, \qquad i=r+1, r+2, \ldots, n \quad \text{(adjoint final conditions)} \tag{11-20e}$$

For all t, $0\leq t\leq T$, and all $\mathbf{v}\in U$.

$$H(\boldsymbol{\lambda}(t), \mathbf{x}(t), \mathbf{v})\leq H(\boldsymbol{\lambda}(t), \mathbf{x}(t), \mathbf{u}(t)) \quad \text{(maximum condition)} \tag{11-20f}$$

where H is the Hamiltonian

$$H(\boldsymbol{\lambda}, \mathbf{x}, \mathbf{u})=\boldsymbol{\lambda}^T\mathbf{f}(\mathbf{x}, \mathbf{u})+\lambda_0 l(\mathbf{x}, \mathbf{u})$$

The presence of the constant λ_0 in this version of the maximum principle is to account for the degeneracy situation discussed earlier. These degenerate situations (where the terminal constraint is overwhelmingly imposing) correspond to $\lambda_0=0$. In these cases the objective does not enter the conditions. In well-formulated problems, however, $\lambda_0>0$, and without loss of generality one may then set $\lambda_0=1$. In practice, therefore, one always tries to apply the maximum principle with $\lambda_0=1$. Indeed, for all examples and problems in this chapter this procedure will work—except for Problem 13.

Example 1 (Shortest Path between Two Points). Consider the classic problem of finding the curve $x(t)$ with $x(0)=0$, $x(1)=1$ with minimum total length. For this problem we write

$$\dot{x}=u$$

$$x(0)=0, \qquad x(1)=1$$

$$J=-\int_0^1 \sqrt{1+u^2}\,dt$$

There are no constraints on the value of $u(t)$.

For this problem the adjoint equation is easily seen to be

$$-\dot{\lambda}=0$$

There is no terminal condition on $\lambda(T)$. However, it is clear that $\lambda(t)=\lambda$, a constant.

The Hamiltonian is

$$H=\lambda u-\sqrt{1+u^2}$$

The optimal $u(t)$ must maximize this at each t. However, since all terms in the Hamiltonian (other than u) do not depend on t, it is clear that $u(t)$ is constant. Thus, the slope of $x(t)$ is constant—that is, the best curve is a straight line.

Example 2 (The Classic Isoperimetric Problem). Problems subject to various integral constraints can be converted to problems with terminal constraints by introducing additional variables whose sole purpose is to keep track of how much of the integral constraint has been used. To illustrate this idea we consider the classic problem of determining the curve that connects two fixed points on the t-axis, has fixed arc length L, and encloses the maximum area between itself and the t-axis. (See Fig. 11.4.)

This problem can be defined with

$$\dot{x}=u$$

$$x(0)=0$$

$$x(T)=0$$

$$J=\int_0^T x\,dt$$

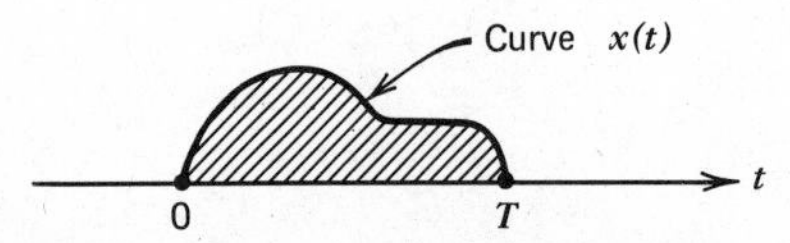

Figure 11.4. The isoperimetric problem.

but, in addition, there is the arc length constraint

$$\int_0^T \sqrt{1+u^2}\, dt = L$$

(with $L > T$). This constraint can be incorporated by introducing an additional state variable $y(t)$ satisfying

$$\dot{y} = \sqrt{1+u^2}$$
$$y(0) = 0$$
$$y(T) = L$$

The overall problem then has a two-dimensional state vector.

The adjoint equations are found to be

$$-\dot{\lambda}_1 = 1$$
$$-\dot{\lambda}_2 = 0$$

Thus, $\lambda_1(t)$ is linear in t and $\lambda_2(t)$ is constant. The Hamiltonian is

$$H = \lambda_1 u + \lambda_2 \sqrt{1+u^2} + x$$

The condition for maximization is obtained by setting the derivative of H with respect to u equal to zero. This yields

$$\lambda_1 + \frac{\lambda_2 u}{\sqrt{1+u^2}} = 0$$

Therefore, substituting $\dot{x} = u$ and using the fact that λ_1 is linear in t, we can conclude that the curve $x(t)$ satisfies an equation of the form

$$\frac{\dot{x}}{\sqrt{1+\dot{x}^2}} = A + Bt$$

for some constants A and B. It can be verified that the arc of a circle

$$(x - x_1)^2 + (t - t_1)^2 = r^2$$

satisfies this equation. The parameters x_1, t_1, and r are chosen to satisfy the constraints.

11.4 FREE TERMINAL TIME PROBLEMS

In some problems the terminal time is not fixed, but is allowed to vary. The terminal time is therefore another variable that can be selected in order to attain the maximum possible objective value. For example, in determining the path of a rocket ship to the moon that minimizes fuel consumption, there is no reason to specify the time of landing. Such problems are treated by yet another simple addition to the maximum principle.

The problem statement in this section is identical to that of the last section, except that the terminal time T is not specified. Explicitly, the problem is: Given

$$\dot{\mathbf{x}}(t) = \mathbf{f}(\mathbf{x}(t), \mathbf{u}(t)) \tag{11-21a}$$

$$\mathbf{x}(0) = \mathbf{x}_0 \tag{11-21b}$$

$$\mathbf{u}(t) \in U \tag{11-21c}$$

$$x_i(T) = \bar{x}_i, \qquad i = 1, 2, \ldots, r \tag{11-21d}$$

find $T > 0$ and $\mathbf{u}(t)$, $0 \leq t \leq T$ so as to maximize

$$J = \psi(\mathbf{x}(T)) + \int_0^T l(\mathbf{x}(t), \mathbf{u}(t))\, dt \tag{11-21e}$$

Clearly if the best T were known, we could fix T at this value and the problem would be one of the type treated in Sect. 11.3. Thus, all of the maximum principle conditions of that section must apply here as well. We require, however, one additional condition from which the unknown value of T can be determined. To find this condition we must go through the procedure of calculating the change in the modified objective function once again, but for this new problem.

The modified objective is

$$\bar{J} = \psi(\mathbf{x}(T)) + \int_0^T [l(\mathbf{x}, \mathbf{u}) + \boldsymbol{\lambda}^T \mathbf{f}(\mathbf{x}, \mathbf{u}) - \boldsymbol{\lambda}^T \dot{\mathbf{x}}]\, dt \tag{11-22}$$

We consider a change to a new control $\mathbf{v}(t)$ with an associated new trajectory $\mathbf{x}(t) + \delta\mathbf{x}(t)$ and a new terminal time $T + dT$. We denote the corresponding new value of the modified objective function by $\bar{J} + d\bar{J}$.

The important new feature of this problem is that the change in the terminal value of the state is *not* $\delta\mathbf{x}(T)$, since the final time itself changes. The new terminal state is actually $\mathbf{x}(T) + \delta\mathbf{x}(T + dT)$. If we denote the new terminal state by $\mathbf{x}(T) + d\mathbf{x}(T)$, then to a first-order approximation

$$d\mathbf{x}(T) = \delta\mathbf{x}(T) + \dot{\mathbf{x}}(T)\, dT = \delta\mathbf{x}(T) + \mathbf{f}(\mathbf{x}(T), \mathbf{u}(T))\, dT \tag{11-23}$$

(See Fig. 11.5.)

As before, we find the change in $\bar{J}$ by considering a first-order Taylor expansion with respect to $\delta\mathbf{x}$. In this case we also consider the change dT. We suppress the details here, but a procedure similar to that used in Sect. 11.1

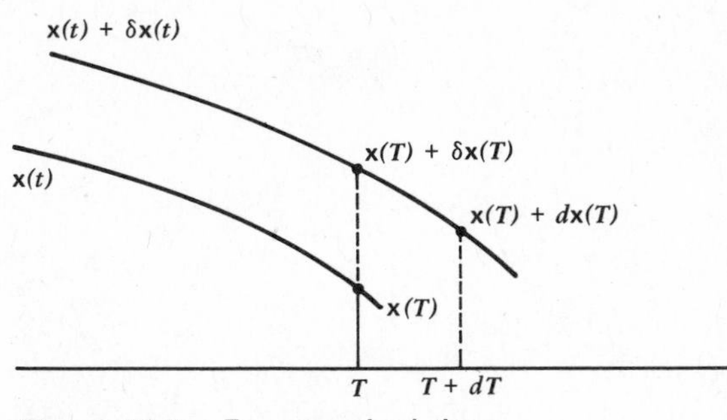

Figure 11.5. Free terminal time.

leads to

$$\begin{aligned} d\bar{J} = {} & \psi_{\mathbf{x}}(\mathbf{x}(T))\, d\mathbf{x}(T) - \boldsymbol{\lambda}(T)^T \delta\mathbf{x}(T) + \boldsymbol{\lambda}(0)^T \delta\mathbf{x}(0) \\ & + l(\mathbf{x}(T), \mathbf{u}(T))\, dT \\ & + \int_0^T [H_{\mathbf{x}}(\boldsymbol{\lambda}, \mathbf{x}, \mathbf{u}) + \dot{\boldsymbol{\lambda}}^T] \delta\mathbf{x}\, dt \\ & + \int_0^T [H(\boldsymbol{\lambda}, \mathbf{x}, \mathbf{v}) - H(\boldsymbol{\lambda}, \mathbf{x}, \mathbf{u})]\, dt \end{aligned} \tag{11-24}$$

By selecting $\boldsymbol{\lambda}^T(t)$ to satisfy the adjoint equation, the first integral term vanishes. The term $\boldsymbol{\lambda}^T(0)\delta\mathbf{x}(0)$ vanishes, since $\delta\mathbf{x}(0) = \mathbf{0}$. Using (11-23), we substitute $\delta\mathbf{x}(T) = d\mathbf{x}(T) - \mathbf{f}(\mathbf{x}(T), \mathbf{u}(T))\, dT$ and find that the only nonzero terms are then

$$\begin{aligned} d\bar{J} = {} & [\psi_{\mathbf{x}}(\mathbf{x}(T)) - \boldsymbol{\lambda}(T)^T]\, d\mathbf{x}(T) \\ & + [\boldsymbol{\lambda}^T(T)\mathbf{f}(\mathbf{x}(T), \mathbf{u}(T)) + l(\mathbf{x}(T), \mathbf{u}(T))]\, dT \\ & + \int_0^T [H(\boldsymbol{\lambda}, \mathbf{x}, \mathbf{v}) - H(\boldsymbol{\lambda}, \mathbf{x}, \mathbf{u})]\, dt \end{aligned}$$

The first term vanishes if conditions are imposed on the components $\lambda_i(T)$, $i = r+1, r+2, \ldots, n$, the same as for the problem with fixed terminal time. Consideration of the integral term yields the maximum condition as usual. Finally, the only remaining term is the second. It is recognized that it is $H(\boldsymbol{\lambda}(T), \mathbf{x}(T), \mathbf{u}(T))\, dT$. It follows that, in order for T to be optimal, $H(\boldsymbol{\lambda}(T), \mathbf{x}(T), \mathbf{u}(T)) = 0$; otherwise a change dT could be found that would improve $\bar{J}$. Thus, the new condition is that the Hamiltonian must vanish at the terminal time.

To summarize, for problems with free terminal time all the usual conditions of the maximum principle of Sect. 11.3 apply, and in addition

$$H(\boldsymbol{\lambda}(T), \mathbf{x}(T), \mathbf{u}(T)) = 0 \tag{11-25}$$

This is the additional condition required to solve for the additional unknown T.

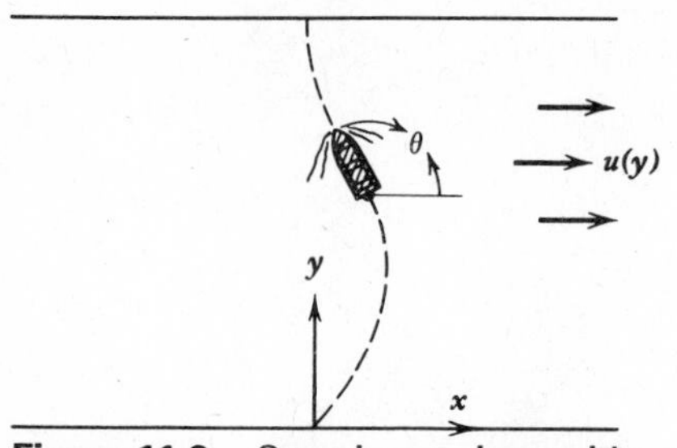

Figure 11.6. Crossing a river with currents.

Example (Zermelo's Problem). We consider the problem of navigating a boat across a river, in which there is strong current, so as to get to a specified point on the other side in minimum time. We assume that the magnitude of the boat's velocity with respect to the water is a constant V. The downstream current at any point depends only on the distance from the bank. (See Fig. 11.6.)

The equations of the boat's motion are

$$\dot{x} = V\cos\theta + u(y) \tag{11-26a}$$

$$\dot{y} = V\sin\theta \tag{11-26b}$$

where x is the downstream position along the river, y is the distance from the origin bank, $u(y)$ is the downstream current, and θ is the heading angle of the boat. The heading angle is the control, which may vary along the path.

Both initial and final values of x and y are specified. The objective function is (negative) total time so we set $\psi = 0$, $l \equiv -1$.

The adjoint equations are easily found to be

$$-\dot{\lambda}_1 = 0 \tag{11-27a}$$

$$-\dot{\lambda}_2 = u'(y)\lambda_1 \tag{11-27b}$$

There are no terminal conditions on the adjoint variables. The Hamiltonian is

$$H = \lambda_1 V\cos\theta + \lambda_1 u(y) + \lambda_2 V\sin\theta - 1 \tag{11-28}$$

The condition for maximization yields

$$H_\theta = -\lambda_1 V\sin\theta + \lambda_2 V\cos\theta = 0 \tag{11-29}$$

and hence

$$\tan\theta(t) = \lambda_2(t)/\lambda_1(t) \tag{11-30}$$

Next we observe that

$$\begin{aligned}\frac{d}{dt}H = {} & \dot{\lambda}_1 V\cos\theta + \dot{\lambda}_1 u(y) + \lambda_1 u'(y)\dot{y} \\ & + \dot{\lambda}_2 V\sin\theta + H_\theta\dot{\theta}\end{aligned}$$

Using (11-26b), (11-27), and (11-29) this is seen to be identically zero. (Also see Problem 7.) Thus, H is constant. Since this is a free-time problem we use the condition derived in this section to infer that this constant is zero. Thus,

$$H \equiv 0 \tag{11-31}$$

Substituting (11-30) into (11-28) we obtain

$$H = \frac{\lambda_1[V + u(y)\cos\theta]}{\cos\theta} - 1$$

Since according to (11-27a) λ_1 is constant, we use (11-31) to obtain the final result:

$$\frac{V + u(y)\cos\theta}{\cos\theta} = A \tag{11-32}$$

for some constant A. Once A is specified this determines θ as a function of y. The constant A is chosen to obtain the proper landing point.

A special case is when $u(y) = u$ is independent of y. Then the optimal paths are straight lines.

11.5 LINEAR SYSTEMS WITH QUADRATIC COST

Problems in which the dynamic system is linear and the objective is quadratic represent an extremely important special family of optimal control problems. This importance stems in part from the fact that in many situations this structure represents a natural formulation; but in actuality the main source of importance derives from the strong analytic results available for this family. The primary feature of these problems is that the optimal control can be expressed in linear feedback form. Thus, the resulting closed-loop system is also a linear dynamic system.

In the standard "linear-quadratic" problem one is given a linear nth-order system

$$\dot{\mathbf{x}}(t) = \mathbf{A}(t)\mathbf{x}(t) + \mathbf{B}(t)\mathbf{u}(t) \tag{11-33a}$$

and a cost function

$$J = \tfrac{1}{2}\int_0^T [\mathbf{x}(t)^T\mathbf{Q}(t)\mathbf{x}(t) + \mathbf{u}(t)^T\mathbf{R}(t)\mathbf{u}(t)]\,dt \tag{11-33b}$$

The cost function is to be *minimized.* In this problem $\mathbf{u}(t)$ is an m-dimensional input function, and it is *not* subject to any constraints. The cost function is quadratic in both the state and the control. The quadratic functions are defined by the matrices $\mathbf{Q}(t)$ and $\mathbf{R}(t)$ that are symmetric matrices of dimension $n \times n$

and $m \times m$, respectively. The matrix $\mathbf{Q}(t)$ is assumed to represent a positive semidefinite quadratic form [so $\mathbf{x}(t)^T\mathbf{Q}(t)\mathbf{x}(t) \geq 0$ for all $\mathbf{x}(t)$; see Sect. 9.11]. The matrix $\mathbf{R}(t)$ is assumed to represent a positive definite quadratic form [and hence $\mathbf{R}(t)$ is nonsingular]. All functions are assumed to be continuous with respect to t.

This problem is a special case of the basic free endpoint problem of Sect. 11.1. Thus, we can easily write down the necessary conditions satisfied by an optimal solution. The adjoint equation is

$$-\dot{\boldsymbol{\lambda}}(t)^T = \boldsymbol{\lambda}(t)^T\mathbf{A}(t) - \mathbf{x}(t)^T\mathbf{Q}(t) \tag{11-34}$$

with terminal condition $\boldsymbol{\lambda}(T) = \mathbf{0}$. The minus sign on the right-hand side of (11-34) is due to the fact that we are maximizing $-J$.

The Hamiltonian is

$$\begin{aligned} H = {} & \boldsymbol{\lambda}(t)^T\mathbf{A}(t)\mathbf{x}(t) + \boldsymbol{\lambda}(t)^T\mathbf{B}(t)\mathbf{u}(t) \\ & -\tfrac{1}{2}\mathbf{x}(t)^T\mathbf{Q}(t)\mathbf{x}(t) - \tfrac{1}{2}\mathbf{u}(t)^T\mathbf{R}(t)\mathbf{u}(t) \end{aligned} \tag{11-35}$$

The condition for maximizing the Hamiltonian with respect to $\mathbf{u}(t)$ is $H_{\mathbf{u}} = \mathbf{0}$; or,

$$\boldsymbol{\lambda}(t)^T\mathbf{B}(t) - \mathbf{u}(t)^T\mathbf{R}(t) = \mathbf{0} \tag{11-36}$$

Therefore,

$$\mathbf{u}(t) = \mathbf{R}(t)^{-1}\mathbf{B}(t)^T\boldsymbol{\lambda}(t) \tag{11-37}$$

This expression can be substituted into the original system equation. If this substitution is made, and if the adjoint equation (11-34) is written in transposed form, we obtain the equations

$$\dot{\mathbf{x}}(t) = \mathbf{A}(t)\mathbf{x}(t) + \mathbf{B}(t)\mathbf{R}(t)^{-1}\mathbf{B}(t)^T\boldsymbol{\lambda}(t) \tag{11-38a}$$

$$\dot{\boldsymbol{\lambda}}(t) = \mathbf{Q}(t)\mathbf{x}(t) - \mathbf{A}(t)^T\boldsymbol{\lambda}(t) \tag{11-38b}$$

with conditions

$$\mathbf{x}(0) = \mathbf{x}_0 \tag{11-38c}$$

$$\boldsymbol{\lambda}(T) = \mathbf{0} \tag{11-38d}$$

In (11-38) there are $2n$ differential equations, $2n$ endpoint conditions, and $2n$ unknown functions. The difficulty, of course, is that the $2n$ conditions are not all at the same endpoint. If they were, say, all initial conditions, the system could be solved (numerically) by integrating forward. Since they are at different ends, a special technique must be developed.

Riccati Equation

Since the system (11-38) is linear, it is clear that both $\mathbf{x}(t)$ and $\boldsymbol{\lambda}(t)$ depend linearly on $\mathbf{x}_0$. Accordingly, $\boldsymbol{\lambda}(t)$ depends linearly on $\mathbf{x}(t)$. This motivates us to try a solution of the form

$$\boldsymbol{\lambda}(t) = -\mathbf{P}(t)\mathbf{x}(t) \tag{11-39}$$

where $\mathbf{P}(t)$ is an (as yet unknown) $n \times n$ matrix. Substituting (11-39) in the system (11-38) yields the two equations

$$\dot{\mathbf{x}}(t) = [\mathbf{A}(t) - \mathbf{B}(t)\mathbf{R}(t)^{-1}\mathbf{B}(t)^T\mathbf{P}(t)]\mathbf{x}(t) \tag{11-40a}$$

$$-\mathbf{P}(t)\dot{\mathbf{x}}(t) - \dot{\mathbf{P}}(t)\mathbf{x}(t) = [\mathbf{Q}(t) + \mathbf{A}(t)^T\mathbf{P}(t)]\mathbf{x}(t) \tag{11-40b}$$

Multiplication of (11-40a) by $\mathbf{P}(t)$ and addition to (11-40b) then yields

$$\begin{aligned}\mathbf{0} = [&\dot{\mathbf{P}}(t) + \mathbf{P}(t)\mathbf{A}(t) + \mathbf{A}(t)^T\mathbf{P}(t) \\ &- \mathbf{P}(t)\mathbf{B}(t)\mathbf{R}(t)^{-1}\mathbf{B}(t)^T\mathbf{P}(t) + \mathbf{Q}(t)]\mathbf{x}(t)\end{aligned} \tag{11-41}$$

This will be satisfied for any $\mathbf{x}(t)$ if $\mathbf{P}(t)$ is chosen so as to satisfy the matrix differential equation

$$\begin{aligned}-\dot{\mathbf{P}}(t) = {} & \mathbf{P}(t)\mathbf{A}(t) + \mathbf{A}(t)^T\mathbf{P}(t) \\ & - \mathbf{P}(t)\mathbf{B}(t)\mathbf{R}(t)^{-1}\mathbf{B}(t)^T\mathbf{P}(t) + \mathbf{Q}(t)\end{aligned} \tag{11-42a}$$

From the endpoint condition $\boldsymbol{\lambda}(T) = \mathbf{0}$ we derive the corresponding condition

$$\mathbf{P}(T) = \mathbf{0} \tag{11-42b}$$

The differential equation (11-42), which is quadratic in the unknown $\mathbf{P}(t)$, is called a *Riccati* equation. The solution $\mathbf{P}(t)$ is symmetric, since $\dot{\mathbf{P}}(t)$ is symmetric for all t. It can also be shown that $\mathbf{P}(t)$ is positive semidefinite.

Feedback Solution

The solution to the control problem (11-33) is now obtained as follows. One first solves the matrix Riccati equation (11-42) by backward integration starting at $t = T$ with the condition $\mathbf{P}(T) = \mathbf{0}$. This solution is usually determined numerically. Once $\mathbf{P}(t)$ is known, it can be used to solve (11-33) in feedback form for any initial state $\mathbf{x}_0$. The control is found by combining (11-37) and (11-39) to obtain

$$\mathbf{u}(t) = -\mathbf{R}(t)^{-1}\mathbf{B}(t)^T\mathbf{P}(t)\mathbf{x}(t) \tag{11-43}$$

or, equivalently,

$$\mathbf{u}(t) = \mathbf{K}(t)\mathbf{x}(t) \tag{11-44}$$

where

$$\mathbf{K}(t) = -\mathbf{R}(t)^{-1}\mathbf{B}(t)^T\mathbf{P}(t) \tag{11-45}$$

The $m \times n$ matrix $\mathbf{K}(t)$ can be computed before operation of the system. Then as the system evolves, the control is computed at each instant on the basis of the current state. This is a feedback solution; and in this case it is linear feedback.

Time-Invariant Case

Suppose now that all matrices ($\mathbf{A}$, $\mathbf{B}$, $\mathbf{Q}$, and $\mathbf{R}$) in the problem definition are constant, independent of time. Suppose also that we consider letting $T \to \infty$ in the problem definition. Then, since the Riccati equation is integrated backward in time, the solution can be expected to approach a constant matrix for t near 0. Accordingly, $\dot{\mathbf{P}}$ approaches $\mathbf{0}$, and the limiting constant matrix $\mathbf{P}$ is a solution to the matrix algebraic equation

$$\mathbf{0} = \mathbf{PA} + \mathbf{A}^T\mathbf{P} - \mathbf{PBR}^{-1}\mathbf{B}^T\mathbf{P} + \mathbf{Q} \tag{11-46}$$

In this case the optimal control is

$$\mathbf{u}(t) = -\mathbf{R}^{-1}\mathbf{B}^T\mathbf{Px}(t) = \mathbf{Kx}(t) \tag{11-47}$$

which is itself a time-invariant linear feedback structure. The overall resulting optimal system is thus governed by

$$\dot{\mathbf{x}}(t) = (\mathbf{A} + \mathbf{BK})\mathbf{x}(t) \tag{11-48}$$

Hence, this optimal control approach provides a sophisticated alternative to the problem of selecting a feedback matrix (compare with Sect. 8.9) to improve the performance of a system.

11.6 DISCRETE-TIME PROBLEMS

Optimal control problems can also be formulated for discrete-time systems. The resulting necessary conditions are quite similar to those for continuous-time systems, although their form is slightly weaker than the maximum principle.

The basic discrete-time optimal control problem consists of a dynamic system (defined for $k = 0, 1, 2, \ldots, N-1$)

$$\mathbf{x}(k+1) = \mathbf{f}(\mathbf{x}(k), \mathbf{u}(k)) \tag{11-49a}$$

an initial condition

$$\mathbf{x}(0) = \mathbf{x}_0 \tag{11-49b}$$

and an objective

$$J = \psi(\mathbf{x}(N)) + \sum_{k=0}^{N-1} l(\mathbf{x}(k), \mathbf{u}(k)) \tag{11-49c}$$

Note that in this formulation there is no constraint on $\mathbf{u}(k)$. Otherwise, the formulation and interpretation are entirely analogous to the continuous-time problem of Sect. 11.1.

The modified objective is in this case taken to be

$$\bar{J}=J-\sum_{k=0}^{N-1}[\boldsymbol{\lambda}(k+1)^T\mathbf{x}(k+1)-\boldsymbol{\lambda}(k+1)^T\mathbf{f}(\mathbf{x}(k),\mathbf{u}(k))]$$

where the $\boldsymbol{\lambda}(k)$'s are yet to be selected. This can be written as

$$\bar{J}=\psi(\mathbf{x}(N))+\sum_{k=0}^{N-1}[H(\boldsymbol{\lambda}(k+1),\mathbf{x}(k),\mathbf{u}(k))-\boldsymbol{\lambda}(k)^T\mathbf{x}(k)]$$
$$-\boldsymbol{\lambda}(N)^T\mathbf{x}(N)+\boldsymbol{\lambda}(0)^T\mathbf{x}(0) \qquad (11\text{-}50)$$

where H is the Hamiltonian

$$H(\boldsymbol{\lambda},\mathbf{x},\mathbf{u})=\boldsymbol{\lambda}^T\mathbf{f}(\mathbf{x},\mathbf{u})+l(\mathbf{x},\mathbf{u}) \qquad (11\text{-}51)$$

This is quite analogous to (11-3).

Following a procedure parallel to that used in Sect. 11.1, the effect of a small change in $\mathbf{u}(k)$ on the modified objective can be determined. The conditions for optimality are then obtained by requiring that this change be zero (at least to first-order). The result of this analysis is expressed in the following theorem.

Theorem (Discrete-Time Optimality). *Suppose $\mathbf{u}(k)$ and $\mathbf{x}(k)$, $k=0,1,\ldots,N$ represent the optimal control and state trajectory for the optimal control problem (11-49). Then there is an adjoint trajectory $\boldsymbol{\lambda}(k)$, $k=0,1,\ldots,N$ such that together $\mathbf{u}(k)$, $\mathbf{x}(k)$, and $\boldsymbol{\lambda}(k)$ satisfy*

$$\mathbf{x}(k+1)=\mathbf{f}(\mathbf{x}(k),\mathbf{u}(k)) \quad \text{(system equation)} \qquad (11\text{-}52a)$$

$$\mathbf{x}(0)=\mathbf{x}_0 \quad \text{(initial state condition)} \qquad (11\text{-}52b)$$

$$\boldsymbol{\lambda}(k)=\boldsymbol{\lambda}(k+1)\mathbf{f}_{\mathbf{x}}(\mathbf{x}(k),\mathbf{u}(k))+l_{\mathbf{x}}(\mathbf{x}(k),\mathbf{u}(k)) \quad \text{(adjoint equation)} \qquad (11\text{-}52c)$$

$$\boldsymbol{\lambda}(N)=\psi_{\mathbf{x}}(\mathbf{x}(N)) \quad \text{(adjoint final condition)}$$

$$H_{\mathbf{u}}(\boldsymbol{\lambda}(k+1),\mathbf{x}(k),\mathbf{u}(k))=\mathbf{0} \quad \text{(variational condition)} \qquad (11\text{-}52d)$$

where H is the Hamiltonian (11-51).

This is quite analogous to the maximum principle of Sect. 11.1, except that instead of requiring that the Hamiltonian be maximized, the condition is that the Hamiltonian must have zero derivative with respect to the control variables.

This general pattern applies to the discrete-time analogs of the continuous-time problems treated in Sects. 11.1 and 11.3. The adjoint equation* is (11-52c). The terminal boundary conditions of the adjoint are determined exactly as in continuous-time problems. Rather than maximization of the Hamiltonian, however, the condition is that the derivative of the Hamiltonian with respect to $\mathbf{u}$ must vanish. This is a slightly weaker condition than maximization.

Example 1 (A Simple Resource Allocation Problem). A special but important class of optimization problems is where a fixed resource must be distributed among a number of different activities. If the resource is money, for example, one may seek an optimal allocation among a number of projects. Suppose an amount A of the resource is given, and there are N activities. If an amount u is allocated to the kth activity, the value accrued by this allocation is $g_k(u)$. The optimization problem is then to maximize

$$g_0(u(0))+g_1(u(1))+\cdots+g_{N-1}(u(N-1))$$

subject to the constraint that

$$u(0)+u(1)+\cdots+u(N-1)=A$$

As a special case we assume $g_k(u)=u^{1/2}$. Thus, the problem is to maximize

$$\sum_{k=0}^{N-1} u(k)^{1/2} \tag{11-53}$$

subject to

$$\sum_{k=0}^{N-1} u(k)=A \tag{11-54}$$

This problem is equivalent to the control problem having system equation

$$x(k+1)=x(k)-u(k)$$

with end conditions

$$x(0)=A, \qquad x(N)=0$$

and objective (to be maximized)

$$J=\sum_{k=0}^{N-1} u(k)^{1/2}$$

This formulation assumes that the allocation is made serially. The state $x(k)$ represents the amount of the resource that is available for allocation to activities k through N.

* To account for the possibility of a degenerate problem, one must include $\lambda_0 \geq 0$ as a coefficient of l in (11-51) and (11-52), and for ψ in the terminal conditions, if there are terminal state constraints.

According to the general theory for discrete-time optimal control, the adjoint equation is

$$\lambda(k)=\lambda(k+1)$$

and there are no endpoint conditions. Thus, $\lambda(k)$ is constant, say, $\lambda(k)=\lambda$. The Hamiltonian is

$$H=\lambda x(k)-\lambda u(k)+u(k)^{1/2}$$

The optimality condition is

$$H_u=-\lambda+\tfrac{1}{2}u(k)^{-1/2}=0$$

Thus, $u(k)=1/4\lambda^2$, and hence $u(k)$ is also constant. The constant is determined so that the constraint is satisfied. Finally,

$$u(k)=A/N$$

11.7 DYNAMIC PROGRAMMING

We now present an alternative approach to optimal control; an approach that in fact exploits the dynamic structure of these problems more directly than the variational approach. The basic concept has a long history, but its scope was broadened considerably by Bellman who coined the term *dynamic programming.* It is an approach that fully exploits the state concept and is therefore quite consistent with the modern approach to dynamic systems.

The Principle of Optimality

The basic observation underlying dynamic programming is the *principle of optimality* that points out a fundamental relation between a given optimal control problem and various other subproblems. Suppose that a dynamic system, either in discrete time or continuous time, is characterized at each time instant by a state vector $\mathbf{x}$. Suppose that, as usual, the evolution of this state is influenced by control inputs. The optimal control problem is to select the inputs so as to maximize a given objective while satisfying various terminal constraints.

Imagine an optimal control problem defined over an interval of time, and suppose the solution is known. Suppose we follow the corresponding trajectory to a time t, arriving at state $\mathbf{x}(t)$. We then consider a new problem, initiated at time t with state $\mathbf{x}(t)$ and for which it is desired to maximize the total objective from that point. Under quite broad assumptions, the solution to that subproblem exactly corresponds to the remainder of the original solution. This observation is stated formally as

The Principle of Optimality: *From any point on an optimal trajectory, the*

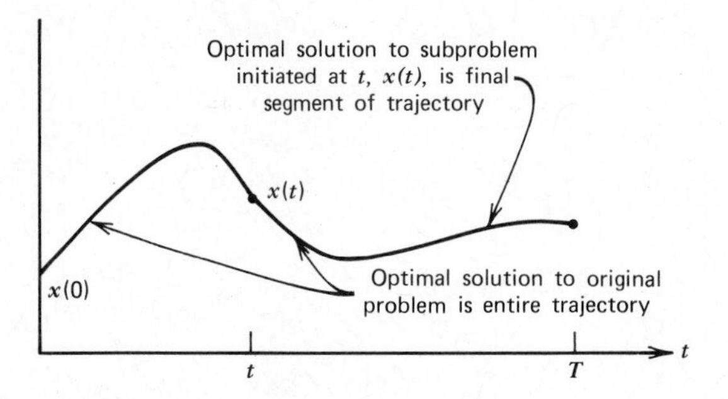

Figure 11.7. Principle of optimality.

remaining trajectory is optimal for the corresponding problem initiated at that point.

This principle, illustrated in Fig. 11.7, allows us to build up solutions by progressing backward in time. It is the basis for a powerful solution procedure.

An example should help clarify this idea. Suppose that from a given location in the city, you pose the optimal control problem of finding the minimum distance path to your home. In this case the objective is path length (which now is to be minimized). Suppose you solve this problem and start out along the optimal path. As you travel, the state is your current position. From any intermediate position, you could formulate the new problem of finding the path that minimizes the distance home from that point. The result would be the same as the remaining path in the old solution.

The Optimal Return Function

The principle of optimality is captured in mathematical terms by introducing the concept of the optimal return function. It is possible to associate with a given state $\mathbf{x}$ at a given time (say t) a value $V(\mathbf{x}, t)$ which represents the maximum value of the objective that could be obtained starting at $\mathbf{x}$ at time t. This function is the *optimal return function.*

For instance, in the case of finding the shortest path to your home, the optimal return function is the shortest distance from each point. [$V(\mathbf{x}, t)$ does not depend on t in this case.]

As a more general example, consider the system

$$\dot{\mathbf{x}}(t) = \mathbf{f}(\mathbf{x}(t), \mathbf{u}(t)) \tag{11-55}$$

with objective

$$J = \psi(\mathbf{x}(T)) + \int_0^T l(\mathbf{x}(t), \mathbf{u}(t))\, dt \tag{11-56}$$

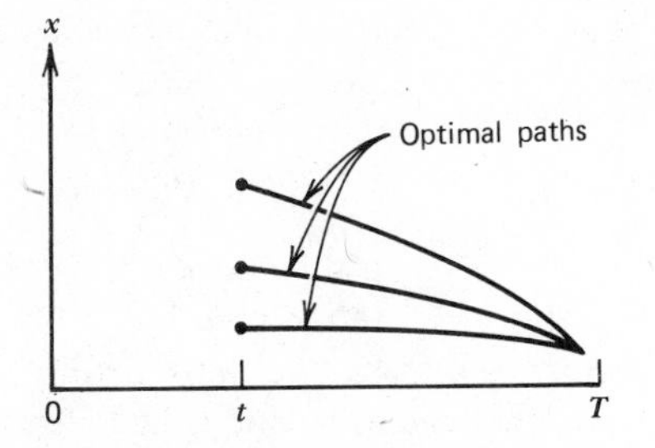

Figure 11.8. Definition of $V(\mathbf{x}, t)$.

To define the optimal return function $V(\mathbf{x}, t)$ one considers the optimal control problem, starting at time t and state $\mathbf{x}$,

$$\dot{\mathbf{x}}(\tau) = \mathbf{f}(\mathbf{x}(\tau), \mathbf{u}(\tau)) \tag{11-57a}$$

$$\mathbf{x}(t) = \mathbf{x} \tag{11-57b}$$

$$J = \psi(\mathbf{x}(T)) + \int_t^T l(\mathbf{x}(\tau), \mathbf{u}(\tau))\, d\tau \tag{11-58}$$

The value $V(\mathbf{x}, t)$ is the maximum possible value of J for this problem. The optimal return function, then, represents the objective value associated with a subproblem of the original problem, initiated at an intermediate time and from an arbitrary state. This is illustrated in Fig. 11.8.

Discrete-Time Systems

Let us apply the dynamic programming idea to the standard discrete-time optimal control problem defined by a system

$$\mathbf{x}(k+1) = \mathbf{f}(\mathbf{x}(k), \mathbf{u}(k)) \tag{11-59a}$$

an initial condition

$$\mathbf{x}(0) = \mathbf{x}_0 \tag{11-59b}$$

control constraints

$$\mathbf{u}(t) \in U \tag{11-59c}$$

and an objective

$$J = \psi(\mathbf{x}(N)) + \sum_{k=0}^{N-1} l(\mathbf{x}(k), \mathbf{u}(k)) \tag{11-59d}$$

For simplicity we take N as fixed.

The corresponding subproblems are exactly the same, except that they are initiated at some $k>0$ and at a specified value of $\mathbf{x}(k)=\mathbf{x}$. The optimal objective value corresponding to such a subproblem is defined as $V(\mathbf{x}, k)$.

The simplest subproblem is the one starting at $k=N$ with a specified $\mathbf{x}(N)$. In this problem the initial point is already the terminal point. No inputs can be selected, and the optimal return is

$$V(\mathbf{x}(N), N)=\psi(\mathbf{x}(N)) \tag{11-60}$$

To determine the other optimal return functions, we work backward a step at a time using the principle of optimality. Suppose the function $V(\mathbf{x}, k+1)$ has been calculated for a given k and all values of $\mathbf{x}$. We wish to calculate the function $V(\mathbf{x}, k)$ for all values of $\mathbf{x}$. Starting with $\mathbf{x}(k)=\mathbf{x}$, we can select a $\mathbf{u}(k)=\mathbf{u}$. This yields an immediate reward of $l(\mathbf{x}, \mathbf{u})$ derived from the summation term in the objective, and it will transfer that state to a value $\mathbf{x}(k+1)=\mathbf{f}(\mathbf{x}, \mathbf{u})$ at the next period $k+1$. However, once the state is known at $k+1$, the remaining objective is determined by the previously evaluated optimal return function. Thus, for a given initial $\mathbf{u}$, the total objective value is

$$J=l(\mathbf{x}, \mathbf{u})+V(\mathbf{f}(\mathbf{x}, \mathbf{u}), k+1) \tag{11-61}$$

assuming an optimal path from $k+1$ to the terminal point N. The optimal return from $\mathbf{x}$ at k is, accordingly, the maximum of (11-61) with respect to all possible choices of $\mathbf{u}$. Thus,

$$V(\mathbf{x}, k)=\operatorname*{Max}_{\mathbf{u}\in U}[l(\mathbf{x}, \mathbf{u})+V(\mathbf{f}(\mathbf{x}, \mathbf{u}), k+1)] \tag{11-62}$$

This is the fundamental recursive expression for $V(\mathbf{x}, k)$. It can be evaluated backward, starting with the condition

$$V(\mathbf{x}, N)=\psi(\mathbf{x})$$

Example (The Allocation Problem). Consider the allocation problem discussed in Sect. 11.6 defined by

$$x(k+1)=x(k)-u(k)$$

$$x(0)=A$$

$$x(N)=0$$

$$J=\sum_{k=0}^{N-1} u(k)^{1/2}$$

The optimal return function $V(x, k)$ is the optimal value that can be obtained by allocating a total of x units of resource among the last $N-k$ project or activity terms. In this example, we may put $V(x, N)=0$. The first nontrivial term is really $V(x, N-1)$. This is the best that can be done by allocating x to the last project. Clearly

$$V(x, N-1)=x^{1/2}$$

For $N-2$ we have

$$V(x, N-2) = \underset{u}{\text{Max}}[u^{1/2} + V(x-u, N-1)]$$

$$= \underset{u}{\text{Max}}[u^{1/2} + (x-u)^{1/2}]$$

The best value of u is $u = x/2$ leading to

$$V(x, N-2) = \sqrt{2x}$$

For $N-3$

$$V(x, N-3) = \underset{u}{\text{Max}}[u^{1/2} + \sqrt{2}\,(x-u)^{1/2}]$$

The best value of u is $u = x/3$ leading to

$$V(x, N-3) = \sqrt{3x}$$

It is clear that in general

$$V(x, N-k) = \sqrt{kx}$$

and the best control is

$$u(N-k) = x(N-k)/k$$

At each stage, one determines the amount of remaining resource, and divides by the number of remaining stages. This determines the allocation to the current stage. Thus, the procedure yields the solution in feedback form—the control is given in terms of the current state. In particular, for the original problem with $x(0) = A$, we obtain $u(0) = A/N$.

Continuous Time

Let us apply the dynamic programming idea to the continuous-time problem defined by

$$\dot{\mathbf{x}}(t) = \mathbf{f}(\mathbf{x}(t), \mathbf{u}(t)) \tag{11-63a}$$

$$\mathbf{x}(0) = \mathbf{x}_0 \tag{11-63b}$$

$$\mathbf{u}(t) \in U \tag{11-63c}$$

$$J = \psi(\mathbf{x}(T)) + \int_0^T l(\mathbf{x}(t), \mathbf{u}(t))\, dt \tag{11-63d}$$

Let $V(\mathbf{x}, t)$ denote the optimal return function; that is, $V(\mathbf{x}, t)$ is the best value of the objective starting at $\mathbf{x}$ at time t. To develop a formula for $V(\mathbf{x}, t)$ we assume that the function is known for $t+\Delta$, where $\Delta>0$ is "small," and then work backward to t.

Following the logic of the discrete-time case, we write

$$V(\mathbf{x}, t) \simeq \underset{\mathbf{u}\in U}{\text{Max}}[l(\mathbf{x}, \mathbf{u})\Delta + V(\mathbf{x}+\mathbf{f}(\mathbf{x}, \mathbf{u})\Delta, t+\Delta)] \tag{11-64}$$

To derive this relation, we have assumed that between times t and $t+\Delta$ a fixed $\mathbf{u}$ is applied. This yields an immediate contribution of (approximately) $l(\mathbf{x}, \mathbf{u})\Delta$ from the integral term in the objective. In addition, it will transfer the state (approximately) to the point $\mathbf{x}+\mathbf{f}(\mathbf{x}, \mathbf{u})\Delta$ at time $t+\Delta$. The optimal return is known from that point.

To simplify expression (11-64) we let $\Delta \to 0$. Assuming that $V(\mathbf{x}, t)$ is a smooth function, we may write

$$V(\mathbf{x}+\mathbf{f}(\mathbf{x}, \mathbf{u})\Delta, t+\Delta) \simeq V(\mathbf{x}, t) + V_t(\mathbf{x}, t)\Delta + V_{\mathbf{x}}(\mathbf{x}, t)\mathbf{f}(\mathbf{x}, \mathbf{u})\Delta \tag{11-65}$$

Substituting this in (11-64) we obtain

$$V(\mathbf{x}, t) = \underset{\mathbf{u}}{\text{Max}}[l(\mathbf{x}, \mathbf{u})\Delta + V(\mathbf{x}, t) + V_t(\mathbf{x}, t)\Delta + V_{\mathbf{x}}(\mathbf{x}, t)\mathbf{f}(\mathbf{x}, \mathbf{u})\Delta] \tag{11-66}$$

Now $V(\mathbf{x}, t)$ does not depend on $\mathbf{u}$, so it can be taken outside the maximization, where it then cancels with the left-hand side. Then $V_t(\mathbf{x}, t)\Delta$ can be taken outside the maximization, and all terms divided by Δ. This yields the final result:

$$0 = V_t(\mathbf{x}, t) + \underset{\mathbf{u}}{\text{Max}}[l(\mathbf{x}, \mathbf{u}) + V_{\mathbf{x}}(\mathbf{x}, t)\mathbf{f}(\mathbf{x}, \mathbf{u})] \tag{11-67}$$

There is also the associated boundary condition

$$V(\mathbf{x}, T) = \psi(\mathbf{x}) \tag{11-68}$$

which is clearly the optimal value obtainable starting at the terminal time.

Equation (11-67) is the *Hamilton–Jacobi–Bellman equation*. It is a partial differential equation for the optimal return function $V(\mathbf{x}, t)$. It is sometimes (but not very often) possible to solve this equation in analytic form. If so, it provides a complete solution to the whole family of optimal control problems defined by the system, the constraints, and the objective. In many situations, however, it must be solved numerically.

A significant advantage of the dynamic programming approach is that it automatically determines the optimal control in feedback form. Once the optimal return function is known, the maximization with respect to $\mathbf{u}$ indicated in (11-67) yields the value of $\mathbf{u}$ that should be employed if the state is $\mathbf{x}$ at time t.

Example (Linear System and Quadratic Cost). Dynamic programming can be applied to the standard "linear-quadratic" problem of Sect. 11.5. Thus, let us consider

$$\dot{\mathbf{x}} = \mathbf{A}\mathbf{x} + \mathbf{B}\mathbf{u} \tag{11-69a}$$

$$J = \tfrac{1}{2}\int_0^T [\mathbf{x}^T\mathbf{Q}\mathbf{x} + \mathbf{u}^T\mathbf{R}\mathbf{u}]\,dt \tag{11-69b}$$

The objective is to be *minimized.* The Hamilton–Jacobi–Bellman equation is (the maximization now becomes minimization)

$$0 = V_t(\mathbf{x}, t) + \underset{\mathbf{u}}{\mathrm{Min}}[\tfrac{1}{2}\mathbf{x}^T\mathbf{Q}\mathbf{x} + \tfrac{1}{2}\mathbf{u}^T\mathbf{R}\mathbf{u} + V_{\mathbf{x}}(\mathbf{x}, t)(\mathbf{A}\mathbf{x} + \mathbf{B}\mathbf{u})] \tag{11-70a}$$

$$V(\mathbf{x}, T) = 0 \tag{11-70b}$$

We hypothesize a solution of the form

$$V(\mathbf{x}, t) = \tfrac{1}{2}\mathbf{x}^T\mathbf{P}(t)\mathbf{x} \tag{11-71}$$

where $\mathbf{P}(t)$ is an $n \times n$ symmetric matrix. Substituting this into (11-70a) yields

$$0 = \tfrac{1}{2}\mathbf{x}^T\dot{\mathbf{P}}(t)\mathbf{x} + \underset{\mathbf{u}}{\mathrm{Min}}[\tfrac{1}{2}\mathbf{x}^T\mathbf{Q}\mathbf{x} + \tfrac{1}{2}\mathbf{u}^T\mathbf{R}\mathbf{u} + \mathbf{x}^T\mathbf{P}(t)(\mathbf{A}\mathbf{x} + \mathbf{B}\mathbf{u})] \tag{11-72}$$

The minimum with respect to $\mathbf{u}$ is obtained by solving

$$\mathbf{u}^T\mathbf{R} + \mathbf{x}^T\mathbf{P}(t)\mathbf{B} = \mathbf{0} \tag{11-73}$$

yielding

$$\mathbf{u} = -\mathbf{R}^{-1}\mathbf{B}^T\mathbf{P}(t)\mathbf{x} \tag{11-74}$$

Substituting this into (11-72) and noting that

$$\mathbf{x}^T\mathbf{P}(t)\mathbf{A}\mathbf{x} = \tfrac{1}{2}\mathbf{x}^T(\mathbf{P}(t)\mathbf{A} + \mathbf{A}^T\mathbf{P}(t))\mathbf{x}$$

yields

$$0 = \mathbf{x}^T(\dot{\mathbf{P}}(t) + \mathbf{Q} + \mathbf{P}(t)\mathbf{A} + \mathbf{A}^T\mathbf{P}(t) - \mathbf{P}(t)\mathbf{B}\mathbf{R}^{-1}\mathbf{B}^T\mathbf{P}(t))\mathbf{x} \tag{11-75}$$

This will be identically true if $\mathbf{P}(t)$ is selected as the solution to the Riccati equation (11-42). Thus, this procedure leads to exactly the same solution as found in Sect. 11.5.

*11.8 STABILITY AND OPTIMAL CONTROL

There is a strong relationship between some optimal control problems and stability theory. Often the optimal return function serves as a Liapunov function.

Consider the optimal control problem with fixed terminal state but free terminal time defined by

$$\dot{\mathbf{x}}(t)=\mathbf{f}(\mathbf{x}(t),\mathbf{u}(t)) \tag{11-76a}$$

$$\mathbf{u}(t)\in U \tag{11-76b}$$

$$\mathbf{x}(T)=\bar{\mathbf{x}} \tag{11-76c}$$

$$J=\int_0^T l(\mathbf{x}(t),\mathbf{u}(t))\,dt \tag{11-76d}$$

The initial state is given. In the present context it is assumed that the problem is to *minimize* J.

We assume that for all $\mathbf{x}\neq\bar{\mathbf{x}}$ and any $\mathbf{u}\in U$

$$l(\mathbf{x},\mathbf{u})>0 \tag{11-77}$$

We assume also that for any initial condition $\mathbf{x}(0)$ there is a unique solution to the problem having a finite objective value. Finally, we assume there is a $\bar{\mathbf{u}}\in U$ such that $\mathbf{f}(\bar{\mathbf{x}},\bar{\mathbf{u}})=\mathbf{0}$.

This problem has an important time-invariant property. The optimal trajectory from a given state $\mathbf{x}$ to the endpoint $\bar{\mathbf{x}}$ is independent of the time t_0 at which $\mathbf{x}(t_0)=\mathbf{x}$. That is, if $\mathbf{x}(0)=\mathbf{x}$ leads to the optimal trajectory $\mathbf{x}(t)$ for $t>0$ with final time T, then the condition $\mathbf{x}(t_0)=\mathbf{x}$ must lead to the trajectory $\mathbf{x}(t+t_0)$ with final time $T+t_0$. Delaying the starting time merely delays the whole solution. This follows because the system, the constraints, and the objective are independent of t. (The time to termination is really some unknown function of the initial state only.)

The optimal control is also a time-invariant function of the state. That is, $\mathbf{u}(t)=\mathbf{u}(\mathbf{x}(t))$. To see this, note that the initial control clearly depends only on the initial state, not on the initial time; then reapply this argument at each instant to the remaining problem. We can assume that $\mathbf{u}(\bar{\mathbf{x}})=\bar{\mathbf{u}}$.

The system, when guided by the optimal control law, is governed by

$$\dot{\mathbf{x}}(t)=\mathbf{f}(\mathbf{x}(t),\mathbf{u}(\mathbf{x}(t))) \tag{11-78}$$

This is a time-invariant (but most likely nonlinear) system. The point $\bar{\mathbf{x}}$ is clearly an equilibrium point. Furthermore, given any initial state, this system eventually reaches $\bar{\mathbf{x}}$. Thus, the system exhibits strong stability properties by its very construction.

Let $V(\mathbf{x})$ be the optimal return function for this problem. That is, $V(\mathbf{x})$ is the *minimum* achievable value of the objective when the system is initiated at $\mathbf{x}$. The optimal return function is also time invariant in this case. (Clearly the minimum value depends only on where the system is initiated—not when.) The function $V(\mathbf{x})$ satisfies

(i) $V(\bar{\mathbf{x}})=0$

since no objective value is accrued if the system is initiated at $\bar{\mathbf{x}}$. Also, the function satisfies

(ii) $V(\mathbf{x})>0$ for $\mathbf{x}\neq\bar{\mathbf{x}}$

since $l(\mathbf{x},\mathbf{u})>0$ for $\mathbf{x}\neq\bar{\mathbf{x}}$. Finally, since

$$V(\mathbf{x}(t))=\int_t^T l(\mathbf{x}(\tau),\mathbf{u}(\mathbf{x}(\tau)))\,d\tau$$

we have

$$\dot{V}(\mathbf{x})=-l(\mathbf{x},\mathbf{u}(\mathbf{x}))$$

Therefore, $V(\mathbf{x})$ satisfies

(iii) $\dot{V}(\mathbf{x})<0$ for $\mathbf{x}\neq\bar{\mathbf{x}}$

Thus, V is a Liapunov function.

The net result of an optimal control problem is to transform a system with inputs to a free system—since the inputs are specified by the optimization. Thus, after the solution of an optimal control problem is found, the behavior of the resulting system can be analyzed using the principles developed throughout this book. If the control is implemented in feedback form, then it may be possible to establish stability as indicated in the above special case. This again is an instance of the theme of Chapter 10—that an appropriate Liapunov function is often directly related to the origin of the system. In this case the connection is the objective function.

11.9 PROBLEMS

1. *Time-Varying Systems.* The theory of optimal control can be easily extended to problems where the system equations, the constraints, and the objective functions are all explicit functions of time. Consider the problem with

$$\dot{\mathbf{x}}(t)=\mathbf{f}(\mathbf{x}(t),\mathbf{u}(t),t)$$

$$\mathbf{x}(0)=\mathbf{x}_0$$

$$\mathbf{u}(t)\in U$$

$$x_i(T)=\bar{x}_i(T),\qquad i=1,2,\ldots,r$$

$$J=\psi(\mathbf{x}(T),T)+\int_0^T l(\mathbf{x}(t),\mathbf{u}(t),t)\,dt$$

with T either free or fixed. Show that by defining an additional state variable $x_{n+1}=t$ this problem can be converted to an equivalent problem without explicit t dependence.

2. *Equivalence of Problem Types.* A continuous-time optimal control problem is sometimes characterized by the form of its objective function as follows:

   (i) $J = \psi(\mathbf{x}(T))$ (problem of Mayer)
   (ii) $J = \int_0^T l(\mathbf{x}(t), \mathbf{u}(t))\, dt$ (problem of Lagrange)
   (iii) $J = \psi(\mathbf{x}(T)) + \int_0^T l(\mathbf{x}(t), \mathbf{u}(t))\, dt$ (problem of Bolza)

   Show that these three types are equivalent. In particular, show that a problem of Bolza can be converted to a problem of Mayer and to a problem of Lagrange.

3. *Pushcart with Friction.* Solve the pushcart problem of Example 2, Sect. 11.2 with the dynamic equation modified to account for friction. Specifically, replace the equation $\ddot{x} = u$ by $\ddot{x} = u - k\dot{x}$.

4. Consider the following optimal control problem:

$$\dot{x} = x + u$$

$$x(0) = 5$$

$$0 \le u(t) \le 2$$

   Minimize

$$J = \int_0^2 (-2x + 3u + \alpha u^2)\, dt$$

   Use the maximum principle to solve for the optimal control in the following two cases:

   (a) $\alpha = 0$
   (b) $\alpha = 1$

5. *The Minimum Principle.* Consider an optimal control problem that is in the form presented in Sect. 11.1, except that the objective J is to be *minimized.* Show that the necessary conditions can be expressed exactly as usual but with the Hamiltonian being minimized.

6. *Optimal Investment.* Consider the problem of determining the optimal investment plan of a production facility. Assume that without any investment the production rate of the facility decreases at a rate proportional to the production rate at that time, but that investment tends to increase the production rate. Specifically, letting P be the production rate and I the investment rate, assume that

$$\dot{P} = -\alpha P + \gamma I, \qquad P(0) = P_0$$

   where $\alpha > 0$, $\gamma > 0$. Assume that the facility operates until time T and is then salvaged at a price proportional to its production rate at that time. Correspondingly, the objective is

$$J = \beta P(T) + \int_0^T [P(t) - I(t)]\, dt$$

where $\beta > 0$. The investment rate is assumed to be positive and bounded above; that is,

$$0 \le I(t) \le \bar{I}$$

(a) Write the necessary conditions for the above problem.
(b) Show that if γ is outside the interval spanned by α and $1/\beta$, the optimal policy is constant for $0 \le t \le T$ [either $I(t) = 0$ or $I(t) = \bar{I}$].
(c) Show that if the optimal policy contains a switch, it occurs at

$$t_s = T - \frac{1}{\alpha} \ln\left(\frac{1/\alpha - \beta}{1/\alpha - 1/\gamma}\right)$$

(d) Show that if the facility were to operate from time T on without any further investment, it could produce a terminal revenue

$$\int_T^\infty P(t)\,dt = (1/\alpha)P(T).$$

(e) Show that if $\gamma > \alpha$ and the salvage value is at least as high as the terminal revenue (that is, $\beta \ge 1/\alpha$), then the optimal policy is to invest at the maximum rate for the entire period.

7. *The Hamiltonian is Constant.* Suppose $\mathbf{u}$ is unconstrained and $\mathbf{f}$ and l do not depend explicitly on t. Let $\boldsymbol{\lambda}(t)$, $\mathbf{x}(t)$, $\mathbf{u}(t)$ satisfy the necessary conditions for optimality (with any form of end-point conditions), then show that $H(\boldsymbol{\lambda}(t), \mathbf{x}(t), \mathbf{u}(t))$ is constant for $0 \le t \le T$.

8. *Geodesics.* Let two points be given on the surface of a sphere of radius R. Show that the curve on the sphere connecting the points and having minimum arc length is the arc of a great circle (that is, a portion of a circle of radius R).

9. *Thrust Programming.* A particle of fixed mass is acted on by a thrust force of constant magnitude. Assuming planar motion, the equations of motion are

$$\dot{u} = A \cos\theta, \qquad \dot{x} = u$$

$$\dot{v} = A \sin\theta, \qquad \dot{y} = v$$

where θ is the angle of the thrust. Show that to maximize some function of the terminal state, the angle of thrust is of the form

$$\tan\theta(t) = \frac{at + b}{ct + d}$$

for some constants a, b, c, d.

10. *The Brachistochrone Problem.* A particle under the force of gravity slides without friction along a curve connecting two fixed points. Consider the problem of determining the shape of the curve that will produce the minimum time path between the two points.

Energy is conserved since there is no friction. Thus, the magnitude of the particle's velocity is $V(y)=\sqrt{y}$, where y is the vertical distance it has fallen. The equations of motion are

$$\dot{x} = V(y)\cos\theta$$

$$\dot{y} = V(y)\sin\theta$$

where θ is the instantaneous angle of the curve. Find an equation relating θ and y. Show that $\dot{\theta}$ is constant.

*11. *Estate Planning.* A man is considering his lifetime plan of investment and expenditure. He has an initial level of savings S and no income other than that which he obtains from investment at a fixed interest rate. His total capital x is therefore governed by the equation

$$\dot{x}(t) = \alpha x(t) - r(t)$$

where $\alpha > 0$ and r denotes his rate of expenditure. His immediate enjoyment due to expenditure is $U(r)$, where U is his *utility function.* In his case $U(r) = r^{1/2}$. Future enjoyment, at time t, is counted less today, at time 0, by incorporation of a discount term $e^{-\beta t}$. Thus, he wishes to maximize

$$J = \int_0^T e^{-\beta t} U(r)\, dt$$

subject to the terminal constraint $x(T)=0$. Using the maximum principle find a complete solution for $r(t)$.

12. *Catenary.* A cable of length L hangs with its two ends fixed at two supports separated by a horizontal distance T. The shape of the curve is $x(t)$. The potential energy of the hanging cable is

$$V = mg\int_0^T x\sqrt{1+\dot{x}^2}\, dt$$

where m is the mass per unit length. The cable hangs so as to minimize the potential energy subject to the condition that its length is fixed.

$$L = \int_0^T \sqrt{1+\dot{x}^2}\, dt$$

(a) Formulate the necessary conditions for this problem and reduce them to a single second-order differential equation in $x(t)$.

*(b) Show that the cable hangs in the shape of a catenary

$$x(t) = a\cosh\left(\frac{t+b}{c}\right) + d$$

where a, b, c, d are constants depending on the parameters of the problem.

13. *Ill-Conditioned Problem.* Consider the problem

$$\dot{x}(t) = u(t)^2$$

$$x(0) = 0, \qquad x(1) = 0$$

$$J = \int_0^1 u(t)\, dt$$

(a) What is the solution?
(b) Show that this solution does not satisfy the necessary conditions of the maximum principle with $\lambda_0 = 1$.
(c) Show that the solution does satisfy the conditions with $\lambda_0 = 0$.

14. *State Variables Constrained to Lie on a Surface.* In a generalization of the free terminal time problem one is given

$$\dot{\mathbf{x}} = \mathbf{f}(\mathbf{x}, \mathbf{u}) \quad \text{(system equation)}$$

$$\mathbf{x}(0) = \mathbf{x}_0 \quad \text{(initial condition)}$$

$$J = \psi[\mathbf{x}(T)] + \int_0^T l(\mathbf{x}, \mathbf{u})\, dt \quad \text{(objective)}$$

$$\boldsymbol{\phi}[\mathbf{x}(T)] = \mathbf{0} \quad \text{(terminal constraints)}$$

where $\boldsymbol{\phi}$ is r-dimensional. The terminal constraints require the state vector to lie on the surface given by $\boldsymbol{\phi}[\mathbf{x}(T)] = \mathbf{0}$ at the unspecified final time T.

To find the necessary conditions for this problem, let $\boldsymbol{\nu}^T$ be an r-dimensional vector of (unknown) *Lagrange multipliers* associated with the constraint $\boldsymbol{\phi}[\mathbf{x}(T)] = \mathbf{0}$. By appending $\boldsymbol{\nu}^T\boldsymbol{\phi}(\mathbf{x}(T)]$ to the objective function one removes the constraint and the problem becomes

$$\text{maximize } \hat{J} = \psi[\mathbf{x}(T)] + \boldsymbol{\nu}^T\boldsymbol{\phi}[\mathbf{x}(T)] + \int_0^T l(\mathbf{x}, \mathbf{u})\, dt$$

$$\text{subject to} \quad \dot{\mathbf{x}} = \mathbf{f}(\mathbf{x}, \mathbf{u})$$

$$\mathbf{x}(0) = \mathbf{x}_0$$

Find a complete set of necessary conditions, including a specification of the terminal constraints on the adjoint variables. Check your condition by verifying that for the constraints $x_i(T) = \bar{x}_i$, $i = 1, 2, \ldots, r$ the condition yields the known results.

*15. It is desired to transfer the state vector of the system

$$\dot{x}_1 = -x_2$$

$$\dot{x}_2 = u$$

from $x_1(0) = x_2(0) = 0$ to the line $x_1(T) + 3x_2(T) = 18$ while *minimizing*

$$J = \tfrac{1}{2}\int_0^T u^2\, dt$$

The final time T is unspecified and u is unconstrained. Solve this problem by the method of Problem 14.

16. *Fish Harvesting.* Consider the problem of determining the optimal plan for harvesting a crop of fish in a large lake. Let $x(t)$ denote the number of fish in the lake, and let $u(t)$ denote the intensity of harvesting activity. In the absence of harvesting $x(t)$ grows exponentially according to $\dot{x} = \alpha x$, where $\alpha > 0$. The rate at which fish are caught when there is harvesting is $r = \beta u^{1-\gamma} x^{\gamma}$ for $\beta > 0$, $0 < \gamma < 1$. (Thus, the rate increases as the intensity increases or as there are more fish to be harvested.) There is a fixed unit price for the harvested crop and a fixed cost per unit of harvesting intensity. The problem is to maximize the total profit over the period $0 \le t \le T$.

    The system is

$$\dot{x}(t) = \alpha x(t) - r(t)$$

$$x(0) = x_0$$

and the objective is

$$J = \int_0^T [pr(t) - cu(t)]\, dt$$

where p is the unit price for the crop and c is the unit cost for harvesting intensity. The objective represents the integral of the profit rate—that is, the total profit.

(a) Write the necessary conditions for this problem using the maximum principle.
(b) Show that u can be written in feedback form as a linear function of x.
(c) Find a differential equation for the adjoint variable that does not depend on x or u.
(d) Show that the optimal return function is of the form $V(x, t) = \lambda(t)x + q(t)$. Find $q(t)$ explicitly.

17. A *Linear-Quadratic Problem.* Consider the problem with system

$$\dot{\mathbf{x}}(t) = \mathbf{A}(t)\mathbf{x}(t) + \mathbf{B}(t)\mathbf{u}(t)$$

and objective to be minimized

$$J = \tfrac{1}{2}\int_0^T [\mathbf{x}(t)^T\mathbf{Q}(t)\mathbf{x}(t) + \mathbf{u}(t)^T\mathbf{R}(t)\mathbf{u}(t)]\, dt$$

with both $\mathbf{x}(0) = \mathbf{x}_0$ and $\mathbf{x}(T) = \mathbf{x}_1$ fixed. The final time T is also fixed. The matrices $\mathbf{Q}(t)$ and $\mathbf{R}(t)$ are as in Sect. 11.5. Write the necessary conditions for this problem. Assume a transformation of the form

$$\mathbf{x}(t) = -\mathbf{P}(t)\boldsymbol{\lambda}(t) + \mathbf{b}(t)$$

and find differential equations for $\mathbf{P}(t)$ and $\mathbf{b}(t)$. Find the control in terms of the state and $\mathbf{P}(t)$ and $\mathbf{b}(t)$.

*18. *Optimal Economic Growth.* A standard aggregate model of the national economy is based on the following three equations:

$$Y = C + I$$

$$\dot{K} = -\mu K + I$$

$$Y = F(K, L)$$

where Y is total economic production, C is consumption, I is investment, K is capital, and L is labor. The first equation is the basic income identity. The second states that capital depreciates but can be replenished by investment. The third equation states that the level of production is a function of the capital and labor employed. If we assume that F exhibits constant returns to scale, then $F(\alpha K, \alpha L) = \alpha F(K, L)$ for $\alpha > 0$. Selecting $\alpha = 1/L$ we find $Y/L = F(K/L, 1) = f(K/L)$. We may then express everything in a per-worker basis. We assume that L grows exponentially according to $\dot{L} = pL$. Then defining $y = Y/L$, $c = C/L$, and so forth, we obtain

$$y = c + i$$

$$\dot{k} = -rk + i$$

$$y = f(k)$$

where $r = \mu + p$. By some easily manipulation this leads to

$$\dot{k} = f(k) - rk - c$$

which is the fundamental equation of growth.

The society selects its growth trajectory by selecting $c(t)$, the consumption per worker. If the society wishes to maximize its discounted aggregate utility, it should determine $c(t)$ to maximize

$$J = \int_0^\infty e^{-\beta t} U(c(t))\, dt$$

subject to $0 \le c \le f(k)$, where k and c are related by the fundamental equation. The function $U(c)$ is the *utility function* of consumption.

(a) Using the maximum principle (without yet worrying about endpoint conditions) express the necessary conditions in terms of a pair of differential equations with variables k and c.

(b) Find a special solution corresponding to $\dot{k} = 0$, $\dot{c} = 0$. Explain why this solution is called *balanced growth.*

19. *Housing Maintenance.* Suppose that the quality of a rental house is characterized by a single variable x. The quality is governed by the equation

$$x(k+1) = \alpha x(k) + u(k) - \frac{u(k)^2}{\bar{x} - x(k)}$$

where $0 < \alpha < 1$, $u(k)$ is the maintenance expenditure in period k, and $\bar{x} > 0$ corresponds to "perfect" condition. The rent is proportional to the quality. A landlord wishes to determine the maintenance policy that maximizes his discounted net profit up to period N, at which point he plans to sell the house. In particular, he wishes to maximize

$$J = \beta^N cx(N) + \sum_{k=0}^{N-1} [px(k) - u(k)]\beta^k$$

where $p > 0$, $0 < \beta < 1$. The quantity $cx(N)$ is the sales price at time N.

(a) Using the variational approach write the necessary conditions and find an equation for $\lambda(k)$ that is independent of $x(k)$ and $u(k)$.
(b) Show that $u(k)$ can be expressed in feedback form.
(c) Find the optimal return function (and show that it is linear in x).
(d) What is a "fair" price c? That is, what value of c would make the landlord indifferent to selling or retaining the house?

20. *Estate-Planning.* Neglect the constraint $x(T) = 0$ in Problem 11. Assume $\beta > \alpha/2$.

(a) Formulate the Hamilton–Jacobi–Bellman equation for the problem.
(b) Find a solution of the form $V(x, t) = f(t)g(x)$ but which does not satisfy the boundary condition. (*Hint*: Try $g(x) = Ax^{1/2}$.)
(c) Find a suitable function $\psi(x, T)$ to append to the objective function so that the solution found in (b) is correct. What is the corresponding feedback control?

21. *Relation of Dynamic Programming to the Maximum Principle.* Let $\mathbf{x}(t)$, $\mathbf{u}(t)$ be an optimal solution to the continuous-time control problem

$$\dot{\mathbf{x}}(t) = \mathbf{f}(\mathbf{x}(t), \mathbf{u}(t))$$

$$\mathbf{u}(t) \in U$$

$$J = \psi(\mathbf{x}(T)) + \int_0^T l(\mathbf{x}(t), \mathbf{u}(t))\, dt$$

Let $V(\mathbf{x}, t)$ be the optimal return function for the problem. Define $\boldsymbol{\lambda}(t)^T = V_\mathbf{x}(\mathbf{x}(t), t)$. Show that $\boldsymbol{\lambda}(t)$, $\mathbf{x}(t)$, and $\mathbf{u}(t)$ satisfy the conditions of the maximum principle. Assume $\mathbf{u}(t)$ is in the interior of U. (*Hint*: $dV_\mathbf{x}/dt = V_{t\mathbf{x}} + V_{\mathbf{xx}}\dot{\mathbf{x}}$.)

22. *Stability.* Consider the optimal control problem (with J to be minimized)

$$\dot{\mathbf{x}}(t) = \mathbf{f}(\mathbf{x}(t), \mathbf{u}(t))$$

$$J = \int_0^\infty l(\mathbf{x}(t), \mathbf{u}(t))\, dt$$

Assume that there is an $\bar{\mathbf{x}}$ and $\bar{\mathbf{u}}$ such that

$$\mathbf{f}(\bar{\mathbf{x}}, \bar{\mathbf{u}}) = \mathbf{0}, \qquad l(\bar{\mathbf{x}}, \bar{\mathbf{u}}) = 0$$

$$l(\mathbf{x}, \mathbf{u}) > 0 \quad \text{if } \bar{\mathbf{x}} \neq \mathbf{x} \text{ or } \mathbf{u} \neq \bar{\mathbf{u}}$$

The functions $\mathbf{f}$ and l are smooth.

(a) Assume also that there is a unique solution to the problem for any initial condition. Show that $\bar{\mathbf{x}}$ is an equilibrium point of the closed-loop optimal system, and that it is asymptotically stable.

(b) Verify this result for

$$\dot{x} = x + u$$

$$J = \int_0^\infty (x^2 + u^2)\, dt$$

NOTES AND REFERENCES

Sections 11.1–11.4. The order of development in these sections deviates substantially from the chronological order, since the maximum principle was a relatively late development. The calculus of variations was initiated with the study by the Bernoulli brothers in about 1696 of the brachristochrone problem. A standard book on the classical approach, which includes an outline of the history, is Bliss [B9]. Also see Bliss [B10]. The classical approach was extended to problems with inequality constraints in about 1930, mainly by McShane [Mc2]. The classical formulation contains no explicit control variable, but the substitution $\dot{x} = u$ can be used to convert these problems to control form. The scope of application was considerably broadened and the notation greatly streamlined by the explicit introduction of the control formulation, and by the general maximum principle of Pontryagin. See [P5]. For an early expository discussion, see Rozonoer [R4], [R5], [R6]. A good introduction to optimal control is Bryson and Ho [B12]. The example on insect colonies is adapted from Macevicz and Oster [M1].

Section 11.5. This problem was originally worked out in detail by Kalman [K2].

Section 11.7. As discussed in the text, the Hamilton–Jacobi approach is a traditional branch of the calculus of variations. Dynamic programming, developed by Bellman [B5] (see also Bellman and Dreyfus [B7]), is now one of the most general and powerful approaches to dynamic optimization. It is applicable to a broad range of problem structures, including several not discussed in this text.

Section 11.8. For an early result along these lines see Kalman and Bertram [K4].

Section 11.9. For more examples similar to those of these problems, and additional theory, see Bryson and Ho [B12], Luenberger [L9], and Intriligator [I1]. The special structure of Problems 16 and 19, which leads to linear feedback laws, was introduced in Luenberger [L11].

References

[A1] Allen, R. G. D. *Macro-Economic Theory: A Mathematical Treatment.* Macmillan, London, 1967.

[A2] Arrow, K. J. and Hahn, F. H. *General Competitive Analysis.* Holden-Day, San Francisco, 1971.

[A3] Arrow, K. J. and Nerlove, M. "A Note on Expectation and Stability," *Econometrica*, **26** (1958), 297–305.

[A4] Aseltine, J. A. *Transform Method in Linear System Analysis.* McGraw-Hill Book Co., New York, 1958.

[B1] Baily, N. T. J. *The Mathematical Theory of Epidemics.* Hafner, New York, 1957.

[B2] Bartholomew, D. J. *Stochastic Models for Social Processes*, 2nd Ed. Wiley, New York, 1973.

[B3] Bartlett, M. S. *Stochastic Population Models in Ecology and Epidemiology.* Methuen, London, 1960; Wiley, New York, 1960.

[B4] Baumol, W. J. *Economic Dynamics*, 3rd Ed. Macmillan, New York, 1970.

[B5] Bellman, R. E. *Dynamic Programming.* Princeton University Press, Princeton, N.J., 1957.

[B6] Bellman, R. E. *Introduction to Matrix Analysis*, 3rd Ed. McGraw-Hill Publishing Co., New York, 1970.

[B7] Bellman, R. E. and Dreyfus, S. E. *Applied Dynamic Programming.* Princeton University Press, Princeton, N.J., 1962.

[B8] Bhat, U. N. *Elements of Applied Stochastic Processes.* Wiley, New York, 1972.

[B9] Bliss, G. A. *Calculus of Variations.* Open Court, LaSalle, Ill., 1925.

[B10] Bliss, G. A. *Lectures on the Calculus of Variations.* Univ. of Chicago Press, Chicago, 1945.

[B11] Braun, M. *Differential Equations and Their Applications.* Springer-Verlag, New York, 1975.

[B12] Bryson, A. E., Jr. and Ho, Y.-C. *Applied Optimal Control.* Blaisdell, Waltham, Mass., 1969.

[C1] Carslaw, H. S. and Jaeger, J. C. *Operational Methods in Applied Mathematics.* Clarendon Press, Oxford, 1945.

[C2] Champernowne, D. G. "A Model of Income Distribution," *Economic Journal,* **63** (1953), 318–351.

[C3] Clark, R. N. *Introduction to Automatic Control Systems,* Wiley, New York, 1962.

[C4] Clarke, B. A. and Disney, R. L. *Probability and Random Processes for Engineers and Scientists.* Wiley, New York, 1970.

[C5] Coddington, E. A. *An Introduction to Ordinary Differential Equations.* Prentice-Hall, Englewood Cliffs, N.J., 1961.

[D1] Derusso, P. M., Roy, R. J., and Close, C. M. *State Variables for Engineers.* Wiley, New York, 1965.

[D2] Donovan, B. and Angress, J. F. *Lattice Vibrations.* Chapman and Hall, London, 1971.

[E1] Estes, W. K. "Toward a Statistical Theory of Learning," *Psychological Review,* **57** (1950), 94–107.

[E2] Ewens, W. J. *Population Genetics.* Methuen, London, 1969.

[F1] Feller, W. *An Introduction to Probability Theory and Its Applications,* Vol. 1, 2nd Ed. Wiley, New York, 1957.

[F2] Fertis, D. *Dynamics and Vibration of Structures.* Wiley, New York, 1973.

[F3] Fischer, J. L. "Solutions for the Natchez Paradox," *Ethnology,* **3** (1964), 53–65.

[F4] Frobenius, G. "Über Matrizen aus positiven Elementen," *Sitzungsberichte, Königl. Preussichen Akad. Wiss,* **8** (1908), 471–76.

[G1] Gandolfo, G. *Mathematical Methods and Models in Economic Dynamics.* North-Holland, Amsterdam, 1971.

[G2] Gantmacher, F. R. *The Theory of Matrices, Vol. I.* Chelsea, New York, 1959.

[G3] Gantmacher, F. R. *The Theory of Matrices, Vol. II.* Chelsea, New York, 1959.

[G4] Gardner, M. F. and Barnes, J. L. *Transients in Linear Systems,* Vol. I. Wiley, New York, 1942.

[G5] Gause, G. F. *The Struggle for Existence.* Hafner, New York, 1964.

[G6] Gilbert, E. G. "Controllability and Observability in Multivariable Control Systems," *J. Soc. Indust. Appl. Math.* Series A: On Control, **1** (1963), 128–151.

[G7] Goel, N. S., Maitra, S. C., and Montroll, E. W. *On the Volterra and Other Nonlinear Models of Interacting Populations.* Academic Press, New York, 1971.

[G8] Goldberg, S. *Introduction to Difference Equations.* Wiley, New York, 1958.

[G9] Goldstein, H. *Classical Mechanics.* Addison-Wesley, Reading, Mass., 1959.

[G10] Gopinath, B. "On the Control of Linear Multiple Input-Output Systems," *Bell System Technical Journal,* Mar. 1971, 1063–1081.

[H1] Hawkins, D. and Simon, H. A. "Note: Some Conditions of Macro-Economic Stability," *Econometrica,* **17** (1949), 245–48.

[H2] Henderson, J. M. and Quandt, R. E. *Microeconomic Theory: A Mathematical Approach,* 2nd Ed. McGraw-Hill, New York, 1971.

[H3] Hoffman, K. and Kunze, R. *Linear Algebra,* 2nd Ed. Prentice-Hall, Englewood Cliffs, N.J., 1961.

[H4] **Homans, G. C.** *The Human Group.* Harcourt Brace Jovanovich, New York, 1950.

[H5] **Housner, G. W. and Hudson, D. E.** *Applied Mechanics Dynamics.* Van Nostrand, Princeton, N.J., 1950.

[H6] **Howard, R. A.** *Dynamic Programming and Markov Processes.* M.I.T. Press, Cambridge, Mass., 1960.

[H7] **Howard, R. A.** *Dynamic Probabilistic Systems, Volume I: Markov Models.* Wiley, New York, 1971.

[I1] **Intriligator, M. D.** *Mathematical Optimization and Economic Theory.* Prentice-Hall, Englewood Cliffs, N.J., 1971.

[K1] **Kalman, R.E.** "On the General Theory of Control Systems," *Proc. of the First IFAC Congress,* Butterworths, London, **1** (1960), 481–491.

[K2] **Kalman, R. E.** "Contributions to the Theory of Optimal Control," *Boletin De La Sociedad Matematica Mexicana* (1960), 102–119.

[K3] **Kalman, R. E.** "Mathematical Description of Linear Dynamical Systems," *J. SIAM on Control,* Ser. A, **1,** No. 2 (1963), 152–192.

[K4] **Kalman, R. E. and Bertram, J. E.** "Control System Analysis and Design via the "Second Method" of Lyapunov I: Continuous-Time Systems," *Journal of Basic Engineering,* June (1960), 371–393.

[K5] **Kalman, R. E. and Bertram, J. E.** "Control System Analysis and Design via the 'Second Method' of Lyapunov II: Discrete-Time Systems," *Journal of Basic Engineering,* June (1960), 394–400.

[K6] **Kane, J.** "Dynamics of the Peter Principle," *Management Science,* **16,** No. 12, (1970), B-800–811.

[K7] **Karlin, S.** *Mathematical Methods and Theory in Games, Programming and Economics, Volume I Matrix Games, Programming, and Mathematical Economics.* Addison-Wesley, Reading, Mass., 1959.

[K8] **Karlin, S.** *A First Course in Stochastic Processes.* Academic Press, New York, 1966.

[K9] **Karlin, Samuel.** *Equilibrium Behavior of Population Genetic Models With Non-Random Mating.* Gordon and Breach, New York, 1969.

[K10] **Kemeny, J. G., Mirkil, H., Snell, J. L., and Thompson, G. L.** *Finite Mathematical Structures,* Prentice-Hall, Englewood Cliffs, N.J., 1959.

[K11] **Kemeny, John G., and Snell, J. Laurie.** *Finite Markov Chains,* Van Nostrand Reinhold, New York, 1960.

[K12] **Kermack, W. O. and McKendrick, A. G.** "Mathematical Theory of Epidemics," *Proceedings of the Royal Society of London,* **115** (1927), 700–721.

[K13] **Keyfitz, N.** *Introduction to the Mathematics of Population.* Addison-Wesley, Reading, Mass. 1968.

[K14] **Kolmogorov, A. N. and Fomin, S. V.** *Elements of the Theory of Functional Analysis,* Vol. 1. Graylock, Rochester, N.Y., 1957.

[K15] **Kordemsky, B. A.** *The Moscow Puzzles: 359 Mathematical Recreations.* Scribner's, New York, 1972.

[K16] **Kwakernaak, H. and Sivan, R.** *Linear Optimal Control Systems.* Wiley-Interscience, New York, 1972.

[L1] **Lanchester, F.** *Aircraft in Warfare, the Dawn of the Fourth Arm.* Constable, London, 1916.

[L2] La Salle, J. and Lefschetz, S. *Stability by Liapunov's Direct Method.* Academic Press, New York, 1961.

[L3] Leontief, W. *Input–Output Economics.* Oxford Univ. Press, New York, 1973.

[L4] Leslie, P. H. "On the Use of Matrices in Certain Population Mathematics," *Biometrika,* **33** (1945), 183–212.

[L5] Lotka, A. *Elements of Mathematical Biology.* Dover, New York, 1956.

[L6] Luenberger, D. G. "Observing the State of a Linear System," *IEEE Trans. Mil. Electron.,* Vol. MIL-8, April (1964), 74–80.

[L7] Luenberger, D. G. "Observers for Multivariable Systems," *IEEE Transactions on Automatic Control,* Vol. AC-11, No. 2, April (1966), 190–197.

[L8] Luenberger, D. G. "Canonical Forms for Linear Multivariable Systems," *IEEE Transactions on Automatic Control,* Vol. AC-12, No. 3, June (1967), 290–293.

[L9] Luenberger, D. G. *Optimization by Vector Space Methods.* Wiley, New York, 1969.

[L10] Luenberger, D. G. "An Introduction to Observers," *IEEE Transactions on Automatic Control,* Vol. AC-16, No. 6 (1971), 596–602.

[L11] Luenberger, D. G. "A Nonlinear Economic Control Problem with a Linear Feedback Solution," *IEEE Transactions on Automatic Control,* Vol. AC-20 (1975), 184–191.

[L12] Luenberger, D. G. "Dynamic Systems in Descriptor Form," *IEEE Transactions on Automatic Control,* Vol. AC-22, No. 2, April (1977).

[L13] Luré, A. I. *Nekotorye Nelineinye Zadachi Teorii Avtomaticheskogo Regulirovaniya.* Gostekhizdat, Moscow, 1951.

[L14] Lyapunov, A. M. "Problème général de la stabilité du mouvement," *Ann. Fac. Sci. Toulouse,* Vol. 9 (1907), pp. 203–474. (Reprinted in Ann. Math. Study No. 17, Princeton Univ. Press, Princeton, N.J., 1949.)

[Mc1] McPhee, W. N. *Formal Theories of Mass Behavior.* Free Press of Glencoe, New York, 1963.

[Mc2] McShane, E. J., "On Multipliers for Lagrange Problems," *Amer. J. Math.* **61** (1939), 809–819.

[M1] Macevicz, S. and Oster G. "Modelling Social Insect Populations II: Optimal Reproductive Strategies in Annual Eusocial Insect Colonies," *Behavioral Ecology and Sociobiology,* **1** (1976), 265–282.

[M2] Martin, W. T. and Reissner, E. *Elementary Differential Equations,* 2nd Ed. Addison-Wesley, Reading, Mass., 1961.

[M3] Meadows, D. L. *Dynamics of Commodity Production Cycles.* Wright-Allen, Cambridge, Mass., 1970.

[M4] Metzler, L. A. "Stability of multiple markets: the Hicks conditions," *Econometrica,* **13** (1945), 277–292.

[M5] Miller, K. S. *Linear Difference Equations.* Benjamin, Reading, Mass., 1968.

[M6] Monod, J. "La Techniqe de Cultures Continues; Théorie et Applications," *Annales de l'Institut Pasteur,* **79,** No. 4 (1950), p. 390.

[M7] Morishima, M. *Equilibrium, Stability, and Growth, A Multisectorial Analysis.* Oxford at the Clarendon Press, Ely House, London, 1964.

[N1] Nikaido, H. *Convex Structures and Economic Theory.* Academic Press, New York, 1968.

[P1] **Papandreou, A. G.** *Fundamentals of Model Construction in Macro-Economics*. C. Serbinis Press, Athens, 1962.

[P2] **Perron, O.** Zur Theorie der Matrizen. *Math. Ann.* **64** (1907).

[P3] **Peter, L. J. and Hull, R.** *The Peter Principle, and Why Things Always Go Wrong*. William Morrow, New York, 1969.

[P4] **Pielou, E. C.** *An Introduction to Mathematical Ecology*. Wiley-Interscience, New York, 1969.

[P5] **Pontryagin, L. S., Boltyanskii, V. G., Gamkrelidze, R. V., and Mishchenko, E. F.**, *The Mathematical Theory of Optimal Processes*, trans. by K. N. Trirogoff, ed. by L. W. Neustadt. Wiley-Interscience, New York, 1962.

[P6] **Prigogine, I.** *Introduction to Thermodynamics of Irreversible Processes*, 3rd Ed. Wiley-Interscience, New York, 1967.

[Q1] **Quirk, J. and Saposnik, R.** *Introduction to General Equilibrium Theory and Welfare Economics*. McGraw-Hill, New York, 1968.

[R1] **Rainville, E. D. and Bedient, P. E.** *Elementary Differential Equations*, 5th Ed. Macmillan, New York, 1974.

[R2] **Rapoport, A.** *Fights, Games, and Debates*. University of Michigan Press, Ann Arbor, 1960.

[R3] **Richardson, L. F.** *Arms and Insecurity*, Boxwood Press, Pittsburgh Quadrangel Book, Chicago, 1960.

[R4] **Rozonoer, L. I.** "Theory of Optimum Systems. I," *Automation and Remote Control*, **20** (10) (1959), 1288–1302.

[R5] **Rozonoer, L. I.** "L. S. Pontryagin's Maximum Principle in Optimal System Theory, II," *Automation and Remote Control*, **20** (11) (1959) 1405–1421.

[R6] **Rozonoer, L. I.** "The Maximum Principle of L. S. Pontryagin in Optimal-System Theory, III," *Automation and Remote Control*, **20** (12) (1959), 1517–1532.

[R7] **Rugh, W. J.** *Mathematical Description of Linear Systems*. Marcel Dekker, New York, 1975.

[S1] **Saaty, T. I.** *Mathematical Models of Arms Control and Disarmament*. Wiley, New York, 1968.

[S2] **Samuelson, P. A.** "Interactions Between the Multiplier Analysis and the Principle of Acceleration," *Rev. Economic Statistics*, **21,** No. 7, May (1939), 75–78.

[S3] **Sandberg, I. W.** "On the Mathematical Theory of Interaction in Social Groups" *IEEE Transactions on Systems, Man, and Cybernetics*, Vol. SMC-4, No. 5 (1974), 432–445.

[S4] **Shannon, C. E. and Weaver, W.** *The Mathematical Theory of Communication*. University of Illinois Press. Urbana, 1949.

[S5] **Simon, H. A.** *Models of Man*. Wiley, New York, 1957.

[S6] **Strang, G.** *Linear Algebra and Its Applications*. Academic Press, New York, 1976.

[T1] **Takayama, A.** *Mathematical Economics*. Dryden Press, Hinsdale, Ill., 1974.

[T2] **Thorp, E. O.** *Beat the Dealer*. Random House (Vintage), New York, 1962.

[V1] **Volterra, V.** "Variazioni e fluttuazioni del numero d'individui in specie animali conviventi," *Memorie della R. Accademia Nazionale dei Lincei*, anno CCCCXXIII, II (1926), 1–110.

[V2] **Von Neumann, J.** "Über ein ökonomisches Gleichungssystem und eine Verallgemeinerung des Brouwerschen Fixpunktsatzes," *Ergebnisse eines Mathematischen Kolloquiums*, **8** (1937), 73–83.

[W1] **Walker, J. and Lehman, J.** *1000 Ways to Win Monopoly Games*. Dell, New York, 1975.

[W2] **Watt, K. E. F.** *Ecology and Resource Management*. McGraw-Hill, New York, 1968.

[W3] **Weinreich, G.** *Fundamental Thermodynamics*. Addison-Wesley, Reading, Mass., 1968.

[W4] **White, D. R., Murdock, G. P., and Scaglion, R.** "Natchez Class and Rank Reconsidered," *Ethnology*, **10,** No. 4, October (1971) 369–388.

[W5] **Wilson, E. O. and Bossert, W. H.** *A Primer of Population Biology*. Sinauer Assoc., Stamford, Conn., 1971.

[W6] **Wonham, W. M.** "On Pole Assignment in Multi-input Controllable Linear Systems," *IEEE Trans. Automatic Control*, vol. AC-12, December (1967), 660–665.

Index